First Edition

Maribeth H. Price
South Dakota School of Mines and Technology

Mastering ArcGIS Pro

McGraw Hill Education

MASTERING ArcGIS PRO

Published by McGraw-Hill Education, 2 Penn Plaza, New York, NY 10121. Copyright © 2020 by McGraw-Hill Education. All rights reserved. Printed in the United States of America. No part of this publication may be reproduced or distributed in any form or by any means, or stored in a database or retrieval system, without the prior written consent of McGraw-Hill Education, including, but not limited to, in any network or other electronic storage or transmission, or broadcast for distance learning.

Some ancillaries, including electronic and print components, may not be available to customers outside the United States.

This book is printed on acid-free paper.

1 2 3 4 5 6 7 8 9 LKV 21 20 19

ISBN 978-1-260-57013-7
MHID 1-260-57013-4

Cover Image: *USGS and Esri*

mheducation.com/highered

Table of Contents

Preface

Welcome to *Mastering ArcGIS Pro*, a detailed primer on learning the latest ArcGIS™ software by Esri®, Inc. This book is designed to offer everything you need to master the basic elements of GIS.

> **Notice:** ArcGIS Pro™, ArcGIS™, ArcMap™, ArcCatalog™, ArcGIS Desktop™, ArcInfo Workstation™, and the other program names used in this text are registered trademarks of Esri, Inc. The software names and the screen shots used in the text are reproduced by permission. For ease of reading, the ™ symbol has been omitted from the names; however, no infringement or denial of the rights of Esri® is thereby intended or condoned by the author.

A new text for a new GIS experience

Although the concepts of GIS have remained fairly constant over time, the software is continually evolving. With the release of ArcGIS Pro, the latest software in the Esri GIS family, a new generation of GIS has arrived. ArcGIS Pro has a 64-bit, multithreaded architecture, uses ribbon-style menus, integrates 2D and 3D applications, and is closely tied to ArcGIS Online.

This text constitutes a major rewrite of *Mastering ArcGIS*, a book that covered GIS concepts and skills using the ArcGIS Desktop programs of ArcMap and ArcCatalog. Although the GIS concepts largely remain the same in both texts, the implementation, and in some cases the terminology, has changed. The new software has also prompted a reorganization of the book in several important ways.

First, the book has been refocused on the basics of GIS. The ArcGIS Pro software capabilities are improving with each new version but have not yet completely matched the capabilities of ArcMap. Partly for this reason, and partly to better match the rhythm of a semester, the book is now presented in 12 chapters, leaving time for instructors to better incorporate exams and projects within the semester. Some of the more advanced and less frequently used skills, such as planar topology and standards-based metadata, have been left for students to explore on their own.

Second, the book includes some new topics. Raster data management has been discussed in a new chapter to acquaint students with compiling and processing raster data sets, supplementing a similar chapter on vector data management. ArcGIS Pro was designed to foster the sharing of GIS data and workflows, and these enhanced capabilities are explored in another new chapter, including how to prepare a database for collecting data using mobile devices.

Third, the chapters and topics have been reorganized to eliminate some repetition and to present the information more logically. The text still roughly follows the project model with data management presented first and analysis second.

The tutorials, questions, and exercises have been rewritten. I have tried to incorporate more open-ended and creative questions in the exercises, rather than relying on cut-and-dried questions and answers. Although this approach makes grading a little harder for instructors, I believe it enhances student learning and makes the exercises more interesting.

I would like to thank the many people who have used and commented on *Mastering ArcGIS*, and I have tried to take those lessons to heart. I hope that this text continues to serve their needs in the rapidly evolving world of GIS.

Previous experience

This book assumes that the reader is comfortable using Windows™ to carry out basic tasks such as copying files, moving directories, opening documents, exploring folders, and editing text and word processing documents. Previous experience with maps and map data is also helpful. No previous GIS experience or training is necessary to use this book.

Elements of the package

This learning system includes a textbook and web site, including

► Twelve chapters on important concepts in GIS data management and analysis

► Comprehensive tutorials in every chapter to learn the skills, with each step demonstrated in a video clip

► A set of exercises and data for practicing skills independently

This book assumes that the student has access to ArcGIS Pro and an organizational account for ArcGIS Online. The Spatial Analyst extension is required for Chapter 11.

Philosophy

This text reflects the author's personal philosophies and prejudices developed from 20 years of teaching GIS at an engineering school. The main goal is not to train geographers but to provide students in any field with GIS skills and knowledge. It is assumed that most students using this book already have a background of discipline-specific knowledge and skills upon which to draw and are seeking to apply geospatial techniques within their own knowledge domains.

► **Concepts and skills are both important.** Students must be able to understand the geographic and database concepts that are integral to spatial data management and analysis. It is not enough to know a series of software steps; one should understand the "why" behind them.

► **GIS is best learned by doing it.** The tutorials constitute a critical component of the book, integrating theory with practical experience and promoting a hands-on, active learning environment.

► **Independent work and projects are critical to learning GIS.** This book includes exercises that encourage students to find solutions independently without a cookbook recipe of steps. A wise instructor will also require students to develop an independent project.

Chapter sequence

The book contains an introduction and 12 chapters. Each chapter includes roughly one week's work for a three-credit semester course. An introductory chapter describes GIS and gives some examples of how it is used. It also provides an overview of GIS project management and how to develop a project. Chapters 1 through 11 follow a roughly project-based sequence: mapping basics, data compilation and management, and spatial analysis. The final chapter explores sharing GIS work with others. These chapters are the core of an introductory GIS class and, by the end of it, students should have little difficulty developing and carrying out an independent GIS project.

Chapter layout

Each chapter is organized into the following sections:

► **Concepts:** provides basic background material for understanding geographic concepts and how they are specifically implemented within ArcGIS. Most chapters have two sections, one (About GIS) covering general GIS concepts and theory, and another (About

ArcGIS) covering the specific implantation of those concepts within ArcGIS Pro. A set of review questions and important terms follows the concepts section.

► **Tutorial:** contains a step-by-step tutorial demonstrating the concepts and skills. The tutorials begin with detailed instructions, which gradually become more general as mastery is built. Every step in the tutorial is demonstrated by accompanying video clips.

► **Exercises:** presents a series of problems to build skill in identifying the appropriate techniques and applying them without step-by-step help. Through these exercises, the student builds an independent mastery of GIS processes. Brief solution methods are included for all exercises, and maps of the results are shown when applicable. A full answer and methods document is available for instructors at the McGraw-Hill Instructor web site.

The book web site, http://www.mhhe.com/Price_pro1e, also contains all the data needed to follow the tutorials and complete the exercises.

Instructors should use judgment in assigning exercises. The typical class would be stretched to complete all the exercises in every chapter. Very good students can complete an entire chapter in 3 to 6 hours, most students would need 6 to 8 hours, and a few students would require 10 or more hours. Students with more computer experience generally find the material easier than others.

Using this text

In working through this book, the following sequence of steps is suggested:

► READ through the Concepts sections to get familiar with the principles and techniques.

► ANSWER the Chapter Review Questions to test comprehension of the material.

► WORK the Tutorial section for a step-by-step tutorial and explanation of key techniques.

► REREAD the Concepts section to reinforce the ideas.

► PRACTICE by doing the Exercises.

Using the tutorial

The tutorial provides step-by-step practice and introduces details on how to perform specific tasks. Students should be encouraged to think about the steps as they are performed and not just race to get to the end.

It is important to follow the directions carefully. Skipping a step or doing it incorrectly may result in a later step not working properly. Saving often will make it easier to go back and correct a mistake in order to continue. Occasionally, a step will not work due to differences between computer systems or software versions. Having an experienced user nearby to identify the problem can help. If one isn't available, however, just skip the step and move on without it.

Using the videos

The web site contains videos of each tutorial step. They are numbered in the text for easy reference. The videos are intended as an alternate learning strategy. It would be tedious to watch all of them. Instead, use them in the following situations:

► When a student does not understand the written instructions or cannot find the correct menu or button

► When a step cannot be made to work properly

> ▸ When a reminder is needed to do a previously learned skill in order to complete a step

> ▸ When a student finds that watching the videos enhances learning

Using video and data components

To view the videos:

The videos are embedded in the electronic version of the text, and they are also available for download from the book's web site. Each chapter may be downloaded separately.

To install the training data:

The gisclass_Pro1e.zip archive contains a folder with the documents and data needed to do the tutorials and exercises. Students must extract this folder to their own hard drives. If more than one person on the computer is using this book, then each person should make a personal copy of the data in a separate folder. The data require approximately 360 MB of disk space.

Extracting the zip file creates a folder named gisclass. It should be extracted to a location on the C:\drive. Do not install it in a user Documents folder or on the Desktop because GIS data often function poorly in those locations.

System requirements

Hardware:

ArcGIS Pro requires a PC 64-bit multiprocessor machine with at least a dual core, and a quad core processor or higher is recommended. Pro requires a minimum of 8 GB of RAM and a good graphics card with at least 2 GB of RAM and 24-bit display color depth.

For a complete description of the system requirements, consult Esri (http://pro.arcgis.com/en /pro-app/get-started/arcgis-pro-system-requirements.htm).

Software:

Windows 10 (64-bit Home, Pro, or Enterprise) or Windows 8.1 (64-bit Pro or Enterprise)

ArcGIS Pro™ 2.1.2 or higher (Basic Level); Spatial Analyst extension required for some exercises

A web browser, such as Chrome, Edge, Firefox, or Internet Explorer

A zip utility such as WinZip or 7zip

A media player that can display the .mp4 video format

Other:

Pro is tightly coupled to ArcGIS Online (AGOL), a cloud-based system built to encourage sharing of GIS data, workflows, and other resources between organizations and users. It is strongly recommended that the users have at least a public account, and a subscription account provides access to significantly more data and tools. A subscription account with publishing privileges is required for some optional exercises in Chapter 12.

Internet access is required for ArcGIS Pro installation and for exercises requiring the use of ArcGIS Online.

For assistance in acquiring or installing these components, contact your system administrator, hardware/software provider, or local computer store.

Acknowledgments

I would like to thank many people who made this book possible. Governor Janklow of South Dakota funded a three-month summer project in 2000 that got the original *Mastering ArcGIS* book started, as part of his Teaching with Technology program. Many students in my GIS classes between 2000 and 2018 tested the text and exercises and helped immensely in making sure the tutorials were clear and worked correctly. Reviewers of previous editions of *Mastering ArcGIS*, including Richard Aspinall, Joe Grengs, Tom Carlson, Susan K. Langley, Henrietta Loustsen, Xun Shi, Richard Lisichenko, John Harmon, Michael Emch, Jim Sloan, Sharolyn Anderson, Talbot Brooks, Qihao Weng, Jeanne Halls, Mark Leipnik, Michael Harrison, Ralph Hitz, Olga Medvedkov, James W. Merchant, Raymond L. Sanders, Jr., Yifei Sun, Fahui Wang, Michael Haas, Jason Kennedy, Dafna Kohn, Jessica Moy, James C. Pivirotto, Peter Price, Judy Sneller, Dave Verbyla, Birgit Mühlenhaus, Jason Duke, Darla Munroe, Wei-Ning Xiang, L. Joe Morgan, Samantha Arundel, Christopher A. Badurek, Tamara Biegas, John E. Harmon, Michael Hass, Nicholas Kohler, David Long, Jaehyung Yu, Sarah Battersby, Gregory S. Bohr, Kelly R. Dubure, Colleen Garrity, Raymond Greene, Joe Grengs, Eileen Johnson, James Leonard, and Tao Tang, provided detailed and helpful comments, and the book is better than it would have been without their efforts.

I especially thank Curtis Price, Bryan Krouse, and their students for testing the first draft of this manuscript in the classroom.

Esri, Inc., was prompt and generous in its granting of permission to use the screen shots, data, and other materials throughout the text. I extend heartfelt thanks to the City of Austin, Texas, for putting their fine GIS data sets in the public domain. I thank George Sielstad, Eddie Childers, Mark Rumble, Tom Junti, and Patsy Horton for their generous donations of data. I am grateful to Tom Leonard and Steve Bauer for their long-term computer lab administration, without which I could not have taught GIS courses or developed this book. I thank the McGraw-Hill team working on this new first edition, especially Michael Ivanov and Jodi Rhomberg. I am grateful to Daryl Pope, who first started me in GIS, and to John Suppe, who encouraged me to return to graduate school and continue doing GIS on a fascinating study of Venus. I thank Charles R. Bacon, with whom I was privileged to work on the geologic map of Crater Lake. I thank my partner, David Stolarz, who provided unfailing encouragement when it seemed as though the writing of this new edition would never end. Last, and certainly not least, I thank Curtis Price and my daughters, Virginia and Madeleine, for their understanding and support during the many, many hours I spent working on this book.

Introduction

Objectives

- Developing a basic understanding of what GIS is, its operations, and its uses
- Getting familiar with GIS project management
- Learning to plan a GIS project
- Finding resources to learn more about GIS

Concepts

What is GIS?

GIS stands for geographic information system. In practical terms, a GIS is a set of computer tools that allows people to work with data that are tied to a particular location on the earth. Although many people think of a GIS as a computer mapping system, its functions are broader and more sophisticated than that. A GIS is a database that is designed to work with map data.

Consider the accounting department of the local telephone company. They maintain a large computer database of their customers, in which they store the name, address, phone number, type of service, and billing information for each customer. This information is only incidentally tied to where customers live; they can carry out most of the important functions (billing, for example) without needing to know where each house is. Of course, they need to have addresses for mailing bills, but it is the post office that worries about where the houses actually ARE. This type of information is called **aspatial data**, meaning that it is not tied, or is only incidentally tied, to a location on the earth's surface.

Employees of the service department, however, need to work with **spatial data** to provide the telephone services. When hundreds of people call in after a power outage, the service department must analyze the distribution of the calls and isolate the location where the outage occurred. When a construction company starts work on a street, workers must be informed of the precise location of buried telephone cables. If a developer builds a new neighborhood, the company must be able to determine the best place to tie into the existing network so that the services are efficiently distributed from the main trunk lines. When technicians prepare lists of house calls for the day, they need to plan the order of visits to minimize the amount of driving time. In these tasks, location is a critical aspect of the job, and the information is spatial.

In this example, two types of software are used. The accounting department uses special software called a *database management system*, or *DBMS*, which is optimized to work with large volumes of aspatial data. The service department needs access to a database that is optimized for working with spatial data, a geographic information system. Because these two types of software are related, they often work together, and they may access the same information. However, they do different things with the data.

A GIS is built from a collection of hardware and software components.

- ***A computer hardware platform.*** Due to the intensive nature of spatial data storage and processing, a GIS was once limited to large mainframe computers or expensive workstations. Today, it can run on a typical desktop personal computer.

▸ **GIS software.** The software varies widely in cost, ease of use, and level of functionality but should offer at least some minimal set of functions, as described in the next few paragraphs. In this book, we study one particular package that is powerful and widely used, but others are available and may be just as suitable for certain applications.

▸ **Data storage.** Some projects use only the hard drive of the GIS computer. Other projects may require more sophisticated solutions if large volumes of data are being stored or multiple users need access to the same data sets. Today, many data sets are stored in digital warehouses and are accessed by many users over the Internet. Compact disc writers and/or USB portable drives are highly useful for backing up and sharing data.

▸ **Data input hardware.** Many GIS projects require sophisticated data entry tools. Digitizer tablets enable the shapes on a paper map to be entered as features in a GIS data file. Scanners create digital images of paper maps. An Internet connection provides easy access to large volumes of GIS data. High-speed connections are preferred because GIS data sets may constitute tens or hundreds of megabytes or more.

▸ **Information output hardware.** A quality color printer capable of letter-size prints provides the minimum desirable output capability for a GIS system. Printers that can handle map-size output (36 in. × 48 in.) will be required for many projects.

▸ **GIS data.** Data come from a variety of sources and in a plethora of formats. Gathering data, assessing their accuracy, and maintaining them usually constitutes the longest and most expensive part of a GIS project.

▸ **GIS personnel.** A system of computers and hardware is useless without trained and knowledgeable people to run it. The contribution of professional training to successful implementation of a GIS is often overlooked.

GIS software varies widely in functionality, but any system claiming to be a GIS should provide the following functions at a minimum:

▸ **Data entry** from a variety of sources, including digitizing, scanning, text files, and the most common spatial data formats; ways to export information to other programs should also be provided.

▸ **Data management** tools, including tools for building data sets, editing spatial features and their attributes, and managing coordinate systems and projections.

▸ **Thematic mapping** (displaying data in map form), including symbolizing map features in different ways and combining map layers for display.

▸ **Data analysis** functions for exploring spatial relationships in and between map layers.

▸ **Map layout** functions for creating soft and hard copy maps with titles, scale bars, north arrows, and other map elements.

Geographic information systems are put to many uses, but providing the means to collect, manage, and analyze data to produce information for better decision making is the common goal and the strength of each GIS. This book is a practical guide to understanding and using a particular geographic information system called ArcGIS, to learn what a GIS is, what it does, and how to apply its capabilities to solve real-world problems.

A history of GIS

Geographic information systems have grown from a long history of cartography begun in the lost mists of time by early people who made sketches on hides or formed crude models of clay as aids to hunting for food or making war. Ptolemy, an astronomer and geographer from the second century B.C., created one of the earliest known atlases, a collection of world, regional, and local maps and advice on how to draw them, which remained essentially unknown to Europeans until the 15th century. Translated into Latin, it became the core of Western geography, influencing cartographic giants, such as Gerhard Mercator, who published his famous world map in 1569. The 17th and 18th centuries saw many important developments in cartography, including the measurement of a degree of longitude by Jean Picard in 1669, the discovery that the earth flattens toward the poles, and the adoption of the Prime Meridian that passes through Greenwich, England. In 1859, French photographer and balloonist Gaspard Felix Tournachon founded the art of remote sensing by carrying large-format cameras into the sky. In an oft-cited early example of spatial analysis, Dr. John Snow mapped cholera deaths in central London in September 1854 and was able to locate the source of the outbreak—a contaminated well. However, until the 20th century, cartography remained an art and a science carried out by laborious calculation and hand drawing. In 1950, Jacqueline Tyrwhitt made the first explicit reference to map overlay techniques in an English textbook on town and country planning, and Ian McHarg was one of the early implementers of the technique for highway planning.

As with many other endeavors, the development of computers inspired cartographers to see what these new machines might do. The early systems developed by these groups, crude and slow by today's standards, nevertheless laid the groundwork for modern geographic information systems. Dr. Roger Tomlinson, head of an Ottawa group of consulting cartographers, has been called the Father of GIS for his promotion of the idea of using computers for mapping and for his vision and effort in developing the Canada Geographic Information System (CGIS) in the mid-1960s. Another pioneering group, the Harvard Laboratory for Computer Graphics and Spatial Analysis, was founded in the mid-1960s by Howard Fisher. He and his colleagues developed a number of early programs between 1966 and 1975, including SYMAP, CALFORM, SYMVU, GRID, POLYVRT, and ODYSSEY. Other notable developers included professors Nystuen, Tobler, Bunge, and Berry from the geography department at Washington University during 1958–1961. In 1970, the US Bureau of the Census produced its first geocoded census and developed the early DIME data format based on the CGIS and POLYVRT data representations. DIME files were widely distributed and were later refined into the TIGER format. These efforts had a pronounced effect on the development of data models for storing and distributing geographic information.

In 1969, Laura and Jack Dangermond founded the Environmental Systems Research Institute (ESRI), which pioneered the powerful idea of linking spatial representation of features with attributes in a table, a core idea that revolutionized the industry and launched the development of Arc/Info, a program whose descendants have captured about 90% of today's GIS market. Other vendors are still active in developing GIS systems, which include packages MAPINFO, MGA from Intergraph, IDRISI from Clark University, and the open-source program GRASS.

What can a GIS do?

A GIS works with many different applications: land use planning, environmental management, sociological analysis, business marketing, and more. Any endeavor that uses spatial data can benefit from a GIS. For example, researchers at the US Department of Agriculture Rocky Mountain Research Station in Rapid City, South Dakota, conducted a study of elk habitat in the Black Hills of South Dakota and Wyoming by placing radio transmitter collars on about 70 elk bulls and cows (Fig. I.1). Using the collars and a handheld antenna, they tracked the animals and obtained their locations. Several thousand locations were collected (Fig. I.1a) for the scientists to study the characteristics of the habitat where elk spend time.

The elk locations were entered into a GIS system for record keeping and analysis. Each location became a point with attached information, including the animal ID number and the date and time of the sighting. Information about vegetation, slope, aspect, elevation, water availability, and other site factors were derived by overlaying the points on other data layers, allowing the biologists to compare the characteristics of sites used by the elk. Figure I.1b shows a map of distances to major roads in the central part of the study area. The elk locations clustered in the darker roadless areas, and statistical analysis demonstrated this observation empirically.

There are many applications of GIS in almost every human endeavor, including business, defense and intelligence, engineering and construction, government, health and human services, conservation, natural resources, public safety, education, transportation, utilities, and communication. In August 2014, the Industries section of the ESRI web site at www.esri.com listed 62 different application areas in these categories, each one with examples, maps, and case studies. Instead of reading through a list here, go to the site and see the latest applications.

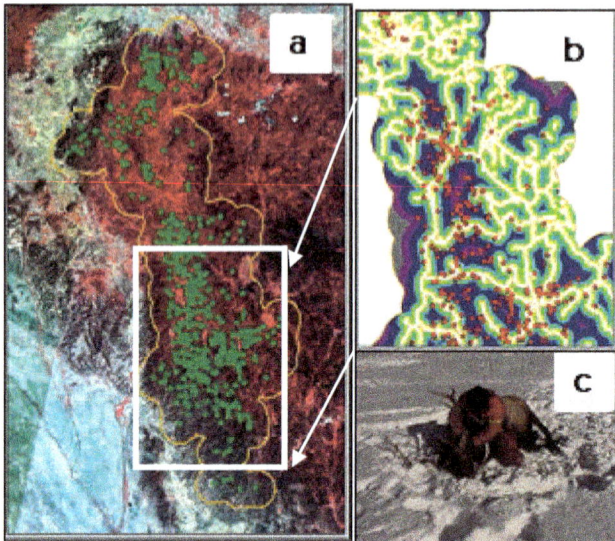

Fig. I.1. Analyzing elk habitat use: (a) elk locations and study area; (b) locations on a map of distance to nearest road; (c) collaring an elk
Courtesy of Mark Rumble

New trends and directions in GIS

The GIS industry has grown exponentially since its inception. Starting with the mainframe and then the desktop computer, GIS began as a relatively private endeavor focused on small clusters of specialized workers who spent years developing expertise with the software and data. Since then, the development of the Internet and the rapidly advancing field of computer hardware have been driving some significant changes in the industry.

Proliferation of options for data sharing

Instead of storing data sets locally on individual computers or intra-organization network drives, more people are serving large volumes of data over the Internet to remote locations within an organization, across organizations, and to the general public. In the past, large collections of data were hosted by various organizations through clearinghouses sponsored by the National Spatial Data Infrastructure (NSDI). The data usually existed as GIS data files in various formats and required significant expertise and the right software to download and use. Today, *GIS servers* are designed to bring GIS data to the general public with a few clicks. Another class of data providers that give wide access to spatial data, although not specifically GIS applications, include web sites such as Google Earth, MapQuest, and Microsoft Virtual Earth.

Like many other computer industries, GIS is heading for the clouds. A cloud is a gigantic array of large computers that customers rent pieces of by the hour, instead of purchasing their own physical hardware. ArcGIS Online is a cloud-based platform for users to collaborate and share GIS data with each other, making it suddenly very easy to share data with a colleague, or with the world, even for those with no particular GIS expertise.

Proliferation of options for working with GIS data

In the early days, people who wanted to do GIS work had to purchase a large, expensive program and learn to use it. Now, a wide variety of scaled applications permits different levels of use for different levels of need. Not every user must have the full program. There are web sites for people who just need to view and print maps, free download software for viewing interactive map publications, and scaled-down versions of the full program with fewer options.

Many organizations are turning to server GIS as a less expensive alternative to purchasing large numbers of GIS licenses. Many workers need GIS, but they only use a small subset of GIS functions on a daily basis. A server GIS can provide both data and a customizable set of viewing and analysis functions to users without a GIS license. All they need is a web browser. Because of more simple and low-cost ways of accessing GIS data and functions, the user base has expanded dramatically.

Expansion of GIS into wireless technology

More people are collecting and sharing data using handheld wireless devices, such as handheld computers, smartphones, and global positioning system (GPS) units. These units can now access Internet data and map servers directly so that users in the field can download background data layers, collect new data, and transmit them back to the large servers. Cell phone technology is advancing rapidly with new geolocation options, web access, and geo-applications arriving daily.

Emphasis on open-source solutions

Instead of relying on proprietary, specialized software, more GIS functions are now implemented within open-source software and hardware. GIS data are now more often stored using engines from commercial database platforms and use the same development environments as the rest of the computer industry. This trend makes it easier to have the GIS communicate with other programs and computers and enhances interoperability between systems and parts of systems.

Customization

With emphasis on open-source solutions, new opportunities have developed in creating customized applications based on a fundamental set of GIS tools, such as a hydrology tool or a wildlife management tool. Smartphone and tablet applications programming is burgeoning. These custom applications gather the commonly used functions of GIS into a smaller interface, introduce new knowledge, and formalize best practices into an easy-to-use interface. Customization requires a high level of expertise in object-oriented computer programming.

Enterprise GIS

Enterprise GIS integrates a server with multiple ways to access the same data, including traditional GIS software programs, web browser applications, and wireless mobile devices. The goal is to meet the data needs of many different levels of users and to provide access to nontraditional users of GIS. The enterprise GIS is the culmination of the other trends and capabilities already mentioned. The costs and challenges in developing and maintaining an enterprise GIS are significant, but the rewards and cost savings can also be substantial.

What do GIS professionals do?

It's getting so easy to create a map these days that one may wonder why anyone should bother to learn GIS. However, the easy solutions are based on the work of experts who provide the data and the software systems to handle them. It may not be that hard to learn to use a smartphone, but engineers and software developers must build the phones and the infrastructure to make them work. GIS is much the same. These days, GIS professionals play a variety of roles. A few broad categories can be defined.

Primary data providers create the base data that form the backbone of many GIS installations. Surveyors and land-planning professionals contribute precise measurement of boundaries. Photogrammetrists develop elevation and other data from aerial photography or the newer laser altimetry (LIDAR) systems. Remote sensing professionals extract all kinds of human-made and natural information from a variety of satellite and airborne measurement systems. Experts in global positioning systems (GPS), which provide base data, are also important.

Applications GIS usually involves professionals trained in other fields, such as geography, hydrology, land use planning, business, and utilities, who use GIS as part of their work. Specialists in mathematics and statistics develop new ways to analyze and interpret spatial data. For these professionals, GIS is an added skill and a tool to make their work more efficient, productive, and valuable.

Development GIS involves skilled software and hardware engineers who build and maintain the GIS software itself, as well as the hardware components upon which it relies—not only computers and hard drives and plotters but also GPS units, wireless devices, scanners, digitizers, and other systems that GIS could not function without. This group also includes an important class of GIS developers who create customized applications, including smartphone apps, from the basic building blocks of existing GIS software.

Distributed database GIS involves computer science professionals with a background in networking, Internet protocols, or database management systems. These specialists set up and maintain the complex server and network systems that allow data services, server GIS, and enterprise GIS to operate.

GIS project management

A GIS project may be a small effort spanning a couple of days by a single person, or it can be an ongoing concern of a large organization with hundreds of people participating. Large or small, however, projects often follow the generalized model shown in Figure I.2, and most new users learn GIS through a project approach. A project usually begins with an assessment of needs. What specific issues must be studied? What kind of information is needed to support decision making? What functions must the GIS perform? How long will the project last? Who will be using the data? What funding is available for start-up and long-term support?

Without a realistic idea of what the system must accomplish, it is impossible to design it efficiently. Users may find that some critical data are absent or that resources have been wasted acquiring data that no one ever uses. In a short-term project the needs are generally clear-cut. A long-term organizational system will find that its needs evolve over time, requiring periodic reassessment. A well-designed system will adapt easily to future modification. The creator of a haphazard system may be constantly redoing previous work when changes arise.

In studies seeking specific answers to scientific or managerial questions, a methodology, or **model**, must be chosen. Models convert the raw data of the project into useful information using a well-defined series of steps and assumptions. Creating a landslide hazards map provides a simple example. One might define a model such that if an area has a steep slope and consists of a shale rock unit, then it should be rated as hazardous. The raw data layers of geology and slope

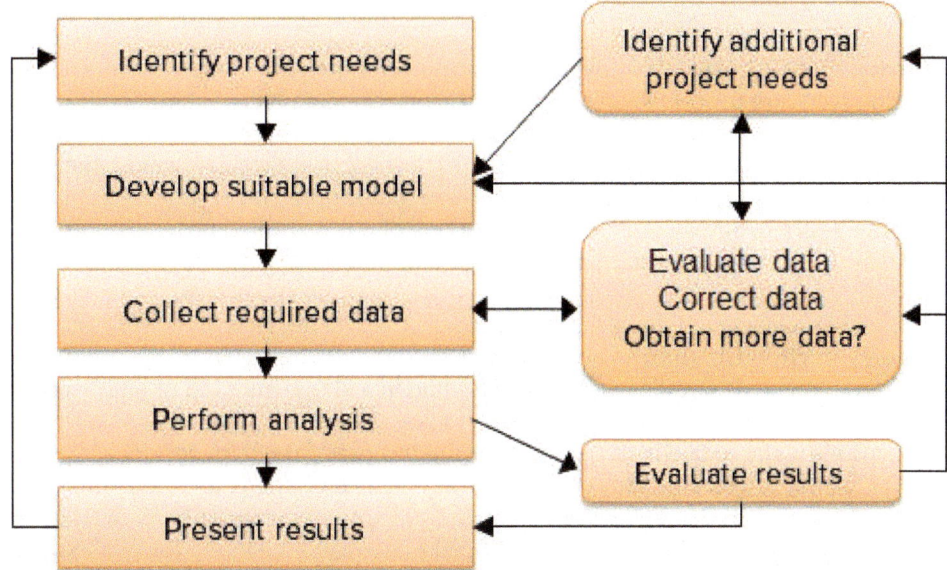

Fig. I.2. Generalized flow diagram representing steps in a GIS project

can be used to create the hazard map. A more complex model might also take into account the dip (bedding angle) of the shale units. Models can be very simple, as in this example, or more complex with many different inputs and calculations.

Once needs are known and the appropriate models have been designed, data collection can begin. GIS data are stored as layers, with each layer representing one type of information, such as roads or soil types. The needs dictate which data layers are required and how accurate they must be. A source for each data layer must be found. In some cases data can be obtained free from other organizations. General base layers, such as elevation, roads, streams, political boundaries, and demographics, are freely available from government sources (although the accuracy and level of detail are not always what one might wish). More specific data must often be developed in-house. For example, a utility company would need to develop its own layers showing electric lines and substations—no one else would be likely to have it.

The spatial detail and accuracy of the data must be evaluated to ensure that they can meet the needs of the project. For example, an engineering firm creating the site plan for a shopping mall could download elevation contours of the site for free. However, standard 10-foot contours cannot provide the detailed surface information needed by the engineers. Instead, the firm might contract with a surveying firm to measure contours at half-foot intervals.

After the data are assembled, the analysis can begin. During the analysis phase, it is not unusual to encounter problems that might require making changes to the model or the data. Thus, the steps of model development, data collection, and analysis often become iterative, as experience gained is used to refine the process. The final result must be checked carefully against reality to identify any shortcomings and to provide guidelines for improving future work.

Finally, no project is complete until the results have been communicated, whether shown informally to a supervisor, published in a scientific journal, or presented during a heated public meeting. Presentations may include maps, reports, slides, or other media.

Web GIS

Since the advent of inexpensive and powerful desktop computers, GIS work has been done mainly by using software programs on these computers. The data have usually resided on the same computer or on a file server on a local area network. The data were often stored in proprietary formats recognized only by the software that created them. However, the urge to share information, the expansion of the Internet, the popularity of smartphones and other mobile devices, and the desire to spread GIS beyond the scope of a few trained professionals have all driven the development of an alternate model, known as Web GIS.

Web GIS comprises a system of components connected by the Internet. It uses the same components as a desktop GIS (hardware, software, stored data, etc.), but those components may be spread across thousands of miles instead of residing in a single building or single office.

Many organizations make data GIS available over the Internet. Free data services, such as Google Earth and MapQuest, provide access to huge volumes of image and map data. These services are tied to specific web applications and mobile apps and cannot be viewed in ArcGIS. Although the data quality and documentation procedures are not designed for professional-level work, the volume and popularity of these sites introduce many people to GIS techniques and data. Other sites, such as the United States Geological Survey's National Map, do allow users to download data.

GIS servers provide geospatial data over Internet connections. Hosting GIS data on a server requires software that responds to requests from users for specific maps or data in the user's current window. Some GIS services are open and free, designed to provide public access to data and maps. Others may be locked down to a particular organization or group of users. Many organizations use servers to make in-house data available to personnel out in the field or to allow employees in different locations to access the same company data sets. Servers provide data in a client-neutral format; as long as the client program knows how to use the service, the data can be used in ArcGIS Desktop, in a web application, on a tablet, or on a smartphone.

Setting up and managing a GIS server is a complex and expensive task requiring suitable equipment, software, and expertise. Cloud-based services are lowering this barrier. A **cloud** consists of warehouses of computers and hard drives managed by a company that rents processing power and disk space to clients. Some companies offer space to individuals to back up their computers or store movies and music files in a place that is accessible to all of their devices.

Many organizations have moved to the cloud to host their GIS data, either as a complement to or even instead of housing their own GIS servers. Advantages of cloud services include the speed and ease of deployment, the ability to scale up and down to meet changing demand, and the security benefits of hosting public information outside the organizational firewall.

ArcGIS Online

ArcGIS Online (AGOL) is a cloud-based platform that provides an environment to create and share GIS data and tools. It also serves the role of a GIS community. Users can create and join groups with common interests. Groups may be open to everyone, such as a group dedicated to people who are interested in bicycling and want to share maps about trips and routes. Or they may be secure and restricted, allowing a group of researchers to share information privately among members invited to be in the group.

ArcGIS Online is designed around the **web map**, an interactive map based solely on GIS services. Web maps perform a restricted set of basic functions, such as zoom and query; however, they are device independent and can be used in web browsers, on mobile devices like smartphones or tablets, and even within social media sites. They are simple to create and share.

Using servers and web maps to collect data on smartphones or tablets has become a widespread activity. It is relatively simple to set up a feature service and configure it for editing. Users with a free account can use the ArcGIS mobile application to access web maps and collect data; some students at the South Dakota School of Mines and Technology used it to map trees on campus and rate damage caused by a blizzard (Fig. I.3). Users with a subscription account can use the more powerful Collector application, which is capable of attaching photos and editing while out of the range of cell or wireless services.

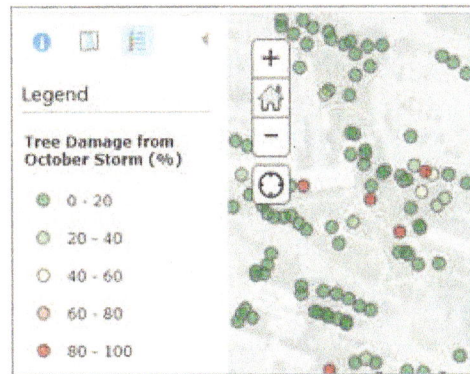

Fig. I.3. Students used smartphones to map damaged trees after a blizzard. Source: Esri

The growing capabilities of ArcGIS Online are turning it into an indispensable tool. ArcGIS Desktop will remain the primary engine for creating, mapping, and analyzing data for quite some time yet because it is far more powerful than the web services, but Desktop users certainly benefit from web services.

Types of accounts

Guests can explore featured maps without creating an account in order to learn about the types of information available. However, an account is needed to access most of AGOL's capabilities. There are two main types of accounts.

A **public account** gives access to some of the data and some of the functionality, allowing users to view data, create map mashups by combining multiple layers of data, save maps, and share maps with others. Purchasers of an Esri product are granted an Esri Account, which is automatically linked to an ArcGIS Online account. Alternatively, one can request a free public account or sign up with an account using a social media login, such as Facebook or Twitter.

Content is categorized as subscriber or premium content. **Subscriber content** is only open to users with an organizational account (as opposed to a public account). **Premium content** requires an organizational account and also consumes service credits, which must be purchased.

Content in AGOL may be shared with others. By default, content is private and visible only to its creator, or it may be shared with a specific group, with everyone in an organization, or with everyone in AGOL.

ArcGIS Pro is designed to integrate directly with the capabilities of ArcGIS Online. An AGOL account is required to use ArcGIS Pro. On program start-up, you will be asked to supply a user account for ArcGIS Online.

Before starting to learn ArcGIS Pro, it is helpful to get familiar with ArcGIS Online in its web browser form. Because it is a live system that is always being improved, it is best to use the online tutorials it provides. Here's how to get started; an account is not required for the first few lessons.

▶ Type arcgis.com into a web browser.

▶ Select Help from the list of links in the upper left corner of the browser.

▶ Scroll down to Take a Lesson.

▶ Click "Get Started with ArcGIS Online."

Learning more about GIS

This book covers many aspects of GIS, but it cannot cover all useful topics. Moreover, the industry evolves rapidly. Completing the lessons in the book is only the first step to mastering GIS, and smart users constantly seek new information and training to build experience and skill. Many methods are available to become more proficient.

▶ **Read the Help.** Maybe you can figure out a cell phone without the manual, but don't turn away from the wealth of information provided in the ArcGIS Pro Help. It doesn't just have instructions for doing things, but it also includes discussions, diagrams, references, and other ways to explore the concepts behind GIS and its implementation.

▶ **Use the Virtual Campus.** Esri has dozens of online courses and seminars available to registered users. Talk to your GIS instructor about getting access to these courses.

▶ **Build a GIS library.** GIS integrates many disciplines, including geography, surveying, cartography, statistics, computer science, spatial statistics, and so on. The more you know of these disciplines, the easier it is to understand not just the *how* of doing things, but also the *why*.

▶ **Join a professional organization.** Many professional societies cater to GIS and remote sensing. They have newsletters and journals, conferences, and lots of professionals who can help answer questions and share their experiences.

▶ **Join GeoNet.** Esri hosts an online GIS community named GeoNet that covers many aspects of GIS. Members can search for answers to questions or ask questions and get answers in a day or two. Groups are available for members with different interests. It is a great place to go when you get stuck and have no one else to consult, and it is a good place to learn about bugs and workarounds.

Chapter 1. What Is GIS?

Objectives

▶ Understanding how real-world features are represented by GIS data

▶ Knowing basic differences between the raster and vector data models

▶ Using and discussing map scales and source scales

▶ Recognizing local and Internet sources of GIS data

▶ Cultivating awareness of the types of shortcomings associated with GIS data

▶ Learning to correctly cite data sets on a map or in a report

▶ Learning the basics of the ArcGIS Pro interface and viewing map data

Mastering the Concepts

GIS Concepts

A geographic information system (GIS) provides data structures and capabilities for storing, analyzing, managing, and publishing map data using a computer. Many different software packages have been developed to accomplish this task. Although this book focuses on a specific software product, many of the concepts are fundamental to all GIS programs, even though the details of implementation may differ. In this book, the first section of each chapter, *Mastering the Concepts*, focuses on these cross-cutting fundamental ideas. The second section, *About ArcGIS*, discusses specific details of ArcGIS Pro, a specific GIS software product published by the company Esri™ (originally the Environmental Systems Research Institute, Inc., or ESRI).

Representing real-world objects on maps

To work with maps on a computer requires storing different types of map data and the information associated with them. Geographic information falls primarily into two categories: *discrete* and *continuous*. **Discrete** data are objects in the real world with specific locations or boundaries, such as houses, cities, roads, or counties. **Continuous** data represent quantities that may be measured anywhere on the earth, such as temperature or elevation. Many different data formats have been invented to encode data for use with GIS programs; however, most follow one of two approaches: the **vector** model, which is designed to store discrete data, or the **raster** model, which is designed to store continuous data. Both models are **georeferenced**, meaning that the information is tied to a specific location on the earth's surface using spatial (x-y) coordinates defined in a standard way. The familiar latitude and longitude values are one example; we will learn more about coordinate systems in Chapter 4. Data and software used to work with georeferenced information are encompassed by the general term **geospatial**.

The vector model uses three basic shapes to represent discrete objects, known as **features**. A **point** feature can represent a single location such as a well or a weather station. **Line** features represent linear objects like roads and rivers. A **polygon** feature represents a closed area such as a county or a state. Vector data sets include two parts: one or more x-y locations that place the feature on the earth's surface (its shape), and a **table** that stores **attributes**, or information about each feature, such as a state's name, abbreviation, population, and so on. A set of similar features stored together, such as the 50 states comprising the United States, is called a **feature class**. A feature class can store only one type of feature: points or lines or polygons, but never a combination, and the features should be the same type of thing. Counties and states, even though both are represented by polygons, should be stored in different feature classes.

The raster model breaks a geographic area into small squares known as **cells** or **pixels**, each one containing a number representing a value being measured or characterized. Some rasters store physical quantities like elevation, temperature, or precipitation. Others may store color and brightness values that make up aerial photographs, satellite images, or a scanned paper map. Even discrete features, such as land use or soil type polygons, may be represented by rasters if needed, even though the model is not optimized for that task.

Figure 1.1 illustrates examples of geospatial data. The green triangle is a point feature representing the location of a fire lookout tower. The blue line features represent streams. The brown shapes are polygons representing the preferred habitat areas for a species of snail. The background shows a raster composed of 30 m × 30 m square cells recording elevation values, with the green values representing the lowest elevations and the white cells the highest elevations.

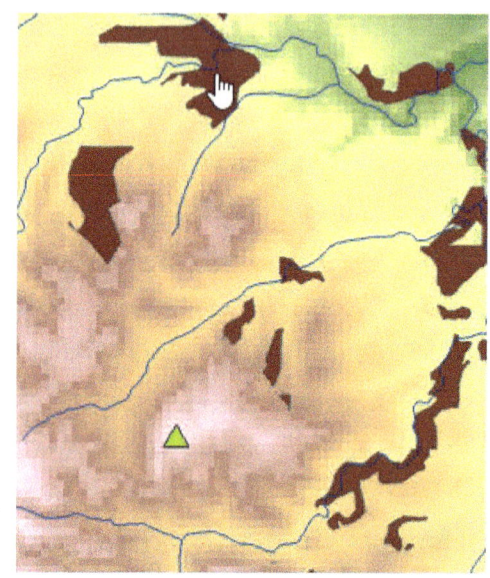

Fig. 1.1. Maps may be composed of multiple layers of different types of information.
Source: USGS

Maps are often constructed from multiple data sets. Each data set is typically called a **layer** once it is placed into a map, and the layers are drawn in a hierarchical order defined by the cartographer, usually with raster or polygon layers on the bottom and point or line layers on top (so that the top layers do not cover up the bottom layers). Figure 1.1 contains four different layers from bottom to top: a raster layer of elevation, a polygon layer of snail habitat (brown), a line layer portraying streams (blue), and a point layer representing the fire tower (green triangle).

Map scale

The act of taking a set of GIS features with x-y coordinate values and drawing them on a screen or printing them on paper establishes a map scale. On a paper map, the scale is fixed at the time of printing. In a digital map configured for interactive display, the scale changes every time the viewer zooms in or out.

What is map scale?

Map scale is a measure of the size at which features in a map are represented. The scale is expressed as a fraction, or ratio, of the size of features represented by the map to the size of the actual objects on the ground. Because it is expressed as a ratio, it is valid for any unit of measure. For the common US Geological Survey topographic map, which has a scale of 1:24,000, one inch on the map represents 24,000 inches on the ground. The map scale can be used to determine the true size of any feature on the map, such as the width of a lake (Fig. 1.2). Measure the lake with a ruler and then set up a proportion such that the map scale equals the measured width over the actual width (x). Then

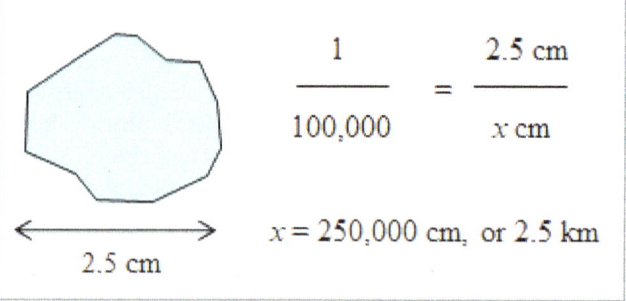

$$\frac{1}{100,000} = \frac{2.5 \text{ cm}}{x \text{ cm}}$$

$$x = 250,000 \text{ cm, or } 2.5 \text{ km}$$

Fig. 1.2. Solving for the size of a lake for a 1:100,000 scale map

solve for x. Keep in mind that the actual width and the measured width will have the same units, which can be converted if necessary.

Often people or publications refer to large-scale maps and small-scale maps. A large-scale map is one in which the *ratio* is large (i.e., the denominator is small). Thus, a 1:24,000 scale map has larger scale than a 1:100,000 scale map. A large-scale map shows a relatively small area, such as a quadrangle, whereas small-scale maps show bigger areas, such as states or countries.

The type of object used to represent features depends on the scale of the map. A river would be represented as a line on a map of the United States because at that scale it is too small for its width to encompass any significant area on the map. If one is viewing a USGS topographic map, however, the river encompasses an area and is represented as a polygon.

Source scale

When data are stored in a GIS, they technically do not have a scale because only the coordinates are stored. They acquire a scale once they are drawn on the screen or on paper. However, a geographic data set has a **source scale**, the original scale or resolution at which it was converted to digital form. Roads on a 1:24,000 quadrangle map are represented in greater detail and with greater accuracy than a highway map for an entire state, such as might be found in an atlas. If the roads were converted to a GIS data set by digitizing or scanning, the detail and accuracy would be carried forward to the digital format. The source scale of a data set is an important attribute and is included when documenting it.

Imagine using data for a project focused on the state of New Jersey. A letter page–size map of the state would have a scale of about 1:2,000,000. A map of a county might have a scale of about 1:300,000. Examine the three different feature classes in Figure 1.3, which shows the state boundary in the Sandy Hook area at a scale of 1:300,000. The purple, orange, and black lines represent three different data sets with source scales of 1:25 million, 1:5 million, and 1:50,000, respectively. The purple feature class was developed for national-scale maps and is a poor representation of New Jersey at this scale. The orange one might be used for statewide mapping, and the black one would be suitable for use at the county level.

One should exercise caution in using data at scales very different from the original source scale. A pipeline digitized from a paper 1:100,000 scale map has an uncertainty of about 170 feet in its location, the approximate thickness of the line on the paper. Displaying the pipeline on a city map at 1:10,000 might look all right, but it could be many feet away from the actual location.

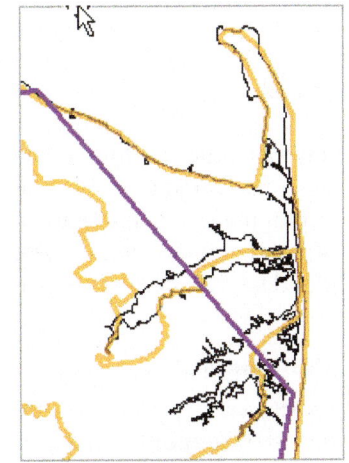

Fig. 1.3. Three New Jersey data sets with different source scales
Source: Esri

From looking at Figure 1.3, one might conclude that we should always obtain and use data at the largest possible scale. However, large-scale data require more data points per unit length, increasing data storage space and slowing the drawing speed. Every application has an optimal scale, and little is gained by using information at a larger scale than needed.

GIS data may be assigned a **scale range**, which specifies the range of scales for which the display of the data is valid so that it will only be shown at those scales. Many web-based maps use scale ranges to present different data sets at different scales, with increasing detail as the user zooms further in.

Local and Internet data sources

The data used in a GIS can come from diverse sources. For many years, the primary source of information was the hard drive of the computer running the GIS software. New data might be downloaded from a data clearinghouse or copied from a CD or DVD. If the computer was part of a local area network set up by a university or company, it might also have access to shared information on a network drive. This capability was especially helpful for large data sets (they were stored only once) or shared data that were often updated (so edits did not have to be propagated to multiple computers). Local and network drives are still important places to store GIS data.

More recently, the Internet has become a major source for geospatial data. Many organizations make data available over the Internet using standard protocols that allow a variety of different devices (**clients**) to access them. Data services such as Google Earth provide access to huge volumes of aerial photography and street and transportation information. Often these services are not accessed directly by GIS programs but instead are used through web pages, smartphone apps, or automotive navigation devices. These services differ from traditional GIS data in that users may freely access information as needed, but they cannot download or store the entire data sets on the local client. Moreover, the developers of clients that use these resources may be charged an access fee to incorporate the services. Creating and updating these giant data sets is an expensive and time-consuming business.

GIS servers

GIS servers provide geospatial data over Internet connections specifically for GIS software to use. Hosting GIS data on a server requires software that responds to requests from users for specific maps or data in the user's current window. Some GIS servers are open and free, designed to provide public access to data and maps. Others may be locked down to a particular organization or group of users. Many organizations use servers to make in-house data available to personnel out in the field or to allow employees in different locations to access the same company data sets. Servers provide data in a client-neutral format; as long as the client program knows how to use the service, the data can be used in ArcGIS Desktop, in a web application, on a tablet, or on a smartphone.

Setting up and managing a GIS server is a complex and expensive task requiring suitable equipment, software, and expertise. Some organizations have moved to the **cloud** to host GIS data, either as a complement to or even instead of housing their own GIS servers. A cloud consists of warehouses of computers and hard drives managed by a company that rents processing power and disk space to members. Some companies offer space to individuals to back up their computers or store movies and music files in a place that is accessible to all of their devices; other customers include companies who have found it less expensive or more convenient than managing their own servers. Advantages of cloud services include the speed and ease of deployment, the ability to scale up and down to meet changing demand, and the security benefits of hosting public information outside an organizational security firewall. **ArcGIS Online** is a cloud service commonly used with ArcGIS Pro.

Data quality

Representing real-world objects as points, lines, polygons, or rasters always involves some degree of **generalization**, or simplifying the data for digital storage, such as turning a house into a rectangular polygon or even a point. No data file can capture all the spatial or attribute qualities of any object. The degree of generalization increases as the map scale decreases. On a standard topographic map, a river has a width and can be modeled as a polygon with two separate banks. A city would be shown as a polygon area. For a national map, however, the river would simply be shown as a line, and a city would be shown as a point.

Even a detailed representation of an object is not always a true characterization. Rivers and lakes can enlarge in size during spring flooding or shrink during a drought. The boundary of a city changes as the city grows. Users of GIS data must never forget that the data they collect and use will contain flaws and that the user has an ethical and legal responsibility to ensure that the data used for a particular purpose are appropriate for the task. When evaluating the quality of a data set, geospatial professionals consider the following aspects.

Geometric accuracy describes how close the x-y values of a data set correspond to the actual locations on the earth's surface. Geometric accuracy is a function of the source scale and of how the data were captured. Surveying is one of the most accurate ways to position features. GPS units have an accuracy that ranges from centimeters to tens of meters. Maps derived from aerial photography or satellite imagery can vary widely in geometric quality based on factors such as the scale of the image, the resolution of the image, imperfections and distortions in the imaging system, and the types of corrections applied to the image. In Figure 1.4, notice that the vector road in white is offset in places from the road as it appears in the aerial photo. These differences can arise from digitizing errors, geometric distortions from the camera or satellite, or other factors.

Fig. 1.4. Aerial photo near Woodenshoe Canyon, Utah
Source: Google Earth and TeleAtlas

Moreover, not every feature can be as precisely located as a road. Imagine mapping the land-cover types: *forest, shrubland, grassland,* and *bare rock* in Figure 1.4. Where does one draw the line between *shrubland* and *grassland*? At what point does the *shrubland* become *forest*? Six people given this photo would create six slightly different maps. Some boundaries would match closely; others would vary as each person made a subjective decision about where to place each boundary.

Thematic accuracy refers to the attributes stored in the table. Some types of data are relatively straightforward to record, such as the name of a city or the number of lanes in a road. Even in this situation, the value of a feature might be incorrectly recorded. Other types of information can never be known exactly. Population data, for example, are collected through a process of surveying and self-reporting that takes many months. It is impossible to include every person. Moreover, people are born and die during the survey process, or they are moving in and out of towns. Population data can never be more than an estimate. It is important to understand the limitations and potential biases associated with thematic data.

Resolution refers to the sampling interval at which data are acquired. Resolution may be spatial, thematic, or temporal. Spatial resolution indicates at what distance interval measurements are taken or recorded. What is the size of a single pixel of satellite data? If one is collecting GPS points by driving along a road, at what interval is each point collected? Thematic resolution can be affected by using categories rather than measured quantities: if one is collecting information on the percent crown cover in a forest, is each measurement reported as a continuous value (32%) or as a classified range (10–20%, 20–30%)? Temporal resolution indicates how often measurements are taken. Census data are collected every 10 years. Temperature data taken at a climate station might be recorded every 15 minutes, but it might also be reported as a monthly or yearly average.

Precision refers to either the number of significant digits used to record a measurement or the statistical variation of a repeated single measurement. Many people confuse precision with accuracy, but it is important to understand the distinction. Imagine recording someone's body temperature with an oral digital thermometer that records to a thousandth of a degree and getting the value of 99.894 degrees Fahrenheit. This measurement would be considered precise. However, imagine that the reading is taken immediately after the person drinks a cup of hot coffee, which causes the thermometer to measure a value higher than true body temperature. Thus, the measurement is precise, but it is not accurate.

Logical consistency evaluates whether a data model or data set accurately represents the real-world relationships between features. In the real world, for example, two adjacent states share a common boundary that is exactly the same (Fig. 1.5). In a database, however, the states might be stored as two separate features with slightly different boundaries. Lines representing streets should connect if the roads they represent meet. A line or a polygon boundary should not cross over itself. A county should not extend past the boundary of its state. It takes special effort to create and manage feature relationships within and between data sets, so not all data sets are logically consistent.

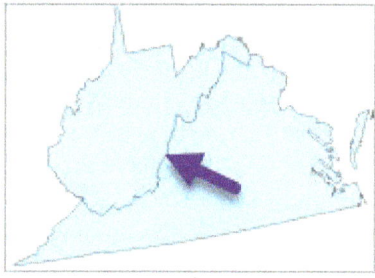

Fig. 1.5. A coincident boundary gets stored twice but is the same for both features.
Source: Esri

Evaluating the quality of a data set can be difficult, especially if the data were created by someone else. Professionals who create data usually also provide **metadata**, which store information about the data set, such as where it came from, how it was developed, who assembled it, how precise it is, and whether it can be given to another person. The user can use the metadata to decide whether a particular data set is suitable for a given purpose. Metadata may consist of a brief description of the most critical information, or they may include many pages of detail.

Citing GIS data sources

Ethical and professional considerations require that any map, publication, or report should cite the data source(s) used and give proper credit to the originators of the data. The metadata or the site from which the data were acquired are good sources of information for citations. The best practice is to record the citation when the data are obtained so the information is available when needed. Generally one cites only data that are publically available (free or purchased). Data created internally within the workplace need not be cited, although often the company name or logo will appear on the map. A data set provided once in response to a personal request should be cited as a personal communication. On a map, only one citation need be included for multiple data sets from the same source, but a report should list all data sets from each source.

Keep in mind that the place where data are found may not be their original source. A GIS administrator may have placed often-used data sets, such as Esri™ Data and Maps, on a university file server for easy access, but the fact that they were obtained internally does not free students from the need to cite them, and the original source should be cited, not the local server.

General format

The purpose of a citation is to allow other people to find and obtain the data. It is not always possible to find all the information needed for a complete citation, but one should do one's best to make it as complete as possible. The following general format for citations may be used:

Data set name (Year published) [source type]. Producer name, producer contact information. *Resource URL: [Date accessed]*.

Data set name. The name is assigned by the creator or provider of the data.

Year published. Some data sets are assembled and provided once or at long intervals, and these are considered to have a publication date. For example, the Esri™ Data and Maps product is released in revised form with each version of the software and carries a publication year. Aerial photography is flown on a particular date (although mosaics such as Google Earth use multiple sources spread over several years). It may take a little hunting to find the publication date. Some data sets are updated at shorter intervals or are even live. These can be assigned the current year in which they were accessed.

Source type. Indicate the format in which the data are available. Types might include physical media (DVD, CD-ROM), a file downloaded from the Internet, or a service that provides live data on demand. Different types of services exist, and new types are being added all the time.

Producer name. Give the name of the person or agency that makes the data available. In some cases, this source may be different from the originator of the data. For example, Esri publishes Data and Maps using data from many different sources. ArcGIS Online serves public data.

Producer contact information. For a company or small agency, the city and state should be included. For large agencies, particularly those with many offices but a unified web presence, the name itself is sufficient. Indicate a clearinghouse name, such as ArcGIS Online, if appropriate.

Resource URL. This entry is optional; include when appropriate. Use only static URLs, even if it means the user has to hunt for the data. (A static URL always has the same form, whereas a dynamic URL is generated automatically based on search strings or other information. Dynamic URLs generally include query control characters such as ? or %.)

Date accessed. This entry is optional because it mainly applies to online data sets. Include the year and month when you accessed or downloaded the data.

Examples of citations

Black Hills RIS Vegetation Database (2008) [downloaded file]. Black Hills National Forest, Custer, SD. URL: http://www.fs.usda.gov/main/blackhills/landmanagement/gis [August, 2010].

Esri™ Data and Maps (2012) [DVD]. Esri™, Inc., Redlands, CA.

National Hydrography Dataset (2015) [downloaded file]. United States Geological Survey on the National Map Viewer. URL: http://viewer.nationalmap.gov/viewer/ [July 23, 2015].

USA Topo Maps (2009) [map service]. Esri™ on ArcGIS Online. URL: http://server.arcgisonline .com/arcgis/services/USA_Topo_Maps/MapServer [January 1, 2012].

EIA Coalbed Methane Field Boundaries (2011) [map service]. US Department of Energy on Arc-GIS Online. URL: http://arcgis.com [August, 2013].

Mineral Operations of Africa and the Middle East (2010) [layer package]. J.M. Eros and Luissette Candelario-Quintana on ArcGIS Online. URL: http://ArcGIS.com [July, 2012].

Badlands National Park GIS Database (2012) [CD-ROM]. Interior, South Dakota: National Park Service—Badlands National Park, personal communication. Airports (2018) [file geodatabase feature class] Price, M.H, Mastering ArcGIS Pro First Edition (tutorial data) mgisdata\Oregon \oregondata.gdb\airports, [01/04/2018]

About ArcGIS

ArcGIS overview

ArcGIS is developed and sold by Esri™. It has a long history and has been through many versions. Originally developed for large mainframe computers, it has metamorphosed from a system based on typed commands to a graphical user interface (GUI). Data models, too, have changed over time,

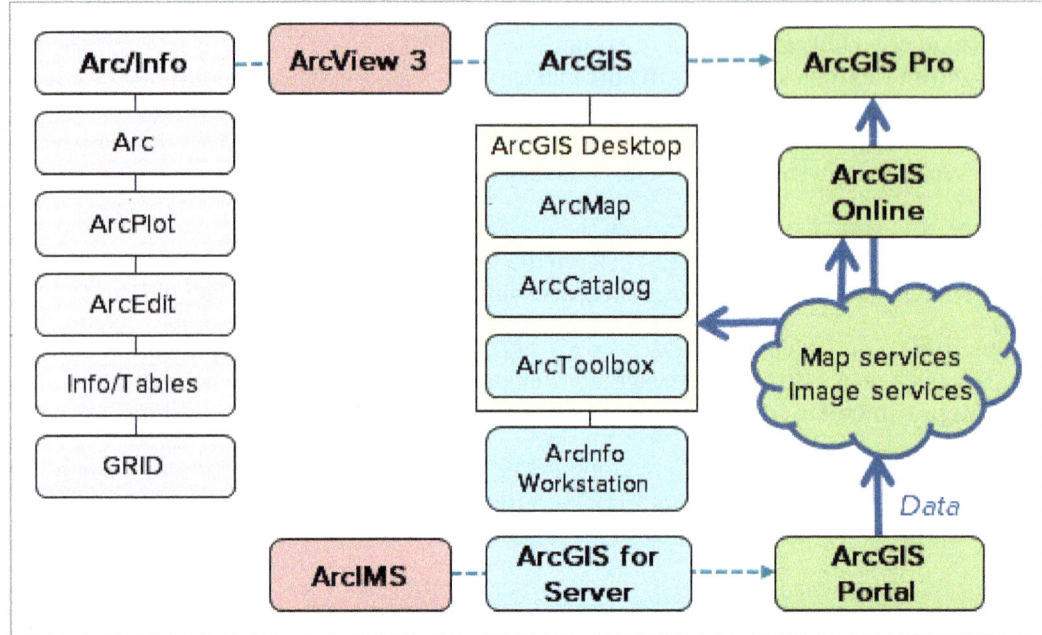

Fig. 1.6. The original Arc/Info program evolved through ArcView and ArcGIS Desktop to ArcGIS Pro, as web GIS servers evolved from ArcIMS and ArcGIS for Server to ArcGIS Portal.

with an increased emphasis on accessing data through the Internet. The following background information should help you to understand the nature of the ArcGIS system and its data.

The old core of the ArcGIS system was initially released in 1982. Called Arc/Info, it included three main programs: Arc, ArcEdit, and ArcPlot (Fig. 1.6), which were built to run on mainframe computers or Unix-type workstations. All of the programs were command based, meaning that the user typed commands to make the program work. The difficulty of learning Arc/Info prompted Esri™ to create ArcView 3 in 1992, a GUI-based program that was easier to use but not as powerful as Arc/Info.

ArcGIS Desktop, released in 2001, combined the power of Arc/Info and the ease of ArcView, and it was ported to the Windows operating system so it could be used on a PC instead of an expensive workstation (earning the name desktop GIS). It contained two programs: ArcMap, which could display, analyze, and edit spatial data, and ArcCatalog, which managed the spatial data files. ArcToolbox contained task-based programs, or tools, that could be used in either program.

Arc/Info, ArcView, and ArcGIS Desktop were originally designed to use GIS data files residing on a local or networked hard drive. However, as Internet-based services expanded, Esri™ began to develop ways to serve GIS data over the World Wide Web. ArcGIS Portal (following its earlier incarnations as ArcIMS [Internet Map Server] and ArcGIS for Server) was built to enable sharing of GIS data. These systems store geospatial data within a commercial database, such as Oracle or SQL Server, and they interact with the World Wide Web to provide data to different clients, including desktop GIS, mobile phones, and web browser applications. These systems can provide both GIS data and GIS analysis tools.

ArcGIS Pro (often shortened to Pro in this text) is the latest version of the desktop GIS. It was designed to take full advantage of multicore 64-bit processors and to easily use GIS services. It has a new interface design and a new system for organizing work. Because of its recent development, it does not yet incorporate all of the functionality provided by ArcMap and ArcCatalog, but new capabilities are added with each version.

Both ArcGIS Desktop and ArcGIS Pro provide different levels of functionality (Basic, Standard, and Advanced) so that users can save money by buying only the level they need. This book primarily uses the functions available in ArcGIS Basic. The platform may be expanded by purchasing **extensions**, or additional software capabilities designed for specific applications such as analyzing raster data (the Spatial Analyst extension) or doing advanced spatial statistics (the Geostatistical Analyst extension). By default, the extensions are installed with the main program but require an additional license in order to be used.

ArcGIS Online (AGOL) is a cloud-based platform that provides many GIS data services as well as an environment to create and share maps and data with other users. AGOL is structured around the **web map**, a map that uses only GIS services. These web maps can be consumed by a variety of clients, including ArcMap, ArcGIS Pro, web pages, smartphones, and tablets.

ArcGIS Pro is designed to integrate directly with the capabilities of ArcGIS Online. The first time the program is started, the user is asked to supply an AGOL user account, which will usually be provided by the course instructor. An account may be a free public account or a subscription account, which provides access to more data and features than a public account. A few of the exercises in this book require a subscription account.

> **TIP:** If you don't have access to a subscription account through your instructor, Esri™ offers a 60-day trial that includes ArcGIS Online and ArcGIS Pro.

The project

Work in ArcGIS Pro is structured within an entity called a **project**, which organizes many types of information needed to do GIS work. Projects are usually developed for a particular geographic area or to accomplish a set of tasks associated with a work project (hence the name). Projects may contain many types of items. Figure 1.7 shows the ArcGIS Pro interface, currently displaying the contents of a project named PacificNorthwestHazards. The center program area portrays **views**, or different objects that the user might work with, such as maps or tables. A **Contents pane** on the left shows different map layers, such as the volcanoes and capitals; the Catalog pane on the right organizes the different items available within the project; they are discussed next in the order in which they appear in Figure 1.7.

Map. Maps are views that display GIS data sets together using specified symbols, labels, and so on. A map may be two dimensional (Hazards in Fig. 1.7), or it may be visualized in three dimensions, in which case it is called a **scene**.

Toolbox. ArcGIS Pro includes many distinct program routines (tools) used to accomplish specific tasks. A user can also create custom tools (models) or write programs (scripts) in a language called Python. Tools created for a specific project can be saved within the project for easy access.

Database. Databases that contain geographic data are often called geodatabases. Every project starts with a **home geodatabase**, a place to store GIS data associated with and generated by work on the project. The geodatabase adopts the name of the project (PacificNorthwestHazards) and is the default location where output data sets are saved.

Geoprocessing history. The project keeps track of every tool run so that one can review previous steps or check what settings were used to run a tool.

Layout. Layouts are formal page designs for printing maps, either on paper or to an electronic format such as PDF. Layouts can portray one or more maps and contain titles, legends, scale bars, north arrows, and so on.

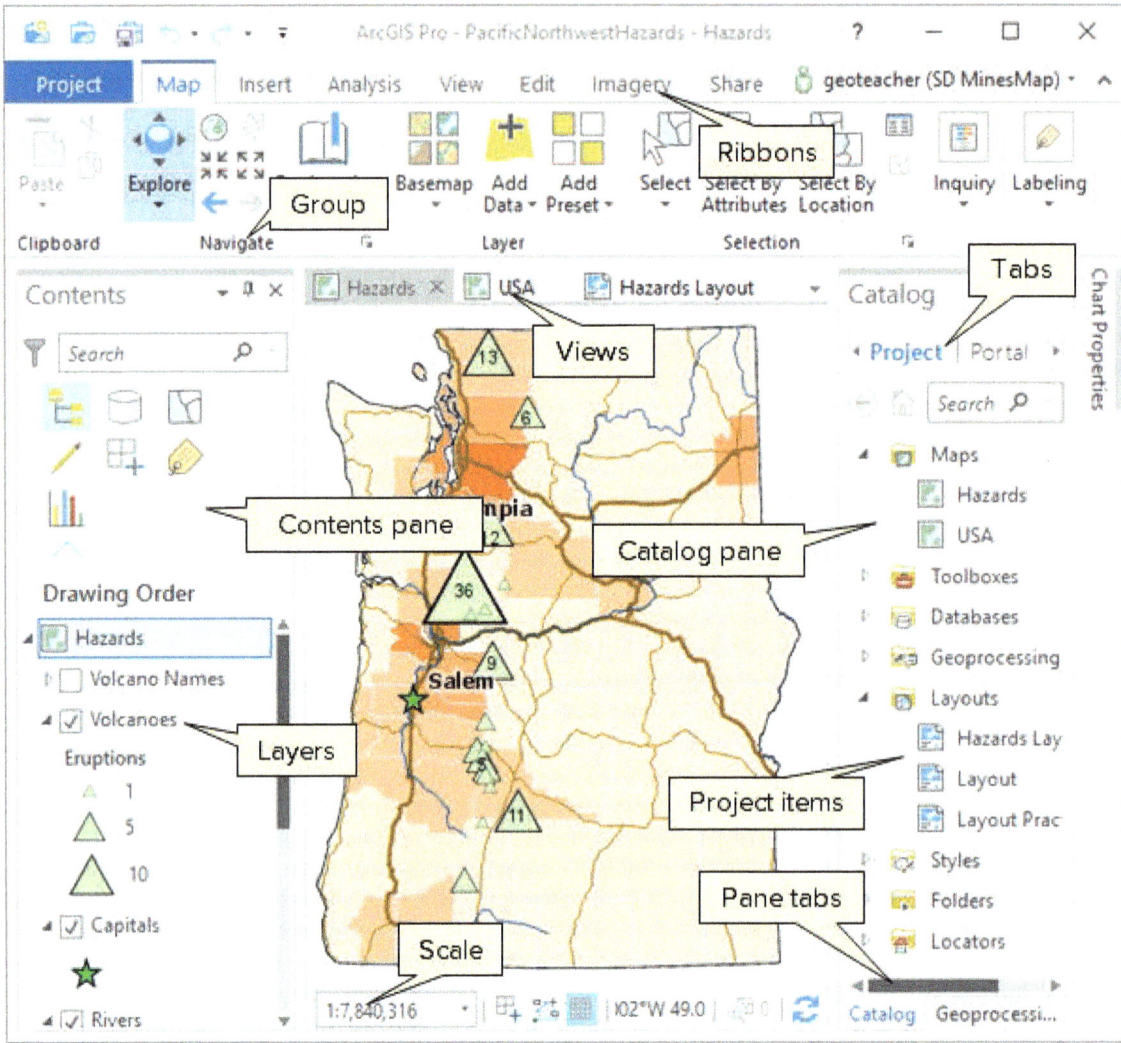

Fig. 1.7. The ArcGIS Pro program interface
Source: Esri

Style. Styles are collections of symbols used to portray map data. ArcGIS Pro comes with default styles (2D and 3D) containing symbols that may be customized individually to meet a specific need. In addition, users or organizations can develop and share their own styles; for example, a set of symbols used to publish geologic maps.

Folder. By default, the **project file** (e.g., PacificNorthwestHazards.aprx) is created inside a folder with the same name as the project, known as the **project folder**. The home geodatabase also resides here. The project folder can be used to store non-GIS data associated with the project, such as word processing documents or spreadsheets. (If the project contains a great deal of additional data, subfolders can be created within the project folder to organize it.) Although some of the data used in a project will be stored in the home geodatabase, it is common to use external data as well, such as information stored elsewhere on the computer or on an organization's file server. A project stores **folder connections** that remember the locations of these data sources for quick access.

Locator. Locators are used to convert street addresses into points on a map by comparing them to data about streets and their address ranges. ArcGIS Online has a default locator, the World

Geocoding Service, which it uses to find a location from a single typed address. Users can create custom locators if they possess the right kind of street data.

Task. Tasks are step-by-step instructions set up to record a certain workflow for future reference or to share with others to make sure that a procedure is correctly performed each time. One can record a series of steps as they are performed and then explain them with detailed instructions.

Portal. Portals provide access to GIS services available over the Internet. Several types of portals are included: (1) data and services hosted in your ArcGIS Online account; (2) public services hosted by others in AGOL; and (3) any other services hosted by organizations to which you belong and have access (a login and password may be required).

The ArcGIS Pro interface

ArcGIS Pro is a powerful program that does many things, so fitting all the tools and commands into a single window on a computer is a daunting task. You must understand the design of the graphical user interface (GUI) to work effectively with the program and to follow the instructions in this book's tutorials.

Pro uses a ribbon-style interface to present access to the functions and commands (Fig. 1.8). It includes **core ribbons**, which are always visible, and **contextual ribbons**, which only appear as you select and work with different parts of the project. For example, when a map layer is clicked, the **Feature Layer** collection of ribbons appears (**Appearance**, **Labeling**, and **Data**) and can be used to modify how a layer looks and behaves (Fig. 1.8).

Within each ribbon, buttons are organized into functional groups. In Figure 1.8, the **Map** ribbon is currently active and shows five groups, including Clipboard, Navigate, Layer, and Selection. Some buttons have a small black arrow underneath (drop-down buttons, like Basemap), which indicates that more than one command is accessible from that button. Simply clicking a drop-down button will access the command currently shown; use the black arrow to select a different button. Some groups have a small boxed arrow in the lower right corner, the Group options button. This button will open properties related to that group so the settings can be modified.

Clicking buttons results in many different actions. Some buttons turn the cursor into a specific tool; others may execute an action (such as zooming in), and some open additional menus, windows, or tools.

There are two basic skills to master in navigating the Pro interface. The first is to learn the terminology for the many program objects (*windows*, *views*, and *panes*) and to learn to

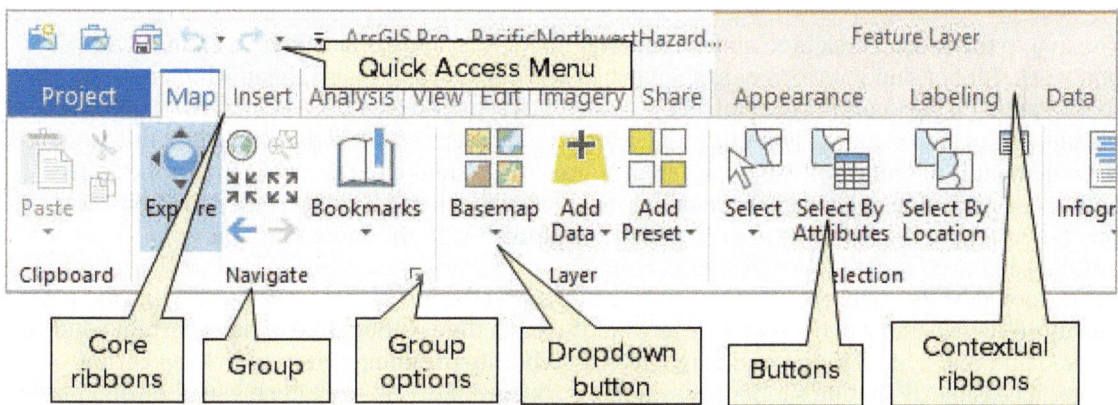

Fig. 1.8. Ribbon organization and terminology
Source: Esri

manage those objects efficiently. The second skill is to get used to the way the interface changes as one works.

Below the ribbon is the main program window. A **window** is a work area that can contain two types of objects, *views* or *panes*. Windows may be floating so that they sit on top of the program area, but more commonly they remain docked within the program area.

A **view** is a window that contains an entity to work with, such as a map, a 3D scene, a table, or a chart. Views can be docked only in the center window of the program, as in Figure 1.7. Multiple views may be open at once, although only one of them can be active at a time.

A **pane** contains commands or settings used to work with the active view (Fig. 1.9). Panes may have multiple tabs used to organize the settings into related sections, icons that present different **panels** containing options or settings, headings that organize settings within a tab, or an Options menu that accesses additional commands or settings. Panes are placed in windows that may be docked around the edges of the program area. Many panes can be open at the same time; they may be placed side by side or stacked on

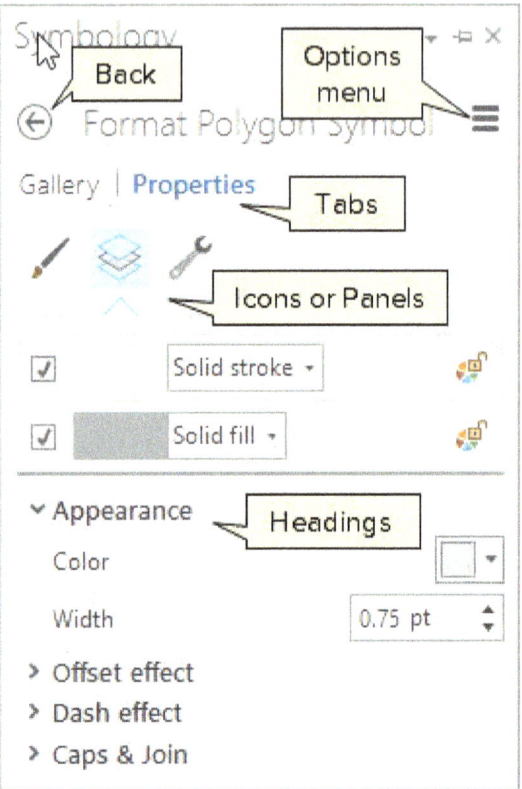

Fig. 1.9. Pane organization and terminology
Source: Esri

top of each other in a window. In Figure 1.7, the right-hand window docked on the side of the program area contains two panes, an active pane (the Catalog pane) and the Geoprocessing pane underneath it (visible as a tab at the bottom of the window).

A single window may contain multiple views or panes (but not both). Within the window, the views or panes may be stacked atop each other, in which case the top one is active. The user switches to a different one by clicking a tab along the top or bottom of the window. Windows also have an auto-hide function, which turns a window into a tab at the edge of the program, keeping it out of the way until it is needed. Clicking the tab opens the window, and moving or clicking elsewhere closes it.

The key to the second task is to understand that the GUI is context sensitive; it changes as you work. Ribbons and buttons only show when it is appropriate to use them, which means that selecting a particular type of object is sometimes required to make the ribbon or buttons visible. For example, a map layer must be selected to change how it looks. Although this flexible arrangement is initially confusing, one soon learns how to make particular menus appear. To make it easier, some objects have a context menu that opens when right-clicked to choose a function directly. Hence, there are often multiple pathways to the same function.

Users soon develop their own preferences for managing the interface. Some people like an uncluttered view and will open and close windows each time. Others like to have certain windows docked and ready at all times, and they develop elaborate grouping schemes to keep certain panes accessible. Sometimes it depends on the workflow: if one is working on symbolizing many map layers, it is handy to keep the Symbology pane open throughout and then close it when ready to analyze data. When editing, it is convenient to group several editing panes in the same window and switch back and forth between them.

TIP: Because the program interface is so flexible, describing how to navigate it in a tutorial is challenging. Learn the meanings and distinctions between the terms *window, view, pane,* and *tab* to better understand the instructions.

Tools and geoprocessing

Many operations in ArcGIS Pro are performed using tools. A **tool** is piece of software designed to perform a specific task, such as deleting a file or calculating statistics. Achieving one's objective may involve executing several tools in sequence; which is called **geoprocessing**. Tools may be accessed several ways: by clicking a button on a ribbon, by choosing an item from a menu, or by selecting a tool in the Geoprocessing pane (Fig. 1.10). Tools are organized functionally into toolboxes (some of them are shown in Fig. 1.10). Examining the contents of the toolboxes is a good way to investigate the many capabilities of ArcGIS, although searching is usually a quicker way to find a specific tool.

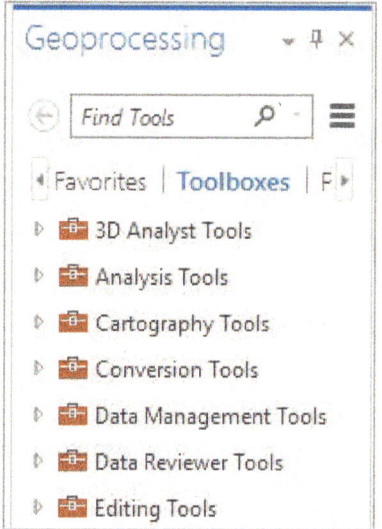

Fig. 1.10. Some of the toolboxes in the Geoprocessing pane
Source: Esri

After opening a tool, the user must set **parameters** that instruct the tool how to proceed. Parameters are different for each tool. Figure 1.11 shows the Buffer tool, which creates polygons showing areas within a certain distance of a set of features. The parameters include the input features (Roads), a name and location for the output polygon dataset (roadbuf300), plus various settings that modify the size and shape of the buffers.

Some common options apply to all tools (shown by numbered squares in Fig. 1.11). (1) The back button clears the settings and returns to the main Geoprocessing pane. (2) The Options button opens a menu with additional choices. (3) The Help button opens a detailed description of the tool. (4) The Environments tab opens additional settings that can be applied to the tool. (5) The information icon appears when the cursor is near a parameter box; it gives information about the parameter. (6) A warning icon appears to notify the user of a potential issue, perhaps that the designated output file already exists and will be overwritten. (7) An error icon appears to indicate that an error is present and the tool will not run until it is corrected. Both warning types will

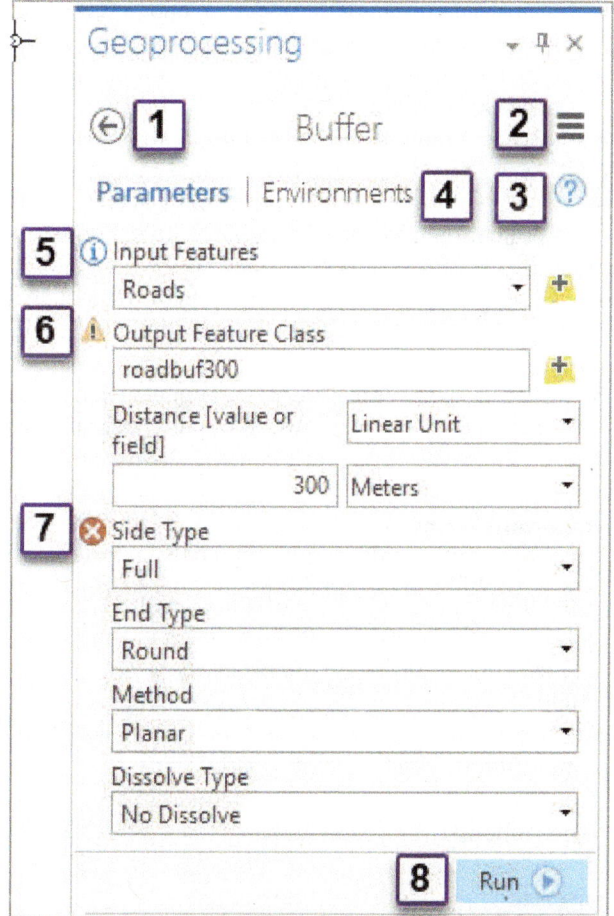

Fig. 1.11. Running a tool
Source: Esri

show a pop-up explaining the issue if the cursor is placed on top of the warning icon. (8) The Run button executes the tool.

TIP: When using a tool for the first time, always read the detailed help information and examine the pop-up information boxes for each parameter. These resources describe the tool and may contain critical information for making it work properly.

ArcGIS Pro includes several ways to automate tools and tasks to make geoprocessing tasks repeatable and more efficient. These techniques will be covered in Chapter 12.

Summary

▶ A GIS is a spatially enabled database that uses map and attribute data to answer questions about where things are and how they are related.

▶ Maps may represent discrete features, such as a road or a lake, or continuous measurements, such as elevation or temperature. The vector data model is ideal for discrete data, and the raster data model excels at storing continuous data.

▶ Map scale is the ratio of the size of objects in the map to their size on the ground. The source scale of GIS data affects their accuracy and precision and the map scales for which they are suitable.

▶ GIS data may be stored locally on a hard drive or on a network server, but data may also be accessed through Internet connections via GIS servers. Server data can be accessed by many clients, including GIS programs, web pages, smartphones, and tablets.

▶ Every GIS user has a responsibility to ensure that data are suitable for the proposed application and an obligation to cite the sources of data used.

▶ Data quality is measured in terms of geometric accuracy, thematic accuracy, resolution, and precision. Metadata store information about GIS data layers to help people use them properly.

▶ ArcGIS Pro uses a flexible and context-sensitive GUI. Understanding the names of interface objects (ribbon, window, view, pane, tab, and heading) helps to navigate the lessons.

Important Terms

ArcGIS Online	folder connection	map scale	resolution
attributes	generalization	metadata	scale range
cell	geometric accuracy	pane	scene
client	geoprocessing	panel	source scale
cloud	geoprocessing history	parameter	style
Contents pane	georeferenced	pixel	table
contextual ribbon	geospatial	point	task
continuous	home geodatabase	polygon	thematic accuracy
core ribbon	layer	portal	tool
database	layout	precision	Toolbox
discrete	line	project	vector
extension	locator	project file	view
feature	logical consistency	project folder	web map
feature class	map	raster	window

Chapter Review Questions

1. Explain the difference between discrete and continuous data sets in your own words. Give three examples of each.

2. A 50-m long swimming pool measures 2-cm on a large scale map. What is the scale of the map?

3. Explain why GIS data sets are considered to have a source scale but not a map scale.

4. In what ways have the sources from which GIS data are obtained changed?

5. Jamie is hiking trails in a forest and using his smartphone to track his position. In this remote location with heavy tree cover, his phone typically records his position within 25 to 50 meters. Discuss the accuracy and precision of the trail map he creates.

6. Judy creates a smartphone app that allows bird-watchers to upload the location and species of birds that they spot. Discuss the potential data quality issues that might affect her final map of bird sightings.

7. Construct an appropriate citation for the data that come with this book.

8. Compare and contrast the terms *window*, *view*, and *pane* as they relate to the ArcGIS Pro user interface.

9. Explain what geoprocessing tools and parameters are.

10. Describe the difference between core ribbons and contextual ribbons.

Expanding your knowledge

Open the ArcGIS Pro Help main page and follow the links to these topics. Explore these sections to learn more about working with aspects of Pro introduced in each chapter. (The organization and content of the Help may change over time. If the specified sections are no longer available, look for similar ones.)

Help > Projects > Catalog pane, catalog view, and browse dialog box

Help > Projects > Terminology for working with projects

Help > Project > Project items

Help > Maps > Interact with maps (down through Infographics)

Mastering the Skills

Preparing to begin

The tutorials in this book use a data set available for download from the book web site at www.mhhe.com/price1e. These data must be installed before starting the tutorials. Each step of the tutorials and all skills learned in this book are illustrated by video clips. See the web site for instructions on downloading and using the tutorial data and videos.

A **TIP** gives useful information about the program or settings that will save time or that are important to know. Pay attention to them.

TIP: In these tutorials, values you must enter are shown in this font: *type this*.

Teaching Tutorial

The following examples provide step-by-step instructions for doing basic tasks and solving basic problems in ArcGIS Pro. The steps to do are highlighted with an arrow →; follow them carefully.

TIP: This book comes with a folder of data called gisclass. Install it on the C:\ drive of the computer and not in your Documents folder, Desktop, or a network drive. Once installed, remember where it is to access it when needed.

Exploring an ArcGIS Pro project

ArcGIS Pro (or Pro, for short) organizes geospatial work in a **project**, a container that remembers the data, maps, and tools being used and organizes them into one folder.

1→ Start ArcGIS Pro and log in with an ArcGIS Online subscription account.

1→ Choose the *Open another project* link. Click Computer and then Browse.

1→ Navigate to the C: drive and then to the gisclass folder (wherever you placed it). It contains two folders, a ClassProjects folder for storing projects related to this book and the mgisdata folder containing additional GIS data.

1→ Navigate into the ClassProjects folder and then into the CraterLake project folder.

1→ Click the project file, CraterLake.aprx, to highlight it. Click OK.

TIP: The four-letter .aprx code is called an extension. Whether it is shown depends on the Windows folder settings. If you don't see it, ignore it and other extensions for the time being.

The center of the program window is occupied by a **map view** of Crater Lake, Oregon. The name of the map, Crater Lake, appears in the upper left corner on a view tab. To its left, the Contents pane shows the map name at the top and the different layers that make up the map (rock types, faults, volcanic vents, etc.).

2→ Click the Lake check box to turn off the Lake layer. The rock units of the floor of the caldera are revealed.

2→ Below the Lake layer, examine the Crater Lake Geology entry, known as a group layer. Uncheck the box next to it to turn all layers in the group off.

2→ Turn on the Hillshade layer, a raster that shows the terrain.

2→ Turn on the Crater Lake Geology group layer again. The Rock Types layer is slightly transparent so that the terrain beneath it can be seen.

2→ Turn the Lake layer back on.

The Contents pane has several icons that open **panels** with different functions (Fig. 1.12).

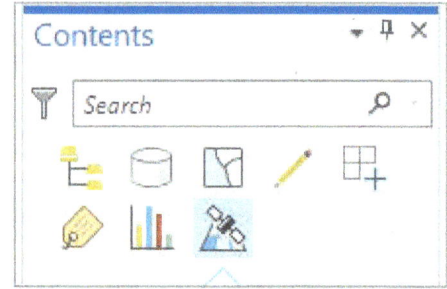

Fig. 1.12. Icons that open panels of the Contents pane

Source: Esri

3→ In the Contents pane, hover over the first icon until its name appears, List By Drawing Order. This is the default view of the pane.

 3→ Click on the List By Data Source icon (the second one), which shows the source of the data used in the map, e.g. the location of the data file. Most of the data comes from the gisclass folder, but the Topographic layer is an Internet-based GIS service.

3→ Click through each of the remaining icons to examine the different panels.

4→ When finished, return to the List By Drawing Order panel (the first one), which can change the order of drawing layers. Layers are drawn in order from bottom to top.

4→ Click and drag the Lake layer above the Vents layer. Notice that the vents in the lake disappear because they are covered by the lake polygon when drawn in this order.

4→ Move the Lake layer back between the Faults and Crater Lake Geology layers.

The top of the window organizes the program functions using ribbons titled **Project**, **Map**, **Insert**, and so on. Each ribbon has different buttons and settings, organized into functional groups.

5→ Click the **Map** ribbon title and examine its groups: **Clipboard**, **Navigate**, **Layer**, and so on.

5→ Click each of the main ribbon titles in turn and examine the groups and buttons.

The ribbon menu is context-sensitive: different ribbons may be visible depending on what is selected in a pane or view. Certain buttons may be accessible or unavailable (dim) based on what the user is doing.

6→ In the Contents pane, click to highlight the Crater Lake map title. Examine the ribbon titles.

6→ In the Contents pane, click on the Vents layer name to highlight it. Notice that a new ribbon group appears, **Feature Layer**, with three additional ribbons: **Appearance**, **Labeling**, and **Data**.

6→ Click on each of the three **Feature Layer** sub-ribbons and examine the groups and buttons.

The **Feature Layer** ribbons are used to control the behavior of a layer: how it appears, whether it has labels, and so on. These buttons only make sense when applied to a particular layer, which is why a layer must be selected to make them visible.

TIP: If a particular ribbon is not visible, check to make sure that the appropriate item is highlighted to make the ribbon appear.

TIP: The tutorial instructions refer to ribbon functions using the format **Ribbon Name: Sub-ribbon Name: Group: Button > Drop-down** button options. The **Sub-ribbon Name** and **Drop-down** portions may not always be present.

7→ In the Contents pane, click the Lake layer to highlight it.

 7→ Click the **Feature Layer: Appearance** ribbon and find the **Effects** group. Set the Transparency slider to about 50%, revealing the rock units underneath.

 7→ Click the Rock Types layer and click Swipe in the **Effects** group. Place the cursor on an edge of the map; then click and drag toward the center to reveal the Hillshade layer.

7→ Click on the **Effects: Swipe** tool again, but notice that it does not turn off.

Some buttons stay in effect until another button function is selected, such as choosing the Explore button to navigate the map.

Navigating 2D and 3D maps

Fig. 1.13. Navigation tools
Source: Esri

8→ Click the **Map** ribbon title and examine the **Navigate** group (Fig. 1.13).

8→ Notice that the **Swipe** tool is still enabled even after switching ribbons.

8→ Hover over the **Explore** button. A button tip appears, explaining how to use the mouse to navigate the map. All buttons have pop-up explanations.

8→ Click the **Explore** button and place the cursor over the map. The **Swipe** triangle is replaced by a hand icon for panning the map.

8→ Use the **Explore** button instructions to learn how to pan and zoom.

8→ Hold the mouse over the other buttons in the **Navigation** group to read about what they do. Experiment and practice with them to understand how each works.

Some button groups have an options menu that accesses additional settings related to that group.

9→ Hover over the small box/arrow icon on the lower right corner of the **Navigate** group, until it shows the pop-up text **Navigation Options**.

9→ Click the **Navigation Options** icon. The Options window opens.

The left side of the Options window lists groups of settings; the Navigation options are already selected. Click on the other sections to explore them, but then return to the Navigation section.

9→ Rolling the mouse wheel forward may be assigned to Zoom in or Zoom out, depending on the user preference. Change the setting now if desired.

9→ Click OK to accept the change.

10→ Click on one of the geologic units in the map (not in the lake). The polygon flashes and a pop-up window appears with information about the polygon.

10→ Click on one of the geologic units in the lake. The pop-up shows information about the lake instead because it is the top layer.

10→ Click the little black arrow near the bottom of the **Explore** button and examine the menu choices. They control which layer(s) will appear in the pop-up.

10→ Select the **Explore** button option: **Selected in Contents**.

10→ In the Contents pane, click on the Crater Floor Geology layer to select it and then click one of the geologic units in the lake in the map. Now the pop-up shows the crater floor geology.

10→ Close the pop-up when finished.

11→ Click the **Full Extent** button to go to the extent of all the layers in the map.

Unfortunately, the Topographic basemap layer has a very large extent (map area). The default Full Extent can be changed using the map's properties.

11→ In the Contents pane, right-click the icon by the Crater Lake map title and choose Properties.

11→ On the left side, click on Extent to see those properties (Fig. 1.14).

11→ Fill the button for Custom Extent. Click the *Calculate from* drop-down underneath it and set it to the Crater Lake Geology layer. Click OK.

11→ Click the **Full Extent** button again and it will zoom to Crater Lake.

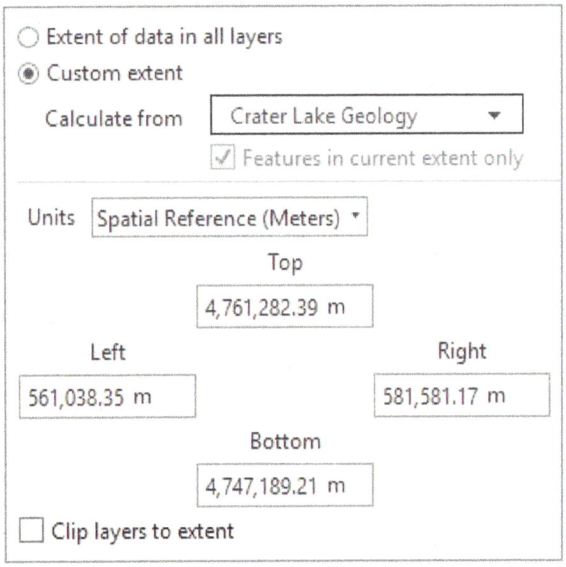

Bookmarks are another way to store a particular map extent and easily return to it. One bookmark has already been created for this project.

12→ Click **Map: Navigate: Bookmarks > Crater Lake**.

12→ Zoom in to the island on the west side of the lake.

Fig. 1.14. Setting the full extent area
Source: Esri

12→ Choose **Bookmarks > New Bookmark**. Enter the name *Wizard Island*. Click OK.

12→ Click the **Bookmarks** drop-down and select the Crater Lake bookmark to return to the view of the lake.

The final navigation tool, Zoom to Selection, is dim because something must be selected before one can zoom to it.

13→ Click **Map: Selection: Select** (just the button, not the drop-down triangle).

13→ Click on Wizard Island. The polygon clicked is highlighted in blue, indicating that it is selected.

13→ Click the **Zoom to Selection** button in the **Navigate** group .

13→ Zoom to the previous extent.

Multiple features may be selected by drawing a box around them.

14→ Click the **Select** button in the **Selection** group. Click and drag a box around several geologic units. Features that pass partially inside the box will be selected.

14→ Click the **Clear** button in the **Selection** group to clear the selection.

15→ Choose **Map: Inquiry: Measure > Measure Distance**.

15→ Examine the instructions in the window that opens; then measure across the lake.

15→ Click the Options drop-down and select Distance Units > Miles.

15→ Explore this tool and the other tools in the Measure drop-down until you can measure distances, areas, and features in a variety of units.

1. What is the area of the lake in hectares? _____

TIP: To hide the Measure window again, open a different tool, such as Explore.

16→ Click the **Map: Inquiry: Locate** tool. It opens the Locate pane.

16→ Type *Timber Crater* in the search box and click Enter. Select Timber Crater, Oregon, USA from the list that appears. The map zooms to Timber Crater.

16→ Search for Olympia and zoom to Olympia, Washington.

16→ Close the Locate pane and click the **Full Extent** button to return to Crater Lake.

Some new symbols probably appeared as you zoomed to Wizard Island.

17→ Click the *Explore* button and zoom in to the rock units southwest of the lake until some different symbols appear. Zoom in more until the labels appear (Fig. 1.15).

17→ Examine the Contents pane and find the Classes and Labels layer. Expand it (by clicking the triangle next to it) to see the volcanic class symbols.

17→ Zoom out until the volcanic class symbols disappear from the map. Notice that the check box by the layer turns light gray, indicating it is inactive and cannot be changed.

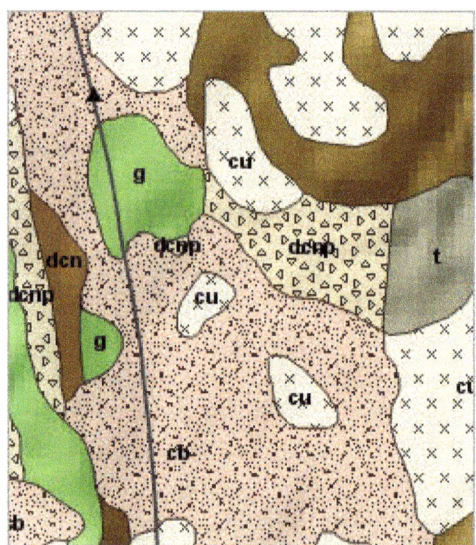

Fig. 1.15. The volcanic unit types and labels don't appear until a preset scale is reached.

Source: USGS

In this map, the Class and Labels layer has been given a **scale range** because the symbols would clutter the map if viewed at smaller scales. Layers can be given many properties to control how they appear and behave.

17→ In Contents, click the Classes and Labels layer to highlight it and reveal the **Feature Layer** ribbon group. Click the **Appearance** ribbon title.

17→ In the **Visibility Range** group, notice that the **Out Beyond** scale has been set to 1:24,000. These symbols appear when the map is zoomed in below that scale.

17→ Click the **Feature Layer: Labeling** ribbon and examine the **Visibility Range** group. A similar setting has been applied to the labels at 1:10,000.

Using the Help

Pro can do much more than can be covered in one textbook. The Help provides an extensive resource for learning more about the software and its capabilities. Getting familiar with the Help, and exploring it often, will greatly enhance your knowledge and skills. The Help is accessed online by default, although it can be downloaded and saved locally.

18→ Click the **Project** ribbon and select the **Help** entry. Wait a moment while a web browser opens to the main Help page.

18→ Examine the blue bar containing the "ArcGIS Pro" heading and the tabs underneath it: Home, Get Started, Help, and so on (Fig. 1.16). Notice that the Help tab is currently active.

18→ Examine the outline of topics on the left: Projects, Maps, Data, . . . Each heading can be expanded or collapsed for more information.

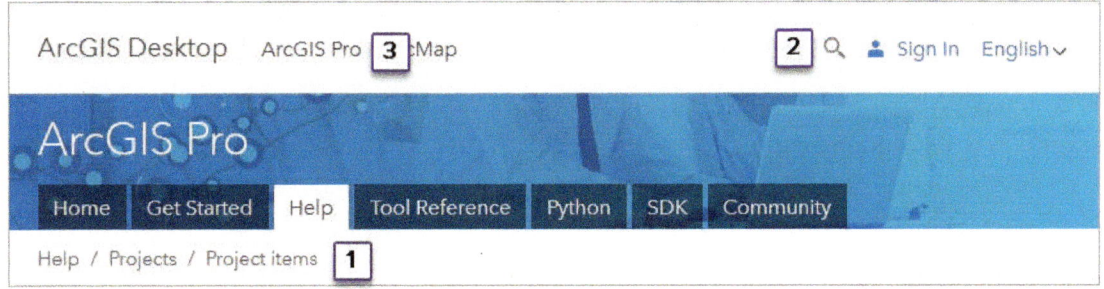

Fig. 1.16. The top of the Help window
Source: Esri

18→ Examine the content presented on the right side of the window.

18→ Expand the Projects heading, and examine the entries.

18→ Click on the *Terminology for working with projects* entry.

18→ Skim through the topics presented in the content for this section.

TIP: The Help organization and content may change over time as the software is updated. If a specific item referenced in this text cannot be found, try looking for a similar topic.

19→ In the outline on the left, click the *Project items* heading.

19→ Notice the expansion of the outline and the many blue links to additional information within the content.

19→ Click on one of the blue links in the content.

19→ Scroll to the bottom of the content and notice additional links in the *Related topics* section.

It is easy to get lost within this information. Use the breadcrumb trail immediately below the blue title bar (1 in Fig. 1.16) to back up to the main sections.

19→ Click Projects in the Help > Projects ... breadcrumb trail.

19→ Click Help in the breadcrumb trail to return to the main Help page.

The SEARCH function provides another method to find information.

20→ Click the Search icon. Be sure to use the Help search tool (2 in Fig. 1.16) and not the search box of the browser, which would leave the Help.

20→ Type symbols on the Search line and click Search.

The left outline now presents filters for different software packages and versions, currently set to ArcGIS Pro. The right side contains different articles. Notice that each article has "ArcGIS Pro" listed after the title. Be sure to view only Pro content for any article.

20→ Click on one of the articles, such as *Modify symbols*.

20→ Briefly examine the outline and the content.

20→ Examine the breadcrumb trail to learn which section of Help is being shown.

20→ Click on each entry in the breadcrumb trail, starting with the last, to back out of the content and return step-by-step to the main Help page.

TIP: In the Search Results page, the main outline and Help tabs disappear. Select an article to get them back, or click ArcGIS Pro at the top of the window (3 in Figure 1.16) and click the Help tab.

20→ Minimize the browser window with the Help, rather than closing it, to facilitate consulting it later.

20→ Return to the ArcGIS Pro program.

Managing windows

Projects can contain more than a single map. The Catalog pane is used to access and manage the different items in a project.

21→ Examine the Catalog pane to the right of the map (Fig. 1.17). Notice that the window has four tabs: Project, Portal, Favorites, and History. (If the Catalog pane is not visible, click **View: Windows: Catalog Pane** to show it.)

21→ The Project tab should currently be highlighted in blue; click it if not. It will show a list of project resources, including Maps and Toolboxes.

21→ Click the triangle next to Maps to expand the section. It contains the Crater Lake map (already open), plus three 3D maps (known as **scenes**).

21→ Double-click the Crater Lake_3D scene to open it.

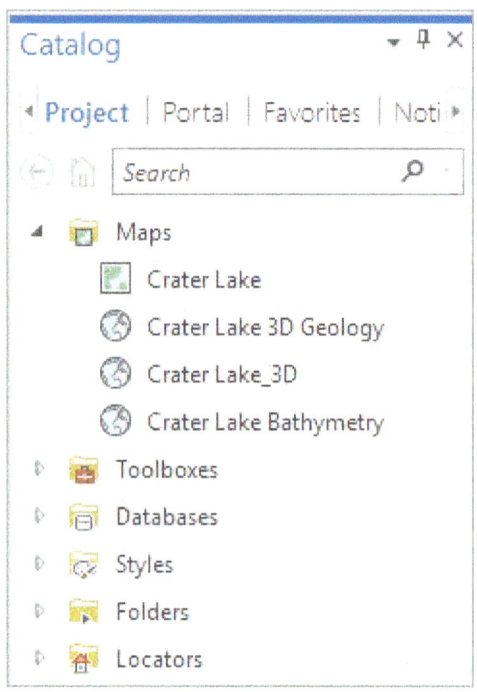

Notice that the previous Crater Lake map view remains open and is visible as a view tab (above the scene), making it easy to switch back and forth. Clicking the x on the tab closes the map, but it remains in the project and can be reopened from the Catalog pane.

Fig. 1.17. The Catalog pane
Source: Esri

22→ Use **Map: Navigate: Bookmarks** to create a new bookmark named *Crater Lake 3D*.

22→ Hover over the **Map: Navigate: Explore** button again and review the instructions for navigating a 3D scene.

22→ Practice zooming, panning, tilting, and rotating the scene until you have mastered the 3D navigation tools.

22→ Return to the Crater Lake 3D bookmark when finished.

23→ In the Contents pane (now showing the layers of the active Crater Lake_3D scene), click the Classes and Labels layer to make sure that the **Feature Layer** ribbon group is visible. Select the **Appearance** ribbon and examine the **Visibility Range** group. In 3D, the scale range is expressed in the viewer's distance from the map.

Managing the placement of windows within the program area is an important skill. Currently a map view and a 3D scene view are open. Both views are docked in the same window, one atop the other, and clicking the view tabs at the top of the window switches between them. Views can also be docked side by side. A docking icon (Fig. 1.18) controls where to place an object within a window: center, left, right, top, or bottom.

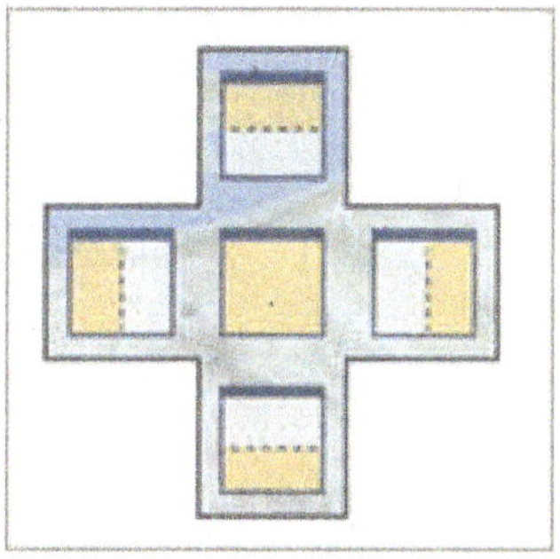

24→ If necessary, resize the program interface to cover only part of the computer screen.

24→ Click the Crater Lake_3D view tab above the scene and drag it down. The docking icon appears.

24→ Drag the scene view to the left option and let go of the mouse button. The map view and scene view are placed side by side.

24→ Click on the scene view tab again, drag it to the map, and choose a different docking option. Try them all.

24→ Drag the scene away until no anchor icon is visible, and let go. The scene becomes a floating window.

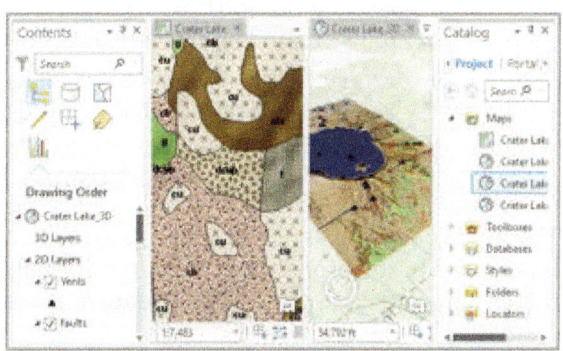

Fig. 1.18. Docking icon
Source: Esri

Any window, whether it contains views (map and scene) or panes (Contents and Catalog), can be docked within the program area or moved to float on top of or outside it.

25→ Click the Crater Lake view title and drag it off the program area so it becomes a floating window as well.

25→ Click the Contents pane title and drag it off the program area, making it float.

25→ Click the Catalog pane title and drag it off the program area, leaving it empty.

26→ Drag the Crater Lake map view back to the program area, and notice that the center of the program area is the only place to dock it.

26→ Drag the Crater Lake_3D scene view back to the program area and use the icon to dock it to the right of the map view.

26→ Drag the Contents pane to the left edge of the program area and dock it there.

26→ Drag the Catalog pane to the left edge and use the icon to dock it in the center, on top of the Contents pane. Use the tabs at the bottom of the pane to switch back and forth between the two panes.

To remove a tabbed pane from the window, it must first be converted to a floating window again.

26→ Click the Catalog tab and drag it to make a new floating window. Then dock the window on the left edge of the program area.

26→ Practice rearranging the windows to master the docking techniques. When finished, leave the map and scene docked side by side. Place the Contents pane on the left and the Catalog pane on the right of the program area (Fig. 1.19).

Fig. 1.19. Place the Contents pane on the left, the Catalog pane on the right, and the two views side by side in the middle.
Source: USGS/Esri

The 2D maps and 3D scenes can be linked together for enhanced visualization.

33

27→ Click **View: Link: Link Views > Center and Scale**.

27→ Navigate in either the map or scene, and the other view updates to match.

27→ Click the Crater Lake view tab at the top of the map. The tab turns blue, indicating it is the active map, and the Contents pane shows its layers.

27→ Click the Crater Lake_3D tab to activate it in order to view or modify its layers.

Rearranging panes can be done the same way as rearranging map and scene views.

28→ Activate the Crater Lake map and click the Vents layer to highlight it.

28→ Click **Feature Layer: Appearance: Drawing: Symbology**.

28→ Experiment with rearranging the Symbology pane and the Catalog pane.

TIP: If two panes are docked in the same window, appearing as tabs, a pane must be separated from the window before it can be docked elsewhere, by dragging the tab away.

⬛ **TIP:** Panes have an Auto-Hide button that turns a pane into a tab along the program edge. When not in use, the pane folds away, conserving real estate for maps in the display area.

TIP: The Contents and Catalog panes are used often and are usually kept open. If needed, use the **View: Windows** group buttons to reopen them.

The flexibility of the interface makes it challenging to write instructions. The tutorials will not dictate how to arrange the windows unless it is important. You may need to independently turn layers on or off, or open and close windows and panes, as you complete the steps. Feel free to experiment to find the arrangements that work best for you.

Exploring project resources

The Catalog pane manages all the resources of a project, including the Maps entry already viewed. Now let's examine some other resources.

29→ Click the Catalog pane tab and make sure that the Project tab is selected, showing the contents of the project.

29→ Expand the Toolboxes entry. Each project has a special toolbox for storing custom tools created by the user. This one is currently empty.

29→ Expand Databases. It shows an icon labeled CraterLake or CraterLake.gdb, the **home geodatabase** that stores data sets related to this project.

29→ Expand the CraterLake database to see the data inside. Mouse over each data set to see a description of the data type, and note the icon used to show it.

The home geodatabase is the default location where the project will save data by default. It is stored in the project folder. The CraterLake home geodatabase contains one table, two raster data sets, and eight vector feature classes. Geodatabases are presented in detail in Chapter 5.

TIP: The .gdb is called a file extension. Whether it appears or not depends on your computer's settings. This book will not normally show the extension unless it is important to do so.

2. List which feature class(es) contain point data and which contain line or polygon data. Also list which ones are rasters.

29→ In the Catalog pane, expand the Styles entry to view the different symbol styles associated with this project.

29→ Expand the Folders entry, which contains shortcuts to data sources used in the project. It currently shows one shortcut, the CraterLake project folder itself.

 29→ Use **Project > Save** or the **Save** icon on the Quick Access menu to save the project. It is smart to save the project periodically.

Vector feature classes, stored as points, lines, and polygons, include a table that stores information about each feature.

30→ Click on the Crater Lake view tab.

30→ In the Contents pane, right-click the Vents layer and choose Attribute Table. (It may appear docked or floating; use your new skills to arrange it as desired.)

30→ Right-click the Faults layer and open its attribute table also.

30→ Close the tables when finished looking at them.

Layers have properties that can be viewed and set. Let's examine some of them.

31→ In the Contents pane, right-click the Vents layer and choose Properties.

31→ Click the General entry, which can be used to set the layer name and to provide another way to modify the visibility range discovered earlier.

31→ Change the layer name to *Volcanic Vents*. This name is a cosmetic name for the layer only; it does not affect the stored data.

31→ Click the Metadata tab. Ideally, a data set includes information about what it is and where it came from.

31→ Click the Source entry. It gives basic information about the storage format and location of the data set (in the project's home geodatabase).

31→ Click OK. Notice that the new name has appeared in the Contents pane.

The source of this data set is the CraterLake home geodatabase. The same data sets for vents, geology, and the lake are used in both the 2D and 3D maps. When a feature class is added to a map, it becomes a **layer**. Each layer has properties, such as the name you just modified, which are unique to the map or scene and can be set independently.

3. Examine the vents' layer name in the 3D scene. Did it update to match the changes made in the 2D map?

Group layers can be used to organize layers with a similar theme and make it easy to turn them on or off together. The Crater Lake Geology entry is an example of a group layer. It contains three individual layers: Crater Floor Geology, Classes and Labels, and Rock Types (Fig. 1.20).

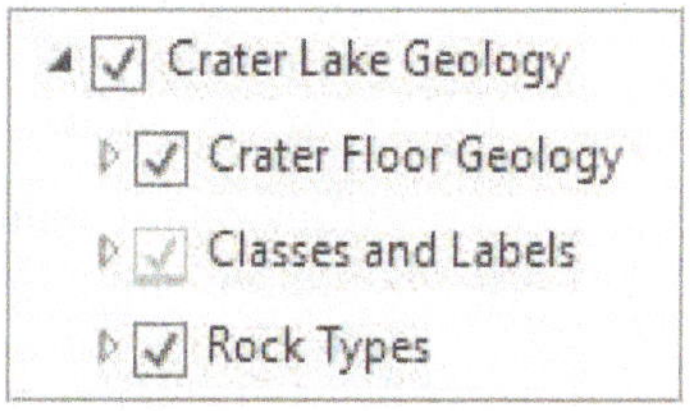

32→ Make sure the Crater Lake map view is active, and turn off the Lake layer.

32→ Use the Crater Lake Geology check box to turn the group layer off, then on, and examine the effect on the map.

32→ Expand each of the layers in the group to see their legends.

Fig. 1.20. The Crater Lake Geology group layer
Source: Esri

32→ Examine the effects as the individual layers are turned off and on.

32→ Collapse the individual layers to hide the symbols, but leave them all checked.

32→ Turn off the Crater Lake Geology group layer and the Hillshade layer.

TIP: To save space, hereafter "In the Contents pane" will be shortened to "In Contents."

The bottom layer in Contents, Topographic, is known as the basemap. It is supplied by ArcGIS Online through an Internet connection, and it covers the whole world.

33→ Choose **Map: Layer: Basemap > Imagery** to switch to the Imagery basemap.

33→ Zoom in. Notice that more detailed imagery with a higher resolution is now displayed. Zoom in to about 1:10,000 until individual trees are visible.

Basemaps use scale ranges to present appropriately detailed imagery at different map scales.

TIP: Hold the Ctrl key and check or uncheck a visibility box to turn all the layers in the Contents off or on at the same time. This shortcut also works to expand or collapse all layers at once.

33→ Switch to the Crater Lake_3D scene and turn off all layers.

33→ Switch to the Imagery basemap and explore the lake and Wizard Island.

We are going to work on one map at a time for a while. Let's unlink this map and scene and fix any rotation present. Maps and scenes have properties and options that can be set for them.

34→ Examine the **View: Link: Link Views** button; it should currently be blue, indicating that the views are linked (note the icons on the view tabs, which also indicate linkage).

34→ Click the **Link Views** button to remove the link.

34→ Close the Crater Lake_3D scene by clicking the x on its view tab.

34→ In Contents, right-click the icon for the Crater Lake map and choose Properties.

34→ Click the General section. Change the Rotation setting to 0. Click OK.

34→ Click **Map: Layer: Basemap > Topographic** to return to the original basemap.

TIP: Click the compass North arrow in a scene to quickly reorient the scene to north up.

Setting map symbols

The Symbology pane modifies the symbols used to display map layers.

35→ In Contents, make sure the Volcanic Vents layer is highlighted, and click **Feature Layer: Appearance: Drawing: Symbology**. The Symbology pane opens.

35→ In the Symbology pane, leave the Symbology set to Single Symbol.

35→ Click on the symbol representation (black triangle) to open the Format Point Symbol pane. Note that it has two tabs: a Gallery for selecting the base symbol and Properties for modifying how the symbol appears.

35→ In the Gallery tab, choose a different symbol, such as Diamond 3. The map updates to the new symbol.

35→ Click on the Properties tab and examine the symbol settings (Fig. 1.21).

The Format Point Symbol pane is an example of packing many options into a small space. It has three different panels, as indicated by the icons (hover over each icon to see its title).

35→ Click the icon for each panel and examine the settings available in each.

35→ Select the first panel, named Symbol. Expand both entries, Appearance and Halo, and view their settings.

35→ Change the color, size, or angle of the symbol, and then click Apply for the changes to take effect on the map.

36→ To start over, click on the Gallery tab to return to the array of symbols.

36→ Use the Gallery and Properties tabs to make the symbols 12-pt. pink stars.

36→ Click the circled arrow to return to the main Symbology pane.

36→ Close the Symbology pane by clicking the x in its upper right corner.

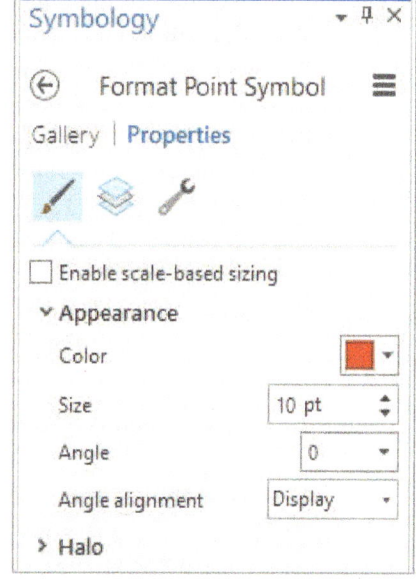

Fig. 1.21. Formatting point symbols
Source: Esri

The symbol settings can be accessed more than one way.

37→ In Contents, right-click the Volcanic Vents layer name to open a context menu. Choose the Symbology entry in the menu to open the Symbology pane.

37→ In Contents, left-click the Vents layer symbol to open the Format Point Symbol tab of the Symbology pane.

37→ In Contents, right-click the Vents layer symbol to select a different color.

37→ Close the Symbology pane. We will learn more about it in Chapter 2.

TIP: Watch for different ways of accessing the same functions. Once a command or pane has been introduced and used several times, the tutorial instructions will simply say to open it, not how to open it.

Working with local GIS data

To work with data outside the project folder, we must create a link, or folder connection, to it. We will add a link to use other data that came with this book.

38→ Click **Insert: Project: Add Folder**.

38→ Click on the C:\ drive and navigate into the gisclass folder (wherever you placed it).

38→ Click the mgisdata folder to highlight it (don't double-click and go inside it) and choose OK.

38→ In the Catalog pane, expand the Folders entry to see the CraterLake project folder and the mgisdata folder. (Collapse the other entries if desired.)

38→ This mgisdata folder will be used many times. Right-click it and choose Add to Favorites to make it easily accessible in other projects from the Catalog Favorites tab.

39→ Expand the mgisdata folder to see subfolders of data, organized geographically.

39→ Expand the Oregon folder and find the oregondata geodatabase. Expand it to see its contents.

39→ Click and drag the counties feature class on top of the Crater Lake map.

39→ Zoom in/out until several of the counties around Crater Lake appear in the map.

40→ In Contents, click the counties layer to highlight it and use **Feature Layer: Appearance: Drawing: Symbology** to open the Symbology pane.

40→ Click the Symbol representation to open the Format Polygon Symbol pane.

40→ In the Gallery tab, type *hollow* in the Search box and click Enter. Choose the Extent Hollow symbol.

40→ Switch to the Properties tab and set the outline color to hot pink. Click Apply.

41→ Zoom in and compare the counties boundary with the basemap (Fig. 1.22).

41→ In Contents, right-click the counties layer and choose Properties (or simply double-click counties).

41→ Read the information in the Metadata section and then close the properties.

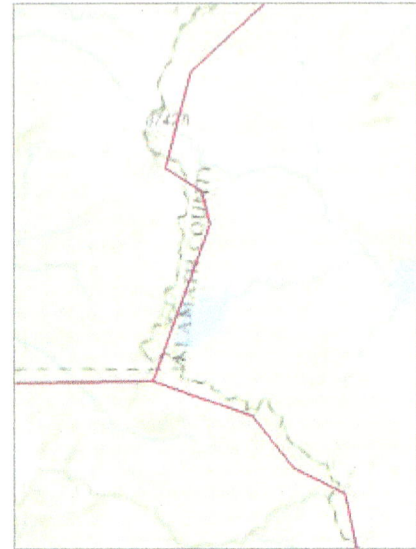

Fig. 1.22. Difference in county source scales

Source: Esri

This example illustrates the importance of **source scale**.
The metadata state that counties came from a generalized data set, optimized for quickly drawing national maps, so it has lower resolution and accuracy than the basemap.

Dragging data from the Catalog pane to the map is one way to add data, but there is another.

42→ Click **Map: Layer: Add Data**. Click the Folders entry to find the mgisdata folder.

42→ Navigate inside mgisdata to find the Oregon\OregonStateGeology\geology feature class. Select it and click OK to add it to the map.

42→ Compare this geology layer with the Crater Lake Geology group by zooming to Crater Lake and turning the geology layer off and on. Leave it off when done.

42→ In Contents, right-click the counties layer and choose Zoom to Layer. Check the scale reading.

Most data sets in the Oregon geodatabase were designed for use at the state scale, about 1:3,000,000. The pink counties appear adequate now. The Crater Lake data sets were compiled at scales around 1:24,000. When viewed at the state scale, the lake is hardly visible. The original source scale of a data set matters and should be considered when searching for data.

Since the Crater Lake and Oregon data sets are not suitable for use together, we will create a new map to continue exploring GIS data.

43→ In Contents, right-click the geology layer and choose Remove. Also remove the counties layer.

43→ Right-click the Crater Lake Geology layer and choose Zoom to Layer.

 43→ Close the Crater Lake map and save the project.

43→ Choose **Insert: Project: New Map**.

43→ In Contents, find the map's icon and click the Map title next to it twice, slowly, to enable editing.

43→ Type *Oregon* for the map name and click Enter.

44→ Use **Map: Layer: Add Data** or the Catalog pane to add the counties feature class from mgisdata\Oregon\oregondata to the map.

44→ In Contents, click twice slowly on the counties layer name and change it to *Counties*.

44→ Use the **Feature Layer: Appearance** ribbon or right-click the Counties layer to open the Symbology pane.

44→ Give the Counties layer a hollow symbol with a green outline color.

TIP: We will be using the Symbology pane extensively, so keep it open. Be sure to have the appropriate layer highlighted before changing symbols.

45→ In the Catalog pane, expand the Transportation "folder" in the oregondata geodatabase.

45→ Click the airports feature class; then hold down the Ctrl key and click on the highways feature class so that both are selected. Drag them to the map.

45→ Symbolize highways with a brown line and airports with an airport symbol.

46→ Add the parks feature class from the oregondata geodatabase. Symbolize it with light green fill and a dark green outline.

46→ Add the hospitals feature class from the oregondata geodatabase. Symbolize it using a thick cross symbol with blue fill and (extra points) a gray outline.

Using ArcGIS Online content

One of the benefits of ArcGIS Pro is its link to ArcGIS Online, which offers extensive data resources, including industry-authored content and information contributed by ordinary users. Online content is accessed through the Portal pane.

47→ In the Catalog pane, switch from the Project tab to the Portal tab by clicking the Portal heading.

47→ Examine the four icons below the heading and hover over each one to view the titles: My Content, Groups, All Portal, and Living Atlas

The first two icons access content saved in your ArcGIS Online account. The All Portal icon searches the entire AGOL holdings. The Living Atlas icon accesses authoritative industry-authored content (for users with an AGOL subscription account).

TIP: Online content is transient, and the specific items mentioned in this tutorial may change or disappear. If an item can't be found, look for something similar and use it instead. The important thing is to learn to search for, evaluate, and use content.

47→ Click the All Portal icon and type *Oregon* in the search box. Click Enter.

47→ Notice the different icons representing different types of data services.

47→ Hover over an item to see a pop-up description.

47→ Note the data location entry in the description. This is the URL of the service providing the data.

TIP: If a URL begins with http://www.arcgis.com/, then it is probably an ArcGIS Online published service. If it begins with anything else, it is probably an ArcGIS Portal installation run by a government or company.

48→ Edit the search term to *Oregon colleges* and click Enter (Fig. 1.23). Look for a ***web map*** icon—for example, Oregon Community Colleges.

48→ Maps cannot be dragged in; they must be opened. Right-click a map icon and choose Add and Open. A new map, linked to the online data, is added to the project.

48→ Choose **Map: Navigate: Explore: Topmost Layer**.

48→ Click on one of the map symbols. A pop-up opens with information about the feature clicked.

48→ In Contents, double-click the map icon to open the map properties. Click the Metadata section.

ArcGIS Online requires a minimum set of metadata for an item to be published. The quality of description is sometimes a clue to the authority of the data set. Like everything else on the Internet, one should be cautious in accepting data at face value.

Layers in a web map are services and can be copied to a different map, including the Oregon map we previously created.

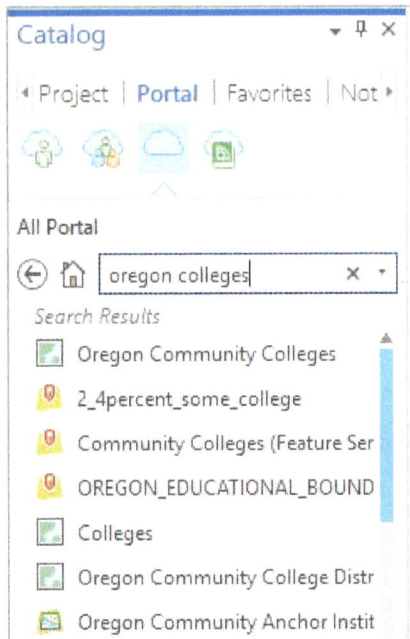

Fig. 1.23. Searching for Oregon colleges
Source: Esri

49→ Close the map properties.

49→ Right-click one of the web map layers and choose Copy.

49→ Switch to the Oregon map, right-click the map icon at the top, and choose Paste.

Labels can provide more information on a map.

50→ In Contents, click the Counties layer to select it.

50→ Examine the **Feature Layer: Labeling: Label Class** group. Note that the label field is currently set to NAME.

50→ Click the **Label** button next to it to make the labels appear. (We'll learn more about setting other label options later.)

50→ Turn off the labels by clicking the **Label** button again.

50→ In Contents, click the airports layer to select it. Turn its labels on.

51→ In the Catalog pane, in the current All Portal search list, look for symbols representing a ***feature layer***, and drag one into the map.

51→ Feature layers are often composed of group layers. In Contents, expand the feature layer, if needed, until the individual layers appear.

51→ Right-click an individual layer and choose Attribute Table. Feature layers behave similarly to feature classes.

Let's explore the Living Atlas (if you don't have a subscription account, skip to Step 55).

52→ In the Catalog pane, click the fourth icon to open the Living Atlas panel.

52→ Type *temperature* in the search box and click Enter.

52→ Find an ***imagery layer*** icon titled USA Mean Temperature and drag it to the map.

Unlike a basemap, feature layers and imagery layers provide actual pixel values, not a pre-rendered snapshot of a map. A user can change the symbols and use the data for analysis.

53→ In Contents, right-click the USA Mean Temperature layer and choose Symbology.

53→ Change the color scheme to a blue-red ramp so it looks more like temperatures.

53→ In Contents, double-click the USA Mean Temperature layer and view the Metadata section. Skim the entire entry. This is well documented data.

4. Construct a citation for the USA Mean Temperature service. Consult both the metadata and the pop-up window in the Portal pane to find all the information needed.

TIP: Internet services can be slow, depending on network connection speed. They can also go down and not work at times. If one service doesn't work, try another.

54→ Close the layer properties window.

54→ In the Catalog pane, change the search term to traffic and click Enter.

54→ Find a *map image layer* icon labeled World Traffic Service and add it to the map.

54→ Turn off the other layers (except the basemap) to see the traffic map better. Zoom into the Portland area.

This layer portrays a real-time map service showing current traffic conditions. Zoom close in to see the local streets light up with information.

54→ Zoom to the extent of Oregon again, and turn on the Counties layer if needed.

54→ Save the project.

We used four types of AGOL services in this exercise (web map, feature layer, imagery layer, and map image layer), but you'll need to learn more about data formats to understand how they differ. We will tackle this material in Chapter 12.

Using geoprocessing

GIS includes more than creating maps; it also provides tools for analyzing map data. Analysis functions and many data management functions are executed as **geoprocessing** tools.

55→ In Contents, turn off all layers except the Topographic basemap and Counties.

55→ Choose **Analysis: Geoprocessing: Tools** to open the Geoprocessing pane.

55→ Examine the three tabs at the top of the window, and click each one in turn.

The Favorites tab keeps a list of frequently used tools. The Toolboxes tab shows all tools organized into toolsets. The Portal pane shows tools that run as services on ArcGIS Online; these tools typically require a subscription account and consume service credits. We would like to gather some statistics about the county populations in Oregon to find the largest, smallest, and average population of the counties.

TIP: The Geoprocessing and Catalog panes are used often. Try docking them one atop the other in a single window. Leave the window open to easily switch back and forth.

55→ In the Geoprocessing pane, click the Favorites panel again.

55→ Enter statistics in the search box.

55→ Mouse over the Summary Statistics tool to learn what it does.

55→ Click the Summary Statistics tool to open it (Fig 1.24)

The Parameters tab provides the information the tool needs to run. Items marked with an asterisk are required; the others will not be used or will contain default values.

55→ Hover over the blue circled question mark to see the short description of the tool. Then click the circle to open the tool Help in a browser window. Minimize the browser.

55→ Hover near a parameter box; a circled "i" will appear. Hover over the "i" to see a description of the parameter.

This tool produces an output table containing the statistics. By default, outputs are placed in the project's home geodatabase. Give all outputs descriptive names that indicate their contents.

56→ For Input Table, choose Counties from the drop-down (the Browse button is only needed if the data are not already in the map).

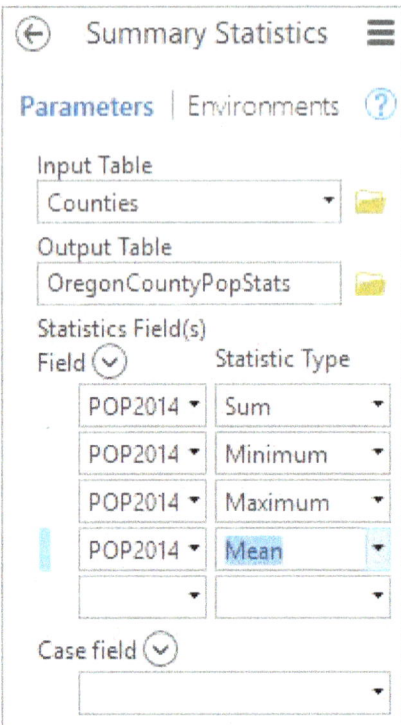

Fig. 1.24. The Summary Statistics tool
Source: Esri

56→ Click in the Output Table box to see the full location and name.

56→ Edit the final part of the name after the last "\" to read *OregonCountyPopStats*. Click Tab to finish entering the name.

56→ Use the Field drop-down box to select the field POP2014. Leave Sum as the Statistic Type.

56→ Use the second Field drop-down and select POP2014 again. Change the Statistic Type to Minimum.

56→ Add entries for the maximum and mean of POP2014 as well.

56→ Leave the Case field blank. The tool should appear as in Figure 1.24.

56→ Click Run at the bottom of the pane to execute the tool.

57→ In Contents, find the new table near the bottom under Standalone Tables.

57→ Right-click the OregonCountyPopStats table and choose Open.

57→ Examine the statistics.

5. How many counties does Oregon have? What is the average population?

We can sort to the original table to find which counties have the smallest and largest populations.

57→ In Contents, right-click the Counties layer and choose Attribute Table.

57→ Right-click the POP2014 field heading and choose Sort Ascending.

6. List the smallest and largest counties with their 2014 populations.

We can use the table data to select features meeting a certain criterion. For example, how many counties have more than 100,000 people?

58→ In the Geoprocessing pane, click the circled back arrow to return to the main Geoprocessing pane.

58→ Search for *select* and open the Select Layer By Attribute tool (Fig. 1.25).

58→ Set the Layer Name to Counties.

58→ Leave the Selection type as *New selection*.

58→ Click Add Clause to enter the condition to be met.

58→ Set the Field to POP2014.

58→ Set the condition to *is Greater Than*.

58→ Type *100000* in the value box.

58→ Click Add to enter the clause in the expression. The tool should look as it does in Figure 1.25.

58→ Click the Run button.

59→ The selected features are highlighted in the table and on the map. Examine the bottom of the table to see the number selected (10).

59→ Click **Map: Selection: Clear** to clear the selected features.

59→ Close all tables.

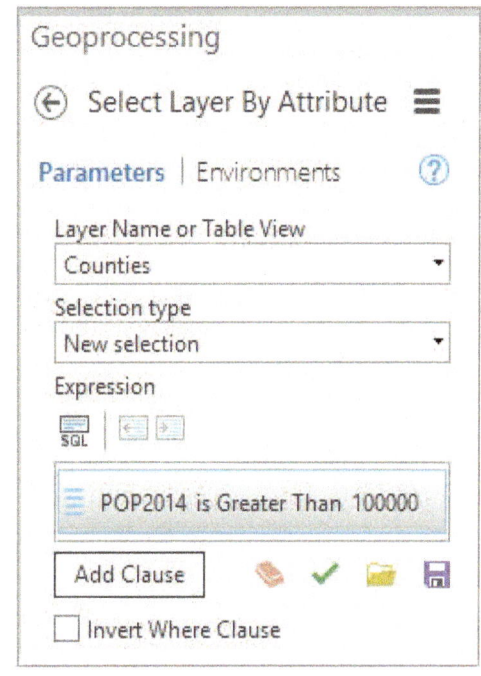

Fig. 1.25. Selecting counties with more than 100,000 people

Source: Esri

This is the end of the tutorial.

→ Save the project and close ArcGIS Pro.

Practice Exercises

TIP: To capture a map or screen shot to submit to your instructor, press the Alt-Prnt Screen keys on the computer, or use a program such as Jing or Snagit. Then paste the capture into a Word document. For help, ask your instructor.

Use ArcGIS Pro to create and explore a map and scene based on your own interests.

1. Start a new project. Name it *ExploreGIS* and save it in the gisclass\ClassProjects folder.

2. Choose a geographic area that interests you, such as your home state or city, your favorite national park, or a place you would like to visit. For best results, it should be at least the size of a moderate city or else you may not find much interesting data for it.

3. Create a map for the area and give it an appropriate name.

4. Search the C:\gisclass\mgisdata folder for one or more data sets that encompass the area chosen. Add them to the map.

5. Search ArcGIS Online for at least four data sets that relate to a theme that interests you and that cover the area chosen. Include a data set or two from the Living Atlas if you have a subscription account.

6. Create a 3D scene and add one or two of the data sets to it.

7. Link the 2D map and the 3D screen together and explore the area.

8. Prepare a brief one- to two-page report. Start with an introduction explaining why you chose the area. **Capture** at least one screen shot each of the map and the scene, number them as figures, and explain what each portrays. Conclude with a discussion of what you learned about the area. At the end, include complete, formal citations for each data set visible in the map and scene.

Chapter 2. Mapping GIS Data

Objectives

▶ Using properties of symbols to differentiate features or rasters in maps

▶ Distinguishing among nominal, categorical, ordinal, and ratio/interval data

▶ Creating maps from attributes using different map types

▶ Selecting appropriate classification methods when displaying numeric attributes

▶ Displaying thematic and image rasters

▶ Understanding the relationship between a layer and its source data set

Mastering the Concepts

GIS Concepts

Geographic Information Systems are used for many purposes, but creating maps is one of the most common. The practice of cartography, or mapmaking, has a long history. Although a computer makes the process easier and offers the ability to explore and edit designs, the cartographer must still understand the basic principles behind portraying spatial data and the aesthetic challenges in designing an effective and attractive way to communicate ideas.

Choosing symbols for maps

Cartographers may choose from many strategies to symbolize features in a map. A layer may simply be portrayed using one symbol, or different features can be assigned different symbols depending on the value of an attribute field. Point data are shown with marker symbols, line features with linear symbols, and polygon features with shaded area symbols (Fig. 2.1).

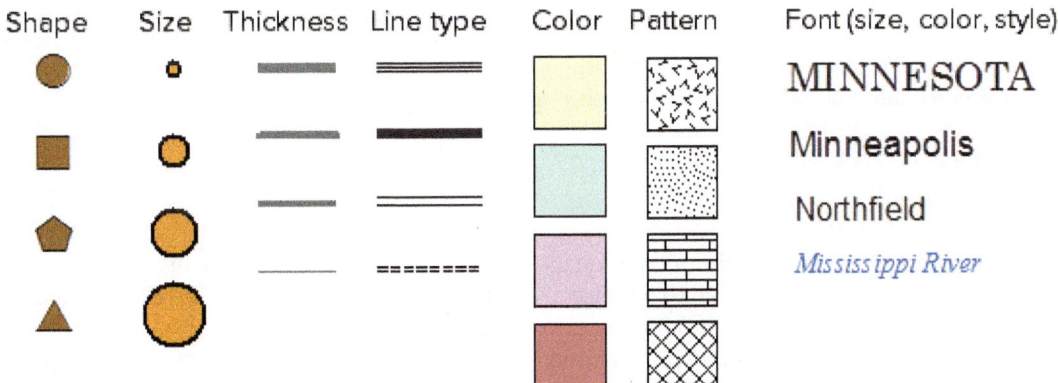

Fig. 2.1. Variations in symbols used to differentiate objects in a map

Cartographers have many ways to signify differences between features in a map: shape, size, thickness, line type, color, pattern, and font (Fig. 2.1). Traditionally these variations are used to show either changes in category (the type of thing) or changes in quantity. Shape, line type, pattern, and font are typically used to show changes in category, such as different types of wells (points), different classes of roads (lines), or different soil units (polygons). Size and thickness are generally used to indicate increases in quantity, such as the population of a city (points) or the discharge of a river (lines) (Fig. 2.1). Text symbol variations usually indicate categories (towns vs. rivers), although font size can indicate qualitative differences in value, such as town size, as long as not too many different sizes are used.

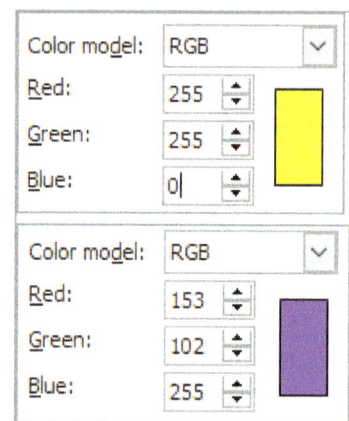

Color may be used to indicate either category or quantity. Colors may be designated using one of several common methods. The **CMYK** model is often used for printing, and it specifies mixtures of inks used in printers or plates (cyan, magenta, yellow, black).

The **RGB** model is based on mixing different proportions of red, green, and blue light on a scale from 0 to 255. If the value of the red light is high (close to 255) and the other two colors have low values near zero, a bright red color will result. If red and green are both high and blue is low, a mixture of red and green will create yellow. (Mixing light is a different process than mixing paint, and different colors will result.) Figure 2.2 shows two possible mixtures and the resulting colors. The RGB model can specify 256^3 different colors (over 16 million).

Fig. 2.2. Computers store and identify colors as mixtures of red, green, and blue light, each measured on a scale from 0 to 255.

Source: Esri

The **HSV** method is instructive for discussing the use of color in portraying features on a map. HSV stands for **hue, saturation, and value** (Fig. 2.3). **Hue** refers to the shade of color (such as red, blue, or yellow) and is established by the wavelength of the light observed. Typically hue is portrayed as a color wheel so that the color values range in degrees from 0 to 360. **Saturation** corresponds to the intensity of the color and is measured as a percentage. Imagine mixing a can of paint, starting with a white base and adding a single pigment—a small amount of pigment yields a low saturation, but a large amount of pigment results in high saturation. **Value** refers to how light or dark the color is, somewhat like putting pigment into a can of base paint that varies from black through gray to white. Value is also measured as a percentage. Any color can be defined using a combination of the three properties.

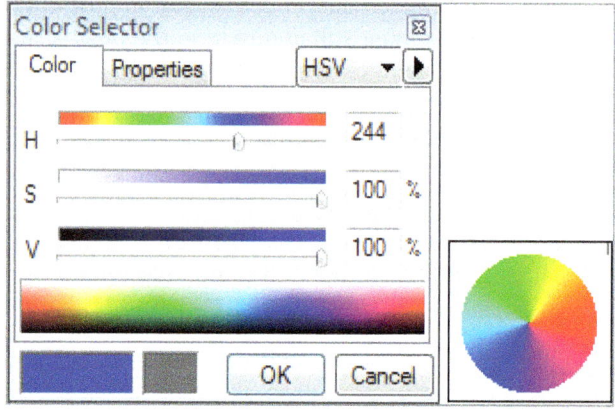

Fig. 2.3. Hue, saturation, and value method of defining color

Source: Esri

A fourth parameter, named **alpha**, may be used with either the RGB or HSV model. It refers to the opacity of a color, or the degree to which it is transparent or see-through, like tinted glass. A zero value indicates that the color is fully transparent, and the highest value, which may be represented either as DN = 255 or 100%, means it is not transparent at all.

The set of colors chosen for a map layer should follow certain guidelines. Figure 2.4 shows three different sets of five colors. A set based on variations in hue should be used to depict changes in category, such as different soil units (Fig. 2.4a). For categories, the saturation and value of the colors should be similar. Quantities are generally indicated using differences in saturation or value, with light or unsaturated colors indicating lower quantities

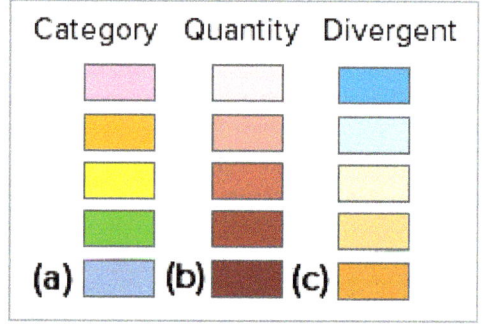

Fig. 2.4. Color ramp variations

and darker or more intense colors indicating higher quantities (Fig. 2.4b). Sometimes a **divergent color set** is helpful in showing variation around a significant middle value (Fig. 2.4c), such as in a climate change map. Areas with no significant temperature change are shown in the middle neutral color, colder temperatures are shown in blue, and warmer temperatures are shown in orange.

The importance of black-and-white symbol schemes cannot be neglected. In commercial printing or copying, color still costs more than black and white. Publishing figures in professional journals or reports may also cost significantly more in color. One should consider, too, how most viewers will see the map. Figures in a master's thesis may look wonderful in color, but some of the people reading it may receive a black-and-white version from inter-library loan. There are many reasons why one might wish to design a map in black and white at the outset. In a black-and-white map, only four or five different shades of gray are typically discernible, so symbol selection relies heavily on variations in shape, size, thickness, line type, or pattern rather than value.

Combinations of these factors may also be used when specifying symbols, such as using points that change in both shape and color, polygons that change color and pattern, or text that has differences in size and font and style. However, combinations should be used sparingly, as the more complex and varied the symbols become, the more difficult it may be to interpret them. In Figure 2.5, both volcanoes and highways are shown using combinations of color, shape, and size (thickness). The combinations serve to accentuate differences within the layer and could be used to emphasize one type of feature, such as using a larger, saturated pink symbol to call attention to the stratovolcanoes. However, this strategy would backfire if multiple point layers were being displayed because it would become difficult to interpret which features belonged to which point layer.

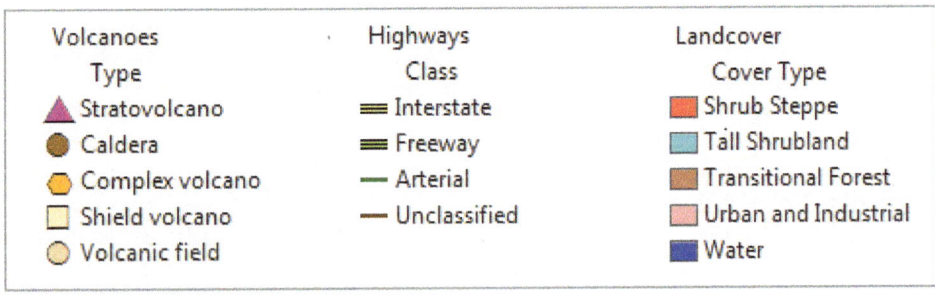

Fig. 2.5. Examples of categorical data representation for points, lines, and polygons

Color choice is complicated by the fact that about 10% of men and 1% of women are color-blind. Different types of color blindness are possible, although red-green color blindness is the most common. If layers are being symbolized for data exploration by one person, then the user may employ any set of colors that aids in the interpretation of the data. However, if the maps are intended for use by many people, it is better to avoid red-green color sets and rely on saturation, value, pattern, and shape variations instead. If divergent colors or multiple hues are needed, combinations such as brown-purple or orange-blue can be interpreted by most color-blind viewers.

Cartographers also need to be sensitive to conventions and connotations associated with different colors or symbols. A **convention** refers to the use of a particular color or symbol in a commonly understood way; for example, using blue to display water, blue and red to contrast cold with heat, or a blue cross to indicate a hospital. Conventional symbols help readers to interpret the map. **Connotation** is an emotional or psychological impact associated with a particular symbol, such as the color red indicating danger, or red, white, and blue evoking feelings of patriotism. However, connotations are often culturally specific. Red may indicate danger in the United States, but it is associated with joy in Thailand and with national pride in China. Making maps for an international audience demands particular care and knowledge in the choice of colors and symbols. Context can matter as well; even in the United States, a red-green map at Christmas might engender more joy than fear.

Types of data and types of maps

Geographers, statisticians, and others characterize measurements or attributes as belonging to one of a set of data types: **nominal, categorical, ordinal, interval**, or **ratio**. The type of data affects how it should be stored in a database and what types of analyses or statistics are appropriate. In mapping, the data type also influences the kind of map used to display the feature attributes.

Nominal data name or identify objects, such as the names of states. Nearly every feature will have a different name. Nominal data are often text values, but they don't have to be: a parcel number or a tax identification number also serves the purpose of uniquely identifying an object. Nominal data are usually portrayed on a map by labels. Each layer typically has one text symbol, but different text symbols may be used to accentuate different layers, such as using all uppercase for state names or italic blue text for rivers.

Categorical data separate features into distinct groups or classes. Figure 2.5 shows examples of categorical data for point features (volcano type), linear features (road class), and polygon features (land cover). Other examples include soil types, ethnic groups, and geological rock formations. Categorical data are often stored as text, but it is also possible to represent the categories with numeric codes, such as Commercial Services = 20 and Industrial = 40. Categorical data are represented by a **unique values map**, which gives each category a different symbol based on shape, line type, color, or pattern (Fig. 2.5) as shown in the geological rock units map in Figure 2.6.

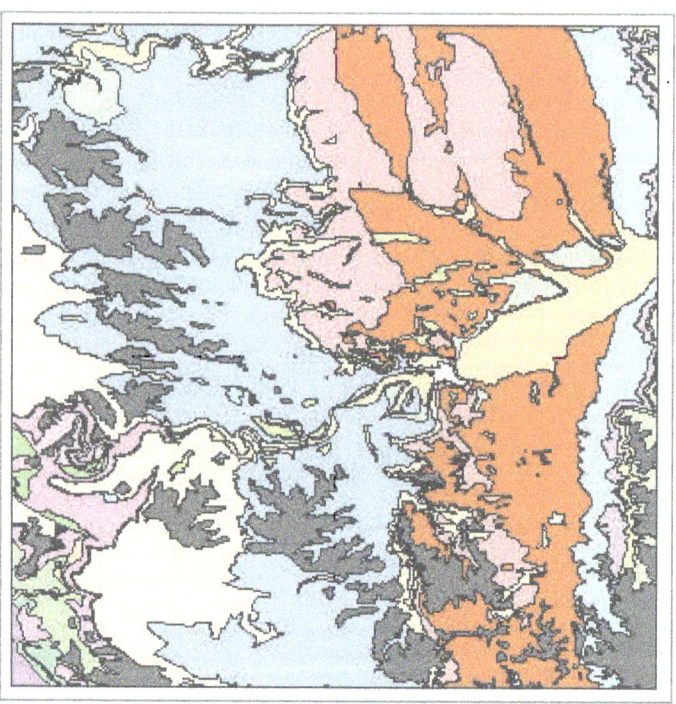

Fig. 2.6. A unique values map based on geological rock types
Source: South Dakota Geological Survey

Ordinal data have categories that are ranked based on some quantitative measure, although the measure may not be linear. For example, urban settlements might be classified as villages, towns, or cities. Students are assigned grades of A, B, C, D, or F. Soils are designated as A, B, C, or D depending on their infiltration. Ordinal data must be represented by unique values maps if the values are text, but value or saturation should be varied instead of hue or pattern so that the quantitative increase can be easily interpreted. Numeric ordinal data may be represented either with unique values maps or graduated color maps.

Quantitative data represent phenomena that fall along a regularly spaced measurement scale, such as distance or rainfall. Equal changes in interval involve equal changes in the quantity being measured. For example, the energy needed to heat a thimble of water from 16 to 17°F is the same as that needed to heat it from 96 to 97°F. **Ratio data** have a meaningful zero point that indicates the absence of the thing being measured. Precipitation is an example of ratio data; zero precipitation corresponds to a total lack of rain, and two inches of rain is twice as much as one inch. Ratio data support all four arithmetic operations of addition, subtraction, multiplication, and division. **Interval data** have a regular scale but are not related to a meaningful zero point. Temperature data measured in the familiar Celsius or Fahrenheit scale are interval data because a temperature of

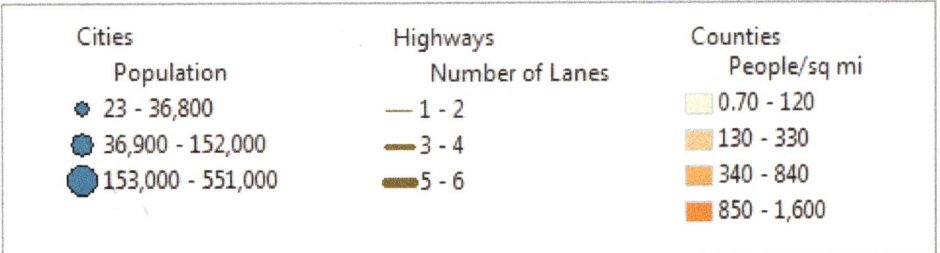

Fig. 2.7. Examples of numeric data representations for points, lines, and polygons

zero does not correspond to a complete lack of temperature (heat energy). Any scale that can have negative values, such as elevation, is an interval scale. Interval scales support only addition and subtraction. Today's high temperature of 80°F might be hotter than yesterday's high of 40°F, but it cannot be said to be twice as hot, except as a figure of speech.

Quantitative numeric data take on values along a continuous scale of possibilities; every state, for example, has its own population value. In order to symbolize numeric data, the values must be partitioned into groups with specific ranges. These ranges are called classes, and the maps are called classified maps. In Figure 2.7, the cities and roads have three classes of values, and population density has four classes. Each class is symbolized using variations in symbol size, thickness, or color saturation or value.

Point or line data are usually displayed by varying symbol size or thickness and are portrayed using a **graduated symbol map**. However, point or line data can also be portrayed using a **proportional symbol map** in which the numeric value is used to proportionally determine the size of the symbol. Instead of a few size classes, the map has a continuous range of symbol sizes. This style is often referred to as an unclassed map. Figure 2.8 compares graduated symbol and proportional symbol maps for the populations of state capitals.

Numeric data classes for polygons are represented using color-shaded symbols (Fig. 2.9); the resulting maps are called **graduated color maps** or **choropleth maps**. The maps are usually symbolized using changes in value or saturation (with a monochromatic color ramp) so that the increase in quantity is clear, as shown in the county population map in Figure 2.9a. Although rainbow color ramps are popular in software packages, using them for graduated color maps is usually a mistake because it is difficult for the mind to interpret hues as representing larger or smaller quantities (Fig. 2.9b). Generally the eye cannot distinguish between more than seven or eight levels of the same color, so no more than that number of classes should be used.

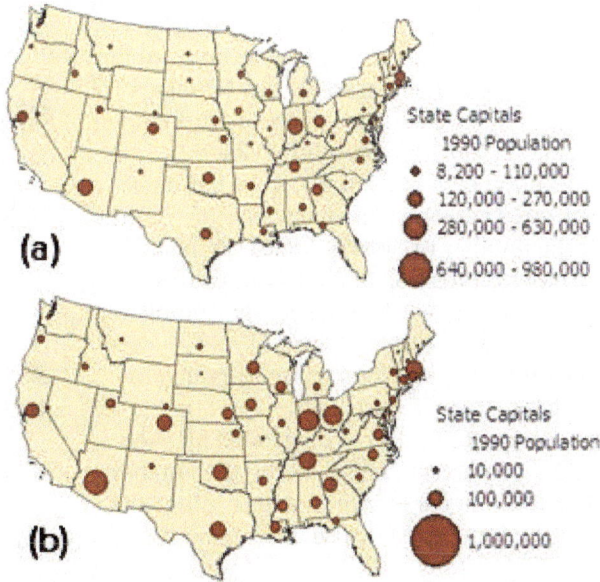

Fig. 2.8. Comparison of (a) a graduated symbol map and (b) a proportional symbol map showing the populations of state capitals

Source: Esri

The modifiable areal unit problem

Data are often measured over one set of areal units, such as a census block, but then combined, or **aggregated**, into a set of larger areal units, such as block groups, tracts, and counties. When the measurement and aggregation units are arbitrary shapes and sizes instead of a regularly spaced grid, the size and shape of units will influence the recorded values. For some measurements, such as the number of farms or lightning strikes, units with larger areas tend to have higher values just because they are larger. In Figure 2.10a, the number of farms is greater in the bigger states like Texas and California. Certain variables, such as the number of home vacancies, are strongly linked to population, and maps made from them reflect population rather than giving insight into underlying patterns. In Figure 2.10b, the largest and most populous states have the most vacancies.

Another issue that can arise is that larger polygons receive greater prominence in the map simply because of their size. In Figure 2.10, the large western states attract more attention than the eastern states. Both of these issues are examples of a phenomenon known as the **modifiable areal unit problem**, or **MAUP**, which occurs when measurements are being aggregated over arbitrarily defined areas. Analyzing spatial or statistical patterns becomes more difficult because the patterns are obscured by the aggregation scheme. However, methods to reduce the effects of MAUP on maps are available.

One approach is to **normalize** the data, dividing each value by a specified variable. If one is concerned that the size of the aggregation area is influencing the magnitude of values, one can divide by the feature area as shown in Figure 2.11a, where the number of farms is divided by state area (compare to Fig. 2.10a). If the values are being influenced by population, then one can divide by population or other suitable variable, as in Figure 2.11b, where the vacancies are divided by the number of housing units, giving a far more interesting and informative map than Figure 2.10b. Another normalization method, less commonly used, shows percentages of the total quantity falling in each feature; Figure 2.11c shows the percentage of Congress controlled by each state, accomplished by showing the number of districts in each state as a percentage of the total number of districts.

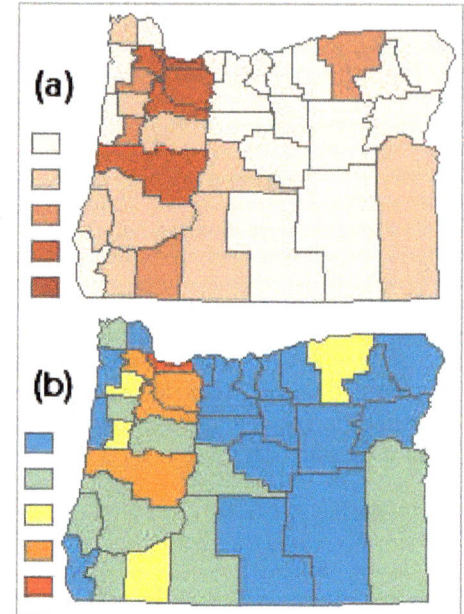

Fig. 2.9. A choropleth map showing population by varying (a) saturations and/or value, and (b) hue
Source: Esri

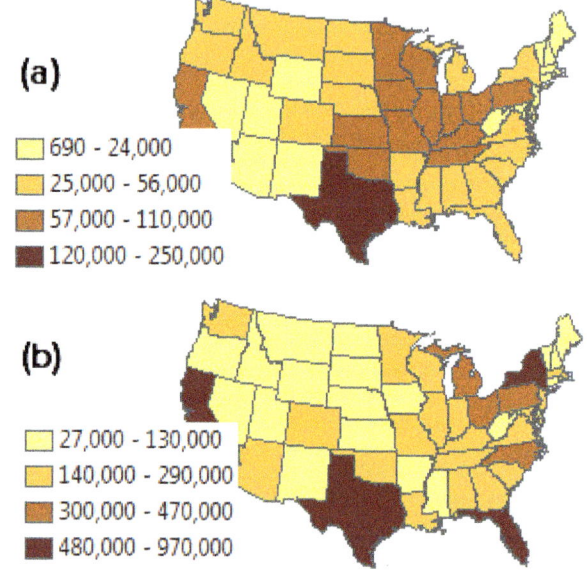

(a)
| 690 - 24,000 |
| 25,000 - 56,000 |
| 57,000 - 110,000 |
| 120,000 - 250,000 |

(b)
| 27,000 - 130,000 |
| 140,000 - 290,000 |
| 300,000 - 470,000 |
| 480,000 - 970,000 |

Fig. 2.10. (a) Number of farms; (b) number of vacant housing units
Source: Esri

However, not all attributes need normalization. In a precipitation map, each value already represents inches of rain falling at any location. The median rent for a county does not depend directly on the area of the county. These attributes would not be normalized. Before mapping any attribute, it is important to stop and consider how the aggregation unit might affect the map and determine whether the values should be normalized, and by what.

The visual MAUP issue, which occurs when large polygons dominate the map, can be addressed by changing the map type. A graduated symbol map applied to polygons places a symbol at the center of each polygon, minimizing the visual imbalance, as shown in a graduated symbol map of the density of farms (Fig. 2.12a; compare to Fig. 2.11a). A **dot density map** uses randomly placed dots to show the magnitude of a value in the attribute table (Fig. 2.12b). Each dot represents a certain number; in this map, 2000 farms. Note that the locations of the dots, which are randomly placed, do not necessarily represent the distribution of farms in the state.

Chart maps

A **chart map** expands the number of attributes that can be displayed on a map by replacing a single symbol with a chart representing several attributes. The chart could be a pie chart, bar chart, or stacked bar chart. Figure 2.13 shows a pie chart map with the proportions of Caucasians, African Americans, and Hispanics in each state. The pie sizes can be all the same (Fig. 2.13a) or proportional to the sum of the three categories, thus showing the relative number of people in the state as well (Fig. 2.13b). Notice how the pie colors are chosen to facilitate recognition of the classes: cream for Caucasians, dark brown for African Americans, and reddish for Hispanics. Such details help make maps easier for the reader to interpret, and they lessen the need for the reader to look back and forth between the legend and the map.

Displaying rasters

Recall from Chapter 1 that the raster model is a cell-based data model in which an array of cells or pixels store numeric values relating to some feature or quantity on the earth's surface. Rasters may be broadly grouped into three main types, each of which is displayed differently: thematic rasters, image rasters, and indexed rasters.

Thematic rasters

A **thematic raster** represents features or quantities, such as roads, geology, elevation, or vegetation density (Fig. 2.14). The raster model is designed to store **continuous** values that may occur anywhere on the earth's surface, such as elevation. However, they may also store **discrete** objects, such as roads or land use polygons. Thus we can speak of a raster as being a continuous or a discrete raster.

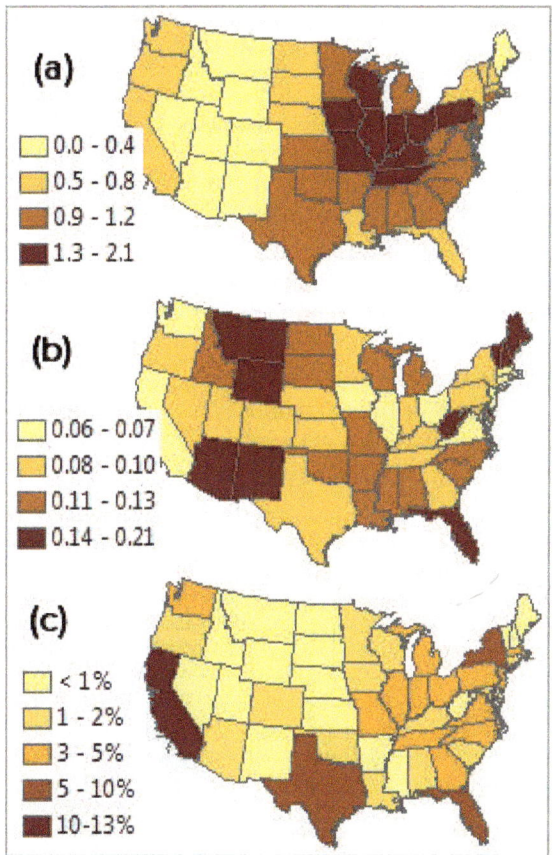

Fig. 2.11. (a) Farms normalized by state area; (b) vacancies normalized by housing units; (c) percentage of Congress controlled by each state
Source: Esri

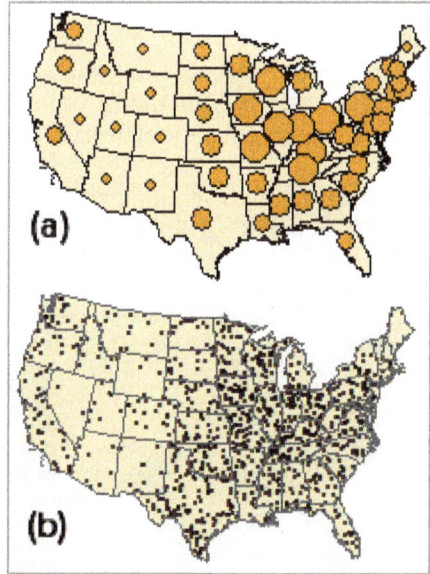

Fig. 2.12. Number of farms shown using (a) an area-normalized graduated symbol map or (b) a dot density map
Source: Esri

Values in a raster are always stored as numbers and may be characterized as categorical, ordinal, or interval/ratio. (Nominal data are not stored in rasters—it would be inefficient.) The methods used to display thematic rasters are similar to those used for displaying vector data and also depend on the data type.

A discrete raster representing categorical or ordinal data is displayed using a unique values map. Just like the unique

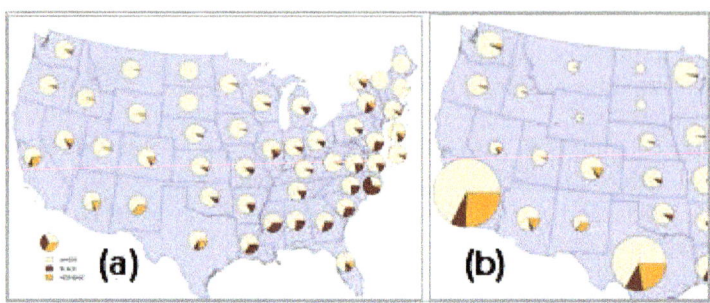

Fig. 2.13. (a) A chart map can represent several attributes, such as the proportion of ethnic groups. (b) The pie size can represent the total population.
Source: Esri

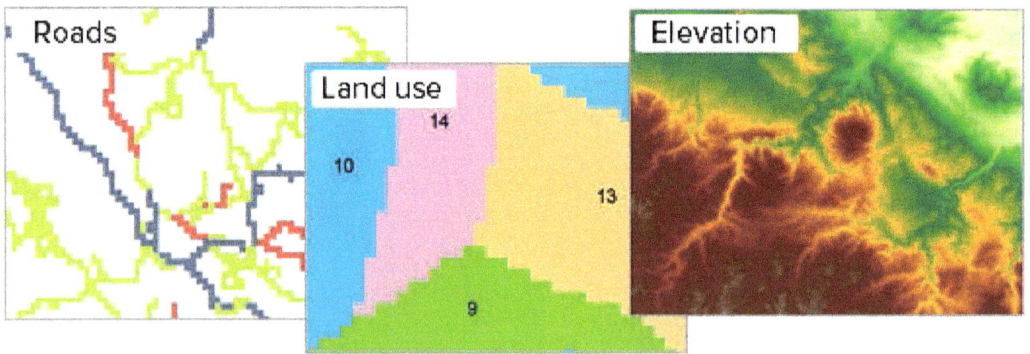

Fig. 2.14. Thematic rasters. Discrete rasters store feature data, such as road types or land use polygons. Continuous rasters store values that exist everywhere, such as elevation.
(Roads): Source: Black Hills National Forest; (Land use and Elevation): Source: USGS

values map used for data sets, each value of the raster receives its own color, as in the geology map shown in Figure 2.15a. Color schemes for unique values maps have 32 possible values and work best with data that have relatively few categories. When representing ordinal data, a monochromatic color scheme is used to communicate a sense of increase.

A thematic raster containing interval or ratio data must be classified into ranges in one of two ways. The **classified** display method divides the values into a small number of bins, similar to

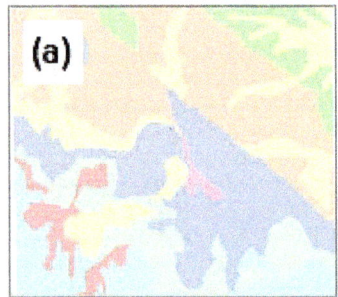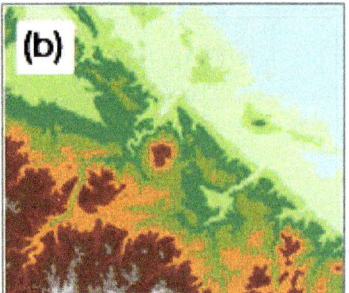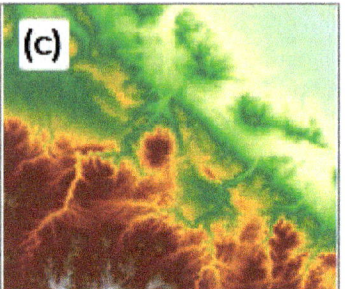

Fig. 2.15. Display methods for thematic rasters: (a) unique values geology; (b) classified elevation; (c) stretched elevation
(a): Source: South Dakota Geological Survey; (b-c): Source: USGS

a graduated color map. A color ramp is chosen to assign the colors to each bin. The elevation map in Figure 2.15b has 12 classes represented by 12 colors from a color ramp. The **stretched** display method scales the image values to a color ramp with 256 shades (Fig. 2.15c). The raster is first subjected to a **slice**, which rescales the elevation values (ranging from 800 to 1600 m) into 256 bins, then matches the bins to the 256 shades in the color ramp. A slice is essentially a classification that creates 256 classes.

The colorful ramps, like those in the elevation rasters in Figure 2.15, ignore the guideline to use saturation or value rather than hue to represent quantities, yet the eye has no trouble interpreting the elevation increase. Continuous rasters typically represent surfaces with an underlying structure, such that low values occur near low values and high ones near high. This structure imposes a visual order and allows the eye to interpret the greater number of colors. Such ramps are not suitable for features or all rasters, but they can be effective when appropriate.

The property that values close to each other tend to be more similar than values farther apart is present, to a greater or lesser extent, in many geographic data sets. It is so common, in fact, that this property is often called the first law of geography, or **Tobler's law**. Not only does it play a role in data visualization, as discussed previously, but it is also a fundamental factor in spatial analysis.

Image rasters

Image rasters include aerial photography and satellite data. The pixels represent degrees of brightness caused by light reflecting from materials on the surface. The brightness values are usually placed on a scale of 0 to 255 DN (digital numbers). If one brightness value is given to each pixel, the image is usually displayed with a grayscale ramp; a dark shadow would have a low brightness and a low DN, and a white cement road would have a high brightness and a high

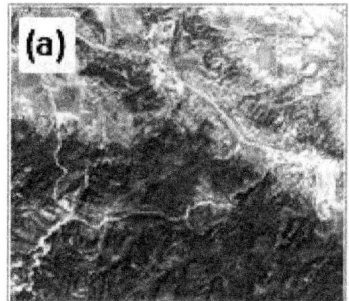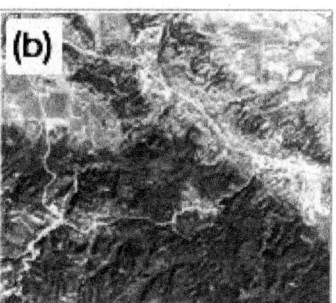

Fig. 2.16. Raster display methods for images: (a) stretched image; (b) RGB composite image.
Source: USGS

DN (Fig. 2.16a). Such images are displayed using the same **stretched** method that is applied to continuous thematic rasters. Images based on the 0 to 255 scale do not need to be sliced, but some images contain larger values and do require slicing. To display a color image, each pixel is given three DN values representing red, green, and blue light, thereby specifying a unique color combination, as discussed earlier. This display method is called an **RGB composite** (Fig. 2.16b).

Stretching image values

Image DN values are often normally distributed. Figure 2.17a shows a typical **histogram** of cell values, with the horizontal axis showing the range of values with the corresponding grayscale color ramp and the vertical axis showing the number of cells for each value (color). The image has few values near 0 or 255, with most of the pixels occurring in the middle gray range. The image is shown in Figure 2.17b, and it appears dim and featureless with little contrast.

The **stretched** display method can improve the display of normally distributed values by ignoring the tails of the distribution. One method assigns the lowest and highest image values to 0 and 255, respectively, and stretches the remaining colors along the ramp. This is called a minimum-maximum stretch. Even greater contrast can be developed using a standard-deviation stretch, which uses only the values within two standard deviations of the mean (Fig. 2.17c).

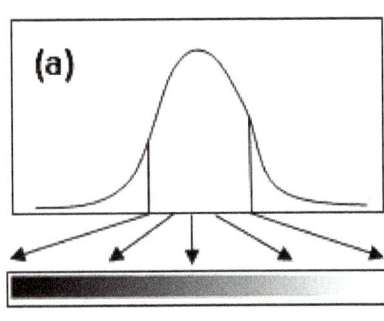

Fig. 2.17. The effects of stretching: (a) stretching an image histogram; (b) no stretch applied; (c) a standard-deviation stretch

(b–c): Source: USGS

Stretches can be applied to thematic continuous rasters as well as images and can be helpful when the image values are not evenly distributed.

Indexed color rasters and colormaps

The RGB color model is capable of storing millions of different colors and is best for fine rendering of images. However, it uses large amounts of memory and storage space. An alternate method for storing color identifies a restricted set of colors that appear in the image. It assigns an integer from 0 to 255 to each color and stores the RGB proportions needed to make it. This list of colors is called the **colormap** and is stored with the image. Each value in the raster is portrayed using its assigned color. Figure 2.18 shows part of a digital raster graphic (DRG). A DRG is a scanned US Geological Survey

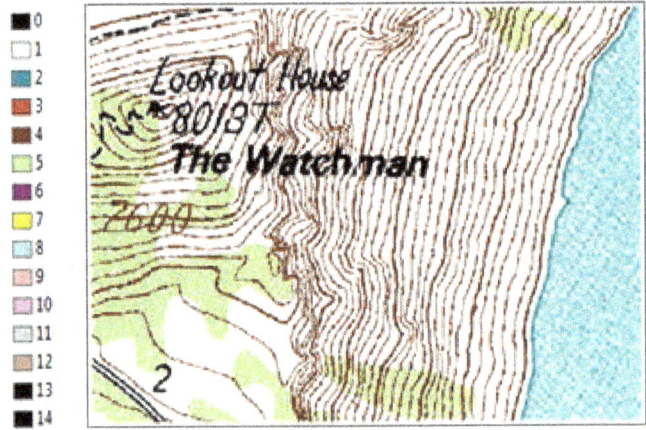

Fig. 2.18. An indexed-color raster stores a colormap that assigns a color to each value in the raster.

Source: USGS

topographic map. The contours are represented by pixels with the value 4 and displayed using the assigned brown color. Every pixel value and defined color used in the map is shown in the accompanying colormap: green for vegetation, black for text, and so on.

Indexed color rasters are commonly produced when scanning large color maps. Storing them in RGB could use hundreds of megabytes or more. An indexed-color map typically uses less than a tenth of the space needed for an RGB map, and the eye cannot easily distinguish between the two.

Classifying numeric data

Mapping numeric data, raster or vector, requires classifying a range of values into a small number of groups, each of which can be represented by a different color or symbol size. This process is called **classification**. There are many ways to classify data, and the choice of method affects the appearance of the map and the message that it portrays. Some common methods are compared in Figure 2.19 using the same data set of average farm size by state.

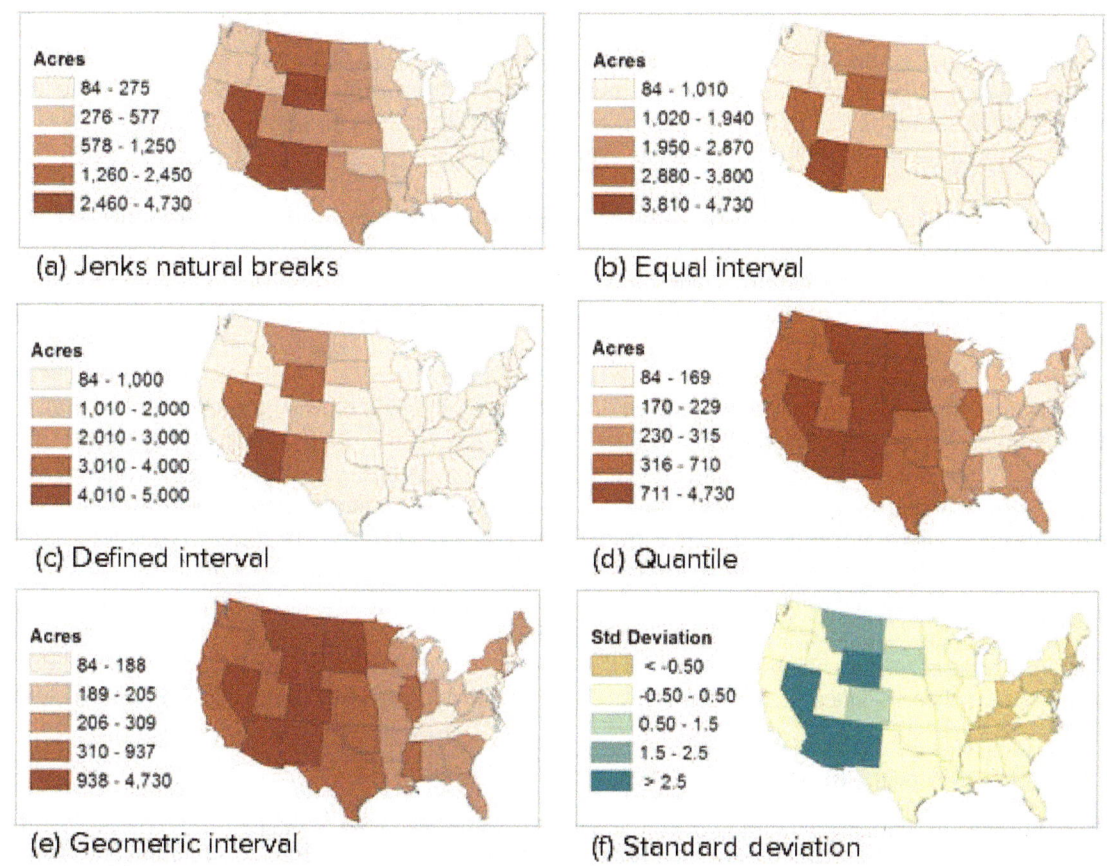

Fig. 2.19. Comparison of different classification methods using the same data set of average farm size in acres by state

Source: Esri

The **Jenks method** sets the class breaks at naturally occurring gaps between groups of data (Fig. 2. 19a). Each class interval can have its own width, and the number of features in each class will vary. The Jenks method works well on unevenly distributed data, such as populations of the capitals shown in Figure 2.8. There are many low-population and medium-population capitals and a few very high-population capitals. The Jenks method works well with almost any data set, making it a natural choice for the default classification scheme in ArcGIS Pro.

The **equal interval** classification divides the values into a specified number of classes of equal size (Fig. 2.19b). This method is useful for ratio data, such as income or precipitation, because it gives a sense of regularity to the observed increases. However, it is hard to predict how many features will end up in each class. Notice in the farm size example that nearly all of the states fall into the first class. Compare this map to Figure 2.19a.

A **defined interval** classification is similar to an equal interval one, except that the user specifies the size of the class interval, and the number of classes then depends on the range of values (Fig. 2.19c). This method will create rounded values in the classes that are easy to interpret. Defined interval maps are ideal when comparing classes composed of percentages, dollars, temperatures, and other values when specific break values are desired (100, 200, etc.). It does, however, suffer the same disadvantages as the equal interval classification.

A **quantile** classification puts about the same number of features in each class (Fig. 2.19d) and enables the display of groups, such as quartiles, commonly of interest in statistics. This method will create a balanced map with all classes equally well represented, but some of the features in the same class could have very different values, and features in different classes could have similar values. Quantile classifications are best applied to uniformly distributed data. Notice that a quantile classification highlights differences between the eastern states that are hidden in the Jenks, equal interval, and defined interval classifications.

A **geometric interval** classification (Fig. 2.19e) bases the class intervals on a geometric series in which each class is multiplied by a constant coefficient to produce the next higher class. A geometric interval classification is designed to work well with continuous data like precipitation and to provide about the same number of values in each class range. It works especially well with positively skewed data distributions.

A **standard deviation** classification apportions the values based on the statistics of the field. The user selects the class breaks as the number of standard deviations, and the data range determines the number of classes needed. This method excels at highlighting which values are typical and which are outliers, especially since a divergent color set is used to accentuate below- versus above-normal values. In Figure 2.19f, the yellow states are close to the mean farm size, several eastern states appear to have smaller-than-average farms, and some western states have much larger farms than average. This map is best applied to normally distributed data.

Finally, if none of the preceding options gives the desired map, the user can manually set class break points to any chosen values. This option works for assigning a logarithmic or exponential scale as well.

Choosing the classification method

The choice of classification method depends on the mapmaker's purposes and on the type of data. Jenks shows the "nearest neighbors" in a distribution, whereas a defined interval or equal interval map does a better job of portraying relative magnitude. Notice the difference between the maps in Figures 2.19a and 2.19b. The Jenks map gives the impression that many states have large farms because the class sizes increase slowly; an equal interval map shows that farms in most states actually average 1000 acres or less. Figure 2.19b makes this observation clear at first glance.

Some data possess the quality that the magnitude of the values has an intrinsic meaning. Percentage data, for example, have a physical and a psychological meaning—people have an intuitive understanding of the difference between 50% and 100%. Differences in median rent occur in dollar values to which people can attach meaning. When dealing with such data, it is wise to use a defined interval map to choose classes with logical break points, such as 20%, 5 inches of rain, or $200, rather than 12.6%, 1.47 inches of rain, or $187. The reader can more effectively interpret the classes.

The distribution of values has impact also. Jenks and geometric interval classifications are designed to work well with unevenly distributed data. Equal interval, defined interval, and quantile maps can be used with any data but show better results with evenly distributed data. The statistics behind standard deviation maps assume that the data are normally distributed.

About ArcGIS

Layers

When a data set is added from a folder to a map in ArcGIS Pro, we call it a **layer**. Although we tend to think of these layers as data sets, there is actually an important distinction. When a picture is inserted in a Word document or a slide show, a *copy* of the picture is stored inside the document, so it is always there if the document gets copied or e-mailed. Even if the original picture is deleted from the hard drive, the copy remains in the document. We are accustomed to this treatment of information, so it is difficult to realize that Pro does NOT treat map data this way. We will learn more about the reasons in Chapter 5.

When data are added to a map, no copy is made. The information remains in its original location, and the map layer simply points to it. The original data on which the layer is based are known as the **source data**. When a layer's properties are changed, such as giving it labels or displaying it with different colors, the source data set does not change. The layer finds the source data each time and manipulates them within the map to produce the desired effect. A layer is like a cooking recipe—it describes where to find the data file (the ingredients) and how to display it (the cooking instructions). The information in the layer, such as which symbols should be used and whether the features have labels, is called the **layer properties**. Layers are held in memory and stored when the project is saved. If the source data set is deleted, renamed, or moved, the map will be unable to display it.

When working in Pro, it is important to always remember that a layer and the geospatial data it references are different objects. The data set may be stored on a disk, or it may come through a portal from ArcGIS Online or another server. *A layer points to the location of a geospatial data set and stores information about how to display and use it*. The same data set can be accessed by multiple layers and maps to present different views of the same information, whether within a single map, within the same project, or across many maps created by different users. Figure 2.20 shows a counties data set stored on a network drive; the same data set has been used by several students in their projects.

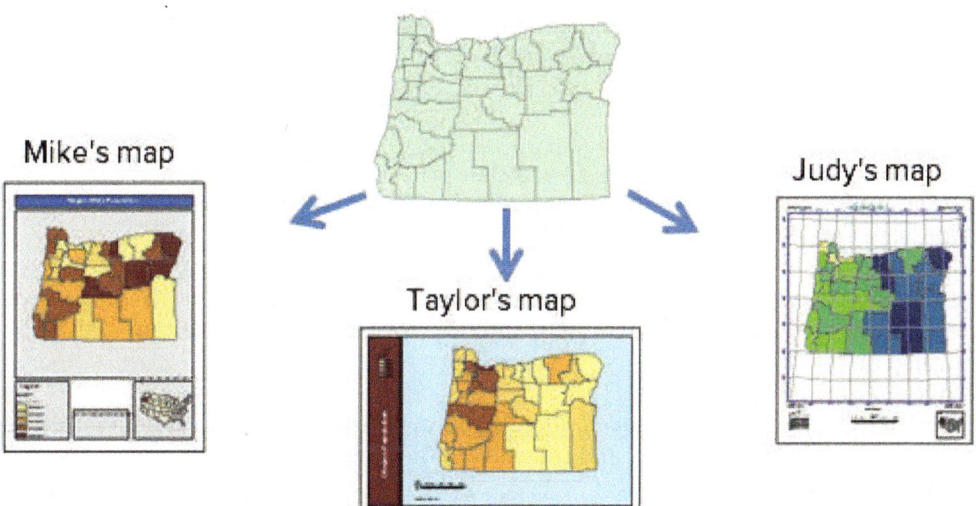

counties feature class
F:\Dept\gisdata\Oregon\oregondata.gdb\counties

Mike's map

Taylor's map

Judy's map

Fig. 2.20. The same feature class can be used in many maps, even by different people.
Source: Esri

A layer may be saved as a **layer file**. This file stores the location of the referenced data set and the properties of the layer. Saved layer files can be used to quickly add a layer with predefined symbols to a map document. Group layer files can organize multiple layers with predefined symbols into thematic maps or base maps. Geologic map symbols, for example, use many specific colors and patterns and can take hours to set up.

In Figure 2.21, a user has combined different sizes, colors, and even shapes to produce a specific classification that shows the arsenic values for water wells, with x indicating values below the detection limit, blue circles to show values less than the Environmental Protection Agency (EPA) standard, and orange circles for values above the EPA standard. By saving the finished layer as a layer file, we can quickly transfer the classification scheme to new layers and preserve the consistency of the symbols.

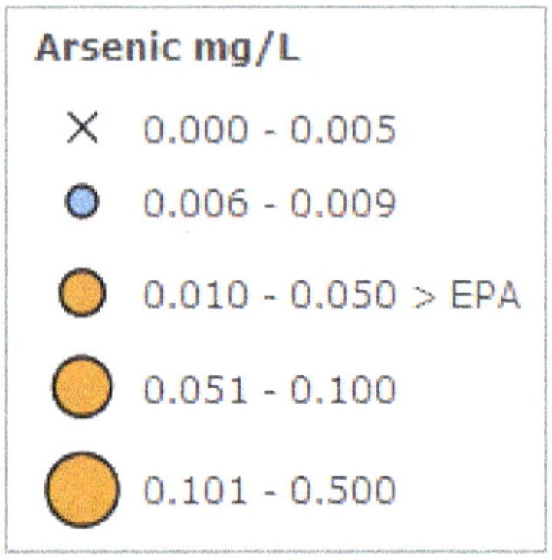

Styles

ArcGIS provides an extensive set of tools for symbolizing features and rasters. Within the Symbology pane (see Fig. 2.22), the user can choose from a variety of predefined symbols and modify the properties of those symbols, such as the fill color, outline width, and outline color, when the available ones are not suitable. The software provides complete flexibility to modify individual symbol layers and create virtually any symbol desired.

Symbols are organized into groups for ease of use. A **style** contains a set of symbols with a related theme and typically includes different symbol types, including point markers, line styles, polygon shades and patterns, color sets, text styles, and even north arrow or scale bar symbols. Pro has two default styles named ArcGIS 2D and ArcGIS 3D. The ArcMap program included many additional styles customized for specific industries or purposes, with themes such as Geology 24K, Environmental, Civic, Crime Analysis, or Hazmat. Although these styles do not appear within Pro, they can be imported if needed.

Arsenic mg/L	
✕	0.000 - 0.005
◉	0.006 - 0.009
●	0.010 - 0.050 > EPA
●	0.051 - 0.100
●	0.101 - 0.500

Fig. 2.21. Custom arsenic classification combining shape, color, and size

Users with very specific symbol needs can create their own styles and fill them with symbols copied from other styles, modified from other styles, or newly created. A search function in the Symbology pane allows the user to search by keyword for matching symbols in any of the installed styles. Searching for *tree*, for example, yields various symbols from different styles.

Summary

▶ Differentiating between features on a map requires variations in symbol shape, size, thickness, line type, color, pattern, or font.

▶ Attribute data have a data type designation: nominal, categorical, ordinal, interval, or ratio. The data type determines the kind of map and even the types of analysis that may be performed on that attribute.

▶ Nominal data name things or uniquely identify them and may be text or numbers. Each feature usually has its own value, and repeats are the exception.

▶ Categorical data group objects into smaller sets identified by a unique value. Numbers may be categorical when they are used as codes. Ordinal data consist of categories that are ranked in some way.

▶ Interval data are measured on a regular scale, and ratio data are measured on a regular scale with a meaningful zero point.

▶ Single symbol maps and labels are used to map nominal data. Unique values maps are used for categorical or ordinal data. Interval or ratio data are displayed using graduated color, graduated symbol, proportional symbol, dot density, or chart maps.

▶ Rasters may contain thematic or image data. Discrete thematic rasters can be displayed using a unique values method. Continuous thematic rasters may be classified or stretched. Image rasters use a stretched or RGB composite display.

▶ Continuous numeric data are classified before being mapped. Classification methods include Jenks natural breaks, equal interval, defined interval, quantile, geometric interval, standard deviation, and manual. The best classification method depends on the type of data and the data distribution.

▶ Data sets become layers when added to a map. Layers reference the original data and store information on how to display them.

▶ In ArcGIS, symbols are grouped into similar themes called styles. Tools are available to allow the user to create virtually any symbol needed for a map.

Important Terms

aggregated	discrete	interval data	ratio data
alpha	divergent color set	Jenks method	RGB composite
categorical data	dot density map	layer	saturation
chart map	dynamic labels	layer file	slice
choropleth map	equal interval	layer properties	source data
classification	geometric interval	modifiable areal unit	standard deviation
classified	graduated color map	problem (MAUP)	stretched
CMYK	graduated symbol map	nominal data	style
colormap	histogram	normalize	thematic raster
connotation	HSV	ordinal data	Tobler's law
continuous	hue	proportional symbol map	unique values map
convention	image	quantile	value
defined interval			

Chapter Review Questions

1. For each of the following types of data, state whether it is nominal, categorical, ordinal, interval, or ratio. Explain your reasoning.

   arsenic concentration in mg/L   elevations of climate stations
   vegetation type   football team rankings
   annual precipitation in inches   number of students in universities
   customer zip codes   letter grades given to students
   social security numbers   college student majors

2. For each of the following attributes, state whether a single symbol, graduated color, or unique values map would be most appropriate. Explain your reasoning.

   political party chosen by voting districts   river flow volumes (cfs)
   lung cancer rates by county   restaurant locations
   flow rates of wells in gallons/minute   soil class, such as loam, clay, etc.

3. If mapping the following attributes for counties, indicate which ones would generally be normalized, and discuss what attribute field(s) should be used.

   average annual snowfall   median home price
   home vacancy rate   number of crimes reported
   Native American population   number of car accidents

4. State whether a unique values, classified, stretched, or RGB composite display method should be used for each of the following rasters, and justify the choice.

land cover classes   black-and-white aerial photo

7-band satellite image   tree canopy percent

elevation   slope in degrees

5. Explain the merits of unclassed maps as compared to classed maps.

6. Describe the difference between a geospatial data set and a layer.

7. Explain the difference between thematic rasters and image rasters.

8. For each of the following map types, find a map from the Internet that uses this type of symbolization: graduated color, graduated symbol (lines), graduated symbol (points), unique values. Turn in a screenshot of each map with a citation. For each one, explain how the differentiation between features is achieved, and critique the choice of symbols used.

Expand your knowledge

Expand your knowledge of ArcGIS Pro by exploring these sections of the Help.

Help > Maps > Author maps > Symbology > Symbolization

Help > Maps > Author maps > Text

Help > Projects > Project Items > Styles

Help > Data > Data types > Imagery and raster > Appearance and Symbology

Mastering the Skills

Teaching Tutorial

The following examples provide step-by-step instructions for doing basic tasks and solving basic problems in ArcGIS. The steps you need to do are highlighted with an arrow →; follow them carefully. Click on the video number in the VideoIndex to view a demonstration of the steps.

Setting basic symbols

For this tutorial, we will start by creating a new, blank project. By default, a project is created inside a folder with the same name, which also includes a home geodatabase.

1→ Start ArcGIS Pro. Under the *Create a new project* heading on the right, click Blank to create a new blank project.

1→ In the Name box, type OregonMaps.

 1→ Click the Browse button for the Location and click the C:\ drive to open it. Navigate into your gisclass folder.

1→ Click the ClassProjects folder to select it as the location (don't go inside it). Click OK.

1→ Keep the *Create a new folder for this project* box checked and click OK.

A blank project will open with the Catalog view visible in the display area. The Catalog view is a larger and more versatile version of the Catalog pane. The home geodatabase is initially empty, of course. We will be using some data from the external mgisdata folder, so we need to add a folder connection to access it.

2→ Click **Insert: Project: Add Folder** and click the C:\ link on the left.

2→ Navigate into your gisclass\mgisdata folder. Click on the Oregon folder to highlight it (don't go inside it) and choose OK.

2→ In the Catalog pane (make sure the Project tab is active), expand the Folders entry to see OregonMaps (the project folder) and Oregon (the folder just added).

2→ Expand the Oregon folder to see two geodatabases and two Excel files.

3→ Click **Insert: Project: New Map**.

3→ In Contents, click the map name (Map) twice, slowly, and rename it *Geography*.

3→ Save the project, and remember to save periodically.

> **TIP:** It is helpful to keep the Catalog and Symbology panes open throughout this tutorial. Dock them in the same window to enable switching back and forth easily, as shown at the bottom of Figure 2.22.

Begin by adding some data sets from the Oregon geodatabase.

4→ Click the **Map: Layer: Add Data** button. Click on the Folders entry on the left side of the window to show the project folder connections.

4→ Double-click the Oregon folder to open it. Double-click the oregondata geodatabase to open it.

4→ Hold the Ctrl key down while selecting first the counties and then the volcanoes feature classes from the Oregon\oregondata geodatabase, so that both are selected. Click OK to add them to the map.

Remember, the Catalog pane can also be used to add data to the map.

5→ In the Catalog pane, expand the oregondata geodatabase and then expand the Transportation container.

5→ Hold the Ctrl key to select both the airports and highways feature classes. Then click and drag them into the map.

5→ Add the rivers data set from the Water container.

> **TIP:** From here on, use whichever method is easiest to add data to the map.

We'll make sure that the full extent of the map is set to the Oregon counties layer.

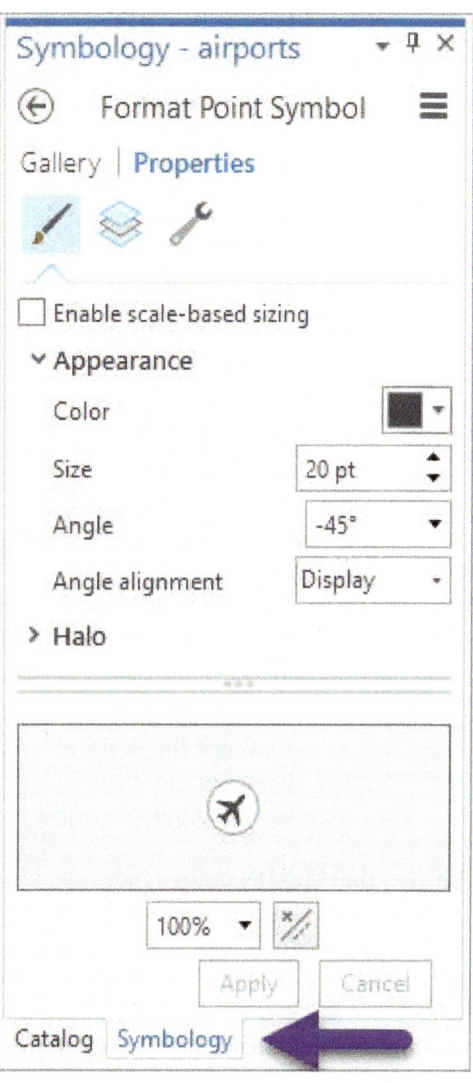

Fig. 2.22. Dock the Catalog and Symbology panes in the same window.
Source: Esri

6→ In Contents, double-click the Geography map icon to open the map properties.

6→ Examine the Extent settings. Fill the button to use a Custom extent and set the drop-down to calculate it from counties. Click OK.

7→ In Contents, click on the airports symbol. It opens to the Format Point Symbol pane.

7→ Click the Gallery tab and examine the symbols to select from.

7→ Type *airplane* into the search box and click Enter.

7→ Change the drop-down box for Project Styles to All Styles to see more symbols.

7→ Choose the largest Airport symbol from the ArcGIS 2D style.

 7→ Click the Properties heading and select the icon for the Symbol panel.

7→ Set the Size to *20* pt. and Angle to *-45* degrees. Click Apply.

Recall that color may be viewed as a combination of three variations: hue, saturation, and value. Let us examine how these variations affect the color produced.

8→ In the Format Point Symbol panel, hover over the Color swatch to view the RGB combination values for the gray color (R:88 G:89 B:91)

8→ In Contents, right-click on the symbol for counties and choose Color Properties.

8→ In the Color Editor, click in the multicolor box on the right to select a color. Examine the Red-Green-Blue DN values. Choose several more colors, and examine the values.

8→ Set the Red Green Blue values each to 255. What color results?

Notice that the DN values increase from the bottom to the top of the box, and the color proportions change from left to right. Now let's examine the HSV color model.

8→ In the Color Editor, change the Color Model from RGB to HSV.

8→ Use the sliders to vary the hue, saturation, and value of the current color to understand how these vectors affect the color produced. Click Cancel.

9→ In Contents, right-click the line symbol for rivers and change the color to deep blue.

9→ Right-click on the symbol for counties and change the color to light yellow.

9→ In the Symbology pane, click the Symbol representation to open the Format Polygon Symbol pane. Click the Properties heading and change the Outline Color to Gray 50% (hover over a color to see its name). Click Apply.

The rivers seem to be cluttering the map at this scale, so they would benefit from a scale range.

10→ In Contents, select the rivers layer and open the **Feature Layer: Appearance** ribbon.

10→ In the **Visibility Range** group, click the **Out Beyond** drop-down and select 1:1,000,000 from the list (if there is no list, type 1000000 in the box and click Enter).

10→ Zoom in until the rivers appear. Then click the **Map: Navigate: Full Extent** button.

TIP: Click in the Out Beyond dropdown and choose Customize to load a default set of scales to choose from, such as ArcGIS Online/Bing Maps/Google.

Labeling features

Nominal data are usually portrayed using labels. **Dynamic labels** can be created for a layer using an attribute field and can be turned on or off for the entire layer. They use an auto-placement function to avoid overlaps, and not all labels may be shown. Dynamic labels are redrawn each time the map is redrawn, and they may change as the scale changes or when the map is printed.

11→ In Contents, select the airports layer and open the **Feature Layer: Labeling** ribbon.

11→ Click the **Label** button in the **Layer** group to turn on the default labels.

11→ In the **Label Class** group, verify that the **Field** is set to NAME.

11→ In the **Text Symbol** group, click the **Text Symbol Style** drop-down to view a collection of preset styles, but do not select any.

11→ In the **Text Symbol** group, set the font to Arial 9-pt. bold.

11→ In the **Label Placement** group, click the **Label Placement Style** drop-down and set it to the **Basic Point** option.

Basic labels are easy to set, but they often are not quite what is wanted. Additional options to control dynamic labels, such as adding a halo to help them stand out, are available. Recall that the small arrow in the lower right corner of a ribbon group opens advanced settings.

 12→ Click the **Feature Layer: Labeling: Text Symbol** group **Options** arrow.

12→ In the Label Class pane, on the Symbol tab, expand the Halo entry.

12→ Click the Halo symbol drop-down and choose White fill 50% transparency (third from left). Increase the Halo size to 2 pt. Click Apply.

12→ Take a moment to examine the other settings on the Symbol tab.

Label classes can be created to assign different symbols to groups within a layer. For example, we could display the busier airports with larger symbols.

13→ Click the **Feature Layer: Labeling: Label Class: Class 1** drop-down and select **Create label class**.

13→ Enter a new label class name of *Busiest Airports* and click OK.

 13→ Click the SQL button next to the Class box.

13→ In the Label Class pane, click Add Clause. Enter the expression: TOT_ENP is Greater than 300000. Click Add. Click Apply.

14→ Click **Feature Layer: Labeling: Label Class: Class > Create label class**. Name the class *Less Busy Airports*.

14→ In the Label Class pane, click the Class drop-down to see or assign which class is currently being modified. It should read Less Busy Airports.

14→ Click Add Clause and enter the expression: TOT_ENP is Less Than or Equal to 300000. Verify the expression and click Apply.

14→ Leave the Label Class pane open; we will use it again shortly.

TIP: The Label Class pane contains many different settings for customizing the appearance and placement of labels. Experiment and explore the Pro Help on labeling to learn more.

We can use the Contents pane to manage the new label classes and easily set their properties. When a label class is selected in Contents, the ribbon settings are applied to that class.

15→ In Contents, click the icon for the Labeling panel.

15→ Under the airports layer, uncheck the Class 1 box to turn off the default labels.

15→ Click the Busiest Airports label class.

15→ On the **Labeling** ribbon, change the symbol to Tahoma 10-pt. bold.

15→ In Contents, click the Less Busy Airports class to select it.

15→ On the **Labeling** ribbon, change the symbol to Tahoma 8-pt. regular.

The labels should appear similar to those shown in Figure 2.23.

TIP: If you have trouble finding ribbons, buttons, or settings, check the Contents pane and make sure that the appropriate item is selected.

16→ In Contents, still in the Labeling panel, select rivers to work with those labels now.

16→ Check the box for the rivers layer to turn the labels on. They don't show yet because the rivers have a display scale set, and the labels initially use the same one.

16→ Zoom to ~1:1,000,000 to make the rivers and labels appear.

16→ On the **Labeling** ribbon, change the symbol to Arial 8-pt. dark blue italic.

It is traditional to curve labels along rivers instead of using straight text.

17→ On the **Labeling** ribbon in the **Label Placement** group, click the Label Placement Style drop-down and select **Water (Line)**.

17→ Click the **Feature Layer: Labeling: Map: View Unplaced** button. With this setting on, labels that don't fit are shown in red.

17→ Zoom and pan around the map to find some unplaced labels. Dynamic labels may change as the map scale and extent change.

17→ Click the **View Unplaced** button again to turn the unplaced labels off.

Symbols and labels may be combined when labeling.

18→ Use the **Map: Navigate: Full Extent** button to zoom to all of Oregon.

18→ In Contents, still in the Labeling panel, right-click the highways layer and choose Symbology. In the Symbology pane, click the symbol.

18→ In the Format Line Symbol pane, click the Gallery tab.

18→ Enter *point* in the search box and click Enter. Choose the 2.0 Point symbol. Switch to the Properties tab and change the color to light brown. Click Apply.

19→ In Contents, select the highways layer and open the **Feature Layer: Labeling** ribbon.

19→ In the **Label Class** group, change the **Field** to HWY_SYMBOL.

19→ In the **Text Symbol** group, click the **Text Symbol Style** drop-down, then scroll down to find and select the Shield 8 symbol.

19→ In the **Label Placement** group, click the **Label Placement Style** drop-down and select the Shield option.

19→ In Contents, check the box beside highways to turn on the labels.

19→ In the **Labeling: Visibility Range** group, set the **Out Beyond** limit to 1:1,000,000.

19→ Zoom in to 1:1,000,000 to see the labels. Then return to the Full Extent of the map.

20→ Give the counties Arial 8 pt. bold labels in a 50% Gray color with a 2-pt. halo. Don't assign a scale range. The map should look similar to Figure 2.23.

20→ In Contents, return to the List By Drawing Order panel.

Creating maps from attributes for points

Currently, each data set is displayed with a single symbol. Maps can also be symbolized from attributes containing categorical, ordinal, interval, or ratio data. Let's examine the table fields.

Fig. 2.23. Oregon map labels and symbols
Source: Esri

21→ In Contents, right-click the volcanoes layer and choose Attribute Table.

1. What is the data type of these two attribute fields in the volcanoes data set?

ELEVATION _Interval_ TYPE _Categorical_

The ELEVATION field contains numeric data, so it must be portrayed using a quantities map, and the values must be classified.

21→ Close the table and select the volcanoes layer in Contents.

21→ Click the **Feature Layer: Appearance: Drawing: Symbology** button.

21→ In the Symbology pane, change the Symbology from Single Symbol to Graduated Colors (Fig. 2.24).

21→ Set Field to ELEVATION if needed. A set of five classes appears with the default Jenks Natural Breaks classification.

21→ Set the Color scheme to Yellow-Orange-Red (5 classes) as shown in Figure 2.24.

The symbols show which volcanoes have lower elevations and which have higher ones. However, the small symbol size makes the colors difficult to see—a common problem with graduated color maps for point data. We can edit the base symbol to make the map easier to interpret.

22→ In the Symbology pane, click the More drop-down and choose Format All Symbols.

22→ In the Format Point Symbols pane, click the Gallery tab and select the Triangle 1 symbol. The symbols will all turn black, but don't worry about that now.

22→ Click the Properties heading and change the symbol size to 12 pt. Click Apply.

22→ Click the back arrow in the Format Point Symbols pane to return to the main Symbology pane. Reset the Color scheme.

TIP: Choosing a triangle symbol that mimics a volcano's shape helps readers to more easily interpret the map.

23→ Turn off the airports and highways layers.

23→ Right-click the counties layer and choose Label to turn off the labels. The settings are stored in the layer, and the same labels can be turned on again at any time.

Observe that the eastern volcanoes are generally lower in elevation, and the highest volcanoes lie in the north. Let's try a graduated symbol map instead.

24→ In Contents, select the volcanoes layer.

24→ In the Symbology pane, change the Symbology drop-down to Graduated Symbols. Keep the Field set to ELEVATION.

24→ Click the Template symbol, click the Gallery tab, and choose the Triangle 3 symbol.

24→ Click the Properties heading and set the color to a reddish brown. Click Apply.

24→ The smallest class symbol is barely visible. Click the circled arrow to return to the main Symbology pane. Change the number of classes to four. Set the minimum size value to 6 pt. and the maximum size to 20 pt.

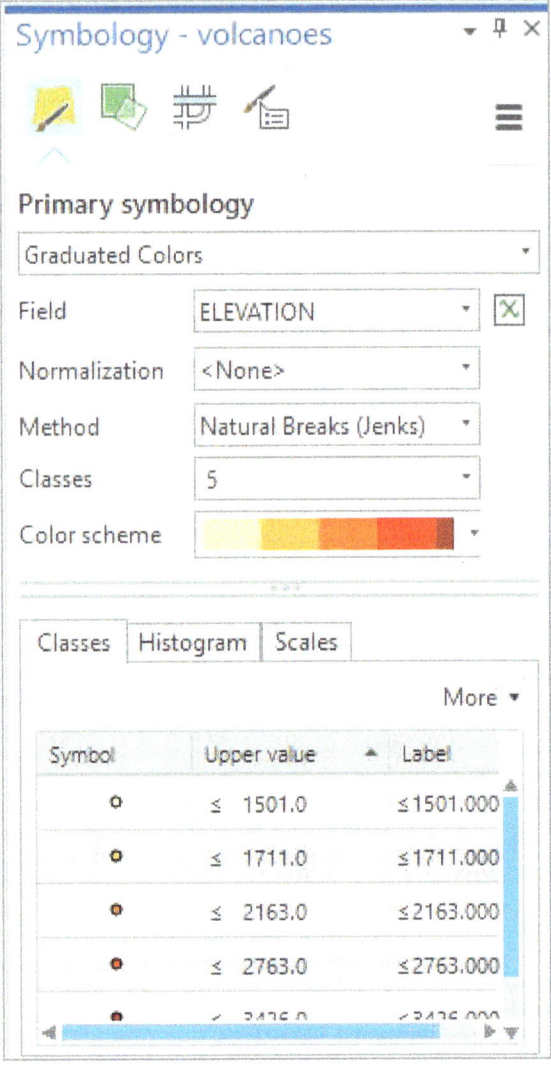

Fig. 2.24. Symbolizing volcano elevations with a graduated color map

Source: Esri

The same data can be presented in different ways, and the mapmaker must choose the method that best communicates the intended message. In your opinion, did the graduated color or graduated symbol map better portray the differences in elevation?

25→ In Contents, rename the volcanoes layer *Volcano Elevation*.

The elevation legend in the Contents pane shows many decimal values, which is undesirable for two reasons. First, they are pointless because they are all zeros. Second, the significant figures used for labels should be consistent with the precision of the values being reported.

 25→ In the Symbology pane, open the Advanced symbol options panel.

25→ Expand the Format labels heading.

25→ Under Rounding, fill the button for *Significant digits* and set the number to 2.

25→ In Contents, click twice slowly on the ELEVATION heading in the Volcano Elevation legend and type (*meters*). The legend should appear as in Figure 2.25.

25→ Save the project.

Fig. 2.25. Elevation legend
Source: Esri

TIP: Evenly rounded class labels give a more professional look to a map.

Next, we will symbolize the volcanoes based on type. However, we don't want to lose the elevation map, so we will create a copy of the layer. One advantage of layers is that we can create different views of the same data set.

Fig. 2.25. Elevation legend
Source: Esri

26→ In Contents, right-click the Volcano Elevation layer and choose Copy.

 26→ Right-click the Geography map icon and choose Paste.

26→ Rename the new Volcano Elevation layer (above the old one) *VolcanoType*. Drag it below the airports layer.

26→ Turn off the Volcano Elevation layer.

Consider how to symbolize the volcano TYPE field. It contains repeating entries like "Stratovolcano" and "Volcanic field," which represent categorical data. The proper type of map for categorical data is a unique values map.

27→ In Contents, select the Volcano Type layer.

27→ In the main Symbology pane, change the Symbology drop-down to Unique Values (see [1] in Fig. 2.26).

27→ Change the Value field to TYPE (2).

TIP: Widen the Symbology pane, if needed, to see all of it. If it is docked, place the cursor over the shared edge with the map window until it becomes a double-sided arrow; then click and drag the boundary to the desired width. If it is undocked, resize it like any other window.

Notice the *<all other values>* entry at the bottom of the categories list. This offers the option of grouping small categories together. Currently, no features appear in this class.

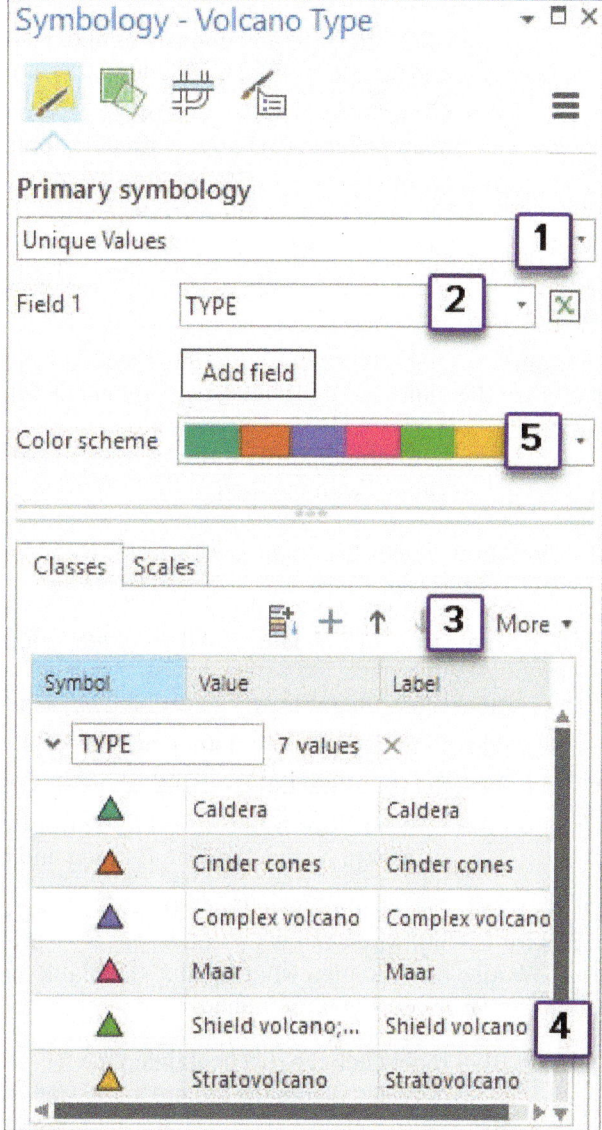

Fig. 2.26. Symbolizing the volcano types
Source: Esri

27→ Click the More drop-down (3) and select *Show all other values* to uncheck it and turn this category off.

Notice the two classes for Shield volcano(es). A data entry error separated these two groups that belong together. We could edit the source data to fix the error, but for now we can work around it.

28→ Click the Shield volcano entry (4), then hold down the Ctrl key and click the Shield volcanoes entry so that both rows are highlighted.

28→ Right-click in the Shield volcano label box (4) and choose Group Values.

28→ Click in the Label box containing the double entry, edit it to simply *Shield volcano*, and click Enter.

TIP: Changing the label here only changed the display properties of the layer. The data entry error is still present in the stored data set referenced by this layer.

29→ Click More (3) and choose Symbols > Format All Symbols. Click the Gallery tab and set the symbol to Triangle 3.

29→ Click the circled back arrow to return to the main Symbology pane. Change the color scheme (5) to one with dark, bold colors that will show up well.

29→ Examine the map and the final Symbology pane (Fig. 2.26).

2. List two ways to open the Symbology pane.

Editing legend details

Portraying a layer effectively sometimes requires modifying the default legend properties. We symbolize the highway types next, but first check the attribute fields available in the table.

30→ In Contents, turn on the highways layer; then right-click it and choose Attribute Table. Examine the HWY_TYPE field.

3. What three values are found in this field? Examine the other fields for clues and list what the three values represent.

Again we have categorical data, and the unique values map type is the appropriate choice.

30→ Close the table and select highways in the Contents pane.

30→ In the main Symbology pane, select the Unique Values map type.

30→ Set the Value field to HWY_TYPE.

30→ Choose More > Show Count to view the number of features in each category. Notice the lone H value and the 58 unlabeled roads.

It is reasonable to assume that "I" stands for interstates, "S" stands for state highways, and "U" stands for US highways, but what could the "H" represent? With only one, it may be a data entry error. We also have no idea what to label the blank highways. However, we can edit the legend categories to create a clearer legend.

31→ Right-click the row containing the "H" value and choose Remove. It removes the category and places the feature in the <all other values> entry.

31→ Right-click on the row with the unlabeled features and remove it also.

31→ Click twice in the "U" box in the Label column to highlight it. Edit the label and change it to *US Highways*.

31→ Replace the "S" label with *State Highways*, the "I" label with *Interstates*, and the <all other values> label with *Unclassified* (Fig. 2.27).

31→ Above the classes, edit the HWY_TYPE box and change it to *Road Class*.

 31→ Click the US Highways row to select it. Click the Up arrow to move it below the Interstates row.

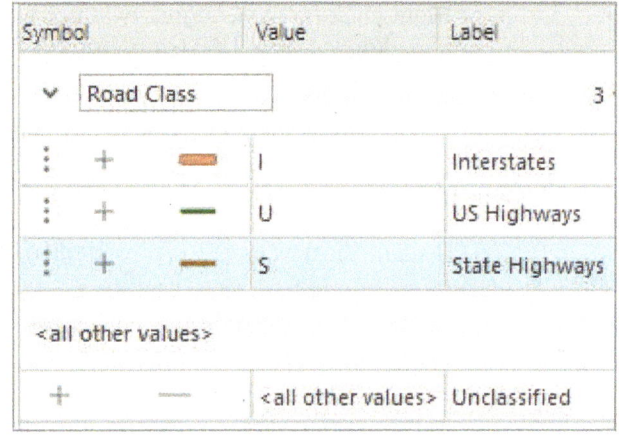

Symbol			Value	Label
⌄	Road Class			3
⋮	＋	▬	I	Interstates
⋮	＋	▬	U	US Highways
⋮	＋	▬	S	State Highways
<all other values>				
	＋	—	<all other values>	Unclassified

Fig. 2.27. Final legend for the highways
Source: Esri

Now we have symbols based on highway type, but the thin colored lines do not distinguish the types well. This classification would benefit from using differences in symbol, thickness, and color to distinguish the different road classes.

32→ Click the symbol for the Interstates to open the Symbol Gallery.

32→ Choose the first Highway symbol and click the back arrow.

32→ Click the symbol for the US Highways. Select the 2.0 Point symbol. Click the Properties heading in the Format Line Symbol window and change the color to dark green. Click Apply and click the back arrow.

32→ Change the symbol for the State Highways to match the U.S. Highways but with a light brown color.

32→ When finished, the legend should look as shown in Figure 2.27. Examine the map and the legend in the Contents pane.

Creating maps for attributes of polygons

Let's create maps based on population using the counties layer.

4. What kind of data does population represent? What kind of map should be used to display it?

33→ In Contents, click the counties layer symbol.

33→ In the main Symbology pane, change the Symbology to Graduated Colors.

33→ Set the Field to POP2014 if necessary.

Remember, population data are usually normalized to the area of the aggregation unit.

33→ Choose SQMI as the Normalization field (near the bottom of the list).

33→ Change the color ramp to a monochromatic one, such as light to dark orange.

33→ Open the Advanced symbol options panel and expand the Format labels heading.

33→ Give the labels two significant digits and check the box to *Show thousands separators*. Return to the Primary panel in the Symbology pane.

We will create a new layer for every symbology, to keep each one for reference.

34→ In Contents, rename the counties layer *Population Density*.

34→ Right-click on the Population Density layer name and choose Copy.

34→ Right-click on the Geography map icon and choose Paste. Leave it on top for now.

Next we will map the median ages of people in the counties. Unlike population, median age is not affected by the size of the area being measured. This attribute should not be normalized.

35→ Rename the new layer (on top) *Median Age*.

35→ In the Symbology pane, change the Field to MED_AGE. Change the normalization field to <None>.

35→ Use the Advanced symbol options panel to format the labels to two significant digits (Fig. 2.28a). Return to the Primary symbology panel.

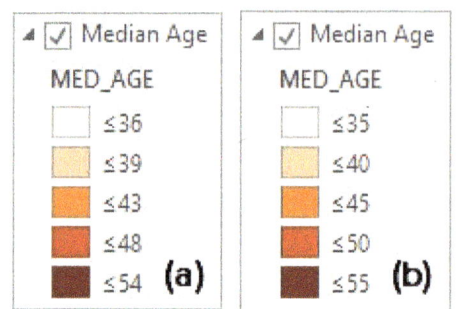

Fig. 2.28. Median age classification: (a) Jenks; (b) defined interval

Source: Esri

Classifying data

We have been using the default Natural Breaks classification so far, but it is not always the best method. Examine the uneven class ranges for the Median Age layer. The legend might make more sense if the age ranges increased by five years each time.

36→ In the Symbology pane for the Median Age layer, change the Method to Defined Interval. Ignore any warning, set the Interval Size to 5, and click Tab.

36→ Reset the value format to two significant digits, and then return to the Primary symbology panel. Examine the legend in the Contents pane (Fig. 2.28b).

Most readers would find this classification easier to understand than the uneven classes produced by the Jenks method. Next, symbolize the population of the cities.

37→ Add the majcities data set from the oregondata geodatabase.

37→ In the Symbology pane for the majcities layer, set the Symbology to Graduated Symbols and set the Field to POP2014.

37→ Notice the tabs in the lower half of the Symbology pane: Classes and Histogram. Click the Histogram tab (Fig. 2.29).

37→ Examine the spacing of the class breaks and horizontal bars showing the number of cities in each class.

37→ Hover over the X symbol to see the mean value. Click More > Show Statistics to see more information. Click it again to turn them off.

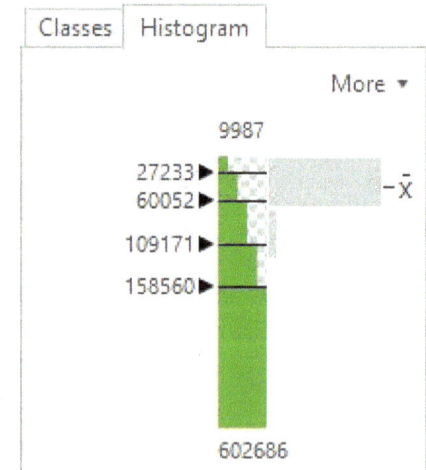

Fig. 2.29. Legend Histogram view

Source: Esri

TIP: Slide the class break lines up or down to create a manual classification, or click the number and type a class break value.

This data set is skewed, with many small cities and one very large one. The default Jenks Natural Breaks is probably the best classification choice. Five classes are a little difficult to distinguish based on size, though, so let's reduce the number to three.

38→ Set the number of Classes to 3.

38→ Click the Template symbol and set the color to a light one that will show up well when the symbols overlap. Click Apply. Return to the main Symbology pane.

38→ Use the Advanced symbol options panel to format the labels to three significant digits with thousands separators; then return to the Primary symbology panel.

Let's look at the activity of the volcanoes now.

39→ Make a copy of the Volcano Elevation layer and paste it into the Geography map. Name it *Known Eruptions* and turn it on.

39→ In the Symbology pane for Known Eruptions, keep the Graduated Symbols map type, but change the Field to KNOWN_ERUP. Examine the class values.

The -999 values are flags to indicate a missing value, and they are messing up the classification statistics. We can specify values to exclude from the classification.

40→ Open the Advanced symbol options panel and expand the Data Exclusion entry.

 40→ Click Add Clause and enter the following expression: KNOWN_ERUP is Less Than 0. Click Add and use the green check mark symbol to verify the expression. Click Apply.

To see the difference, reset the symbols in some way.

41→ Return to the Primary symbology panel. Set the classification method to Defined Interval and set the Interval Size to 3.

41→ In the Label column, change the <excluded> label to *No Info*.

41→ Examine the map and the legend in Contents.

An alternative to exclusion is to make a new layer that only shows the features of interest, such as showing only the active volcanoes with at least one known eruption.

42→ Click **Analysis: Geoprocessing: Tools** to open the Geoprocessing pane.

42→ Search for the Make Feature Layer tool and open it.

42→ Set the Input Features to the *Volcano Elevation* layer.

42→ Name the Output Layer *Active Volcanoes*.

42→ Click Add Clause and enter the expression: KNOWN_ERUP is Greater Than 0.

The appearance of the table is also part of a layer's properties. Let's turn off some fields to better focus on the ones of interest. Remember, these changes are just part of the layer properties, and they do not modify the source data.

43→ In the Field Info section, uncheck all fields except NAME, TYPE, and KNOWN_ERUP.

43→ Click Run to run the tool. Close the Geoprocessing pane when it finishes.

43→ In Contents, turn off the other volcano layers. Right-click the Active Volcanoes layer and choose Attribute Table.

Notice that the table contains only the fields specified and only volcanoes with at least one eruption. Now, before we finish with the volcanoes, let's try an unclassed map, which holds the advantage that no classification is required, and the view of the data is unbiased.

44→ Close the table.

44→ In Contents, select the Active Volcanoes layer and open the Symbology pane.

44→ Set the map type to Proportional Symbols. Set the Field to KNOWN_ERUP.

44→ If the map symbols don't appear in different sizes (possible software bug), change the Unit to Miles and then back to <Unknown>.

44→ Click the Template symbol and change it to the Triangle 3 symbol with a light color. Click Apply and click the back arrow.

44→ Set the Maximum size to 40.

44→ Click the Classes tab and set the Legend count to 3. Examine the map.

Displaying thematic rasters

Thematic rasters portray map data, such as soil types, roads, elevation, or precipitation. The current map is crowded with layers already, so we'll create a new map to experiment.

45→ Save the project.

45→ Click **Insert: Project: New Map** and wait for the new map to open.

45→ In Contents, rename the map Rasters.

We will begin with a digital elevation model (DEM), a raster that portrays an elevation surface. This DEM has a pixel size of one kilometer.

46→ Add the gtopo1km raster from the oregondata geodatabase to the map.

46→ In Contents, double-click the gtopo1km layer to open its properties. Examine the Source settings. Note that the vertical units (elevations) are in meters. Close the properties window.

46→ In Contents, right-click the gtopo1km layer and choose Symbology.

5. What kind of data type is elevation? What kind of raster display method(s) could be used to display it?

Pro chooses an initial display method based on the raster type. In this case, it chose the Stretch method with a grayscale color ramp. Notice the minimum and maximum elevation values, from 1 to 3124 meters. The Stretched method slices these values into 256 bins for use with the 256-grayscale ramp.

47→ In the Rasters map view, click on the raster to view values for a pixel.

Figure 2.30 shows the results when a gray pixel is clicked. The Pixel Value is the elevation, 1453 meters. The Stretched value indicates the bin number, 173. This bin is displayed as light gray. The COUNT field tells how many pixels have this value.

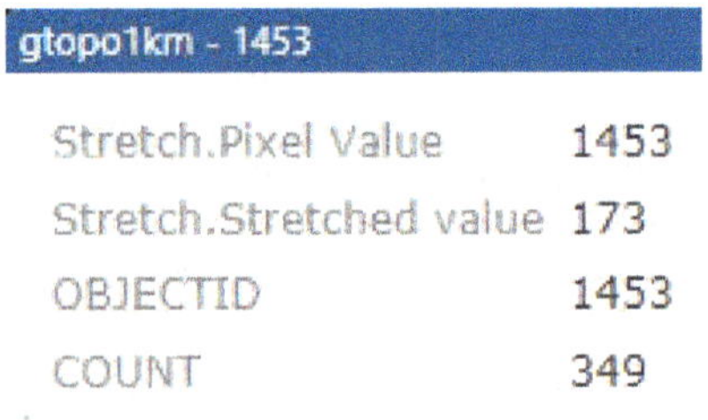

gtopo1km - 1453	
Stretch.Pixel Value	1453
Stretch.Stretched value	173
OBJECTID	1453
COUNT	349

Fig. 2.30. Examining pixels in the DEM
Source: Esri

47→ Click on more pixels, choosing black, white, and gray ones. Compare the pixel values and the stretched values for each one. Close the information pop-up when done.

48→ Examine the Symbology pane. A Standard Deviation stretch is being used with 2 standard deviations.

48→ Change the Stretch type to None and click Apply. Examine the map.

48→ Try the Minimum-Maximum stretch; then change the Stretch type back to Standard Deviation.

49→ Try choosing different color schemes.

49→ End by choosing the color scheme shown in Figure 2.31 (Elevation #1).

The Stretch method uses 256 equal interval classes. The classified display method gives control of the number of classes and the classification method.

49→ Rename the gtopo1km layer *Elev-Stretched*.

49→ Copy the Elev-Stretched layer and paste it into the Rasters map. Rename the new layer *Elev-Classified*.

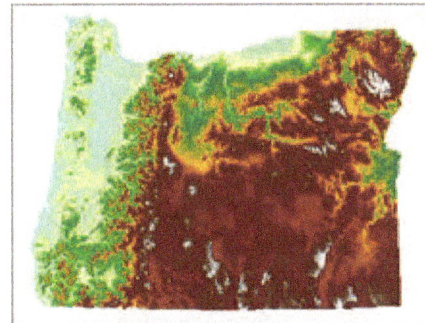

Fig. 2.31. Elevation terrain color ramp, stretched
Source: USGS

50→ In the Symbology pane for Elev-Classified, change the Symbology method to Classify. Find the same Elevation #1 color scheme and select it.

50→ Set the number of classes to 10. Continuous data, because they often have underlying spatial patterns, can use more classes and colors.

50→ In the map window, click on the map to examine the pixel values and the class values (which go from 0 to 9 rather than 1 to 10). Close the information pop-up.

This display method permits the classification parameters to be modified. The classification defaulted to Jenks Natural Breaks. However, elevations are best displayed in even classes with rounded numbers.

51→ In the Symbology pane for the Elev-Classified layer, change the Method to Defined Interval. Set the Interval size to 200 meters.

Notice how few white pixels show now (Fig. 2.32). The Stretched method and the Jenks classification overemphasized the areas with high elevations.

52→ Add the gtoposhd raster from the oregondata geodatabase to the map.

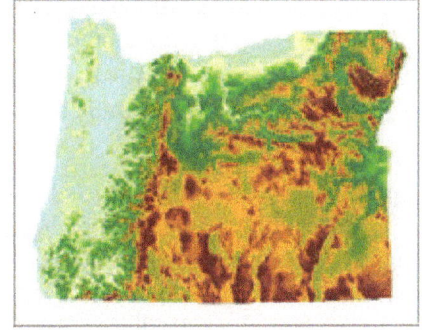

This hillshade raster is derived from a DEM by modeling how the illuminated surface might appear, and it contains brightness values from 0 to 255. Hillshade rasters provide an intuitive and detailed view of the surface and make a useful base for transparent overlays.

Fig. 2.32. Elevation terrain color ramp, classified
Source: USGS

52→ Click on the Elev-Classified layer and drag it above the gtoposhd layer in the Contents pane. Remember, layers are drawn from bottom to top.

52→ With the Elev-Classified layer still highlighted, open the **Raster Layer: Appearance** ribbon. In the **Effects** group, use the **Transparency** slider to set the layer transparency to 50% (Fig. 2.33).

53→ In Contents, collapse the legends for all layers

53→ Add the oregondata\slopeclass raster to the map. It shows areas of low, medium, and high slope indicated by the values 1, 2, and 3.

6. What type of data does this raster contain? What type of map should be used to display it?

> 53→ In the Symbology pane, change the Color scheme to a monochromatic ramp.

Fig. 2.33. Transparent elevation displayed over a hillshade raster

Source: USGS

> 54→ Add the oregondata\landcover raster to the map.
> 54→ Examine the Symbology pane for landcover. Scroll down and examine the color symbols and labels in the list. A total of 150 different colors are represented.
> 54→ In the map, click on different colors of the landcover raster to view the habitat codes and names. Close the pop-up when finished.

The Symbology defaulted to the Colormap method, which assigns preselected colors to each raster value using an alpha-red-green-blue combination. If a raster file includes a stored colormap, then this display method will be initially used.

In the information pop-up, the first row shows the pixel value (57 in Fig. 2.34). The second row contains the color mixture values. The raster table includes a field called HABNAME, which groups the cover classes into fewer categories. Let's display the land cover using this field.

landcover - Ponderosa Pine Forests and Woodlands	
Colormap.Pixel Value	57
Colormap.Color (a,r,g,b)	255,115,0,76
OBJECTID	56
Count	673976
GAPCODE	4240
ORCODE	4240
DISPLAY	4240 Ponderosa Pine
HABNAME	Ponderosa Pine Forests
HABCODE	33
LCCODE	420

> 55→ In the Symbology pane for landcover, change the display method to Unique Values.
> 55→ Change the Value field to HABNAME and choose a color scheme.

Images represent brightness of land features as measured by a camera or spectrometer, usually using the RGB color model.

Fig. 2.34. Identifying land cover pixels

Source: Esri

> 56→ Turn off all layers in the Rasters map.
>
>
>
> 56→ In the Catalog pane, click the Portal tab and select the All Portal icon.
> 56→ Type naip in the search window. Find the imagery layer icon labeled USA NAIP Imagery: Natural Color, and add it to the map.
> 56→ Examine the Symbology pane for the NAIP layer, noticing that it defaulted to the RGB symbology method. The legend in Contents shows the RGB symbols also.

Typically, an imagery layer gets more detailed at larger scales. This feat is accomplished using scale ranges showing different sets of imagery depending on the map scale.

This is the end of the tutorial.

> → Save the project and close ArcGIS Pro.

Practice Exercises

In this exercise, you will create four new maps in your OregonMaps project, each one using a different theme: Volcanic Hazards, Agriculture, Housing, and Physiography. Use the data in the oregondata geodatabase. Make each map both aesthetic and legible.

In the Volcanic Hazards map:

1. Show the population density of the counties. Also show the hospitals, marked with blue crosses.

2. Create a proportional symbol map of the volcanoes based on the KNOWN_ERUP field.

In the Agriculture map:

3. Create a map of counties showing the density of farms. Label the county names.

4. Create a map showing the transportation routes, symbolized by type (road type, rail, and air).

In the Housing map:

5. Create a map showing the vacancy rate of the counties. Which field is most appropriate to normalize this data? Label the county names.

6. Show the major cities data set (majcities) symbolized by population.

In the Physiography map:

7. Add the state boundary (statedtl) and the county boundaries, and then add the World Imagery service from ArcGIS Online. Use a hollow symbol and a contrasting outline color for the state and county boundaries so that they show up well against the imagery.

For all the maps:

8. Examine the legends in the Contents pane. Be sure they use good classification schemes, rounded label values, and appropriate significant digits or decimal places. Be sure that each layer has an informative capitalized name.

9. Capture a screenshot of each map. Make sure that the screenshot includes the Contents pane showing the corresponding legends for each map.

Chapter 3. Presenting GIS Data

Objectives

▶ Learning basic principles of map design

▶ Creating text and labels on maps

▶ Creating map layouts and printed maps

Mastering the Concepts

GIS Concepts

Symbolizing spatial data is a powerful analysis technique, but sooner or later the information will need to be shared with others as a formal map. Every map has a purpose established by the cartographer, and the success of a map is judged by how well the map fulfills that purpose. This chapter introduces some basic design principles for maps, but it cannot cover the full range of concepts included in the art and science of cartography. Taking a cartography class or reading map or graphic design texts is a good way to improve such skills.

Map design can be envisioned as a six-step process (Fig. 3.1), but it is not linear. It begins by developing a vision of the map's objective, and this vision must inform each step: selecting the data, arranging the page, symbolizing the information, reviewing the map, and editing to improve it. Each step must constantly refer back to the main objective to ensure that the final map supports the objective in every possible way.

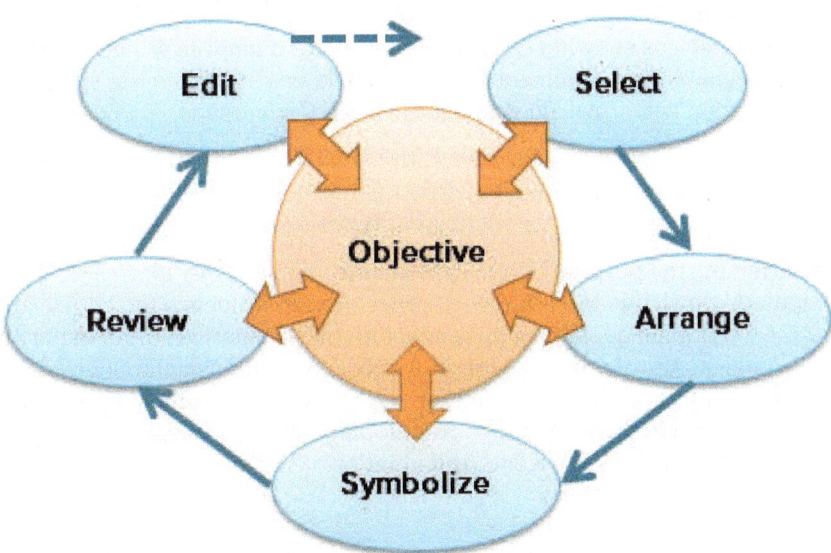

Fig. 3.1. The map design process—every step must return to the map objective.

TIP: To illustrate this process, the map used in the tutorial will be used as an example of map design. Consult these orange boxes to follow through the design considerations.

Determine the map objective

Developing a clearly defined objective for a map is a critical but sometimes neglected design step. It determines the form and content of the map, and it provides the context in which to review a map and ask whether it works. Four issues should be considered: the purpose of the map, the audience, the medium, and the conditions of use.

Purpose and audience

Maps serve different functions, and the function strongly influences all steps of the design. So the first question to answer is, "What is the purpose?" A complete list of every map purpose is impossible, but a few of the common ones include:

▶ *Locating*. Some maps show the location of some place or places of interest, such as showing the location of a store or a choice of hotels on a street map, perhaps with directions to get there. A walking or biking tour map would also fall into this category. Such maps need little more than the point(s) of interest and a street matrix with labels.

▶ *Navigating*. In a navigation map, the location of interest is not already known, so the maps tends to be richer in data to allow the user to identify and travel to different types of locations. A highway atlas, for example, shows many cities, connecting roads, and other features of interest such as parks, rivers, or mountain ranges.

▶ *Compiling*. Some maps gather information, sometimes very detailed and complex information, to meet the specific needs of experts. Geologic maps, utility network maps, and construction or landscaping plans are examples. The information will be specific to the task at hand.

▶ *Convincing*. Many maps are created to make a point or to illustrate the conclusions of a study, such as showing the potential sites for a landfill, telling the story of historical battles, making an argument for a new park, or identifying hot spots of crime. The information will primarily support the intended message.

▶ *Comparing*. Some designs use multiple maps to compare information. A map might show the historical and modern street maps, decadal differences in population or climate, or the spatial distributions of different types of cancers.

Part of determining the purpose of a map is identifying its *target audience*. The characteristics and anticipated knowledge level of the audience greatly influence the choice and presentation of the data. A traditional geologic map is a wealth of information for an expert but means little to a general citizen. A professional might not appreciate whimsical touches designed to appeal to elementary scholars. A map produced for residents of a retirement community could have larger text and symbols than usual to accommodate vision impairments. The best maps take into account the likes, dislikes, knowledge, and abilities of the target audience and produce the best possible experience for them.

Medium and conditions of use

Maps are designed with a specific medium in mind. For a paper map, the page size and map scale are a significant concern. Maps should be designed at the size intended for production; scaling generally introduces unwanted artifacts or affects the artistic balance. The properties of the available printers—the size of paper it can handle, the type of ink used, and the printing resolution—will also influence the final product. If many copies will be produced, the creator must factor in the costs of printing in choosing the medium, the paper quality, and whether color should be used.

Today, cartographers have exciting new media to design for: electronic files like PDF or presentation slides, web sites, smartphones, and tablets. Electronic maps have different requirements than printed maps and present unique design challenges. The author of a web map has little control over the size of the product; it could be shown on a conference screen many feet across or reduced to a few inches on a mobile device. The maps compete for attention in a flood of content, and most users will move on if they don't find what they need in less than a minute. Web map users rarely take time to explore complex content at a leisurely pace. The maps must be adaptable, obvious, and attractive.

Web maps can be interactive, which adds a welcome new dimension to cartography. Users can zoom, turn layers on and off, measure objects, and even perform analysis functions like getting directions or viewing an elevation profile. However, simplicity is still important. It is tempting to turn a web map into a baby GIS with dozens of buttons, but it is better to design a web map to show or do one thing really well rather than to try to pack it with data and functions.

The *conditions of use* can be considered part of the medium as well. Posters viewed from a distance require larger fonts and symbols than maps intended to be spread on a table. Maps that would be used under dim lighting conditions should avoid complex backgrounds like imagery and use larger symbols and better contrast than a map viewed under normal room light. Maps designed for outdoor data collection on a smartphone must use large symbols and intense colors to contend with the tiny screen and bright sunlight. Being aware of special physical conditions enables the cartographer to design the map to function properly in its environment.

The purpose of the tutorial map is to portray the potential dangers of volcanoes to populations living in the Pacific Northwest. Its audience is the average adult citizen, and it is intended to be printed in color on a letter-size page.

Select the data

Once the purpose of the map is clear, the cartographer selects the data it will contain. Think of the map as telling a story. A good story has main characters that move the action and minor characters that must not outshine the lead players. A mapmaker must know which data play the starring roles and which layers must support without intruding.

The key term in this process is *selection*. Geodatabases often contain a wealth of information that could be put on maps; more often the trick is to remove what isn't needed. Simplicity is one of the guiding tenets of cartography. Every piece of data should work to support the map objective.

When selecting data for a map, one should also be aware of potential privacy issues or unintended economic consequences of presenting certain types of information. For example, laying a fault map on top of the city parcels layer might have consequences for the property values of parcels near faults. The public may not be aware that most faults are inactive and do not pose a threat. Maps of fossil sites or archeological digs must ensure that the exact locations of fossils or artifacts remain out of the public domain to protect them from poaching. One should also be cautious about putting private information on public maps, such as the names of parcel owners.

The scale of the map is an important consideration, and the data layers selected should be appropriate. Coarse-scale data should not be placed on large-scale maps. Occasionally large-scale data may be needed on a small-scale map, although often some type of simplification must be applied to individual layers. This process is called **cartographic generalization**. Eight different techniques are commonly employed (Fig. 3.2). With the exception of refinement, which can be performed using queries and selection layers, most techniques will involve creating a new simplified feature class to be used in place of the original.

TIP: These generalization functions can be found in the Cartography Tools toolset.

To present information from a larger-scale layer at a smaller scale, one might do the following:

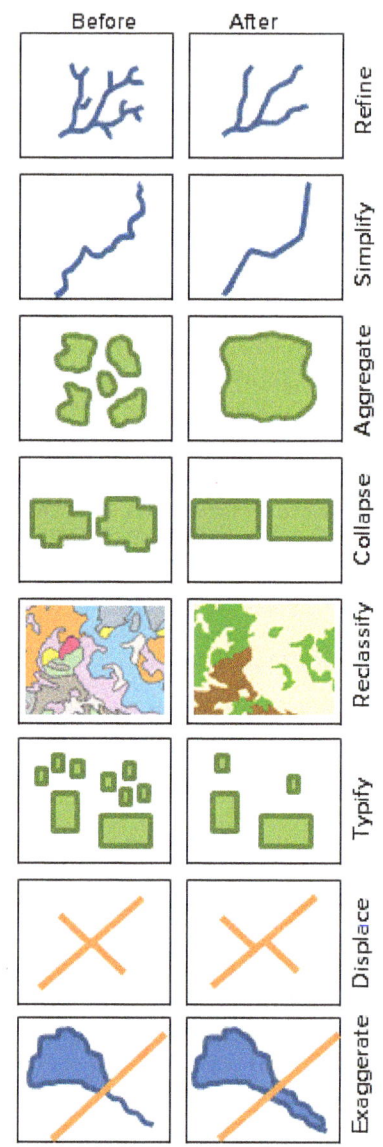

▶ *Refine* a feature class by omitting certain features, especially smaller or more detailed ones, such as removing higher-order streams and keeping the main trunk rivers.

▶ *Simplify* lines or polygons by removing vertices or *smoothing* to give them less detailed shapes.

▶ *Aggregate* many small features into larger ones, such as turning many close-lying small wetland polygons into a generalized marsh area.

▶ *Collapse* features to simpler forms, such as turning detailed building footprints into simple rectangles.

▶ *Classify* a detailed set of attributes into a simpler set, such as reducing geologic formations to a classification of igneous, metamorphic, or sedimentary rocks.

Some types of changes involve "falsifying" the underlying spatial data and should be used sparingly. A note on the map should indicate that such techniques have affected the locations or shapes of certain layers.

▶ *Typify* features using representative rather than actual features. For example, replace the point markers for actual homes with a smaller number of "fake" points representing scattered houses.

▶ *Displace* features slightly so they do not run over each other. For example, move a road to avoid the appearance of an intersection with another road.

▶ *Exaggerate* the importance of features to give them greater prominence, such as enlarging the width of a water body to prevent its appearing as a line.

Fig. 3.2. Cartographic generalizations

Graticules and grids

A **graticule grid** shows latitude and longitude markings on a map (Fig. 3.3a). Two other types of grids are often used on maps. A **measurement grid** shows the distance units in meters for feet based on the x-y values of the map (Fig. 3.3b). Both types of grids are helpful for locating known coordinates on a printed map. A **reference grid** shows letters and numbers delineating each grid square (Fig. 3.3c). A reference grid is commonly found in a telephone directory in conjunction with a street index, allowing streets to be located within squares of the map.

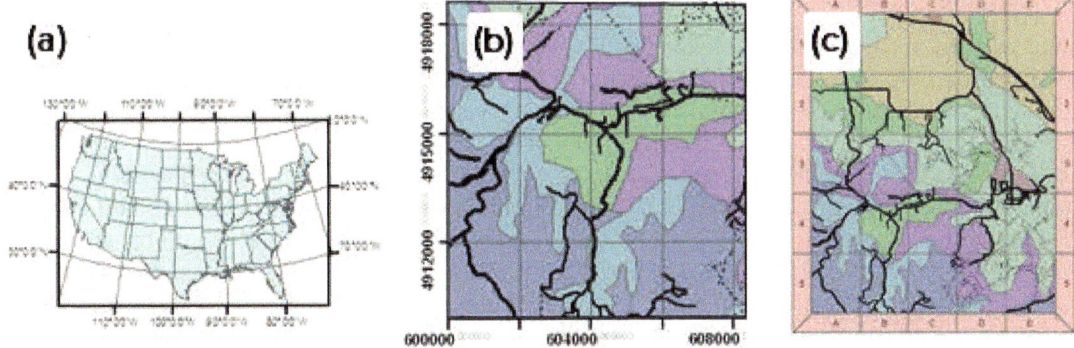

Fig. 3.3. Data frame grids: (a) graticule grid; (b) UTM measurement grid; (c) reference grid
(a) Source: Esri; (b) Source: Black Hills National Forest; (c) Source: South Dakota Geological Survey

Selecting map elements

Map layouts traditionally contain certain **map elements** (sometimes called map surrounds) in addition to the map itself. A location map is used to show where the main map extends, especially when the main map is not easily recognizable to the readers. A legend explains the symbols used. A scale bar indicates distances, and a north arrow shows direction. A graticule grid shows latitude-longitude lines or tics. A **neatline** can be used to enclose the entire map or groups of elements within it. Titles and text explain what the map is about and cite the data used to create it. Images or graphic shapes can be added to support the story.

Common practice asserts that all maps must have a title, legend, scale bar, and north arrow, but this rule is not necessarily true. Some maps are so simple and well symbolized that a legend is not needed. Scale bars and north arrows may not be appropriate for some maps, as we will see in Chapter 4, and they may be superfluous even when correctly used. A map portraying teacher salaries in the United States, for example, could omit both items; most viewers already know where north is, and distances have no bearing on the message of the map. It is better to think of this list of elements as tools, not rules, and use them only if they support the map objective.

The Volcanic Hazards map will focus on Washington and Oregon and will have a scale of about 1:4,000,000 in an equidistant projection. Volcanoes are the main point; counties and capitals will portray populations (census tracts would be too detailed at this scale). Highways are needed to show escape routes. Rivers are needed to show potential lahars (volcanic mudflows).

Arrange the map elements

The next step is to decide how the map will be arranged on the page. At this point it is a good idea to walk away from the computer and use a paper and pencil to sketch a mock-up of the design and decide how the parts of the map should be arranged, instead of succumbing to what is currently shown on the screen.

Return to the analogy of the map as a story. A story has main characters, supporting ones, and a defined path through the events. In a map, the direction of the story and the relative importance of the elements is achieved by establishing a **visual hierarchy**, which refers to the order in which the reader perceives the different objects in a design. A good map establishes a clear pathway through the map objects by using carefully arranged elements and appropriate symbols.

Balancing the elements

A map with good balance and a clear visual hierarchy is achieved by paying attention to several important guidelines: the size of the elements, the visual center of a page, the rule of thirds, the alignment of elements, and the treatment of negative space.

The relative *size* of the map elements should be considered; larger objects will naturally command more attention. The scale of the map should be as large as possible to enhance the legibility of the map features. Scale bars, north arrows, and legends should be unobtrusive. Use large fonts to draw attention to titles and key points, and smaller text for informational details. The viewing distance must be considered; a poster map requires a minimum of 18-pt. font to be comfortably read from several feet away, and it will generally need larger symbols as well.

The **visual center** of a page is about 5% higher than the geometric center in the exact middle (Fig. 3.4a). Centering the map on the geometric center makes the map appear unbalanced and heavy.

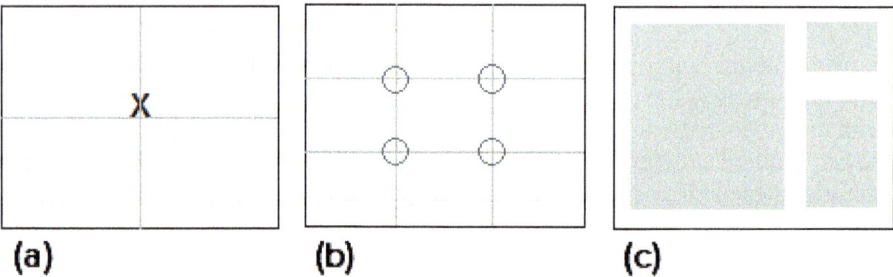

Fig. 3.4. Ways to balance elements: (a) place the map on the visual center; (b) place important elements at these intersections; (c) use aligned rows and columns.

The **rule of thirds** is a graphic arts principle used to enliven a design; although it is primarily used as a guide to composing photographs, it can also be helpful in map design. Imagine the page cut into horizontal and vertical thirds (Fig. 3.4b). Important elements placed on these intersections will have greater impact than if placed in the center of the map.

The *alignment* of the map elements is also important. Alignment can be achieved in two ways. First, arrange the map elements into columns or rows, rather than letting them wander all over the page. Second, be sure that the edges of elements align with each other.

A rough mock-up is helpful in evaluating the balance and visual hierarchy of a design. In Figure 3.4c, the gray boxes represent placeholders for the main map elements. Notice that two columns have been envisioned, one for the main map on the left (which is centered near the left line of thirds), and a second column for the other elements. The exterior edges of all three elements are exactly aligned at the top, left, and right.

The visual center and the rule of thirds influence the view path because the eye is typically drawn to these areas first, other things being equal. In addition, the eye tends to travel from left to right because it is used to reading that way (although this tendency would be reversed in countries using right-to-left languages). A good map design uses these influences rather than working against them. Which way does the eye move in Figure 3.4c? Probably you saw the large square on the left first, then moved to the upper right and down. Placing the map, title, and legend in these three areas, respectively, would create a logical flow through the map.

The treatment of **negative space**, or the blank areas on the page, is more important than it might at first appear. Large irregular blank areas make a map look unbalanced, yet too little space can make a design appear crowded. Negative space can actually be used as another design element that can separate or group elements more subtly than neatlines. Maps often look better when the designer resists the temptation to put a box around every element. In Figure 3.4c, the negative space serves to separate the three main map elements. Page margins are an important type of negative space.

Maps often show irregularly shaped geographic regions that require creativity in achieving balance. Imagine placing California in the layout shown in Figure 3.5. The map area is relatively small, limiting the scale and the legibility of the features. Moreover, the state's shape creates irregular negative space inside the map frame. Figure 3.5 shows three alternative treatments based on a portrait layout that maximizes the scale. The non-map elements are placed inside the data frame (a space solution that may not occur to beginning map designers, but which can be highly effective). Give each design a swift glance. Which one seems the most balanced?

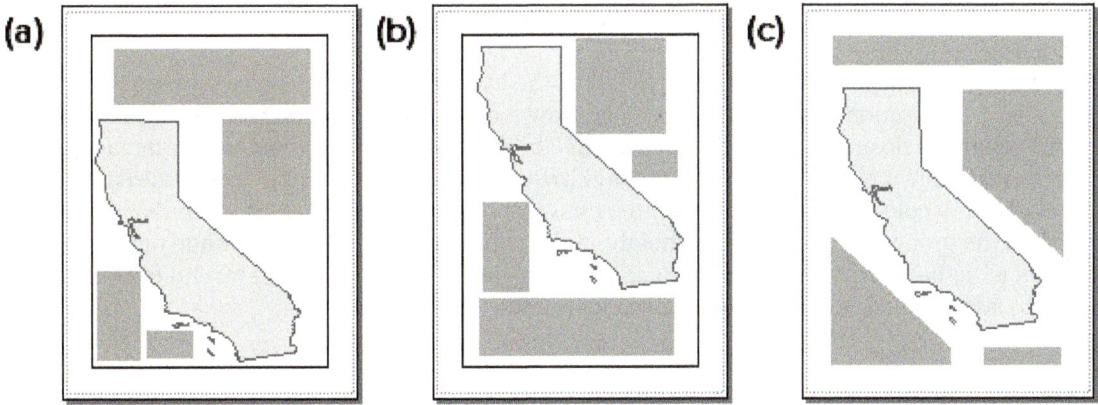

Fig. 3.5. Managing the visual balance and negative space in a layout

Figure 3.5a ignored the visual center and placed California near the bottom, creating a heavy sensation exacerbated by two more blocks added to fill the negative space. The visual center is blank in this layout. The upper elements draw the eye first and then force it left to the map, ignoring the left-to-right preference. Some parts of the map appear crowded, others empty. The edges are not aligned, and the neatline crowds the map.

In Figure 3.5b, the state is now too high, although the negative space is more evenly balanced. The edges are irregular, and the elements appear haphazardly arranged. They crowd the state in some spots and leave open areas in others. The eye is drawn to the map first and travels left-to-right across it, which is fine, but then is unsure what to view next.

Figure 3.5c is the best arrangement so far. In this case, the placeholders for the elements have been reshaped to echo the state's shape, which leaves uninterrupted negative space around California and draws attention to it. The north and south sections of the state fall roughly in the target areas advocated by the rule of thirds and encourage the eye to scan the state left to right. The exterior edges of all elements fall on the boundary of the data frame, which subtly encloses the map and allows the neatline to be removed.

In the tutorial map, Washington and Oregon form a roughly rectangular shape with little negative space, so the supporting elements can go outside the data frame. The map can be placed on the left to draw the eye first. The elements will include a location map, a legend, and some titles and supporting text. An equidistant projection and a scale bar will be used in case the reader wishes to estimate distances from the volcanoes to the vulnerable populations.

Symbolize the data

Symbolizing the features in the layout has enormous impact on the final result. Chapter 2 introduced ways in which symbols may be manipulated by changing their shape, size, thickness, line type, font, and color. It also discussed the best color choices for different data types, managing issues such as color blindness, and employing convention and connotation to enhance map interpretation. All of these principles remain important, but in a layout we must also consider the interaction of symbols between multiple layers. Symbolization is a powerful tool for manipulating the visual hierarchy in a layout through color, contrast, foreground/background relationships, size, symbol type, patterns, and branding.

Color is one of the strongest influences in a map. Dark, saturated colors attract attention and help indicate the focus of a map; they should be used sparingly. Pastel colors are perfect for the supporting layers. Color can also be used to group features or map elements—objects that are shown in similar colors will be perceived to belong with each other.

Another important aspect of color is **contrast**, or the juxtaposition of colors with very different hue, value, or saturation. Consider Figure 3.6a: How many squares do you see? Did you neglect the fifth square enclosing the four obvious ones? The dark squares stand out and command the **foreground**, while the gray fades into the **background**. This effect is achieved by the saturation and value of the colors and even more by the contrast between them. The blue stands out more than the orange, and the outlines heighten the contrast, especially the orange outline that contrasts with the interior orange as well as the gray exterior. The visual hierarchy encourages the eye to start in the lower right and move clockwise.

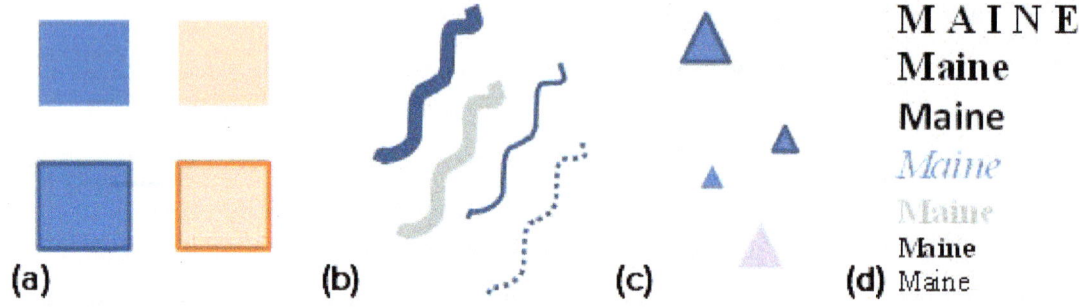

Fig. 3.6. Tools for establishing visual hierarchy: (a) color and contrast; (b) color, thickness, and line style; (c) color, size, and contrast; (d) size, spacing, font, color, contrast, and style

The effects of color and contrast are repeated in Figure 3.6b, with the dark blue lines appearing more prominent than the gray one. *Line thickness* and *symbol style* also affect the hierarchy, with thicker and more continuous lines dominating thinner or discontinuous ones. (Which is more prominent, the thick gray line or the thin blue one?) In Figure 3.6c, the additional attribute of *size* is varied; the larger something is, the more attention it commands. Text symbols have font and style, in addition to size and color, to rank them in the hierarchy (Fig. 3.6d).

Humans do not simply perceive objects on a page; the brain tries to organize the objects into *patterns*. Consider the four different panels in Figure 3.6—what makes them into different panels? First, the negative space between the gray squares separates them. Second, the similarity of the symbols within each panel (rectangles, lines, points, text) encourages the brain to group them together. The separation is balanced, however, by the use of repeated colors to tie the entire figure together. What is the effect of having the single purple triangle? Does it stand out more than if it were orange like the square?

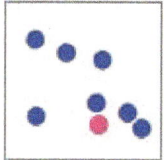 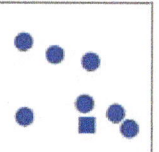   

Fig. 3.7. Size, color, and shape in patterns

Look at each box in Figure 3.7 and pay attention to how your brain tries to group the markers. Which variation (size, color, or shape) makes the points easiest to group? Which sets are easier to separate? Which ones emphasize that the points belong to one map layer? When evaluating symbols for layers, take note of how your mind tries to group or separate objects, and use these effects to help support the visual hierarchy.

Symbols in a map should bring the important elements to the foreground and relegate less important features to the background. In Figure 3.8a, the world map has no clear hierarchy; the ocean and graticules grab as much attention as the country populations. Three slight adjustments make the hierarchy clear: (1) the oceans are displayed in light blue to displace them to the background; (2) the graticules are colored light gray to decrease the contrast; and (3) the graticules are placed behind the countries to further enforce them as background elements (Fig. 3.8b).

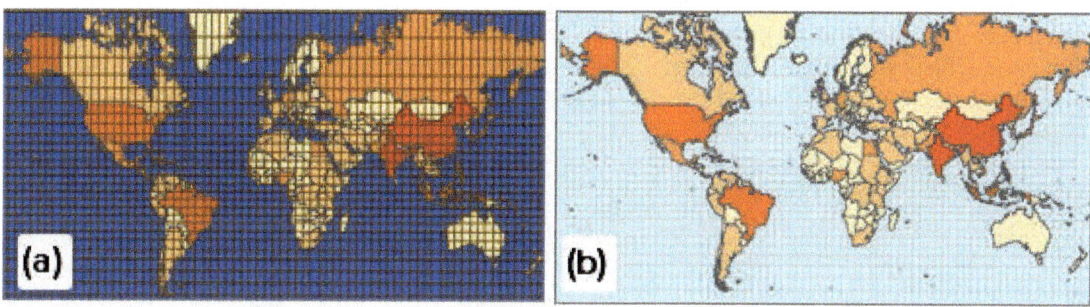

Fig. 3.8. Establishing a clear visual hierarchy by manipulating color value and contrast
Source: Esri

Of course, establishing the hierarchy is more difficult when many layers (and text) are involved. When in doubt about how to treat a layer, one should always return to the objective of the map and ask what role the layer plays—should it be emphasized, downplayed, or even removed to better achieve the map objective? Examine the different ArcGIS Online basemaps and study how they achieve visual hierarchy. The *Terrain With Labels* basemap is an especially good example of differentiating between multiple background layers without swamping user data placed in the foreground (Fig. 3.9).

Another factor in choosing symbols for the map is the desire to achieve **branding**. An organization may wish to impart a certain look or feel to its maps across all of its publications through the consistent use of certain colors, fonts, marker symbols, line types, or images. It is not difficult to select symbols to honor these choices. For an example of branding, examine the *National Geographic* basemap in ArcGIS Online. *National Geographic* readers would find a familiar face in this map.

In the tutorial map, the volcanoes and population information are most important, followed by the transportation. The county populations will cover the most map area and will be symbolized in shades of orange; other colors will be chosen to develop the appropriate contrast.

Review the draft

The review step should not be skimped, any more than a student should neglect to proofread a research paper before turning it in. However, we are looking for more than typographical or grammatical errors. We must return once again to the map objective and evaluate how well the map succeeds in telling its story.

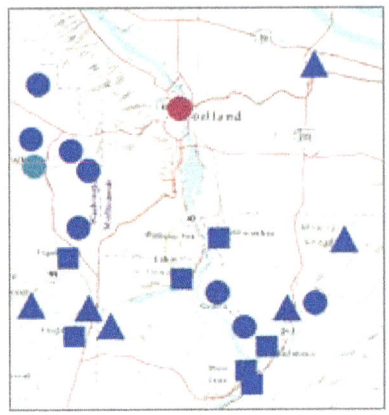

Fig. 3.9. A basemap is the epitome of designing for the background.
Source: Esri

Author's first review

Start with your own review, applying the principles in this chapter. Is it easy to determine what the map is about? Is the map balanced? Is it clear what is foreground and what is background? Does the eye go first to the more important elements, or does it dart all over, unsure what to focus on? Do the symbols clearly portray the message? Are the colors and arrangement pleasant? Even when good principles have been cultivated all through the design, one must still evaluate whether they all work together.

Don't neglect the details. Make sure the margins are sufficient for the printer being used. Check for errors in spelling and grammar. Evaluate the legend for common mistakes, such as using file names rather than real words, using inappropriate significant figures, or allowing the neatline to crowd the text (Fig. 3.10). Many cartographers feel that the title "Legend" is superfluous and do not care to see it on maps.

Legend	(a)
cities.shp	
POP2007	
·	0 - 122212
●	122213 - 486861
●	486862 - 1502129
●	1502130 - 3908521
●	3908522 - 8323732

Legend	(b)
US Cities	
2007 Population	
·	0 - 122,000
●	123,000 - 487,000
●	488,000 - 1,500,000
●	1,510,000 - 3,910,000
●	3,920,000 - 8,320,000

Fig. 3.10. (a) A poor legend; (b) a better legend
Source: Esri

Be aware of significant figures in all publications. Not only do excessive figures make the classes hard to interpret (Fig. 3.10), but they announce that someone does not understand the data. Population, for example, is always an estimate, and it is ridiculous to state it down to the last place (or worse, with decimals). Classifying already generalizes data to a small set of bins; variations in the hundreds and thousands make little difference.

Do not assume that every layer needs to go in the legend. If a set of features is familiar to all map viewers (such as state boundaries on a United States map) or is so apparent from the title or purpose of the map (labeled paths on a city walking tour map), consider leaving them out of the legend. Consider those basemaps again—they generally have no legends, yet most viewers have no difficulty recognizing what is being shown (Fig. 3.9).

Let's see how an intial map review is conducted by the author. The map on the left in Figure 3.11a shows a first draft based on software defaults. (Hopefully a first draft would be better than this one, but unfortunately the author has seen similar student maps turned in for an assignment.) Assume that the map objective is to communicate areas with larger-than-usual households. Cover the map on the right and identify ways that the draft map might be improved. Then read on.

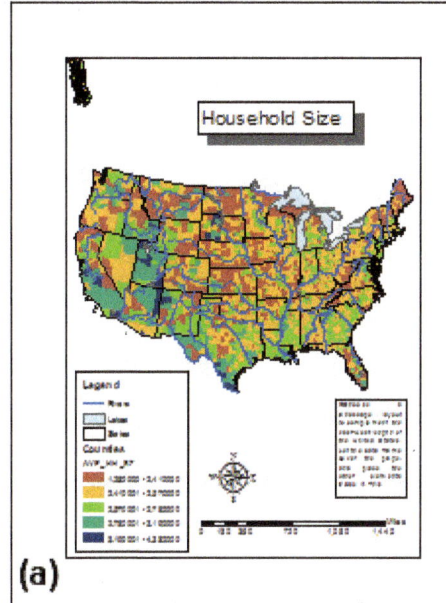

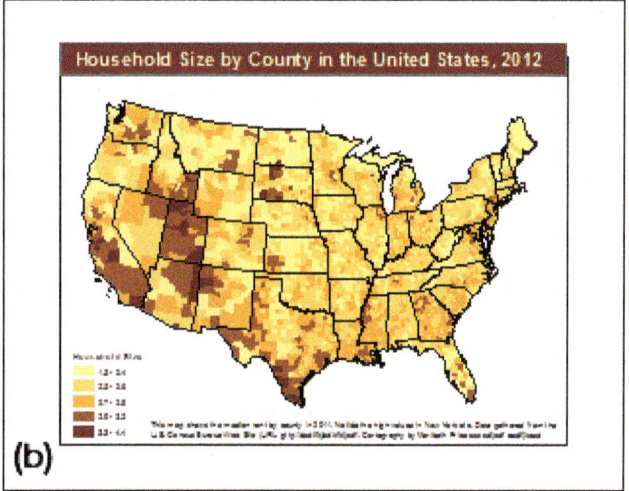

Fig. 3.11. Improving a draft layout: (a) original draft; (b) revised layout

Source: Esri

Evaluate the balance first. The map is placed near the visual center, but everything else is poorly arranged. The elements are crowded on the bottom and leave a large negative space above, from which Alaska dangles down, neither in nor out of the map. The cartographer has tried to fill the space with a box and shadow around the title, but that is a gimmicky solution. The legend is large compared to the map and contains too many significant figures. The neatlines around the legend and text boxes distract from the map. The scale and north arrow are too large, the north arrow is inappropriate in a conic projection, and the scale bar is immaterial to the map objective. The element edges are not aligned. The visual path is not clear; the eye is drawn first to the map, which is fine, but then it tends to bounce around between the other elements.

The symbolization in Figure 3.11a is also ineffective. The map uses a color scheme that resembles natural geographic features rather than human ones and, because it is based on hue rather than saturation or value, it does not clearly indicate quantity. The red hues draw attention to the low values in the visual hierarchy—the opposite of the map objective. The rivers and lakes are irrelevant to the topic and distract from the message.

The revision on the right seems like a different map (Fig. 3.11b). A landscape orientation better fits the shape of the United States and supports a larger map scale for improved legibility. The color scheme shows quantities and draws attention to the high values using saturated colors. The irrelevant lakes and rivers have been removed. The legend has been reduced to the single layer that needs explaining, the significant figures have been corrected, and the redundant "Legend" heading was deleted. The map region is familiar to most readers, so the scale and north arrow are superfluous and have been deleted. The distracting neatlines around the legend and text are gone, and the text has been anchored across the lower neatline. The new colors and expanded width help the title appear as part of the map rather than floating aimlessly above it. The visual hierarchy is simple and clear. Notice how taking elements OUT of the map improved the focus and legibility.

External review

Equally important is to have others look at the map to make sure that the author's intended message is clear to the readers. These reviewers need not be experts; in fact, if one is designing the map for a general audience, then those are the people who should review it.

The first impression made by a layout is extremely important. Maps are not just rational, logical documents like spreadsheets—they can have emotional impacts, and the author must know what kinds of emotions the map elicits from the target audience. Start by giving the reviewer about three to four seconds to look at the map, then take it away and ask: What were the first things that you noticed? Was the overall impression positive or negative? What made you feel that way?

Let them look at the map again for a few more minutes. Then ask them what they think the map is about and what it is telling them. Ask if any part of the map is confusing or too complicated. This feedback will help you to evaluate whether the map achieves its objective. At this point one can also get more detailed feedback on a design if the reviewers are knowledgeable users, qualified in cartography or in the field the map represents.

Keep in mind that design principles sometimes conflict with each other in a particular situation, and a choice must be made based on which guideline will best achieve the map objective. On occasion, a map author may want to break a guideline on purpose to achieve a particular effect, as some novelists may sometimes use sentence fragments. Just be sure to do it for a good reason and not out of ignorance or laziness. One should also remember that a map is not correct or incorrect, any more than a novel or a painting can be. However, some maps are better than others. When reviewing maps and considering alternative strategies to achieve the map objective, keep asking, "Would this make the map better?"

Observing other people's maps is a great way to become a better cartographer. Give a map the first impression test by looking at it for a few seconds. Notice what stood out and whether the first impression was positive or negative, and analyze why it felt that way. Practice analyzing the visual pathway through a map and how it is achieved (or ignored). Then evaluate the map's balance and symbology. What works well? What doesn't? How could it be improved? Watch for effective techniques to incorporate in your own maps.

We will review the Volcanic Hazards map as part of the tutorial.

Edit and improve

The review step will usually highlight some things that need changing. Sometimes they will be minor edits, but at other times one may need to return to the selecting, arranging, or symbolizing steps for a more drastic redesign. Regardless, there are a few principles to keep in mind when editing.

Concentrate the message. Make sure that every part of the design is working toward the map objective in the clearest way possible. Include the minimum number of layers and elements needed to accomplish the map's purpose. You would be surprised at how often removing a layer or reclassifying a complex symbol scheme will improve a map.

Simplify, simplify. Many maps are needlessly complicated, with too many different kinds of text, symbols, or fonts. Just because the software provides default neatlines, fill colors, and shadows doesn't mean that they are always a good idea—they should serve a purpose, such as a neutral background to separate a scale bar from the map behind it. Remember that negative space is often a great way to separate or group map elements. Stay away from complicated cartoon symbols—they may be popular for web maps, but they can hamper interpretation. The less junk placed on the map, the clearer the intended message will be.

Experiment. If a design doesn't seem to be working well, experiment. Save a copy of the map with a new name and try something daring, like changing the page orientation and completely rearranging the elements. Rethink, resymbolize. That's the beauty of computer cartography—it is easy to try something different with no risk.

Test it. Before pronouncing the map complete, test it under the conditions of actual use. For paper maps, print a test copy on the same printer to be used for production. Colors on screen may differ significantly from the printed colors and may need adjusting to produce the desired result. Margins may shift, causing the map to appear uncentered on the page. If it is a poster, print it full size and hang it on the wall, then step back and make sure that the arrangement and fonts work well from a viewing distance of two meters. If it is a web map, try it on several computers, on a tablet, and on a smartphone to make sure that it works for all media on which it might be viewed.

About ArcGIS

Maps in ArcGIS

ArcGIS provides an environment for cartography that spans the range from creating a simple map for a report to preparing sets of maps for commercial printing. This chapter covers the basic steps to create a printer-ready map, but the reader should be aware that many advanced cartographic tools are available, as well as tools for developing graphs and database report printouts; although they are not covered here, the Help files have more information about them.

In ArcGIS, the arrangement of one or more maps intended for printing or distribution is called a **layout**. The layout mode in ArcGIS Pro facilitates map design by incorporating map elements, such as legends, north arrows, scale bars, text, and more. Figure 3.12 shows a basic layout illustrating the geology of Crater Lake, Oregon. It includes a main map, a location map showing the location of Crater Lake within Oregon, a legend, a scale bar, and a north arrow. Simple layouts like this one can be prepared in fifteen minutes, or one can spend hours or weeks creating complex and detailed layouts with many maps and elements.

> **TIP:** The terminology used in this chapter is quite specific. Be sure to understand the differences in the definitions for **map** (or project map), **map frame**, **layout**, and **layout file**.

A **map** (or project map) is a container for sets of data within a project; you have used them extensively already. Layouts incorporate one or more maps from a project. A layout contains **map frames**, which are derived from the project's maps and contain the same information. The layout shown in Figure 3.12 contains two map frames: the main one showing aquifer depths, and a small one beneath it showing the location of the main map within South Dakota. Each map frame refers to a project map. Turning data off and on in the project map affects how it appears in all views, including the map frames on a layout.

A project may contain multiple layouts incorporating different groups of maps in the project. Layouts may also be saved outside the project as a **layout file** (.pagx), which can be used

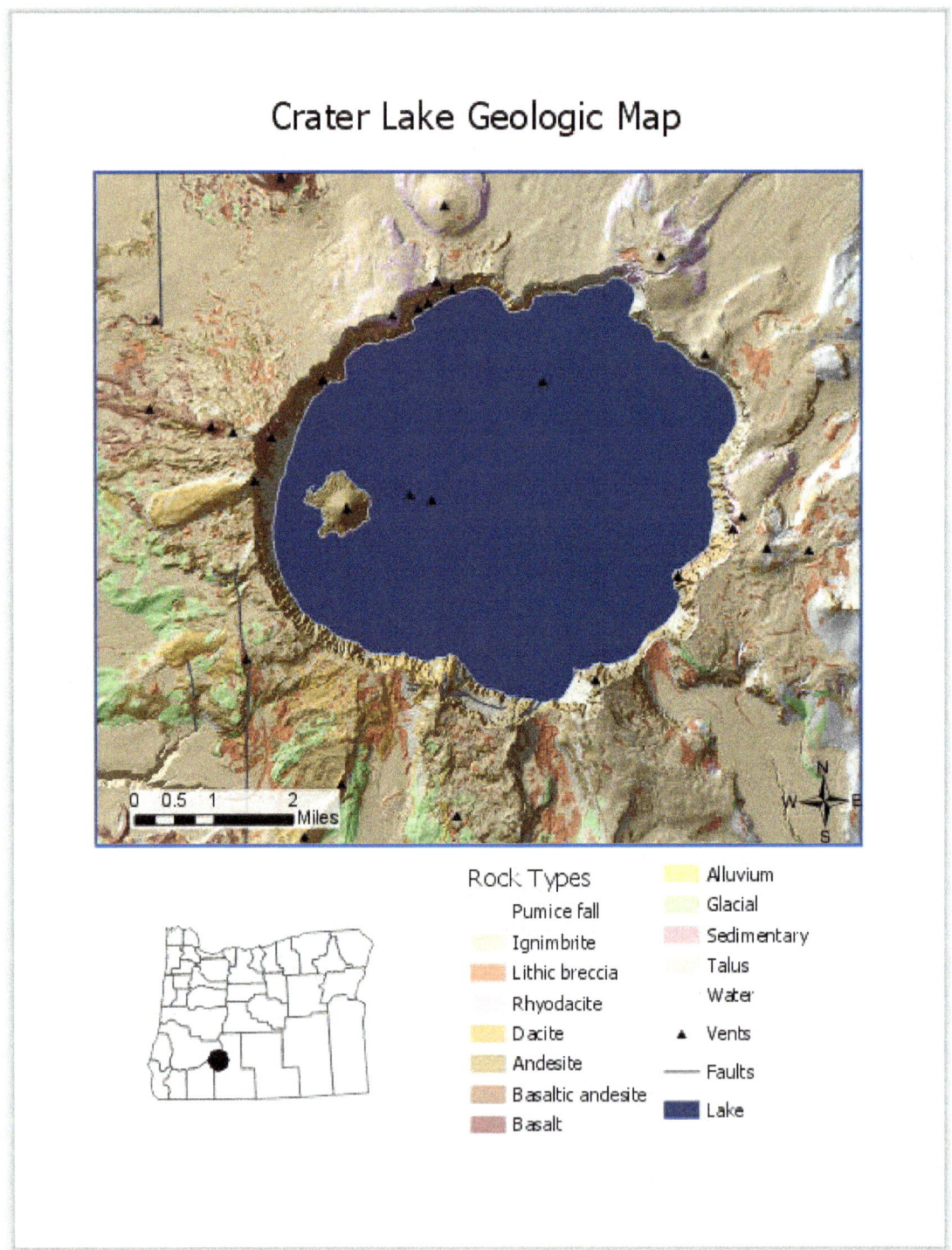

Fig. 3.12. A map layout of Crater Lake
Source: USGS/Esri

to re-create a design or to give an organization's maps a common format and brand. It is also possible to create map books, in which the sample map design is applied to multiple tiles covering a larger area, with the map extents, legends, titles, and other aspects generated automatically for each map using a feature called Map Series.

Assigning map scales

GIS spatial data acquire a scale as soon as they are drawn on a computer screen, printed on paper, or saved in a PDF. The first step of creating any layout is generally to define the size of the layout page, which will be influenced by the page-size capabilities of the available printers. Once the page size is set, the map frames are arranged on the layout page, and the user must decide how much of the map data will be shown within each map frame. The **map extent** is defined as the range of x-y values displayed in the map frame. Zooming in reduces the map extent; zooming out enlarges it. The printed scale is determined by the map extent and the size of the map frame on the paper.

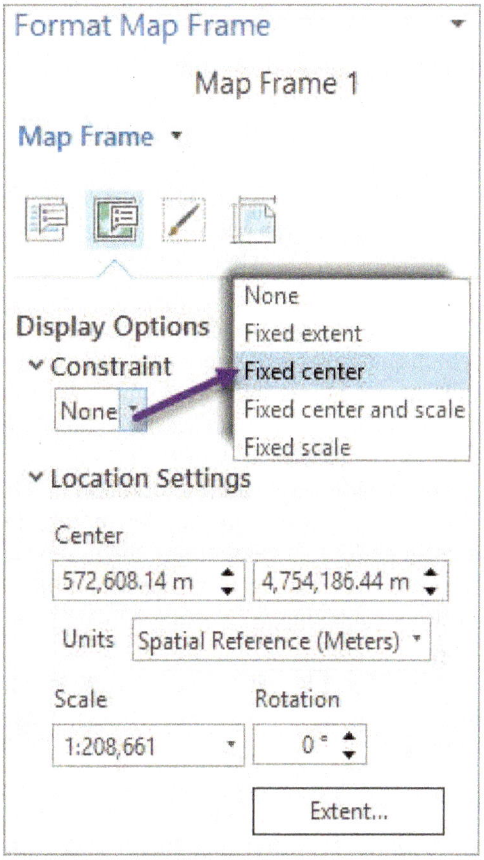

The data shown in a map frame is linked to the project map, but the map extents of the two are independent. The map extent of the map frame is set from the layout and persists even if the user opens the project map and zooms to a different extent. This feature of the software allows one to zoom and manipulate the project maps without affecting the extents in any of the layouts. As with project maps, the views of map frames may be linked to show the same extent and scale.

By default, the map extent can be modified within the map frame by zooming or panning, but the user can set optional constraints to prevent certain

Fig. 3.13. Scaling options for map frames
Source: Esri

navigation actions. For example, if designing a layout to show information at a particular scale like 1:24,000, the author can disable zooming so that the scale remains fixed, preventing anyone from accidentally changing the map design. The following types of constraints can be assigned to a map frame (Fig. 3.13).

Fixed extent sets the exact extent (or position of the camera for a 3D scene). The map frame may be resized but must conform to the same aspect ratio. All navigation is disabled. The extent may be set to the current extent of the frame, from one or more of the layers in the frame, or from a rectangle derived from specified coordinates.

Fixed center fixes the center point of the view, permitting zooming and rotating (3D) around the center of the map extent. The panning function is disabled.

Fixed center and scale fixes the center point of the view as well as the scale of the map frame. Zoom and pan are both disabled, but rotation is allowed. This option is similar to Extent, but it allows the map frame itself to be resized to a different aspect ratio.

Fixed scale fixes the map scale to a specific value such as 1:24,000. The frame may be resized, which will increase the geographic area of the map displayed. Zooming is disabled, but panning is permitted.

Rotation changes the angle of the map in degrees.

Setting up scale bars

Some maps should contain a scale bar or a statement of the scale, such as "Map scale is 1:62,500." ArcGIS Pro can automatically construct a suitable scale and resize it when the map scale changes. The user has to set several options, which requires some knowledge of terminology.

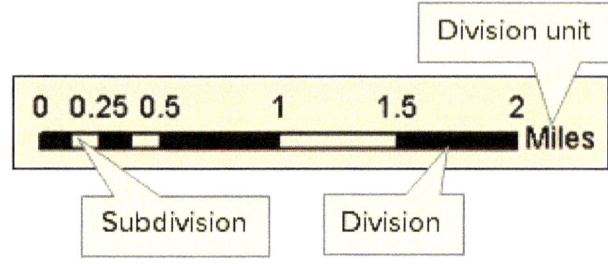

Fig. 3.14. Terminology of a scale bar

The **divisions** are the sections into which the scale is divided. The scale bar in Figure 3.14 has four divisions. The first division may be subdivided, yielding what are called **subdivisions**. This scale bar also has four subdivisions. The **division units** are the units of measurement used to divide the scale bar (miles in this case). The **division value** is the length of each division, expressed in division units. This scale bar has a division value of 0.5 mile. Divisions should be expressed in rounded, easily divisible numbers to make them easy to use for estimating distances. The division units, the map scale, the division size, and the length of the scale bar are interrelated. Zooming changes the map scale and requires the scale bar to change also. Three options control how the scale bar responds to changes in map scale.

Adjust Width keeps the division size and number of divisions constant. If the map scale changes, the entire scale bar gets larger or smaller. This method keeps nicely rounded numbers for the divisions but may result in an unacceptably small or large scale bar if the scale changes dramatically.

Adjust Division Value keeps the scale bar roughly the same length on the page and with the same number of divisions but changes the size of the divisions. If the map scale changes, the division size of the scale in Figure 3.14 might change to a less desirable interval such as 0.4 or 0.6 miles. This option preserves the width of the scale bar but may result in awkward division sizes and labels.

Adjust Number of Divisions leaves the division value constant and adds or subtracts full divisions if the scale changes. For minor changes in scale, the bar will increase or decrease slightly in size. For dramatic changes, the scale bar will remain about the same size but may have more or fewer divisions. The bar in Figure 3.14, for example, might become 1 mile long or 3 miles long, but the division size would stay 0.5.

Labeling, text, and annotation

ArcGIS offers several ways to place text on maps. You have already learned how to create *dynamic labels*, which are derived from a feature attribute and applied to an entire layer at once. *Graphic text* is used to place a few labels manually. *Annotation* is used to gain exact control of labels. We briefly summarize the key features of each type.

Graphic text items are constructed by the user, one at a time. Different tools are provided to create standard, splined, callout, or multiline text or to label a feature using the value from its attribute table (Fig. 3.15). All graphic text exists on the map as simple graphic elements. Items remain where they are placed but are tedious to create when many labels are needed. Graphic text can only be placed on a layout, not in a project map.

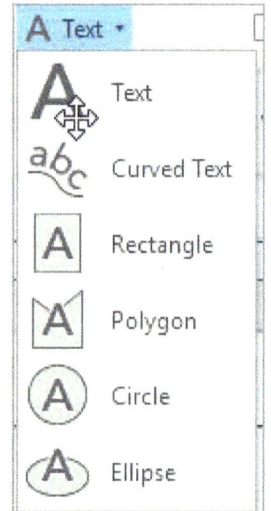

Fig. 3.15. Tools for creating graphic text
Source: Esri

Dynamic labels are created from an attribute and are redrawn every time the user changes the map view. They can be turned on or off as needed. Dynamic labels use a placement algorithm that omits overlapping labels, known as the labeling engine. By default, ArcGIS Pro uses the advanced Maplex Label Engine, which allows the user to set rules about placement. These rules govern three label issues: (1) position, or defining the allowed or preferred locations of the label relative to its feature; (2) fitting strategy, or the types of alterations allowed to the label position to fit more of them in a congested area; and (3) conflict resolution, which defines which labels are more important should labeling conflicts occur. The engine tries to place as many labels as possible within the guidelines set by the user. Which labels are displayed can vary, depending on the zoom scale and label size.

Because the labels are redrawn each time, printing maps with dynamic labels can bring surprises. What shows on the screen does not necessarily appear the same way on the printed map. For many purposes a few extra or missing labels do not matter. If they do, however, converting labels to annotation provides greater control over label placement.

Annotation refers to labels that are stored individually as features so that each one can be independently placed and modified, allowing the maximum degree of control. Annotation is often created from dynamic labels and then saved within a geodatabase, which allows it to be reused in many maps. ArcGIS Pro 2.1 has the ability to create or edit simple geodatabase annotation, and it can display annotation features created and edited in ArcMap, which has more advanced and flexible editing capabilities for annotation than Pro 2.2.

Reference scale

When labels and annotation are created, they are assigned a size in points. For a given map, the size of the text relative to the features, such as the counties shown in Figure 3.16a, is determined by the text point size and the current map scale. But what happens if the scale changes? By default, the labels (and symbols) stay the same size when the user zooms in or out (Fig. 3.16b); 12-pt. text remains 12-pt. text regardless of the scale of the map.

This behavior changes if the user sets the **reference scale**, which is a property of the project map. The reference scale is the scale at which the labels and symbols appear at their assigned size. If a user places 12-pt. labels on a map with a scale of 1:100,000, then the labels will appear 12 points in size only as long as the map scale remains 1:100,000. If the user zooms out, the labels and symbols decrease in size. If the user zooms in, they increase in size (Fig. 3.16c).

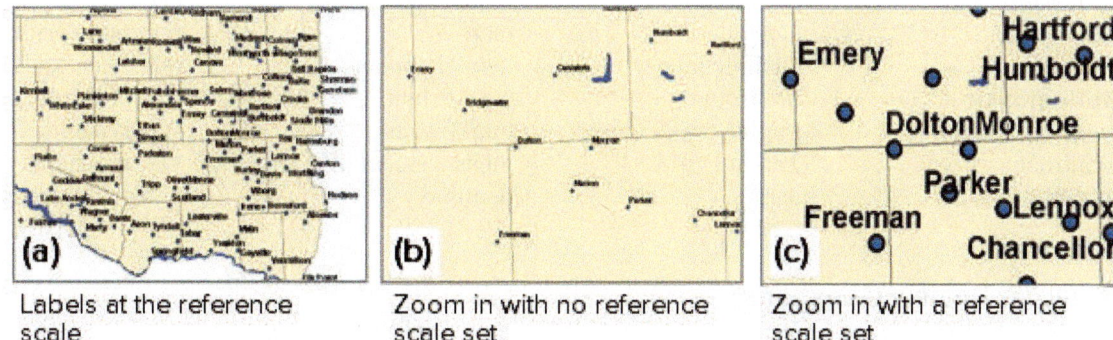

Labels at the reference scale   Zoom in with no reference scale set   Zoom in with a reference scale set

Fig. 3.16. (a) Text and symbols appear at their assigned size at the reference scale. (b) If the reference scale is not set, the symbols will always appear the same size. (c) If the reference scale is set, then zooming in or out changes the size of the text and symbols.
Source: Esri

TIP: Be sure to understand the differences among the terms *map scale, reference scale,* and *visible scale range*. Map scale determines the ratio of features on the map to features on the ground. The reference scale is the scale at which symbols and text appear at their assigned size. The visible scale range determines the range of scales at which a layer will appear.

Summary

▶ Designing a map consists of six interlinked steps: determine the objective, select the data, arrange the elements, symbolize the elements, review the map, and edit/improve the map.

▶ Maps serve many purposes, each of which influences the content, medium, target audience, and conditions of use.

▶ Every set of data included should serve the purpose of the map and be appropriate to the map scale. Layers that are too detailed can be simplified through cartographic generalization.

▶ Visual hierarchy refers to the order in which the reader perceives the different objects in a design; it is manipulated using the arrangement and symbolization of map elements.

▶ Balanced arrangement of map elements uses the size, visual center, rule of thirds, alignment, and negative space to establish a visual path through the map.

▶ Symbolization affects the visual hierarchy through size, color, contrast, foreground /background relationships, font, symbol style, and pattern.

▶ Maps should be reviewed by the author and also by other to ensure that it meets its objective and is free of petty errors. First impressions are important.

▶ When editing and improving the map, work to concentrate the message and simplify the presentation. Be sure to test the map under the conditions of actual use.

▶ Three types of labeling are available. Graphic text labels are placed on layouts individually. Dynamic labels are automatically generated for entire layers at a time. Annotation offers the greatest control; it is created from dynamic labels and stored in a geodatabase.

Important Terms

annotation	division units	layout file	negative space
background	division value	map	reference grid
branding	dynamic labels	map elements	reference scale
cartographic	foreground	map extent	rule of thirds
generalization	graphic text	map frame	subdivisions
contrast	graticule grid	measurement grid	visual center
divisions	layout	neatline	visual hierarchy

Chapter Review Questions

1. Go online and find an example for each of the map purposes listed in the first section. **Capture** a screenshot of each one and then explain why you chose it as a good example.

Find another map to critically review, provide a large screen shot so it can be easily seen, and answer the following questions about it.

2. Describe your first impression of this map. Was it positive or negative? What aspects of the map caused you to react that way?

3. State the map objective and explain how each layer does or does not support it. Are all of layers necessary or could some have been omitted?

4. Describe the visual hierarchy that you see (or don't see) in this map and explain how it is created (or ignored).

5. Critique the balance and alignment of this map and describe how well it uses negative space.

6. Critique the symbols used for this map and describe how the symbols support or detract from the map objective.

7. Critique the legend of the map, if it has one.

8. Could simplifying this map improve it? List the ways, or explain why there are none.

Additional reading on cartography

MacEachren, Alan M. 1995. *How Maps Work: Representation, Visualization, and Design.* New York: The Guilford Press.

Monmonier, Mark. 1996. *How to Lie with Maps.* Chicago: The University of Chicago Press. Second edition. First published 1991.

Robinson, Arthur H., Joel L. Morrison, Phillip C. Muehrcke, A. Jon Kimerling, and Stephen C. Guptill. 1995. *Elements of Cartography.* New York: John Wiley & Sons. Sixth edition. First published in 1953.

Expand your knowledge

Expand your knowledge of ArcGIS Pro by reading these sections of the Help. Use the breadcrumb trail to navigate to each topic, or search for a key term in the section title to find it.

> Help > Layouts (all sections)

Mastering the Skills

Teaching Tutorial

The following examples provide step-by-step instructions for doing basic tasks and solving basic problems in ArcGIS. The steps you need to do are highlighted with an arrow →; follow them carefully. Click on the video number in the VideoIndex to view a demonstration of the steps.

> → Start ArcGIS Pro and open the PacificNorthwestHazards project from the gisclass\ ClassProjects folder.

> → Remember to save the project frequently.

Setting up the map page

This project map shows volcanic hazards in Washington and Oregon. Before creating the layout, we should decide on the page size and orientation. This map will be placed in a research paper, so a letter-size page is best, with a landscape layout similar to Figure 3.26.

> 1→ Click the **Insert: Project: New Layout** drop-down and select the ANSI Landscape Letter 8.5 × 11 template (first icon on the second row). A Layout view opens.

1→ In Contents, change the Layout name to *Hazards Layout*.

1→ In a layout view, the Contents pane has different panel icons. Hover over each icon to view its title, but stay in the first icon, List by Drawing Order.

1→ Examine the **Insert** ribbon. In a layout view, additional button groups are available, such as **Map Frames**, **Map Surrounds,** and **Graphics**.

TIP: Always design and edit a map at the page size intended for printing.

When creating a layout, it is important to provide suitable margins so that printing does not crop the map. Rulers and guides can aid the map designer. Rulers measure distances on the page in inches or centimeters. A guide is a placed line use to align the map elements.

2→ Click the **Layout** ribbon and examine its groups and commands.

2→ In the **Layout: Show** group, the boxes for Rulers and Guides should be checked. No guides will be visible yet because they must be created first.

2→ On the top (horizontal) ruler, position the cursor at about 0.75 inches. Right-click and choose Add Guide.

It is not easy to position the guide exactly, and multiple margins are needed, so a menu can place multiple guides. Let's set all margins to exactly 0.75 inches.

3→ Right-click the blue guide line on the map and choose Remove Guide.

3→ Right-click on the top ruler at 0.75 inches again and choose Add Guides.

3→ For orientation, choose Both.

3→ For Placement, choose *Offset from edge.*

3→ Edit the margin value, if necessary, to read exactly 0.75 inches. Click OK.

Next, we need to create a map frame containing the Hazards map.

1. Describe the difference between a map and a map frame. If this distinction is not clear, you will have trouble following the tutorial directions.

4→ In the **Insert: Map Frames** group, click the drop-down on the **Map Frame** button and choose the second Hazards map. (The Default choice opens with the default map extent; the second button uses the current map extent.)

4→ A Map Frame appears in Contents. Rename it *Hazards Map Frame.*

4→ In the layout view, click the resizing handle on the upper left corner and align it with the margin guide. Notice that the handle snaps to the guide when it gets close.

4→ Click the handle in the lower right corner and align it with the margin guide.

Snapping is controlled by the icon in the lower left corner of the map area (Fig. 3.17); hover on the icon to make the menu pop up. Snapping may be turned on or off at need, and two types of snapping are available: to rulers and guides, and to temporary lines created for other map elements.

Next, we will set up the layout design. Although it is simple to arrange map frames by clicking, dragging, and resizing, sometimes precise sizes and locations are needed. Set these options using the map frame properties.

Fig. 3.17. Layout snapping controls
Source: Esri

5→ In Contents, right-click the Hazards Map Frame and choose Properties. The Format Map Frame pane for the Hazards Map Frame appears.

5→ The drop-down, currently labeled Map Frame, controls which item is being modified. Use the drop-down to click and examine each choice; then set it back to Map Frame.

5→ Examine the panel icons and set it to the Placement panel.

5→ Verify that the X and Y positions are set to 0.75 (inch).

5→ Set the rame size to 6 inches wide and 7 inches high, clicking Tab after each change to set it and see the result.

TIP: Notice that opening the Format Map Frame pane also opened a **Map Frame: Format** ribbon. Settings may be altered in either location, but the pane offers additional settings.

Now the map frame is set, but Washington and Oregon do not fill the area of the frame (Fig. 3.18). The map frame extent needs to be adjusted.

6→ Open the **Layout** ribbon and examine the **Navigate** group (Fig. 3.19). The icons appear different because they affect the layout page rather than the map.

6→ Click the **Navigate** tool and try to zoom and pan the map.

In a layout view, the navigation tools affect the entire layout page. To change the extent of the map inside the frame, the frame must first be activated.

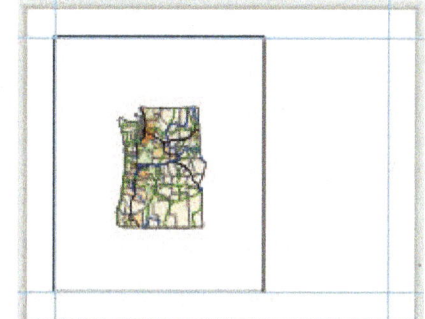

Fig. 3.18. Initial map frame and map extent

Source: Esri

6→ In Contents, right-click the Hazards Map Frame and choose Activate. Notice the addition of the **Activated Map Frame** ribbon.

6→ Click the **Map** ribbon and use the Explore tool or the **Navigation** tools to enlarge the Hazards map so that Washington and Oregon are centered in the frame and fill the space. (Recall that right-click-drag provides a continuous zoom function.)

6→ When satisfied, choose **Activated Map Frame: Layout: Map: Close Activation**.

Fig. 3.19. Layout navigation tools

Source: Esri

This map activation process may seem tedious, but it serves an important function by making the extent of the map frame in the layout independent of the project map.

7→ Switch to the Hazards map view.

7→ Zoom way out of the map.

7→ Click the Hazards Layout view tab to switch back to layout view. The map extent set in the previous step has not changed.

However, the content of the map frame is not independent of the map.

7→ Switch to the Hazards map view and turn off the Volcanoes layer.

7→ Switch back to the Hazards Layout view. The volcanoes are no longer visible.

7→ In Contents, expand the Hazards Map Frame entry below the Hazards Layout name, if necessary. The Hazards map and its layers appear.

7→ Turn the Volcanoes back on.

Notice that the Hazards Layout contains the Hazards Map Frame, which contains the Hazards Map, which contains its layers, making it easy to modify symbols or labels in the map while remaining in the layout.

Let's take a moment to examine the ribbon changes when working in a layout. The ribbons contain the tools we need to work with each element of the layout.

8→ In the Contents pane, click the Hazards Layout entry. Click the **Layout** ribbon to see its contents.

8→ Click the Hazards Map Frame entry. A **Map Frame: Format** ribbon appears, as well as a Format Map Frame pane. Click the ribbon and examine its contents.

8→ Click to highlight the Volcanoes feature layer. Examine the familiar **Feature Layer** ribbon set when it appears.

TIP: In the Contents pane for a layout view, changes to the map and feature layers affect both the layout and the map. Only the map extent is independent. Changes to the layout are primarily made by selecting different elements in the Contents and modifying their properties.

Now that we have established the desired map extent, let us prevent any accidental change to it.

9→ In Contents, right-click the Hazards Map Frame and choose Properties.

9→ In the Format Map Frame pane, click the Display Options icon.

9→ Under the Constraint heading, change the drop-box choice to Fixed Extent.

Creating a location map

Many maps, especially if they show a place that is not commonly known, include a location map to inform the viewer where the map is. To add a location map, we need another map frame.

10→ In Contents, click the Hazards Layout entry to highlight it.

10→ Click the **Insert: Map Frames: Map Frame** drop-down and select the second (not default) USA map.

10→ In Contents, rename the new map frame *USA Map Frame*.

11→ Click and drag the new map frame to the right corner of the layout, snapping to the right and bottom margin guides.

11→ On the top ruler, right-click at about 7 inches and choose Add Guides...

11→ Choose a Vertical guide in a Single location at exactly 7.0 inches. Click OK.

11→ Use the resizing handles to snap the left side of the USA map frame to the new guide.

As before, we need to adjust and lock the map extent within the map frame. It is a little tricky because the optimal size of the frame depends in part on the map scale, so we will learn how to adjust both at once.

12→ In Contents, right-click the USA Map Frame entry and choose Activate.

12→ Use the **Map: Navigate** group buttons to center the lower 48 states in the map frame, making the map as large as possible. Now the frame is too tall, though.

12→ Choose the **Map Frame: Format** ribbon. In the **Size & Position** group, set the **Height** to 2.0 and click the Tab key to enter the value.

12→ Click the **Map: Navigate: Explore** button and make final adjustments to the position of the map within the frame.

12→ Click the **Activated Map Frame: Layout: Map: Close Activation** button.

13→ In Contents, right-click the USA Map Frame entry and choose Properties.

13→ In the Format Map Frame pane, click the Display Options panel icon and set the Constraint to Fixed Extent.

A location map often includes a box showing the extent of the main map. We can add this as a property of the USA data frame.

14→ In Contents, expand the USA Map Frame to show the USA map, if necessary, and click to select it (Fig. 3.20).

14→ Click the **Insert: Map Frames: Extent Indicator** drop-down and select the Hazards Map Frame. The Format Extent Indicator pane opens and an Extent of Hazards Map Frame item appears in the Contents pane, currently selected.

14→ In the Format Extent Indicator pane, click the Symbol representation and then switch from the Gallery to the Properties heading. Change the Outline color to dark orange and the Outline width to 3-pt. Click Apply.

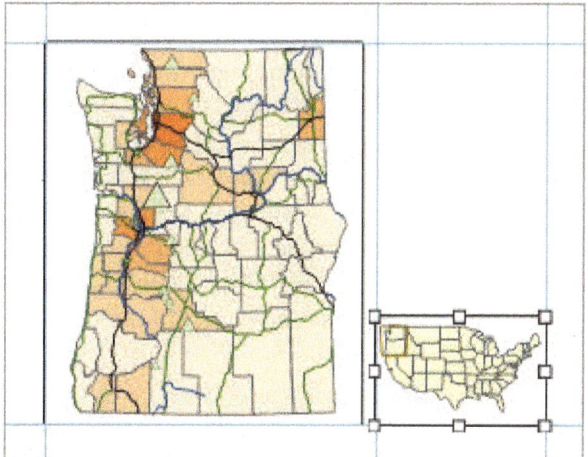

Fig. 3.20. The final map frames
Source: Esri

14→ At this stage, the layout should look similar to Figure 3.20.

TIP: Notice that the Format pane appears any time you right-click a layout element in Contents and select Properties, customized each time to the object that was right-clicked. It is convenient to keep the Format pane docked and open, allowing for rapid format as needed.

15→ In Contents, click the second icon to open the List by Element Type panel. Expand the Hazard Map Frame and its contents if necessary.

15→ Collapse the legends for the individual layers in the Hazards map, as shown in Figure 3.21, to see all of the elements more easily.

Take a moment to examine the Contents pane in the List by Element Type panel (Fig. 3.21). As we add elements to the layout, they appear in the Contents. Selecting an element provides access to additional ribbons, and the properties of any element can be opened and modified. We will use this mode of the Contents pane extensively as we continue working on the layout.

Using graphic text on layouts

One can create **graphic text** on the layout using the text tools on Insert ribbon.

16→ Click the **Insert: Text: Text** drop-down and examine the available tools. Click the **Text** tool.

16→ In the layout, click offshore of the states, which creates a small label, Text. Notice that a Text entry also appears in the Contents pane.

16→ The text is highlighted, ready to replace. Type Pacific Ocean.

Once created, the text label can be edited in many ways.

16→ In the map, place the mouse on the text to show an insertion bar, used to edit the text.

16→ Place the mouse near the text to show a black arrow, used to select the text box.

16→ Select the text box. Click and drag one of the square handles to resize the text.

16→ Find the rotate symbol under the middle bottom of the text. Click and drag to rotate the text to vertical.

16→ Click and drag on top of the text box to move it.

17→ In Contents or on the map, double-click the Text entry to open its properties in the Format Text pane. Explore the settings under each tab, icon, and heading in the pane.

17→ In the Text Symbol tab, under the General icon, expand the Appearance heading and give the text an Arial 18-pt. blue italic symbol. Click Apply.

17→ Keep the Format Text pane open for a while, docked where convenient.

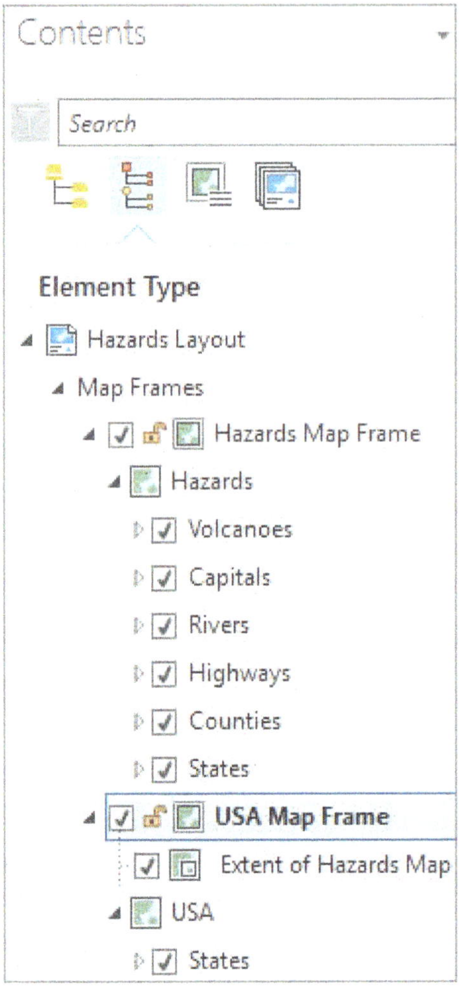

Contents

Search

Element Type

▲ Hazards Layout
  ▲ Map Frames
    ▲ ☑ Hazards Map Frame
      ▲ Hazards
        ▷ ☑ Volcanoes
        ▷ ☑ Capitals
        ▷ ☑ Rivers
        ▷ ☑ Highways
        ▷ ☑ Counties
        ▷ ☑ States
    ▲ ☑ **USA Map Frame**
        ☑ Extent of Hazards Map
      ▲ USA
        ▷ ☑ States

Fig. 3.21. The Contents pane in View By Element Type mode
Source: Esri

TIP: To locate items in this complex pane, we will refer to them using the format Tab > Icon > Heading, e.g. Text Symbol > Formatting > Appearance. (Hover over an icon to see its name.)

18→ Let's try a different text tool. Move the Pacific Ocean text to the margin for now.

18→ Click the **Insert: Text: Text** drop-down and select the **Curved Text** tool. Note that this icon is now the one shown on the ribbon.

18→ Click to enter vertices of a line that curves along the coast of Oregon and Washington from south to north. Double-click to end the line.

18→ Type Pacific Ocean. The text appears splined along the entered line. A second Text 1 element appears in the Contents pane, selected.

18→ Click on the map to finish entering the text.

Use either the format ribbon or the format pane to set text options.

19→ Open the **Text: Format** ribbon and use the **Text Symbol** group settings to give the text an Arial 24-pt. blue italic symbol.

19→ In the Format Text pane, use the Text Symbol > Formatting > Formatting settings to change the Letter spacing to 100%. Click Apply.

Let's keep the second label and delete the first.

19→ In Contents, right-click the Text element (the first label) and choose Delete.

19→ To make it easier to identify, click on the Text 1 element name in Contents and edit it to Pacific Ocean.

Let's give the map a title and author credit. Instead of manipulating the text symbol properties afterwards, we can specify a symbol beforehand.

20→ Click the **Insert: Text: Default Text Symbol** drop-down (looks like Aa) and select the **Title (Sans Serif)** symbol under the **Layout** section, near the bottom.

20→ In the **Text** group, select the **Text** drop-down (now showing as **Curved Text**) and select the **Text** tool.

20→ Click to the right of the main map frame. Type Pacific Northwest, press Enter, and type Volcanic Hazards on the next line.

20→ In Contents, click on the new Text entry and rename it Title.

20→ In the Format Text pane, use the Text Symbol > General > Position settings to set the *Horizontal alignment* to Centered. Click Apply.

20→ In the layout view, move the cursor close to the new text and select it. Move it to the center top of the space (consult Figure 3.26 if needed) using the vertical and temporary center guides to snap it in place.

TIP: At times, a black arrow tool will appear as the mouse moves over the layout, used to select items. Select them from the Contents pane if selecting them from the layout is difficult.

21→ Click the **Insert: Text: Default Text Symbol (Aa)** drop-down and select the **Subtitle (Sans Serif)** symbol.

21→ Repeat the previous steps to add the text by Your Name, centered under the title.

21→ In Contents, rename the new text element Author. Save the project.

Working with dynamic labels

To label the state capitals, we will use dynamic labels, introduced in Chapter 2.

22→ In Contents, expand the Hazards Map Frame and Hazards map. Click the Capitals layer to highlight it.

22→ Click the **Feature Layer: Labeling: Layer: Label** button to turn on the labels.

22→ Use the **Labeling** ribbon to set the label properties to 10-pt. Tahoma bold text.

22→ In the **Text Symbol** group on the ribbon, click the Options button (tiny arrow in the lower right corner) to open the Label Class pane.

22→ In the Label Class pane, use the Symbol > General > Halo settings and give the labels a 2-pt. halo. Close the Label Class pane.

TIP: You may wonder why we are now using different panes and settings to modify this text. It is because the capital names are dynamic labels associated with a map layer, whereas the author and titles are graphic text associated with the layout.

Next we will label the volcanoes with the number of eruptions, but to avoid clutter, we want labels only for the volcanoes with at least five eruptions.

23→ In Contents, select the Volcanoes layer and open the **Feature Layer: Labeling** ribbon.

23→ Set the label field to KNOWN_ERUP, the font to Arial 8 pt., and turn on the labels.

23→ In the **Label Placement** group, choose the **Centered Point** style.

23→ Click the **Options** button for the **Text Symbol** group to open the Label Class pane.

23→ Open the Symbol > General > Position settings and set the Offset Y to −3 pt. to center the labels better in the symbol. Click Apply.

Next we set a query so that only the most active volcanoes have the eruption labels.

24→ Click the **Feature Layer: Labeling: Label Class: SQL** button.

24→ In the Label Class pane, click Add Clause, enter the expression: KNOWN_ERUP is Greater Than 4. Click Add and click Apply.

24→ The labels for the smaller volcanoes disappear. Close the Label Class pane.

Now we also want to label the same volcanoes with names, but to do that we need to create another layer. The Make Feature Layer tool can create a layer that is a subset of another.

25→ If needed, click **Analysis: Geoprocessing: Tools** to open the Geoprocessing pane.

25→ In the Geoprocessing pane, search for and open the Make Feature Layer tool.

25→ Set the Input Features to Volcanoes and the Output Layer to *Volcano Names*.

25→ Click Add Clause and enter the expression: KNOWN_ERUP is Greater Than 4. Click Add and run the tool.

Next, we'll create dynamic labels for the Volcano Names layer. Since we are only using the layer for labeling, we will make the symbols invisible.

26→ In Contents, select the Volcano Names layer and use the Symbology pane to specify the Circle 1 symbol with 1-pt. size and no color. Click Apply.

26→ Switch to the **Labeling** ribbon and turn on the labels. Change the **Text Symbol** to Arial 10-pt. bold font. Use the **Text Symbol** group options to create halos.

26→ In the **Map** group, click **View Unplaced** to see any unplaced labels in red.

The volcano names are interfering with the numbers in the triangles. We can move the volcano names slightly to eliminate interference.

27→ In the **Labeling: Label Placement** group, set the style to **Upper Right Only**.

27→ Click the **Label Placement** group **Options** button to open the Label Class pane. Use the Symbol > General > Position settings to set the Offset X to 4-pt.

27→ Close the Label Class pane.

Before going on, take a moment to analyze the visual hierarchy of the symbols in the map. The complex interstate symbols are prominent, and we don't need four different road classes for this map. We can create a simpler classification.

28→ In Contents, select the Highways layer and choose **Feature Layer: Appearance: Drawing: Symbology**.

28→ Hold the Ctrl key and click the Interstate and Freeway rows so that both are highlighted.

28→ Right-click a highlighted row and choose Group Values.

28→ Edit the Label column for the new class and rename it *Interstates*. Use the up arrow to move it to the top of the legend.

29→ Repeat the previous step to group the Arterial and Unclassified classes together. Edit the label to rename it *Highways*.

29→ Click the Interstates symbol and set it to a 2-pt. simple brown line. Set the Highways symbol to a 1-pt. line of the same color. The legend classes should now appear as in Figure 3.22.

Symbol			Value	Label
∨	HYW_TYPE			
⋮	+	—	Interstate; Free...	Interstates
⋮	+	—	Unclassified; Art...	Highways

Fig. 3.22. Editing the highway classes
Source: Esri

The county lines are too similar to the highways, causing confusion. We can reduce the contrast to bring the highways forward and relegate the county lines to the background.

30→ In Contents, click the Counties layer to show its settings in the Symbology pane.

30→ Click on the More drop-down and select Format All Symbols. Change the Outline color to Gray 10%. The final result should appear as in Figure 3.23.

30→ Save the project.

Adding a legend to the map

Now that the map is in good shape, it is time to start creating the map surrounds, like the legend.

31→ In Contents, click to highlight the Hazards Map Frame element. (The legend will be created for the selected frame.)

31→ Click **Insert: Map Surrounds: Legend**. Draw a box in the empty space below the titles and above the location map, snapping to the vertical guides.

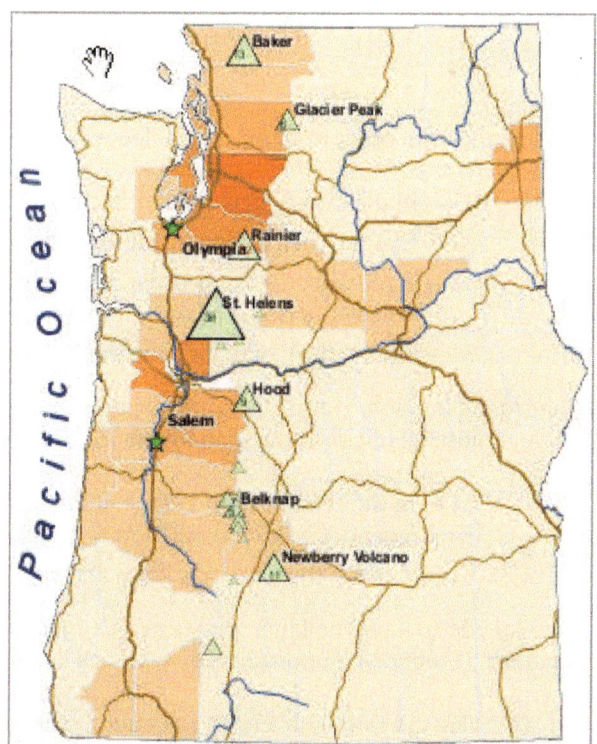

Fig. 3.23. Final map labels and symbols
Source: Esri

31→ The legend appears, and a Legend element appears in the Contents pane.

31→ In Contents, expand the Legend element to show the layers. You will use these entries to make changes in the legend.

First of all, the Volcano Names layer has only labels. It doesn't need to go in the legend.

31→ In Contents, under the Legend element, right-click the Volcano Names element and choose Remove.

Let's take a moment to get acquainted with the legend editing procedures because they are complex and can be confusing.

32→ In Contents, click the Legend element and open the **Legend: Format** ribbon.

32→ Examine the **Current Selection** group with its single drop-down box. This box controls which aspect of the legend we are working with.

32→ Click the drop-down and try some of the other choices. Some tools in the ribbon may be turned off, depending on what is selected.

For example, let's choose a different font for the entire legend.

32→ Make sure the **Legend: Format: Current Selection** drop-down is set to **Legend**.

32→ In the **Text Symbol** group, change the font to Arial. (Leave the size setting blank; the legend includes multiple text sizes, and we don't want to alter them.)

TIP: Depending on what is selected in Contents and in this drop-down, you may change each item individually or multiple items at once. Be aware of what is selected, and what the properties ribbon is referring to, before making changes.

Let's change the order of layers in the legend.

33→ In the Contents pane, select the first icon to open the List by Drawing Order panel.

33→ Expand the Legend element again, if needed, to see the layers.

33→ To change the layer order, if necessary, click and drag a layer to its new position: Volcanoes first, Capitals next, then Rivers, Highways, Counties, and States.

33→ Switch the Contents pane back to the List by Element Type panel.

The legend has some unprofessional aspects, including the use of cryptic abbreviations. We can change these headings in the map layers for a more understandable legend.

34→ In Contents, expand elements under the Hazards Map Frame to view the Volcanoes legend. Click on the KNOWN_ERUP heading and edit it to read Eruptions.

34→ Expand the Counties layer and replace POP10_SQMI with Persons/sq.mi.

It looks odd to have the layer names (Volcanoes, Counties) in nearly the same font size as the headings (Eruptions, Population/sq. mi). Let's change this default.

35→ In Contents, click the Legend element to highlight it.

35→ In the **Legend: Format: Current Selection** group, change the drop-down to **Layer Names**. Change the font size to 14 pt.

35→ Change the **Current Selection** drop-down to **Headings**. Set the font size to 12 pt.

The HWY_TYPE heading under the Highways layer name does not provide any useful information and can be deleted. We can control which parts of a legend item (Layer Name, Heading, Label, or Description) are shown.

35→ In Contents, under Legend, right-click the Highways layer and choose Properties.

35→ In the Format Legend Item pane, under the Show heading, uncheck the box for Headings. Leave the pane open for now.

TIP: The Format Legend pane provides advanced settings not available in the ribbon. You can open it from the Options buttons in the different groups on the ribbon, or you can open it by right-clicking a legend element and choosing Properties.

We may want to resize the legend at some point, but we don't want this action to change the font sizes. We can control this resizing behavior.

36→ In Contents, click the Legend element to switch to the Format Legend pane. Find the Legend > Legend Arrangement Options > Fitting Strategy settings.

36→ Change the Fitting Strategy drop-down to *Adjust columns.*

36→ Check the box to *Set equal widths.*

You may notice that the Highways legend item is split between columns. We can explicitly set the column break.

37→ In Contents, under the Legend element, select the Highways layer. The Format Legend Item pane appears.

37→ Under the Arrangement heading, check the box to *Keep in single column.*

37→ In the layout, resize the Legend width to fit around the actual content. If made too small, part of the legend will disappear. Just size it larger to bring it back.

37→ Adjust the legend position, centering it between the guides and just below the title, leaving space underneath. The final legend should appear as in Figure 3.24.

37→ Save the project.

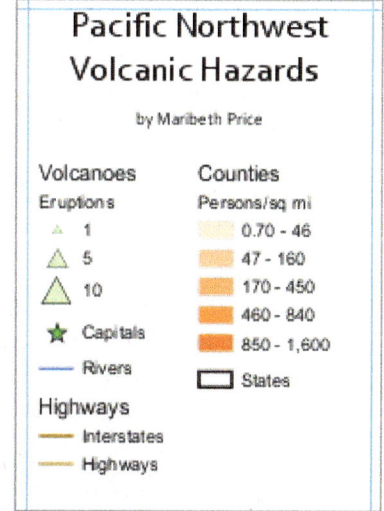

Fig. 3.24. The final legend
Source: Esri

Placing a scale bar on the map

Hazards are affected by distance from the volcanoes, so a scale bar is appropriate on this map. The scale bar is placed in the frame selected in the Contents.

2. To review the scale bar terms, write the correct terms in the boxes of Figure 3.25.

38→ In Contents, select the Hazards Map Frame element.

38→ Click the **Insert: Map Surrounds: Scale Bar** drop-down and select **Alternating Scale Bar 1**. The scale bar appears on the map.

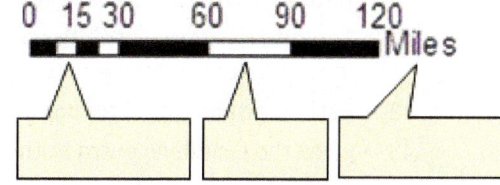

Fig. 3.25. Terminology of a scale bar

38→ Move the scale bar to the lower right corner of the Hazards map frame, on top of Oregon.

38→ Use the *Layout: Navigate: Navigate* tool to zoom in and examine the scale bar.

The bar may be too long or too short, and it may have uneven divisions, such as 110 miles. If it appears unsuitable, try resizing the scale bar to improve it.

39→ Switch to the *Layout: Elements: Select* tool.

39→ Select the Scale Bar element and drag the right boundary of the scale bar to the left to decrease the length of the scale bar.

39→ Repeat until the scale bar is about 120 miles long.

The scale bar may no longer show nice, round numbers. We can modify the scale bar to get exactly the properties we want.

40→ In the **Scale Bar: Design: Divisions** group, change *Adjust division value* to *Adjust number of divisions*.

40→ Set the division value in the top left box to *50* miles and set the Subdivisions to *2*.

40→ Examine the new scale bar. Resize it to 100 miles and place it neatly in the lower right corner of Oregon.

41→ Switch to the **Scale Bar: Format** ribbon. In the **Current Selection** group, set the drop-down to **Background**.

41→ In the **Background** group, set the **Fill** to Sand.

41→ Zoom to the entire layout page and save the project.

Adding other map elements

Many other elements can be added to the map using the Insert menu. Keep in mind that the elements will refer to the data frame that is currently active. Let's add a north arrow based on the *Hazards* map frame.

42→ In Contents, select the Hazards Map Frame element.

42→ Click the **Insert: Map Surrounds: North Arrow** drop-down and choose a slender, unobtrusive symbol like ArcGIS North 5. The north arrow appears on the map.

42→ Click and drag the north arrow to the lower-right corner of the Hazards frame. Keep it small and unobtrusive.

42→ Open the **North Arrow: Design** ribbon and examine the settings.

A grid can be created to place graticules marks around the map.

43→ In Contents, select the Hazards Map Frame element.

43→ Click the **Insert: Map Frames: Grid** drop-down and select the **Black Vertical Label Graticule** (the second option).

43→ The gridlines are distracting and unnecessary. In Contents, right-click the Black Vertical Label Graticule element and choose Properties.

43→ In the Format Map Grid pane, click the Components icon.

43→ Click the Gridlines entry in the Components box. Underneath it, click the line symbol to edit it.

43→ In the Properties tab, use the Gridlines > Properties > Symbol > Appearance settings to change the color to No color. Click Apply and close the Format Map Grid pane.

A professional map cites the data sources used. We will use a wrapped text box so that it can be resized to fit the available space easily.

44→ Zoom to the section of layout below the legend.

44→ Click the **Insert: Text: Text** drop-down and select the **Rectangle** tool.

44→ Click and drag to create a box inside the blank area below the legend.

44→ Type the following citation into the box without using any Enter keys: *Data source: Mastering ArcGIS Pro tutorial data, 1st edition [Download]. (2019) McGraw-Hill Higher Ed, Dubuque, IA [October, 2019].*

45→ Click outside the text box to finish the data entry. Click to select the text box.

45→ Open the **Text: Format** ribbon and set the font to Arial 10 pt.

45→ Adjust the text box to fit neatly in the space available.

45→ Zoom back to the full layout page. The layout should look similar to Figure 3.26.

At this point, a test print can be made to ensure that the colors and symbols work as well on paper as they do on screen.

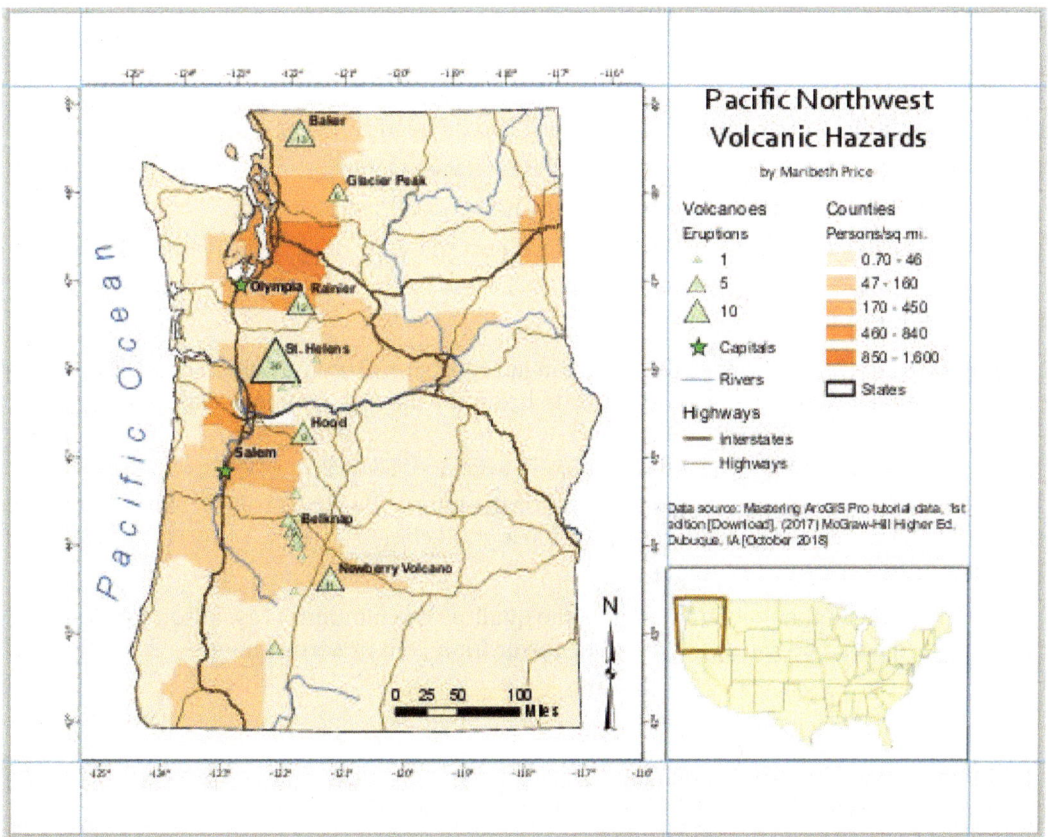

Fig. 3.26. The final layout
Source: Esri

46→ Save the project before printing (always a wise move).

46→ Click **Share: Print: Layout**.

46→ Select the printer or adjust the print settings, if needed.

46→ When ready, click Print.

TIP: Examine the margins of the print. Some printers shift the map off-center, and the map margins must be adjusted for that printer.

Reviewing and editing the layout

Now for the final review. Go take a break, then return and look at the map at full scale, both on paper and on screen. You might notice the following issues.

The visual hierarchy established by the symbols could be clearer, especially for the line features, which draw attention from the volcanoes. The highways and rivers are difficult to tell apart. The Pacific Ocean label is distracting.

While evaluating the design, return to the map objective: to draw attention to the fact that population centers in Washington and Oregon lie close to historically active volcanoes. Perhaps the roads and rivers are not as important as originally decided. We can edit the map to make it more attractive and clarify its message, which will be the task in the first exercise.

It takes practice to create and edit layouts. Let's create a practice copy of this layout to experiment with later.

47→ In the Catalog pane, expand the Layouts heading.

47→ Right-click the Hazards Layout and choose Copy.

47→ Right-click the Layouts heading and choose Paste.

47→ Click twice slowly on Hazards Layout1 (the new one) and rename it Layout Practice.

Exporting a layout

Maps can be saved as image files for inclusion in reports or presentations. The PDF format is a popular way to distribute maps to others or to bring the map to a commercial printing center.

48→ Click **Share: Export: Layout**.

48→ Choose the location and name for the image file. The project folder (but not inside the project geodatabase) is often a good choice.

48→ Set the Save As type to PDF.

48→ Check the resolution and image quality. The minimum resolution for printing should be 300 dpi. It could be lower for a web or screen image.

48→ Click Export to save the map.

TIP: If planning to place the map in another document, check the box to *Clip Output to Graphics Extent* and save as a JPEG or GIF. Check that the resolution is at least 300 dpi.

This is the end of the tutorial.

→ Save the project and exit ArcGIS Pro.

Practice Exercises

1. In the Catalog pane, create another copy of the Hazards Layout and name it *Hazards Final*. Edit the copy to produce the version in Figure 3.27, including removing the highways and rivers, as well as increasing the sizes of the symbols and labels. When done, add the World Terrain basemap from ArcGIS Online to the main map frame to provide context (but turn off its reference layer of features and names). Make the counties about 30% transparent so that the basemap shows through slightly. Save the map as a PDF to turn in.

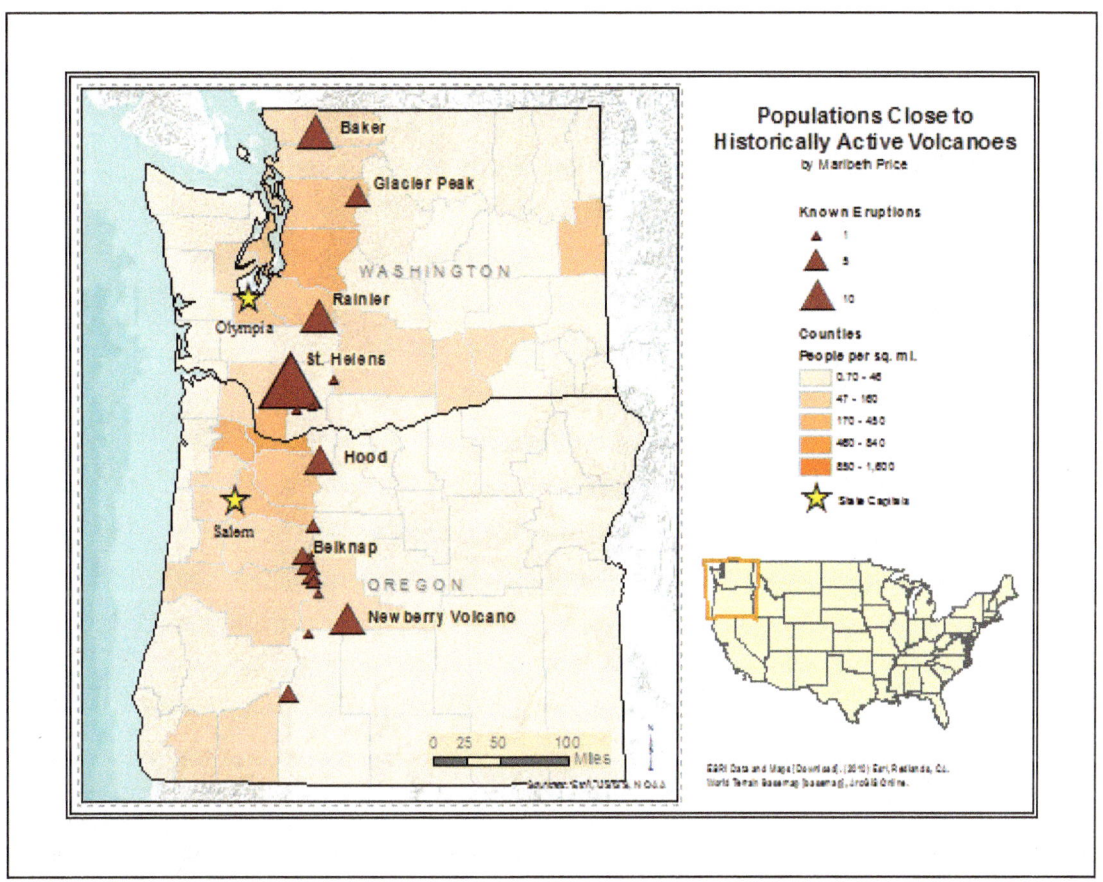

Fig. 3.27. Revised map from the tutorial
Source: Esri

2. Using the *CraterLake* project or the project you created for the Chapter 1 exercises, design and create an attractive map layout. Pick any map objective you like, but make sure the layout communicates it clearly. Export the map as a PDF file and print it. Share it with some classmates, and help each other critique the maps. Write the suggestions on the back of the PDF and turn it in.

Chapter 4. Coordinate Systems

Objectives

▶ Learning the basic properties and uses of coordinate systems

▶ Understanding the difference between geographic and projected coordinate systems

▶ Understanding different projections and the distortions they cause

▶ Choosing appropriate projections for a map or geodatabase

▶ Managing and troubleshooting coordinate systems of feature classes

Mastering the Concepts

GIS Concepts

About coordinate systems

A successful GIS system depends on managing coordinate systems correctly, and a person's skill (or lack of it) can make the difference between usable and trashed data. Coordinate systems are also a vast and complex topic. Experience working with coordinate systems and regular review of these concepts will ultimately provide a sound basis for effectively working with coordinates.

In order to display and analyze maps on the screen, a GIS system uses **coordinate pairs**, which specify the location and shape of a particular feature. A point is described by a single x-y coordinate pair located in space; a line is a series of x-y coordinates. A **coordinate space** is an agreed-upon range of coordinates used to portray the features.

Imagine that a surveyor, Carlos, is mapping an industrial site using typical surveying instruments that measure angles, directions, and distances. He may begin at a fence corner at the lower right of the complex and assign that point the coordinates (0, 0). This is the **origin** of the coordinate space (Fig. 4.1). Carlos then determines that the southwest corner of building A is 175 meters east and 200 meters north of the origin, and he assigns that corner the coordinates (175, 200). He uses meters to record the distances and to determine the coordinates, thereby establishing meters as the **map units**. The **extent** is the range of x-y values stored (in Fig. 4.1, about 0–800 for x and 0–700 for y.) The combination of origin and map units becomes the coordinate space of this map. In this case the coordinate space is arbitrary and chosen for convenience.

Now imagine that the surveyor's colleague Jasmin is surveying another site across town. She also chose the corner of her site as the origin, but she is measuring distances in feet. Carlos and Jasmin take their data back to the office and decide to create a map showing the location of both sites, each transferring the coordinates into a GIS. When they plot the sites, however, the two different sites show up in exactly the same spot, even though in real life they are several miles apart. This problem occurs because both surveyors used an arbitrary (0, 0) origin unrelated to the other

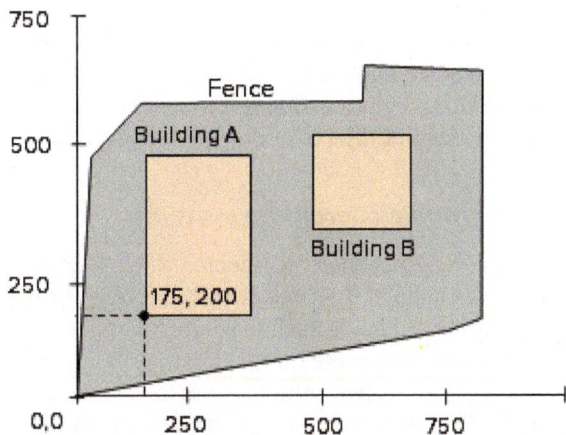

Fig. 4.1. An arbitrary coordinate space used for surveying a site

site. Moreover Jasmin's buildings appear about three times as large as Carlos's buildings because she is measuring in feet instead of meters. To plot these together, the two surveyors would need to define a single common origin and convert both sites to the same map units.

To avoid problems such as this, geographers and map creators usually employ a standard **coordinate system (CS)**, a coordinate space whose characteristics are defined and established according to cartographic standards and tied to a specific location on the earth's surface. The surveyors could have used a global positioning system (GPS) instrument to establish the x-y coordinates of their origins in a common coordinate system, such as Universal Transverse Mercator (UTM), and then entered their surveyed points relative to that base. Then their maps would have been located correctly with respect to each other.

Standard coordinate systems

Both raster and vector data rely on x-y values to locate data to a particular spot on the earth's surface. The x-y values of the coordinate pairs can vary, however. The choice of values and units to store a data set is called its **coordinate system**. Consider a standard topographic map, which actually has three different coordinate systems marked on it. The corners are marked with degrees of latitude and longitude, which is termed a *geographic coordinate system (GCS)*. Another set of markings indicates a scale in meters representing the **UTM**, or Universal Transverse Mercator, coordinate system. A third set of markings shows a scale in feet, corresponding to a **State Plane** coordinate system. Any location on the map can be represented by three different x-y pairs corresponding to one of the three coordinate systems (Fig. 4.2).

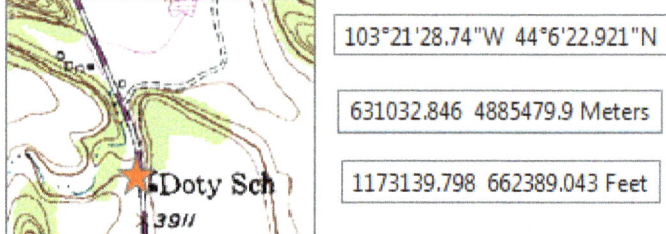

103°21'28.74"W 44°6'22.921"N

631032.846 4885479.9 Meters

1173139.798 662389.043 Feet

Fig. 4.2. A location can be stored using different coordinate systems and units. The x-y location of this school is shown in degrees, UTM meters, and State Plane feet.
Source: Esri

A global positioning system (GPS) unit also has this flexibility. It can be set to record a location in GCS degrees, UTM meters, State Plane feet, or other coordinate systems.

When creating a vector or raster data set, one must choose a coordinate system and units for storing the x-y values. It is also important to label the data so that the user knows which coordinate system has been selected and what units were chosen for the x-y values. Thus, every GIS data set should have a label that records the type of coordinate system and units used to store the x-y data inside it. Coordinate systems fall into two categories, geographic coordinate systems and projected coordinate systems.

Geographic coordinate systems

One commonly used coordinate system is based on measuring angles from the center of the earth and has units of **degrees** (Fig. 4.3). **Longitude** measures horizontal angles east or west of the **Prime Meridian**, which is the line where longitude equals 0 that passes through Greenwich, England. Longitudes fall in the range –180 to +180. **Latitude** measures vertical angles above or below the equator, which has latitude equal to 0, and values range from –90 at the South Pole to +90 at the North Pole. This system of measurement is called a **geographic coordinate system**, or **GCS**.

A GCS location is determined when the vector defined by a latitude-longitude pair intersects the surface of the earth. But what surface? Because the earth's shape is irregular, its surface can

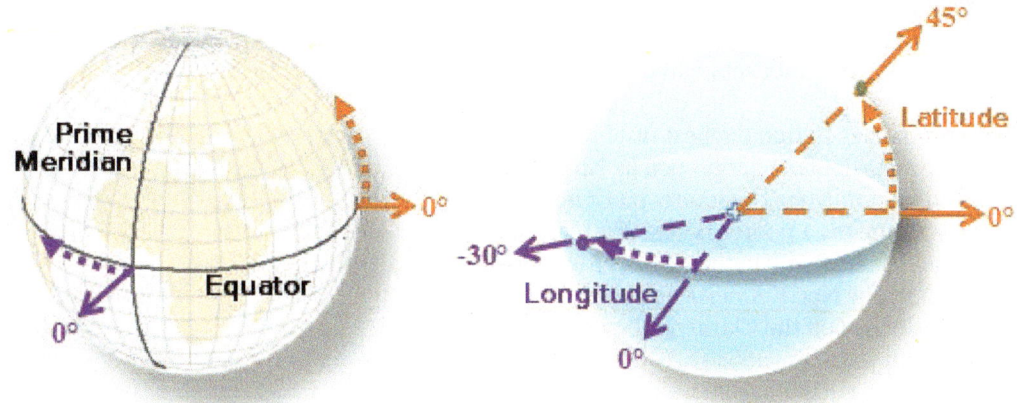

Fig. 4.3. Longitude measures E-W angles in the plane of the equator; latitude measures N-S angles above or below the plane of the equator.

only be approximated, and the difference between the true shape and the approximate shape is a source of positional errors. For this reason, different approximations have been developed to meet specific mapping needs, each one representing a slightly different shape for the earth. Such an approximation is called a **datum**.

How is a datum defined? The earth's rotation causes it to bulge at the equator, so the first task is to model the earth's surface using a smooth figure called a **spheroid**, an oblong sphere with a major and a minor axis. (The bulge in Fig. 4.4 is highly exaggerated.) The chosen spheroid becomes a part of the datum definition. (Some people use the term **ellipsoid** rather than spheroid.) Different spheroids have been used in the past because estimates of the earth's major and minor axes have varied over time. Clarke 1866 was a common spheroid used in North America, but it is being replaced by more recent and accurate satellite-measured spheroids.

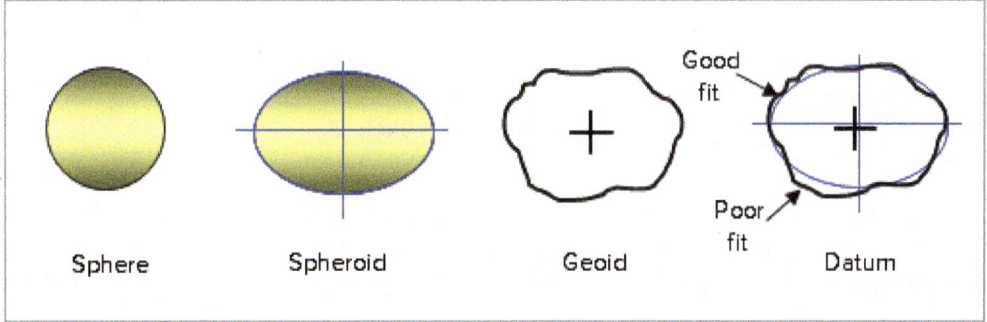

Fig. 4.4. Relationship among the spheroid, geoid, and datum

However, a spheroid does not adequately represent the shape of the earth, either, so a second correction is made. The closest approximation to the earth's shape is the **geoid** (Fig. 4.4), an irregular, equipotential surface based on gravity measurements. Sea level varies from place to place due to rotational, topographic, and compositional differences in the earth's mantle, and the geoid is often described as what the shape of the earth would be if there were only seas and no continents. The geoid can be modeled mathematically, but the equations are too computationally intensive to serve well as the basis for mapping. Instead, the spheroid is shifted relative to the geoid until a best-fit solution is obtained (Fig. 4.4). A datum, then, consists of the chosen spheroid and the location of its center relative to the geoid center.

A datum that finds the best fit overall is called an *earth-centered datum*. The World Geodetic Survey of 1984 is an earth-centered datum based on the WGS 1984 spheroid and is commonly used for international data sets and web maps. It is the default datum for most GPS units.

Another approach is to find the best fit of the spheroid to the geoid for a particular region, such as North America, to produce a *local datum*. The improved fit in one place, however, often means a poorer fit elsewhere, so a local datum should be used only in the regions for which it was developed. The North American Datum of 1927 (NAD 1927) is based on the Clarke 1866 spheroid. The North American Datum of 1983 (NAD 1983) is based on the more accurate GRS 1980 spheroid. Both datums use a network of surveyed benchmarks to further optimize the fit and provide more accurate positions; this approach is often employed for local datums. Coordinates stored using NAD 1927 and NAD 1983 can differ by several hundred meters, which, although not significant for a national map, can cause serious registration issues for larger-scale maps. Figure 4.5 shows the offset between roads stored using NAD 1927 and an image stored using NAD 1983.

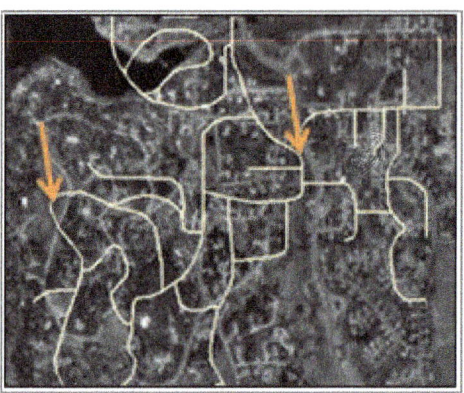

Fig. 4.5. Offset between road lines in NAD 1927 and image in NAD 1983
Source: USGS

At times it is necessary to convert from one datum to another; the process is called datum **transformation**. It does not always use exact mathematical formulas, and it may require localized estimates and fitting. Often, multiple methods are available. The transformation itself may introduce new errors in coordinate locations, on the order of several meters, so it is not a good idea to transform GIS data back and forth repeatedly and thus multiply these errors. Usually, one picks the datum to be used for a project (or organization) and sticks with it, converting data once when they are brought into the geodatabase using the best available transformation.

Adjustments of NAD 1983, based on improved surveying of the benchmarks, take place periodically and may be used for very high-precision surveying applications when differences on the order of centimeters are important. Such adjustments have names such as CORS96, HARN, and NSRS2007 (Fig. 4.6). However, unless you have been told to use one of these specialized NAD adjustments, it is best to choose the standard datum, NAD 1983. Most data downloaded from other sources will use the standard NAD 1983; choosing a different one will require that transformations be made, potentially introducing new positional errors into the data.

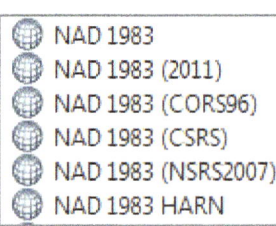

Fig. 4.6. NAD options
Source: Esri

The datum forms the base definition of a geographic coordinate system (GCS) used to store locations in degrees of latitude and longitude. The same place will have slightly different values of latitude and longitude depending on which datum was used. The GCS typically takes its name from its defining datum, and to many, the terms *datum* and *GCS* are interchangeable.

Map projections

A GCS is a three-dimensional coordinate system, but maps need to be flat. Therefore, cartography requires converting the locations on a globe to locations on a piece of paper. This process is called **projection**. One starts with a defined GCS and applies a set of mathematical formulas that convert degrees of latitude and longitude into planar *x-y* coordinates on the paper. During this conversion, the three-dimensional GCS becomes a two-dimensional representation, and degrees are converted into feet or meters. The chosen earth shape (the GCS) influences where locations end up on the paper, so the GCS is inherited by the projection and becomes a part of its definition.

One can imagine the projection process visually as a clear globe with the graticules of the earth printed in black. Placing a light bulb at the center of the earth and wrapping a piece of paper around the equator in the shape of a cylinder (Fig. 4.7) will cast the graticules as shadows on the paper. Tracing them yields a map that can be unrolled and laid flat.

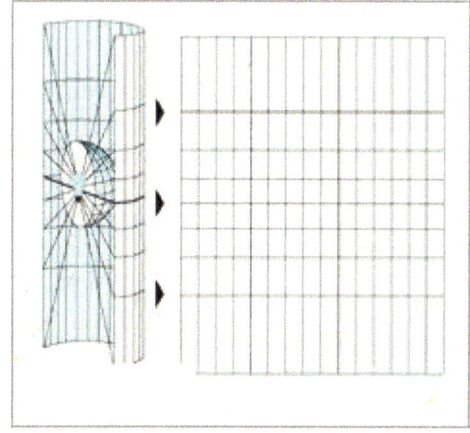

Imagine the globe in Figure 4.7 as a flexible sphere, like a beach ball, and consider what happens if it is slightly compressed. The projected lines will shift where they intersect the cylinder, and therefore the locations on the paper will also change. This visualization illustrates the effect of using a different earth shape (datum) as the basis of the projection, and it helps explain why maps based on different datums may be offset even if they are in the same projection.

Fig. 4.7. Projecting a map from 3D to planar coordinates

Source: Esri

Projections are grouped into three major classes, depending on the shape of the surface onto which the GCS locations are projected. A **cylindrical projection** uses a cylindrical surface that lays tangent to (touches) the earth at the equator along a great circle (Fig. 4.8). Rotating the cylinder sideways and making it tangent to the earth along a line of longitude produces a **transverse projection**. An **oblique projection** places the tangent at an angle. In each case, distortion is absent along the tangent and increases away from it. Cylindrical projections typically preserve shape or direction at the expense of area and distance.

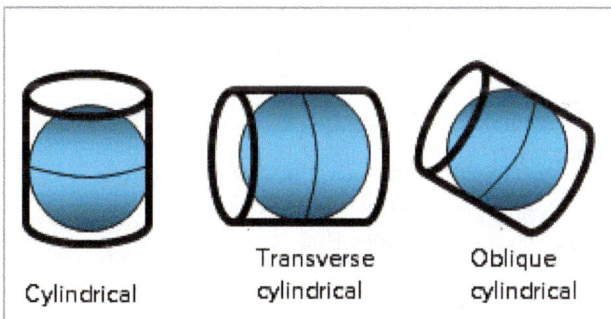

Fig. 4.8. Cylindrical projections

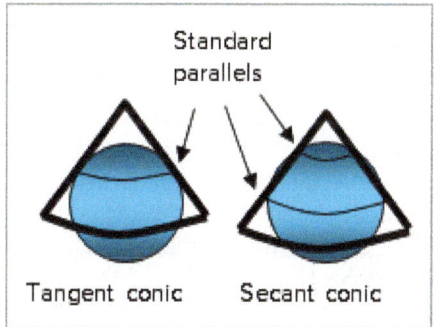

Fig. 4.9. Conic projections

A **conic projection** is based on setting a cone on the sphere (Fig. 4.9). If the cone is tangent to the globe along a line of latitude, then the projection is a **tangent projection**. The line of tangency is called a **standard parallel**. One can also place the cone *through* the sphere so that it touches in two places—a **secant projection**. Such projections have two standard parallels. (Cylindrical projections may also be tangent or secant.) One can also place a plane tangent or secant to the sphere, producing an **azimuthal projection** (Fig. 4.10). Other

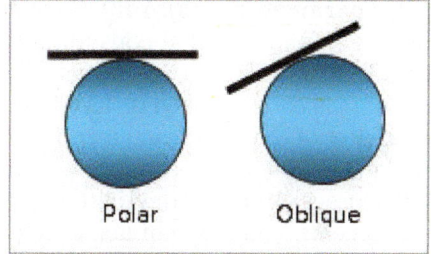

Fig. 4.10. Azimuthal projections

names for this method are **stereographic**
and **orthographic projection**. These
projections are often used for displaying the
earth's poles, and for that reason they are
sometimes called polar projections.

Projecting the earth to a plane always
incurs distortion of the map. Four properties
of map features may be distorted: *area,
distance, shape,* and *direction.* Usually
map projections reduce or eliminate certain
distortions at the expense of the others.
Maps based on cylindrical projections
(Fig. 4.11a) typically preserve direction
and shape at the expense of distance and
area. Notice that the longitude lines point
north-south, indicating the correct direction,
but Alaska and Greenland are enlarged.
Conic projections typically preserve area
or distance at the expense of direction and
shape (Fig. 4.11b). The longitude lines no
longer point north-south, but the areas are
better preserved. Maps based on azimuthal
projections also typically preserve areas or
distances (Fig. 4.11c). The inside front cover
of the text summarizes some of the more common map
projections along with the properties and uses of each; full
descriptions are provided in the ArcGIS Pro Help.

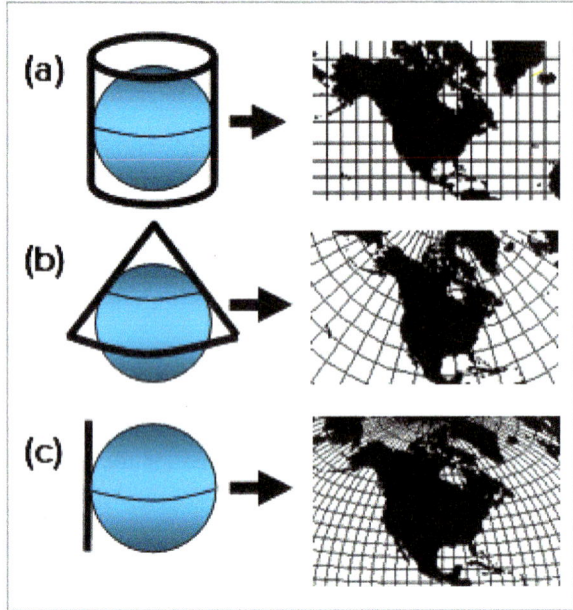

Fig. 4.11. Comparison of distortions caused by
(a) cylindrical; (b) conic; and (c) azimuthal
projections

Source: Esri

In any projection, no distortion occurs where the globe
touches the cylinder, cone, or plane. The standard
parallels, or the center of an azimuthal projection, are thus
undistorted in all four properties, and the distortion will
increase away from these locations. The standard parallels
of a projection are typically placed below the upper sixth
and above the lower sixth of the area of interest. For
example, the region of Turkey covers approximately 36 to
42 degrees, so the standard parallels might be set to 37 and
41 degrees (Fig. 4.12).

Fig. 4.12. Choosing standard
parallels for Turkey.

Source: Esri

Map projections have certain properties that are
used to customize the projection for a particular
geographic area and purpose; these properties are
called **parameters** (Fig. 4.13). One parameter is
the line of longitude that constitutes the **central
meridian**, or $x = 0$ line on the map. The central
meridian is chosen based on its position near
the center of the area being mapped. A map of
England would have a central meridian close
to 0 degrees. A map of the United States would
have a central meridian of about −100 degrees
because that meridian is close to the center of the
continental United States. The **latitude of origin**,

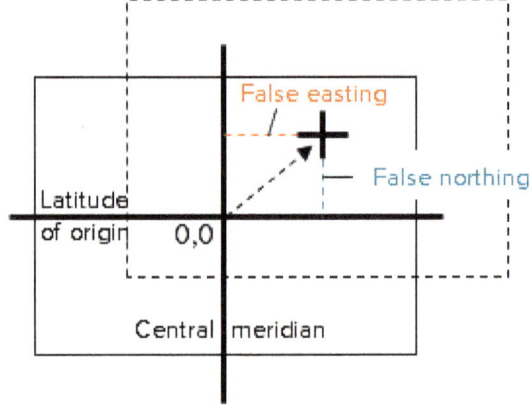

Fig. 4.13. Some parameters of map projections

also sometimes called the **reference latitude**, is the $y = 0$ line. The equator is often used as the reference latitude. The **standard parallels** are also parameters. Maps may also have a **false northing** or **false easting** (Fig. 4.13). These are arbitrary numbers added to the x-y coordinates in order to translate the map to a new location in coordinate space. Most often, the false easting and northing are used to ensure that all the coordinates of the map are positive.

Common projection systems

Several projections deserve special description because they are so commonly used. One, the GCS, is not actually a projection but is often treated as one. UTM is a family of coordinate systems used worldwide. State Plane is another such family used within the United States, and other countries have similar national or state systems used for large-scale maps.

Geographic coordinate systems

As a three-dimensional coordinate system, a GCS cannot, strictly speaking, be displayed on a flat map or screen. However, in practice a simple equirectangular projection can be applied that treats degrees as planar distances rather than as angles (Fig. 4.14). Unfortunately, distortions are introduced by this practice. On the globe, the latitude lines get shorter at higher latitudes, going to zero at the poles, yet in the planar display, these lines are all the same length. The further north or south an area is from the equator, the more the map gets distorted. Analysis functions also treat a GCS data set as if it were planar, and if a function uses area or distance in its calculations, then the results may be incorrect. The GCS is primarily used for distributing data; once it is chosen, the user should project it before incorporating it into databases for mapping or analysis.

TIP: It is inappropriate to use a GCS to portray maps or to do spatial analysis, and doing so may cause incorrect results.

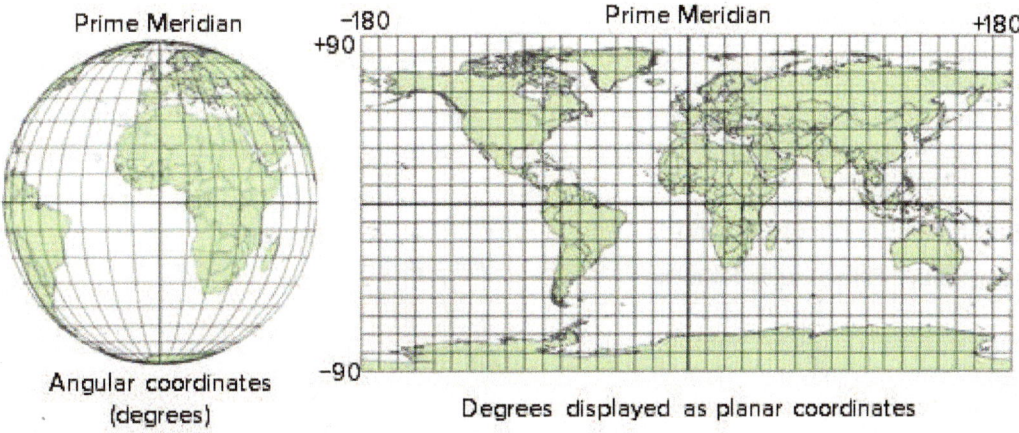

Fig. 4.14. A GCS is projected flat for display in a GIS, causing distortion
Source: Esri

The Universal Transverse Mercator (UTM) system

The **Universal Transverse Mercator (UTM)** system is based on a secant transverse cylindrical projection. Two lines of tangency of the cylinder to the sphere fall about 180 km to each side of the central meridian (Fig. 4.15a). A zone is defined to include 3 degrees on each side of the central meridian between the 80S and 84N latitude lines. Since the projection is true along the lines of tangency, the distortion inside the zone is minimal. To map different locations, the cylinder is

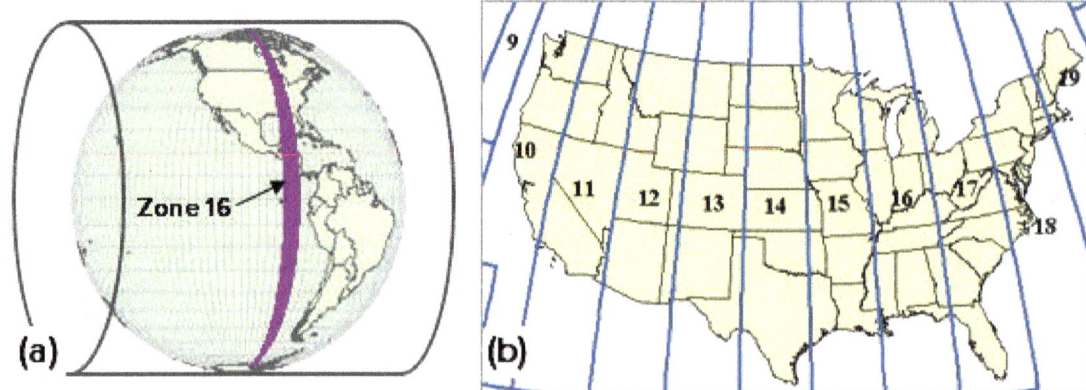

Fig. 4.15. (a) The UTM projection system has 60 north-south zones, each 6 degrees wide. (b) This map shows the UTM zones of the conterminous United States.
Source: Esri

rotated about the globe in 6-degree intervals, producing 60 zones around the world. Areas within a single zone have negligible distortion in all four properties of shape, direction, distance, and area.

Each UTM zone has its own central meridian and splits at the equator into a north and south component. In order to eliminate negative x-y coordinates, southern zones have a false northing of 10 million meters, and both northern and southern zones have a false easting of 500,000 meters. UTM is convenient because users need only know the zone number and hemisphere. In the United States and in ArcGIS, each zone is indicated by a number and the designation N for the Northern Hemisphere or S for the Southern Hemisphere—for example, UTM Zone 14N. (The worldwide UTM system uses letters C to X to represent different latitudinal zones.) Figure 4.15b shows the UTM zones of the conterminous United States.

The UTM system is used extensively for US Geological Survey 1:100,000 and 1:24,000 scale publications, for county maps, for maps of states within single zones, and for maps covering small areas. If the region of interest lies on the boundary between two zones, the user can choose the zone containing the larger area to represent the map. This choice may be acceptable if only a small area extends out of the zone. If not, State Plane zones may provide a better alternative.

The State Plane Coordinate System

The **State Plane Coordinate System (SPCS)** includes an assortment of projections developed in the 1930s by the US Coast and Geodetic Survey for large-scale mapping in the United States. To maintain the desired level of accuracy, most states are broken into zones with different parameters used to minimize distortion in each zone (Fig. 4.16). Like the UTM system, an SPCS is identified by its name and zone, and distortion is negligible within a single zone.

The SPCS uses three different projections: Lambert Conformal Conic, Transverse Mercator, and Oblique Mercator. North-south oriented zones use a Transverse Mercator

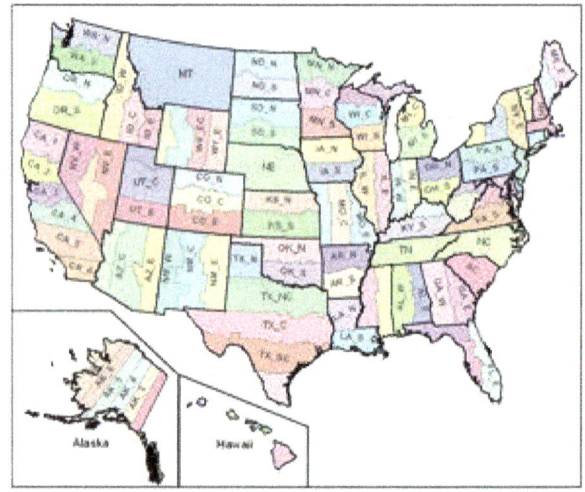

Fig. 4.16. State Plane zones of the United States
Source: Esri

projection because it provides the least distortion. East-west oriented zones use a Lambert Conformal Conic projection. A state may use more than one projection, depending on the size and shape of its zones. Alaska has 10 zones and uses all three projections. Zones are formally identified using a FIPS code composed of a two-digit state code and a two-digit zone code (e.g., FIPS 0407 for Zone 7 of California). Users may refer to them informally by state name and zone, for example, California State Plane Zone 7.

The original SPCS was based on NAD27 and used units of feet. Improvements in surveying necessitated the revision of the SPCS in 1983. At that time, it adopted the NAD83 datum and metric units. Certain zones were also revised and in some cases eliminated. Even though the official units changed from feet to meters, users may specify either unit for both versions of SPCS. Additional information about both the State Plane and UTM systems can be found in the ArcGIS Pro Help.

State and national grids

Some large states define a special coordinate system for statewide maps because a single State Plane zone cannot represent it accurately. The Oregon Statewide Lambert coordinate system is one example. Many countries define their own coordinate systems for national maps or their own systems of projections (similar to State Plane) for local and regional maps. Examples include the New Zealand National Grid and the Canada CSRS98 system. A variety of projections, datums, and spheroids are employed. ArcGIS Pro maintains a collection of these predefined coordinate system definitions for use with state and country data.

Choosing projections

At the beginning of every GIS project, one of the first tasks is to establish the projection to be used because best practices dictate that all data sets in a project share the same projection and datum. The choice affects which map properties are preserved, and a poor choice may cause incorrect analysis results. Some guidelines for choosing projections for a project are described next. Preference should be given to one of the standard coordinate systems for the sake of interoperability, but if none of the existing systems provides the right properties, a user can define a custom projection.

Maps of the world. Distortions cannot be avoided in small-scale maps, so they must be managed. Consider the applications. If distances or areas are to be analyzed as part of the project, then a conic equidistant or equal area map should be chosen. If navigation or wind and water currents are important, then preserving direction is critical, and a Mercator projection works well. If portraying attractive maps is the main goal, then a compromise projection like Robinson might be best.

Maps of countries or continents. Many countries or regions have one or more projections defined for general mapping purposes, such as USA Contiguous Equidistant Conic or Europe Albers Equal Area Conic. Some countries have their own defined mapping systems. Like world maps, the application determines which properties of the map are important to preserve.

Maps of states. Smaller states within a single UTM or State Plane zone can use them. Larger states require selecting one of the zones; if desired, one can modify the central meridian and/or parallels of the zone to work better for the entire state. Some states have a statewide projection defined, such as Oregon Statewide Lambert.

Local and regional maps. Any region that fits inside a single UTM or State Plane zone can use either. Many countries, such as Canada and Australia, have defined their own mapping systems similar to the State Plane system in the United States. The preferred system would place the region to be mapped close to the center of the zone, where the distortion is least. However, other factors may take precedence, such as the existing coordinate system of one's data sources, or the decision of the agency that oversees the project.

Sometimes the region to be mapped falls close to a UTM or State Plane zone boundary or spans two zones. In this case, a customized coordinate system is desirable. Typically, one starts with a UTM or State Plane zone and adjusts the central meridian and/or the standard parallels to minimize the distortion. A good rule of thumb is that the central meridian should bisect the map extent, and the parallels for secant maps should be placed one-sixth of the north-south extent from the edges.

Impact of coordinate systems

The choice of the coordinate system can have a significant impact on the portrayal of a map or the analysis of spatial data, and it must be considered carefully.

Portraying maps and layouts

The choice of projection used for a map, and especially for a layout, depends on the objective of the map. A map that requires distances to be measured must use a projection that preserves distance. A map that compares information across states or counties will be influenced by size, and so equal area maps are usually preferred to minimize MAUP issues. A geographic coordinate system is almost never appropriate for a map because of the significant distance and area distortions, and it should be avoided.

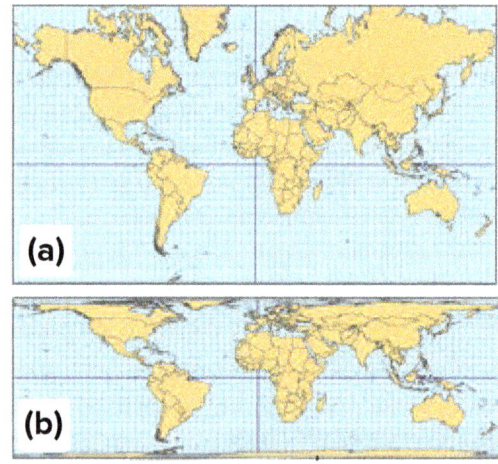

Sometimes the projection can influence the map story in unintended ways. The cylindrical Mercator projection has been popular for world maps and is often used in web maps (Fig. 4.17a), but critics have charged that this projection enlarges countries in Eurasia and North America, sending a subliminal message that the Northern Hemisphere is more important than developing countries near the equator. Attempts have been made to popularize equal-area world projections that don't exaggerate the Northern Hemisphere (Fig 4.17b).

The projection may also influence the elements placed in a map. A conic projection does not preserve direction, and a single arrow cannot represent north everywhere on the map. A map that distorts distance should not have a simple scale bar.

Fig. 4.17. (a) A Mercator projection and (b) an equal-area cylindrical projection
Source: Esri

Spatial analysis

Many types of spatial analysis can be adversely affected when the input data sets are not stored in an appropriate coordinate system. Figure 4.18 shows a distance analysis in which tourist attractions are being linked with the closest hotels, indicated by the color of the stars. In Figure 4.18a, the data sets were stored in a GCS. The three attractions in the northeast corner were assigned to Hotel C based on the distances in decimal degrees. In Figure 4.18b, the feature classes were projected to a UTM coordinate system, and the attractions were assigned instead to Hotel B. Since a UTM projection preserves distance within the zone, we know that the second example gives the correct distances and is the valid result.

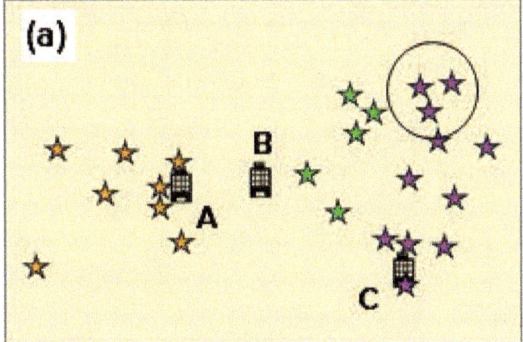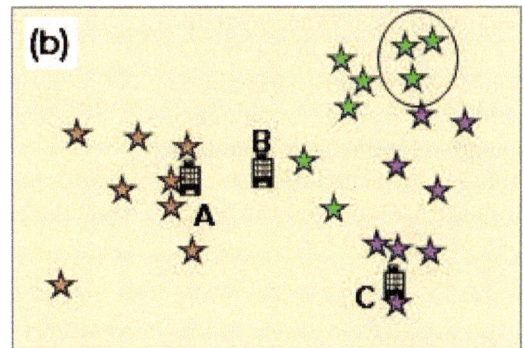

Fig. 4.18. The effect of the coordinate system in a distance analysis. (a) Three attractions are incorrectly assigned to Hotel C (purple stars) when the input feature classes have a GCS. (b) When the feature classes are projected to UTM, the attractions are correctly assigned to Hotel B (green stars).

Overlay and spatial join tools will accept any coordinate system or combination of coordinate systems for input, but rules of precedence dictate the coordinate system of the output. Outputs will be projected according to the following rules:

▶ If the output is placed in a feature dataset, the coordinate system will always match that of the feature dataset.

▶ If the coordinate system is set in the Environment settings and the output will not become part of a feature dataset, the Environment settings coordinate system is used.

▶ If the Environment setting is not set, the default rule applies—that the output will match the coordinate system of the first input to the tool.

TIP: Even with these rules, unexpected problems may emerge when mingling coordinate systems during analysis. It is wise to keep a project's data sets in a common projection whenever possible.

About ArcGIS

ArcGIS Pro has many capabilities for managing coordinate systems for data sets. It can display data sets stored in different coordinate systems and convert data sets from one coordinate system to another. It supports a wide variety of different projections and gives users a complete set of tools for defining, converting, and managing coordinate systems.

On the fly projection

In older GIS systems, the coordinate systems of different data sets had to match in order for them to be drawn together in the same map. UTM data could only be shown with other UTM data, State Plane data with other State Plane data, and so on. If data were in different coordinate systems, they would need to be converted to the same coordinate system prior to display.

Although it is still true that coordinate systems must match for data to be displayed together, many GIS systems can now perform the conversion on the fly. This feature allows data to be stored in different coordinate systems yet be drawn together. In ArcGIS Pro, every map has a designated coordinate system. The initial coordinate system is assigned to match the first data set added to the map, but it can be changed by the user. Any data set that is stored in a different coordinate system than the map will be projected to match on the fly (Fig. 4.19).

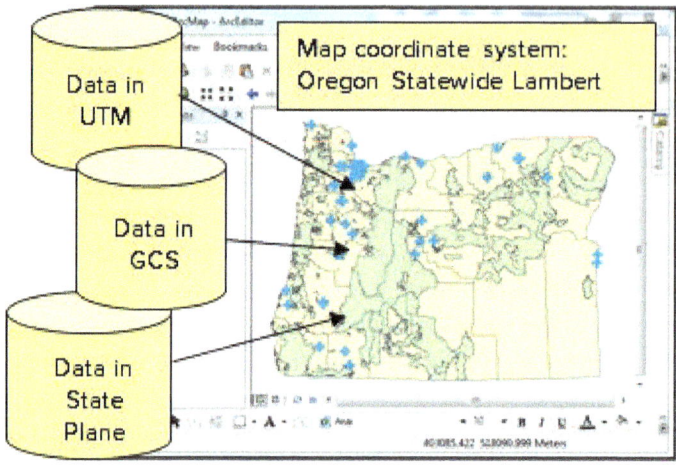

The units defined for the map coordinate system, whether they are meters, feet, or degrees, become the **map units** and may differ from the stored units in the files. In Figure 4.19, the UTM data use meters to store the x-y coordinates, the GCS data use degrees, and the State Plane data use feet. The Oregon Statewide Lambert coordinate system uses meters, so meters become the map units. The user may specify a different set of units, called the **display units**, to be used on the x-y location readout in the map view.

Fig. 4.19. Data in any coordinate system can be displayed together by setting the map coordinate system.
Source: Esri

In order to project on the fly, Pro must know the coordinate system of each data set, as defined by its label, or spatial reference. If the spatial reference is Unknown, then Pro will display the data with whatever coordinates it has. Sometimes it will fortuitously match the other data in the frame, but at other times it will appear totally out of place. If it matches, then one knows that the Unknown layer has a coordinate system that is the same as, or at least very similar to, the data frame, and can create a spatial reference label for it.

Pro will also automatically transform a datum on the fly if a data set uses a different datum than the map. However, a datum transformation is not always exact, and positional errors of several meters are common. For national or state maps, the difference is negligible and can be safely accepted. Large-scale maps below 1:24,000 may be visibly affected. Sometimes it is useful to know when a transformation is being applied, and a user may set Pro to issue warnings when maps contain data in more than one datum.

Labeling coordinate systems

Pro can display data together even when they have different projections. This capability relies on every data set having a coordinate system label, which records the characteristics of the coordinate system, including the datum and projection (Fig. 4.20). This label serves two purposes: it documents this information about the coordinate system, and it helps the GIS display and manage the data. Making sure that all data sets have correct coordinate system definitions is a critical step in building a GIS database. The coordinate system label is stored and copied as part of the data set. A coordinate system may also be stored separately as a file with a .prj extension. These files may be created, saved, copied, and used to define the spatial reference for new data sets.

Data sets may contain unprojected or projected data. If unprojected, the coordinates are stored in decimal degrees of latitude and longitude. The feature class coordinate system label would include simply the GCS (datum), such as GCS North American 1983. The stored x-y coordinates of a data set on the disk would fall in the range −180 to +180 degrees for x and −90 to +90 degrees for y. The map units would be degrees.

Projected data have had a map projection applied to them, converting the GCS latitude-longitude values to feet or meters in a planar coordinate system. The map units become meters or feet. Typically, the x-y values in a projected coordinate system encompass millions or billions of meters. The feature class coordinate system definition includes the GCS/datum plus a description of the projection and its parameters, such as the central meridian, reference latitude, standard parallels, and false easting and northing, as shown in Figure 4.13.

The coordinate system is part of a larger description called the **spatial reference**, which includes the coordinate system, the X/Y Domain, and the resolution. The **X/Y Domain** is the range of allowable x-y values that can be stored in a feature class. Storing a world map in degrees requires a small domain, since degree values range at most between −180 and +180. Storing a world map in feet requires a large domain because there are more than 115 million feet in the circumference of the earth. The **resolution** represents the underlying accuracy of the values; a resolution of 0.001 meter means that coordinate values are stored to the nearest thousandth of a meter.

Spatial Reference	
Projected Coordinate System	North America Equidistant
Projection	Equidistant Conic
WKID	0
Authority	
Linear Unit	Meter (1.0)
False Easting	0.0
False Northing	0.0
Central Meridian	−120.5
Standard Parallel 1	20.0
Standard Parallel 2	60.0
Latitude Of Origin	40.0
Geographic coordinate system	GCS North American 1983
WKID	4269
Authority	EPSG
Angular Unit	Degree (0.0174532925199
Prime Meridian	Greenwich (0.0)
Datum	D North American 1983
Spheroid	GRS 1980
Semimajor Axis	6378137.0
Semiminor Axis	6356752.314140356
Inverse Flattening	298.257222101

Fig. 4.20. The spatial reference
Source: Esri

The **extent** of the data layer is the range of x-y coordinates of the features actually in the feature class. The extent of a data layer is shown in its properties (Fig. 4.21). This information can indicate whether the layer is projected or unprojected. Values between +180 and −180 indicate an unprojected GCS coordinate system with map units in degrees, whereas large values indicate a projected coordinate system with units of feet or meters.

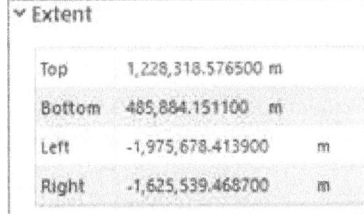

Extent		
Top	1,228,318.576500 m	
Bottom	485,884.151100	m
Left	−1,975,678.413900	m
Right	−1,625,539.468700	m

Fig. 4.21. The extent of the data set in Figure 4.20
Source: Esri

A note on terminology

Many people use the terms *coordinate system* and *projection* interchangeably to refer to the spatial reference (the complete label) of a data set. Using the terms this way is not strictly correct because a coordinate system is only part of the spatial reference, and a projection simply is a mathematical conversion from a 3D to a 2D coordinate system. However, it is a common practice, and usually the intended meaning is clear in context.

Missing coordinate system labels

Occasionally, a user will encounter a GIS feature class or raster without a coordinate system label, and the coordinate system appears as "Unknown". To use the data, the user must determine the actual coordinate system of the data and create a label for it. Finding the coordinate system can be a challenge because it usually cannot be determined by looking at the data set. The user must find documentation supplied by the data provider, such as a web page or document at the site where the data were downloaded or a separate description or document that was supplied with the data. The user might have to contact the data supplier to get this information. As a last resort, the user might have to choose one of the more commonly used coordinate systems and test whether it is the correct one.

Once the coordinate system is known, the user may create the coordinate system label using the **Define Projection tool**. The Define Projection tool labels a data set with a coordinate system and allows the user to specify all the coordinate system properties, including the GCS, datum, central meridian, and so on. Once defined, the data set is ready for use. The same tool is used to define a label for rasters.

Projecting data

Usually, GIS projects choose a single coordinate system and convert all data to match it. An organization will usually establish an official projection and datum for its area of interest and use it consistently. Outside data from different coordinate systems are always converted to the official one at the outset using the **Project tool**.

The Project tool is found under the Data Management section of the Toolbox (Fig. 4.22). It converts the x-y coordinate values in a feature class to a different coordinate system and saves them in a new feature class with a new spatial reference label. The original data remain unchanged. The original data set must have a correctly labeled coordinate system before it is projected. If the label is Unknown, the Project tool cannot work. Rasters have their own set of projection tools.

Troubleshooting coordinate system problems

A coordinate system problem occurs when two layers should map on top of each other but don't. They might be off by a few hundred meters or by thousands of miles. Most of these problems result from mislabeling a data set with the wrong coordinate system. The usual remedy is to determine the correct coordinate system and fix the label.

ArcGIS Pro can be used to troubleshoot such problems. In Figure 4.23, the user is examining the extent and spatial

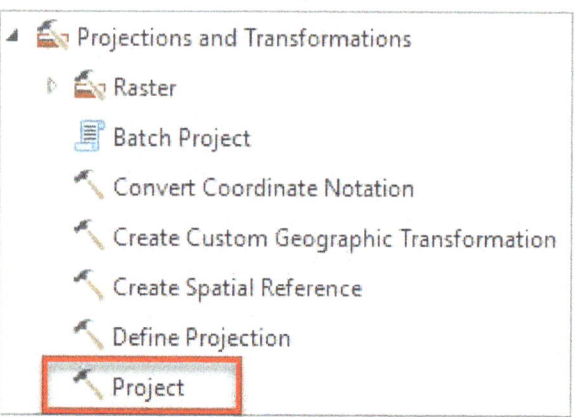

Fig. 4.22. The Projections and Transformations toolset
Source: Esri

Extent		
Top	44.000000	m
Bottom	43.875000	m
Left	-103.500000	m
Right	-103.375000	m

Spatial Reference	
Projected Coordinate System	NAD 1983 UTM Zone 13N
Projection	Transverse Mercator
WKID	26913
Authority	EPSG
Linear Unit	Meter (1.0)

Fig. 4.23. An incorrectly labeled coordinate system
Source: Esri

reference of a layer. The extent shows x-y values in the range −103.375 to −103.500 for x and 44.875 to 44.000 for y. These values are clearly degrees. However, the label underneath lists NAD 1983 UTM Zone 13N as the coordinate system, and the x-y values are listed with meters as the map units. The label does not match the x-y extent values. So which is correct?

The x-y values are always correct because they are obtained from the values stored in the file. This data set really *is* stored in degrees, and the label is incorrect. Imagine adding this feature class to a map set to the UTM coordinate system. Pro reads the label and mistakenly concludes that the x-y values are already in UTM meters and don't need to be projected on the fly. The x-y values are plotted as degrees in the UTM coordinate space, so they don't match the other data and appear in a completely different location.

To fix the problem, one must know the correct coordinate system and use Define Projection to modify the label. Another data set in the same area that is known to be correct will be needed to compare the results and verify that the correct label was assigned. If the data set aligns with the known one, then the label was set correctly. If it is still misaligned, then a different label is needed.

Data in a GCS are relatively easy to correct because the choice of possible datums is limited. Projected data are virtually impossible to guess unless they are in one of the common projections or some outside information is available to narrow the range of possibilities.

Define Projection versus Project

Why would a data set be mislabeled with the wrong coordinate system? The usual culprit is a user who does not understand the tools used to manage coordinate systems and who makes the mistake of using Define Projection when the Project tool is required.

Imagine Frank has a shapefile with GCS coordinates and wants to convert it to a UTM projection to match his other data. He carelessly selects the Define Projection tool and sets the coordinate system to UTM. However, the tool merely puts a UTM label on the data set without changing the coordinates inside. So now the data set is incorrectly labeled UTM and the coordinates inside are still in a GCS. (This situation is analogous to placing an albacore tuna label on a can of cat food, with disastrous results to a dinner recipe.) When the data are added to Pro, it will choose the wrong equations to convert the data to the data frame of the coordinate system, and the feature class will no longer appear where it should.

WARNING: Many people get confused about the functions of the Define Projection tool and the Project tool. This confusion can ruin data sets, so ***pay attention*** to the next two paragraphs, and then be careful to select the correct tool for the task at hand.

The **Project tool** acts on the x-y coordinates of a layer and converts them to a different coordinate system, producing a new feature class and leaving the original feature class unchanged. The new file has a new coordinate system and label. Use the Project tool to convert a layer in one coordinate system to a different one. The Project tool should be used only on layers that already work properly and appear in the right location.

The **Define Projection** tool only changes the coordinate system label of the feature class without affecting the coordinates inside. It should only be used on a data set that has an Unknown coordinate system or on a data set that was previously mislabeled, does not appear in the right location, and needs to be fixed.

Summary

▶ All GIS data have coordinate systems, which define the units and axes used to represent map features as *x-y* coordinates. The complete definition is called the spatial reference.

▶ A geographic coordinate system, or GCS, uses angular measures of latitude and longitude with units of degrees. It is based on an approximation of the earth's shape known as a datum.

▶ A datum includes a spheroid and its location relative to the earth's geoid. It is used to reduce map errors introduced by differences between the spheroid and the earth's actual surface. Common datums in North America include NAD 1927, NAD 1983, and WGS 1984.

▶ Data for a project should be stored in a common GCS. Transforming from one datum to another should be minimized because the process may introduce new positional errors.

▶ Map projections are mathematical equations that convert degrees from a GCS into planar *x-y* coordinates of meters and feet so that the map may be portrayed on a flat sheet of paper. All map projections introduce distortions of area, distance, direction, and shape.

▶ UTM and State Plane are common projection systems in the United States. Each is a family of projections designed to minimize distortion over the area covered (a zone).

▶ Every data set used in ArcGIS should have the correct coordinate system defined and stored with the map features. ArcGIS uses these definitions to display data correctly.

▶ Most coordinate system problems result from missing or incorrect labels on data sets. The Define Projection tool is used to update a data set with an undefined or incorrect label.

▶ Data may be permanently projected from one coordinate system to another using the Project tool or Project Raster tool in ArcToolbox.

TIP: The ArcGIS Help section also has some good material about projections.

Important Terms

central meridian	false northing	projection,	resolution
coordinate pair	geographic coordinate	azimuthal	spatial reference
coordinate space	system (GCS)	conic	spheroid
coordinate system	geoid	cylindrical	standard parallel
datum	latitude	oblique	State Plane
Define Projection	latitude of origin	orthographic	Coordinate System
tool	longitude	secant	transformation
degrees	map units	stereographic	Universal Transverse
display units	origin	tangent	Mercator (UTM)
ellipsoid	parameter	transverse	X/Y Domain
extent	Prime Meridian	reference latitude	
false easting	Project tool		

Chapter Review Questions

1. If a data set's features have x coordinates between -180 and $+180$, what is the coordinate system likely to be? In what units are the coordinates?

2. Explain why the UTM projection would not be a good choice for making a map of the contiguous United States.

3. Explain the difference between a local and an earth-centered datum.

4. Examine Figures 4.8 through 4.10, and explain why conic projections usually conserve area and distance but cylindrical projections typically preserve direction.

5. What is the difference between a central meridian and the Prime Meridian?

6. A shapefile has an Unknown coordinate system, but a file on the web site says that the coordinate system is UTM Zone 13 NAD 1983. What is the next step?

7. A shapefile has a UTM Zone 10 NAD 1983 coordinate system and must be brought into a state database that uses the Oregon Statewide Lambert coordinate system. What is the next step?

8. You need to create a map for the entire state of Idaho. What coordinate system options are there? Which one would yield the most accurate map?

9. Examine the UTM and State Plane Zones for California. Explain why creating an accurate map projection for the whole state would be a challenge. Discuss the options; then select a projection and define a set of reasonable parameters.

10. Using a web browser, find at least three world maps that clearly use different projections. **Capture** a screen shot of each one. Underneath, discuss the likely type of projection (cylindrical, conic, or azimuthal) that each is based upon.

Expand your knowledge

Expand your knowledge of ArcGIS Pro by reading these sections of the Help. Use the breadcrumb trail to navigate to each topic, or search for a term in the section title to find it.

Help > Maps > Properties of maps > Coordinate Systems (all sections)

These articles are readily available and contain good discussions and examples of projections.

Synder, John P. and Philip M. Voxland, 1989, An Album of Map Projections, US Geological Survey Professional Paper 1453, 249 p., URL: https://pubs.usgs.gov/pp/1453/report.pdf

Wikipedia article on Map Projection, URL: https://en.wikipedia.org/wiki/Map_projection

Mastering the Skills

Teaching Tutorial

The following examples provide step-by-step instructions for doing basic tasks and solving basic problems in ArcGIS. The steps you need to do are highlighted with an arrow →; follow them carefully.

Map coordinate systems

For this lesson, we will create a new project that will focus on water quality issues in the Black Hills of South Dakota. To keep things organized, we will create and save the project in the ClassProjects folder of the book data in gisclass.

> 1→ Start ArcGIS Pro. Under *Create a new project,* click on Blank.
>
> 1→ Name it *BHWaterStudy* and click the Browse button for the Location. Save it in the gisclass\ClassProjects folder.
>
> 1→ Keep the *Create a new folder for this project* box checked and click OK.

A blank project will usually open with the Catalog view visible in the display area. Like the Catalog pane, the Catalog view shows the items associated with the project, but it has additional capabilities, such as viewing or editing metadata.

> 2→ Examine the **View: Windows** ribbon and notice the Catalog pane and the Catalog view side by side. These buttons can reopen either one if it gets closed.

To complete this project (and many others), we will need access to data in the mgisdata folder. Since this is a new project, we must add a folder connection to be able to access it. You added the mgisdata folder to the Favorites tab in the Chapter 1 tutorial.

> 2→ In the Catalog pane, switch to the Favorites tab. Right-click the mgisdata folder and choose Add To Project. (If it's not in the Favorites tab, choose **Insert: Project: Add Folder,** navigate into the gisclass folder, select the mgisdata folder, and click OK.)
>
> 2→ In the Catalog pane, switch back to the Project tab. Expand the Folders entry to verify that the mgisdata folder is now listed as a data source for the project.

Before we start to compile data, we will spend some time learning about coordinate systems. Every data set is stored in a specific coordinate system and is labeled with the properties of that coordinate system (the spatial reference). However, maps can be portrayed in any coordinate system using on the fly projection.

> 3→ Click **Insert: Project: New Map** to create a new map for this empty project.
>
> 3→ In Contents, rename the map *World.*
>
> 3→ In Contents, right-click the World map icon and choose Properties (or double-click the icon). This action opens the map properties for World.
>
> 3→ On the left, click on the Coordinate Systems entry and examine the Current XY coordinate system in the blue-highlighted box.

Initially, the map coordinate system is set to WGS 1984 Web Mercator (auxiliary sphere), the preferred coordinate system for data in ArcGIS Online. Not all data in ArcGIS Online are in Web Mercator, but many are, including all the basemaps.

> 4→ Close the Map Properties window by clicking Cancel.
>
> 4→ Click the **Map: Layer: Add Data** button or use the Catalog pane to add the country data set from the mgisdata\World folder
>
> 4→ In Contents, double-click the World map icon to open the map properties again.
>
> 4→ Click the Coordinate Systems entry and examine the map coordinate system again.

1. What is the map coordinate system (CS) now? _____

When the first data set is added to a map, the map coordinate system defaults to the stored coordinate system of the new data. It will remain this way unless it is set to a different one (it does not update when every data set is added, only the first one). WGS 1984 is an example of a geographic coordinate system (GCS) using degrees of longitude and latitude.

4→ Examine the lower box, labeled XY Coordinate Systems Available.

4→ Scroll the box up and down to see all the entries.

This box sets the coordinate system for the map. Notice the Layers heading, which is currently expanded and shows the coordinate systems of all layers in the map: GCS WGS 1984 from the country shapefile just added and Web Mercator from the Topographic basemap. A Favorites folder saves frequently used coordinate systems. The other two folders store pre-configured geographic and projected coordinate systems. We will pick a different CS in a moment.

5→ Close the Map Properties window by clicking Cancel.

5→ In Contents, right-click the country layer name and choose Properties to open the layer properties (or double-click the layer name).

5→ Click the Source entry and examine the information on the right.

5→ Expand the Extent heading and examine the information.

5→ Expand the Spatial Reference heading and examine the coordinate system information.

The spatial reference, or coordinate system, of this data set is GCS WGS 1984. The extent is the range of coordinates present in the file, stored in decimal degrees (dd).

6→ Click Cancel to close the Map Properties window.

6→ Move the mouse cursor over the map and watch the coordinate values in the lower bar of the map view update the x-y location in degrees.

Recall that the Prime Meridian (0° longitude) runs through Greenwich, England. Locations west of this line have negative longitudes, and locations east of it are positive. Locations north of the equator have positive latitudes, and those south of the equator have negative latitudes. However, the X,Y Coordinate box usually shows E/W and N/S tags; W and S indicate negative values.

6→ Zoom in to the tip of Florida in the United States.

6→ Use the cursor to hover over the southeastern tip of Florida and observe the x-y coordinates (Fig. 4.24).

2. What are the coordinates of the SE tip of Florida? _____

6→ Open the World map properties and click the General settings.

3. What are the map units of this frame? _____ What are the display units? _____

The map units can't be set explicitly because they are determined by the map's coordinate system. However, the user can choose the units displayed in in the X,Y Coordinate box.

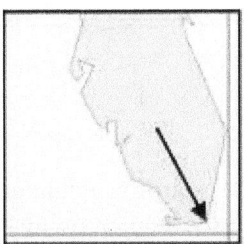

6→ Change the Display units to Degrees Minutes Seconds and click OK.

Fig. 4.24. Hover here.
Source: Esri

7→ Click the **Map: Navigate: Full Extent** button to see the whole world again.

7→ Open the World map properties and click the Coordinate Systems entry.

7→ In the folder tree on the right, collapse the Layers folder.

7→ Expand the Projected coordinate system folder and examine its contents.

7→ Expand the World folder. Click on the Mercator (world) projection.

7→ In the upper right corner, click the Details link to learn more about this Mercator coordinate system. Examine the parameters.

Coordinate System Details	
Projected Coordinate System	World Mercator
Projection 1	Mercator
WKID	54004
Authority	Esri
Linear Unit	Meter (1.0)
False Easting	0.0
False Northing	0.0
Central Meridian	0.0
Standard Parallel 1	0.0
Geographic coordinate system 2	GCS WGS 1984
WKID	4326
Authority	EPSG
Angular Unit	Degree (0.01745
Prime Meridian	Greenwich (0.0)
Datum	D WGS 1984
Spheroid	WGS 1984
Semimajor Axis	6378137.0
Semiminor Axis	6356752.314245

Fig. 4.25. Details of the World Mercator projection
Source: Esri

Notice that a projected coordinate system has two parts (Fig. 4.25): (1) the projection and its parameters, and (2) the underlying geographic coordinate system (GCS) used to approximate the earth's shape in 3D before it is projected to 2D. This World Mercator coordinate system uses GCS WGS 1984 for its GCS, which is also known as the datum.

4. What are the map units for this Mercator projection? _____

7→ Close the Coordinate System Details window.

7→ Click OK to accept Mercator (World) as the coordinate system for the map.

TIP: Changing the coordinate system of the map only changes how it is displayed. It does not affect the coordinate systems used to store the feature classes.

8→ Use the Add Data button or the Catalog pane to add the mgisdata\World\latlong data set to the map.

8→ Open the World map properties and examine the General settings.

8→ Change the display units to meters and click OK.

8→ Use the cursor and coordinate box to locate the approximate coordinate origin, where both x and y equal zero (look for where the coordinates change from W to E and N to S).

Mercator is a cylindrical projection. The central meridian where $x = 0$ corresponds with the Prime Meridian that runs through Greenwich, England. The $y = 0$ line is the equator and is called the latitude of origin. Coordinates west of the central meridian and south of the reference latitude have negative values. Florida is west of the central meridian; its x coordinates are negative.

5. Which continent has primarily negative x AND negative y coordinates in this projection? _____ Which one has primarily positive x and y coordinates? _____

    9→ Add the states feature class from the mgisdata\Usa\usdata geodatabase.

    9→ Open the World map properties again and view the Coordinate Systems settings.

    9→ Find and expand the Layers folder again.

    9→ Expand each of the three coordinate system entries in the Layers folder.

This Layers folder is a handy place to review the coordinate systems for all of the layers in a map. Currently, the countries and latlong layers are in a GCS, the states are in a conic projection, and the basemap is in Web Mercator. Yet all of them are being displayed in the Mercator (World) projection set for the map. We can set the map coordinate system to match any of the layers.

    9→ In the Layers folder, click on the North American Equidistant Conic coordinate system to highlight it.

    9→ Click the Details link in the upper right corner and examine the map parameters. Notice that this coordinate system has a different underlying GCS (NAD 1983) than the other layers (WGS 1984).

TIP: Keep in mind that the *map* coordinate systems in Pro may be different from the coordinate systems in which the data sets are stored.

6. What longitude is the central meridian? _____ What is the latitude of origin? _____ What are the standard parallels? _____ and_____.

    9→ Close the Details window and click OK to accept the coordinate system.

The world map looks bizarre because this coordinate system is designed for North America. It is easy to tell that it is a conic projection in this view. Notice that North America is in the center because the central meridian of the map lies in North America.

    10→ Zoom in to North America (including Central America).

    10→ In Contents, select the latlong layer and open the **Feature Layer: Labeling** ribbon.

    10→ In the **Label Class** group, set the **Field** to Value. Click the **Label** button to turn on the labels.

    10→ Find the central meridian, the standard parallels, and the latitude of origin on the map.

We observed that the states data set uses the NAD 1983 datum, whereas the map and the other data sets use the WGS 1984 datum. The map is transforming states on the fly to match the WGS 1984 datum. At this map scale, the conversion errors are negligible, but at large scales the differences may become significant. Pro can issue a warning when data sets with different datums are placed in the same map.

    11→ Choose **Project: Options** and view the Map and Scene settings.

    11→ Expand the Spatial Reference heading.

    11→ Scroll down and check the box labeled *Warn if transformation between geographic coordinate systems is required to align data sources correctly.*

    11→ Click OK, and click the back arrow to return to the project.

This setting will apply to new maps, as we will discover later.

Understanding map distortion

The circumference of the world at the equator is about 25,000 miles. The circumference at the poles is zero. Latitudes in between have intermediate lengths. Let's look again at the WGS 1984 GCS.

12→ Open the World map properties and view the Coordinate Systems settings.

12→ Find and expand the Layers folder.

12→ Click the GCS WGS 1984 entry to set the data frame to that CS and click OK.

12→ Click the **Map: Navigate: Full Extent** button to show the whole world again.

12→ In Contents, right-click the latlong layer and choose Label to turn the labels off.

When a GCS is displayed, the angular units are drawn in an equirectangular projection, as if the degrees were a planar x-y coordinate system from $x = -180$ to $+180$ and $y = -90$ to $+90$. East-west distances are increasingly distorted away from the equator.

13→ Add the circles shapefile from the mgisdata\World folder.

13→ In Contents, drag the circles above the latlong layer in the Contents window.

Each circle has a radius of 5 degrees, and they would appear identical on a globe, but here the distance distortion of the GCS map is clearly shown. The areas and distances, if determined for these circles, would be incorrect over most of the globe.

TIP: Any GIS analysis based on distance or area measurements would be wrong when data are stored in a GCS. It is best to choose appropriate projected coordinate systems to store data.

14→ Open the map properties of the World map and select the Coordinate Systems settings.

14→ Type Web Mercator in the Search box and click Enter.

14→ Expand the Projected coordinate system and World folders.

14→ Select the WGS 1984 Web Mercator (auxiliary sphere) projection and click OK.

This coordinate system is preferred for data in ArcGIS Online. Area and distance are both distorted in this projection, especially at high latitudes. However, Mercator does preserve direction and shape. The circles remain circles, and the longitude meridians point north. Because it preserves direction and shape, it is a good choice for viewing data all over the world. However, it is NOT suitable for conducting spatial analysis involving areas or distances.

TIP: Projecting rasters on the fly is slower than projecting vector data. If using basemaps from ArcGIS Online, drawing performance can be improved by setting the map's coordinate system to WGS 1984 Web Mercator (auxiliary sphere).

15→ Use the map properties and the search box to change the coordinate system of the map to Robinson (world).

Robinson is called a compromise projection because it minimizes distortion in all four properties but preserves none. Distortion is unavoidable when using small-scale world and national maps.

16→ Change the map coordinate system to Sinusoidal (world).

Sinusoidal is an equal-area projection, meaning that it preserves area and, to some extent, distance. However, it does so at the expense of shape and direction. The longitude lines converge toward the poles, and the circles have equivalent areas now, but most are no longer circles. The distortion gets worse away from the equator and the central meridian.

17→ Change the map coordinate system to Africa Sinusoidal.

17→ Zoom in to Africa.

Although globally the sinusoidal projection shows great distortion, when customized for the region of Africa it fairly well preserves area and distance, although a direction distortion is apparent. For even smaller areas, Universal Transverse Mercator (UTM) is a good choice.

18→ Change the map coordinate system to WGS 1984 UTM Zone 35S. (**Hint:** This one is more difficult to find. Search for UTM and then expand the UTM, WGS84, and Southern Hemisphere folders to find the right one.)

18→ Add the utmzone35 feature class from the mgisdata\World folder.

The narrow strip of UTM Zone 35 runs along the tangent of a transverse cylindrical projection. Within the zone, the circles remain circles, and all four properties of area, distance, shape, and direction are well preserved. Areas above the equator would use Zone 35N.

Converting coordinate systems when compiling data

When compiling data for a project, it is important to ensure that all data are converted to the same suitable coordinate system. Imagine compiling data for a study of groundwater quality in two quadrangles in the Black Hills of South Dakota. A colleague has downloaded the information; it is your task to compile it into a geodatabase. Let's review this data, which is stored in the mgisdata\BlackHills\Downloads folder.

19→ Choose the Catalog tab above the map to switch to the Catalog view, a more versatile version of the Catalog pane.

19→ In the Catalog view, click the Folders entry. Navigate into the mgisdata\BlackHills \Downloads folder.

19→ Click on the first entry, benchn27_hc.shp. The metadata will appear on the right.

19→ Take a few minutes to peruse the metadata for each data set in the folder.

The home geodatabase of this project, BHWaterStudy, will provide the repository for the data, and the project lead has elected to use NAD 1983 UTM Zone 13N as the projection. All the data must be stored in this projection (not just displayed in the map, but actually converted).

20→ Save the project and close the World map.

20→ Use **Insert: Project: New Map** to open a new map.

20→ In Contents, rename the map Study Quads.

20→ Click **Map: Layer: Add Data**. Navigate to the mgisdata\BlackHills\Downloads folder and use Shift-click to add all of the feature classes to the map.

Remember setting the Project options to warn the user if different datums are used in the same map? A Transformation Warning window has appeared as the result of that setting (Fig. 4.26), warning the user

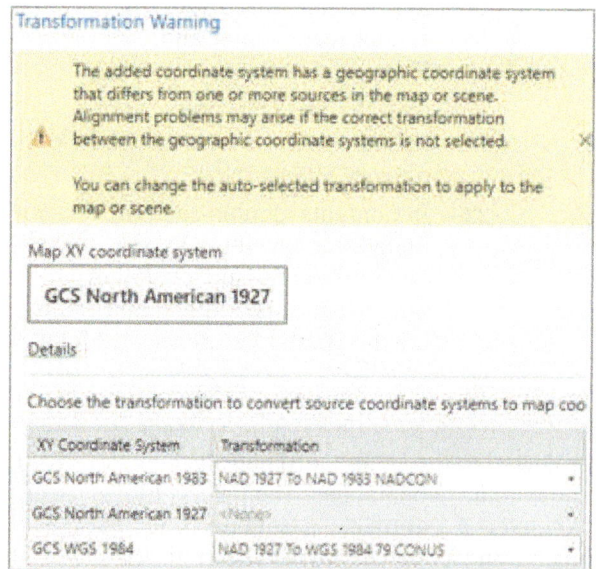

Fig. 4.26. The Transformation Warning
Source: Esri

that a default transformation will be applied. An experienced user might choose an alternate transformation method, but for a beginner, the default is the best choice.

20→ Click OK in the warning window.

20→ Zoom to the extent of the StudyBoundary data set.

It may appear that all the data sets are in the same coordinate system because they all draw together on the map, but this is not true. Remember, on-the-fly projection plots everything in the map coordinate system, even when the stored coordinate system is different.

21→ Open the map properties for Study Quads and examine the Coordinate Systems settings.

21→ Find and expand the Layers folder.

Seven coordinate systems, including three different datums, are listed (Fig. 4.27). The task is to save each feature class to the home geodatabase in NAD 1983 UTM Zone 13. Tackling them one at a time is best.

21→ Click Cancel to close the Map Properties window.

22→ In Contents, turn off all layers except Topographic and StudyBoundary.

22→ Give the StudyBoundary layer a hollow symbol with a thick black outline.

22→ In Contents, double-click the StudyBoundary layer to open the layer properties and click the Source settings. Expand and examine the Spatial Reference carefully.

The StudyBoundary feature class is stored in a GCS with an NAD 1983 datum. It needs to be projected to the UTM coordinate system and saved in the home geodatabase. The Project tool is used to convert a feature class from one CS to another by creating a copy in the new CS (Fig. 4.28).

23→ Close the layer properties window.

23→ In the Geoprocessing pane, search for *project*.

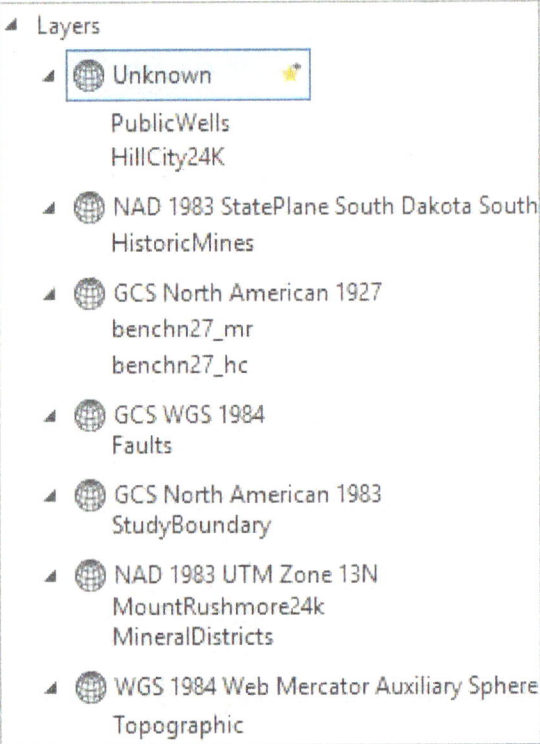

Fig. 4.27. The data are stored in seven different coordinate systems.
Source: Esri

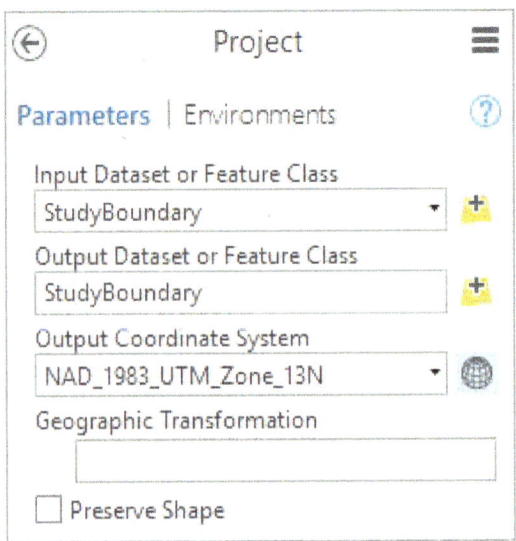

Fig. 4.28. Projecting a feature class to a new coordinate system
Source: Esri

23→ Click the Project tool (NOT the Define Projection tool) to open it.

23→ Set the Input Dataset to StudyBoundary.

23→ Name the Output Dataset StudyBoundary and make sure it is being saved in the home geodatabase BHWaterStudy.

7. Will the new StudyBoundary data set overwrite the old one, since they share the same name? Why or why not?

24→ Click the globe icon to choose an Output Coordinate System. Type UTM in the search box and click Enter. Expand the Projected Coordinate System, UTM, and NAD 1983 entries.

24→ Find NAD 1983 UTM Zone 13N and click to highlight it.

24→ Click the tiny star icon next to the name to add this coordinate system to the Favorites list to make it easier to find again.

24→ Click OK to accept the coordinate system, and double-check that the Output Coordinate System in the tool reads NAD_1983_UTM_Zone_13N.

24→ Click Run to execute the tool.

A second StudyBoundary feature class is added to the Contents window, most likely above the first one. Having two with the same name might get confusing, so let's change the Contents panel to show the source of each data set. We can remove the original and keep the new, projected one.

25→ In Contents, click the second view icon to open the List By Data Source panel.

25→ In Contents, find the StudyBoundary layer stored in the home geodatabase (it will be by itself, since it is the first data set we have saved there).

25→ Find the original StudyBoundary layer saved in the Downloads folder. (Hover over the database entry to see its full name and location.)

25→ In Contents, right-click the StudyBoundary layer from the Downloads folder and choose Remove.

25→ Leave the Contents pane in this panel (List By Data Source) to help keep track of which data sets come from which folder.

Notice that the shape of the study boundary did not change when we projected it. The map coordinate system is still set, and the new data sets will continue to be shown in the map coordinate system unless we modify it. Let's change it to our final desired coordinate system.

26→ Open the map properties for the Study Quads map and view the Coordinate Systems settings.

26→ Scroll to the top of the list and expand the Favorites folder.

26→ Click the NAD 1983 UTM Zone 13N entry to select it, and click OK.

Let's do another one. The instructions are abbreviated, so review the previous steps if needed.

27→ Turn on the HistoricMines layer.

27→ Open the HistoricMines layer properties and examine the Spatial Reference in the Source settings. Close the properties when finished.

8. What coordinate system is HistoricMines stored in? _____

28→ In the Geoprocessing pane, set or change the Input Dataset to HistoricMines (if the tool was left open, this next step is streamlined).

28→ Set or change the Output Dataset to HistoricMines and make sure it is being saved in the home geodatabase. (Click Tab to accept the entry and get rid of the warning.)

28→ Set or leave the Output Coordinate System set to NAD_1983_UTM_Zone_13N.

28→ Click Run.

28→ When the tool finishes, remove the original HistoricMines in the Downloads folder and keep the new one in the home geodatabase.

TIP: By default, all output feature classes and tables are saved in the home geodatabase, which serves an important function of keeping the analysis results and data created during a project in one convenient location. ***Always save output to the home geodatabase unless specified otherwise in the instructions.***

29→ Turn on MineralDistricts, open its layer properties, and examine the Spatial Reference. Close the layer properties when finished.

This data set is already in the desired coordinate system, so all we need to do is copy it to the home geodatabase.

30→ In the Geoprocessing pane, click the circled back arrow to return to the main Geoprocessing pane.

30→ Search for and open the Copy Features tool.

30→ Set the Input Features to MineralDistricts.

30→ Also name the Output Feature Class MineralDistricts. Run the tool.

30→ Remove the original MineralDistricts layer from the map.

30→ Save the project before going on.

Understanding transformations

We are making good progress, but the next few data sets are stored in a different datum (GCS) than the desired coordinate system.

31→ Open the map properties for the Study Quads map and examine the Coordinate Systems settings.

31→ Expand the Layers folder to examine the coordinate systems present in the map.

Notice that three different datums are represented here: NAD 1983, North American 1927, and WGS 1984. When projecting to a new coordinate system with a different datum, the user must also specify a **transformation**, or a conversion to a new datum.

32→ Close the map properties, turn on the Faults layer, and examine its spatial reference in the layer properties.

32→ Open the Project tool in the Geoprocessing pane.

32→ Set the Input Dataset to Faults and save the output as Faults.

32→ Select the Output Coordinate System NAD_1983_UTM_Zone_13N, as before.

Notice that a transformation automatically appears when one specifies an output coordinate system in a different GCS than the input coordinate system. Choosing the best transformation method is beyond the scope of this book, but the top entry recommended by the software is usually the best one.

32→ Carefully click the drop-down arrow next to the listed transformation and examine the other choices, but do not change it.

32→ Run the tool and remove the original Faults layer from the Contents window.

Since we are changing datums now, it is best to close (click the back arrow) on the Project tool to clear it each time. To choose the output coordinate system easily, click the drop-down and choose Current Map [Study Quads], since we know that we have set it to the desired CS.

33→ Run the Project tool to project the benchn27_hc feature class, including the datum transformation from NAD 1927 to NAD 1983. Name the output *bench_hc.*

33→ Remove the original benchn27_hc feature class from the map.

34→ Run the Project tool to project the benchn27_mr feature class, including the datum transformation from NAD 1927 to NAD 1983. Name the output *bench_mr.*

34→ Remove the original benchn27_mr feature class from the map.

TIP: Turn off the Transformation Warning now, if desired, in the Project > Options > Map and Scene settings. However, leaving it on is a good way to cultivate awareness of datum transformations and potential issues with them.

Troubleshooting projection problems

Sometimes the spatial reference label of a data set is missing or has been compromised by improper practices. In some cases, the data set can be salvaged if the correct coordinate system of the data can be identified and the label corrected. This section shows how.

35→ Save the project.

35→ Turn on the PublicWells layer and examine its spatial reference.

35→ In the Geoprocessing pane, clear and reopen the Project tool.

35→ Set the Input Dataset to PublicWells and click Tab. A red X appears.

According to the Spatial Reference, this data set has an unknown coordinate system. As a result, it cannot be projected yet. The tool must know what coordinate system to start with before it can project a data set to a new one. You need to figure out what the coordinate system should be and set the label to match.

35→ Turn off all layers except StudyBoundary and PublicWells.

35→ Examine the map. The public wells all appear inside the study area.

The fact that the wells appear in the right place on the map is an important clue. When the label is missing, Pro cannot project the data set on the fly, so it simply plots whatever *x-y* values are stored in the data set. Since they appear correctly, the actual coordinate system of the wells is the same as the map coordinate system, NAD 1983 UTM Zone 13N. Let's demonstrate this.

36→ Open the map properties for Study Quads and view the Coordinate Systems settings.

36→ Change the map coordinate system to WGS 1984 Web Mercator (auxiliary sphere). (It is easier to expand the Layers folder and select it there, instead of searching.)

The public wells are no longer visible in the map. Where did they go?

36→ In Contents, right-click the PublicWells layer and choose Zoom to Layer.

36→ Turn on the Topographic basemap and zoom out until the world location can be identified.

We must give PublicWells a spatial reference to make it a complete data set that will function properly in all maps. Use the Define Projection tool to assign a spatial reference label to a data set without modifying the stored coordinates.

37→ In the Geoprocessing pane, open the Define Projection tool.

37→ Set the Input Dataset to PublicWells.

37→ Set the Coordinate System to NAD 1983 UTM Zone 13N (pick it from the *Favorites* folder) and click Run.

37→ Zoom to the PublicWells layer again; they are now in the right place.

When the proper coordinate system has been set, the wells are plotted in the right location regardless of the map coordinate system. The wells are already in the desired coordinate system, so they just need to be copied to the home geodatabase.

38→ Set the map coordinate system back to NAD 1983 UTM Zone 13N (use the entry in the Favorites folder).

38→ In the Geoprocessing pane, open the Copy Features tool and copy PublicWells to the home geodatabase using the same name.

38→ Remove the original PublicWells from the map.

39→ Turn on the HillCity24K layer. It does not appear anywhere on the map.

39→ Zoom to the HillCity24K layer and zoom out (way out!) until its world location is apparent. Then zoom back to the StudyBoundary layer.

39→ Open the layer properties for HillCity24K and examine the spatial reference.

This feature class also has an unknown coordinate system, but in this case it did not plot in the right location, which reveals that its coordinate system does NOT match the map's coordinate system. The actual coordinate system needs to be identified, and the **extent**, or range of x-y values stored in the data, sometimes provides a clue.

39→ Expand the Extent heading in the layer properties.

The Extent shows the range of x-y coordinates as small numbers, less than 180, which is a reliable indication that the coordinate system is in degrees. So a GCS label must be assigned, but which datum? There is no way to know for certain, but NAD83 is the most common datum for US data, so try it first.

40→ In the Geoprocessing pane, open the Define Projection tool.

40→ Set the Input Dataset to HillCity24K (Fig. 4.29).

40→ Set the Output Coordinate system to NAD 1983 (search for NAD, then expand Geographic Coordinate System and North America). Select the plain NAD 1983 choice. Click the star to add it to your Favorites folder.

40→ Run the tool.

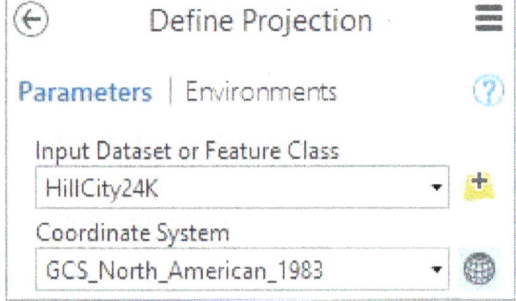

Fig. 4.29. Labeling a data set with a coordinate system

Source: Esri

TIP: Unless specifically told otherwise, always use the basic datum, NAD 1983, and not one of the specialized ones like (2011) or (CORS96) or (HARN). If you picked one of these by mistake, click the button again, select the correct one, and run the tool again.

41→ Datums typically differ by a few hundred meters or less, so zoom way in to the corner of the study area and confirm that the HillCity24K edge aligns closely with the StudyBoundary area edge.

41→ Zoom back to the StudyBoundary layer.

Now the HillCity24K feature class has a correct spatial reference label, but it still must be projected and saved to the home geodatabase.

42→ In the Geoprocessing pane, open the Project tool.

42→ Project the HillCity24K feature class to the NAD 1983 UTM Zone 13N coordinate system and save it in the home geodatabase using the same name.

42→ Remove the original HillCity24K feature class.

43→ Turn on the MountRushmore24k layer. It does not appear on the map.

43→ Check where it plots, and then return to the study area.

43→ Open the layer properties for the MountRushmore24k layer and examine the spatial reference and the extent.

The spatial reference indicates that the coordinate system is NAD 1983 UTM Zone 13N, but it is still not plotting in the right place, which is odd. However, examine the extent carefully and notice that the x-y values are small numbers again, indicating that the data are in a geographic coordinate system. The spatial reference says one thing, but the extent says another. The extent is always correct; therefore, this data set is really stored in a GCS, and the spatial reference label is incorrect.

TIP: The Extent x-y values show the values actually in the file and are the best guide to the true coordinate system.

This example shows what happens when someone uses Define Projection when they should have used Project. This data set has been mislabeled with the wrong coordinate system, and thus it is not plotted correctly on the map. To fix it, set the label to GCS NAD 1983.

44→ Use the Define Projection tool to change the label of MountRushmore24k to the GCS North American 1983 coordinate system. It then appears in the right place.

45→ Use the Project tool to project MountRushmore24k to the NAD 1983 UTM Zone 13N coordinate system and save it in the home geodatabase with the same name.

45→ Remove the original layer from the map.

The four remaining layers from the Downloads folder will be used in the next chapter.

45→ Remove MtRushmoreStructure, HillCityStructure, MtRushmoreGeology, and HillCityGeology from the map.

It is time to take stock and check your work. At this point, the map should contain only the Topographic basemap and layers from the home geodatabase. Moreover, all layers except the basemap should be in the NAD 1983 UTM Zone 13N coordinate system.

46→ Make sure that the Contents pane is set to the List by Data Source panel.

46→ Confirm that all of the layers except the basemap are sourced from the gisclass \ClassProjects\BHWaterStudy\BHWaterStudy home geodatabase.

46→ Open the map properties for the Study Quads map and examine the Coordinate Systems settings. Find and expand the Layers entry.

46→ Confirm that only two coordinate systems are listed, NAD 1983 UTM Zone 13N and WGS 1984 Web Mercator Auxiliary Sphere (from the basemap), as in Figure 4.30.

TIP: If more than these two coordinate systems are listed, you made an error. Return to the relevant steps and figure out what went wrong so you can avoid that error in the future. However, you do not need to fix the problems to finish the tutorial.

47→ The Catalog view tab should still be visible in the display area. (If the tab is not there, **Choose View: Windows: Catalog view.**)

47→ Navigate to the home geodatabase in gisclass\Class Projects\BHWaterStudy.

47→ Click on each of the new, projected data sets to examine it. Note that the metadata was copied to the new files.

47→ Save the project.

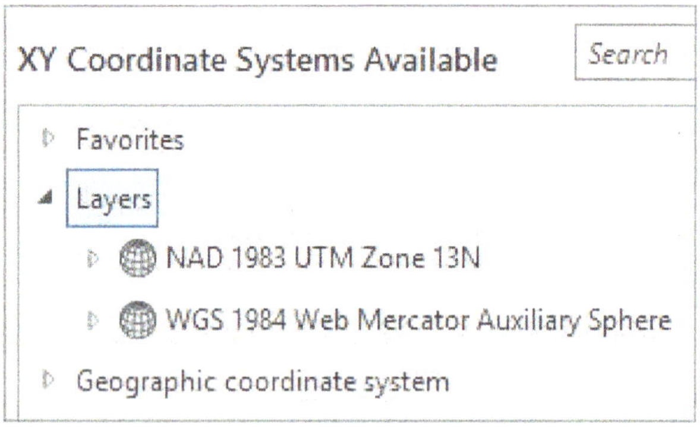

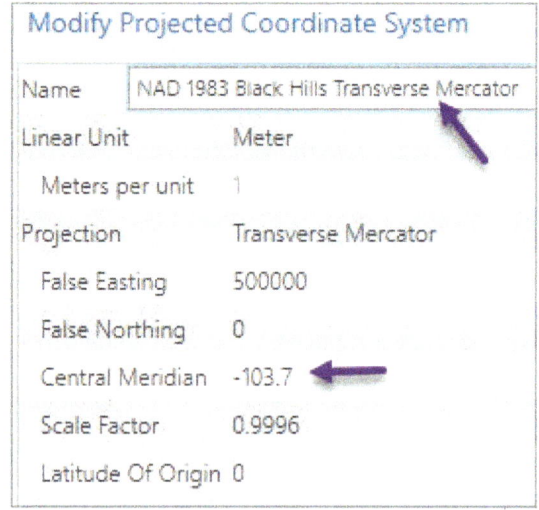

Fig. 4.30. The map properties should show only two coordinate systems in the Layers list.
Source: Esri

Creating a custom coordinate system

Sometimes a predefined coordinate system is not available or does not have the desired accuracy for an area, requiring a custom coordinate system to be defined. Suppose a client in the Black Hills is dissatisfied with UTM Zone 13N because the area is on the edge of the zone, and maps show slight distortions. A custom projection can be defined based on UTM Zone 13N but with a modified central meridian to place the Black Hills at the center of the projection.

48→ Open a new map and rename it *New CS.*

48→ Open the map properties and view the Coordinate Systems entry.

48→ Find the NAD 1983 UTM Zone 13N coordinate system. Right-click it and choose Copy and Modify.

48→ Rename the coordinate system *NAD 1983 Black Hills Transverse Mercator* (Fig. 4.31).

48→ Change the central meridian from −105 to −103.7 and click Save.

Fig. 4.31. Modifying a coordinate system to create a custom one
Source: Esri

Now the map is set to the custom coordinate system, but it would be helpful to save it as a file to use it for other maps or data sets. By default it will be saved in the user's Favorites folder; if others need to use it, then it should be saved in a shared location.

49→ Right-click the new coordinate system and choose Save As Projection File.

49→ Type *NAD 1983 Black Hills Transverse Mercator* for the file name. Note the folder it will be saved to and then click Save.

49→ Click OK to accept the new coordinate system for the map.

For a map of Turkey, one finds that no predefined coordinate system is available. In this case, a new custom coordinate system can be created. Because it is a large area oriented east-west, a secant conic projection would be most suitable. The central meridian should be placed in the center of Turkey, and the standard parallels should divide the top and bottom sixths from the middle.

50→ Create a new map and rename it *Turkey*.

50→ Zoom to Turkey in the map.

50→ Find the approximate central longitude and range of latitudes of Turkey by using the cursor and reading the values from the cursor XY display box.

Central longitude _____
Southern latitude _____
Northern latitude_____

Armed with this information, we can define a custom coordinate system based on the Lambert Conformal Conic projection.

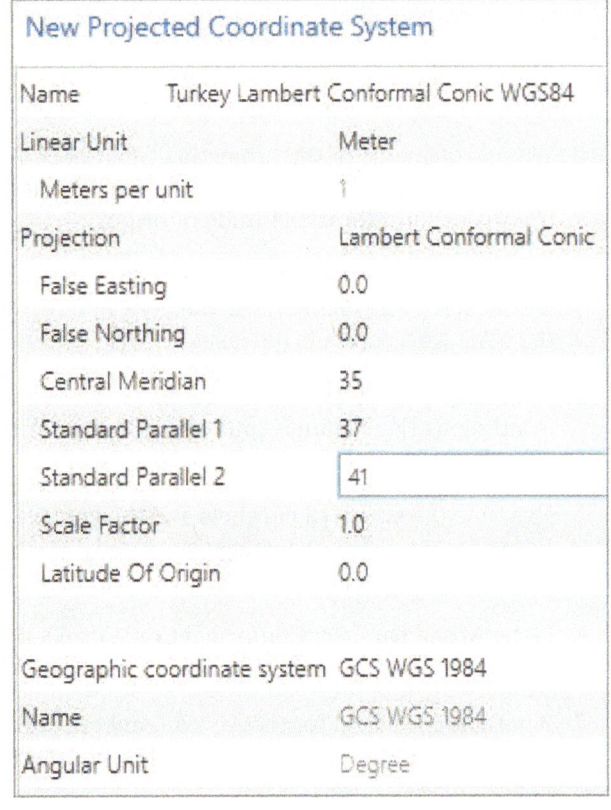

New Projected Coordinate System	
Name	Turkey Lambert Conformal Conic WGS84
Linear Unit	Meter
Meters per unit	1
Projection	Lambert Conformal Conic
False Easting	0.0
False Northing	0.0
Central Meridian	35
Standard Parallel 1	37
Standard Parallel 2	41
Scale Factor	1.0
Latitude Of Origin	0.0
Geographic coordinate system	GCS WGS 1984
Name	GCS WGS 1984
Angular Unit	Degree

Fig. 4.32. Creating a custom projection for Turkey
Source: Esri

51→ Open the Turkey map properties and view the Coordinate Systems settings.

51→ Click the Add Coordinate System drop-down and choose New Projected Coordinate System.

51→ Enter *Turkey Lambert Conformal Conic WGS84* as the name (Fig. 4.32).

51→ For the Projection, choose Lambert Conformal Conic from the drop-down list.

51→ Carefully change the Central Meridian to 35 degrees.

51→ Set Standard Parallel 1 to 37 degrees. Set Standard Parallel 2 to 41 degrees.

51→ Leave the geographic coordinate system as WGS 1984, a good choice. Click Save.

52→ Right-click the new coordinate system and choose Save As Projection File. Retype the name and save it in the Favorites folder.

52→ Click OK in the map properties and view the map.

TIP: To use the saved projection file, click the Add Coordinate System drop-down and choose Import. Open it in Notepad if you want to see what it looks like—it is a text file.

This is the end of the tutorial.

→ Save the project and close ArcGIS Pro.

Practice Exercises

1. Examine the coordinate system for the streets feature class in the Austin geodatabase. What is the name of the coordinate system? Is it projected or unprojected? What are the map units?

2. What is the name of the projection (not the coordinate system) used by the feature classes in the Oregon geodatabase? What are the central meridian and the standard parallel(s)? Does it use the equator for the latitude of origin?

3. Determine the coordinates of the summit of Mount Rainier, Washington in (a) degrees-minutes-seconds, (b) UTM Zone 10 (in meters), and (c) Washington State Plane South (in US feet). Use the NAD 1983 datum in all cases.

4. Compare the projection parameters of the North America Equidistant Conic and the USA Contiguous Equidistant Conic predefined coordinate systems. Explain the reason for the differences.

5. Examine the standard parallels and the latitude of origin of the Africa Lambert Conformal Conic predefined coordinate system and state them. Predict whether any areas of Africa have negative y coordinates in this projection. Why or why not?

6. Is the Africa Lambert Conformal Conic projection a tangent or a secant projection? How can you tell?

7. What datum is used for the Africa Lambert Conformal Conic coordinate system? Why is it different from the one used for the North America Equidistant Conic coordinate system?

8. Create a new map and add the utmzone.shp data set from the mgisdata\World folder and the spcszn83 feature class from the mgisdata\Usa\usdata geodatabase. Examine these locations relative to the UTM and State Plane zone boundaries. Select which zone would be a better (less distorted) choice for the area, and explain why.

   Amarillo, Texas

   Pensacola, Florida

   State of Tennessee

9. You are working on a statewide Mississippi project and decide to define a custom coordinate system. Start with one of the State Plane zones and modify it slightly to make it better for the whole state. Explain the approach, and **Capture** the window showing the custom coordinate system description created.

10. The Austin folder contains two shapefiles showing dog off-leash areas as points and polygons. Both have coordinate problems. Describe the problem for each, then fix them and create a map showing both the points and the polygons with a backdrop of the major transportation arteries in Austin. (**Hint**: Open Pro and load arteries first in order to be able to compare the other data sets against it.) **Capture** the map.

Chapter 5. Managing Vector Data

Objectives

▶ Understanding how GIS data differ from typical computer files

▶ Organizing GIS data sets and folders

▶ Compiling data for geodatabases through importing, exporting, or modifying existing data

▶ Understanding how map documents work with GIS data

▶ Documenting data sources using metadata

Mastering the Concepts

GIS Concepts

Chapter 1 introduced the idea of different data models to handle discrete data (spatial objects like rivers or states) versus continuous data (like elevation). In this chapter we will discuss the details of the vector data model and learn how GIS data can be different from other types of data that you may have handled before. We will also learn about strategies and best practices for creating and managing data collections for a project.

The vector model

The vector data model is designed to handle the storage of discrete objects on the earth's surface. Vector data use a series of x-y locations to store information as points, lines, or polygons (Fig. 5.1). The choice of x-y values used to store the locations is the coordinate system, discussed in Chapter 4. A point consists of a single x-y coordinate pair. A line includes two or more pairs of coordinates—the endpoints of the line are termed **nodes**, and each of the intermediate pairs is called a **vertex**. A polygon is a group of **vertices** that define a closed area.

Fig. 5.1. The vector data model uses a series of x-y locations to represent points, lines, and polygon areas.

Similar features may be stored together as a **feature class**. Information about each feature, its **attributes**, are stored in a table associated with the feature class (Fig. 5.2). A special field, called the **Feature ID (FID)** or **Object ID (OID)**, links the spatial data with the attributes. Each feature's attributes are stored in one row of the table. When a state is highlighted on the map, its matching attributes are highlighted in the table, and vice versa. This live link between the spatial and attribute information has many benefits, such as creating maps from attributes as we did in Chapter 2.

Some GIS systems can also handle **multipart features**, in which a single feature includes multiple spatial objects. For example, Hawaii requires multiple polygons to represent the different islands, but in a states feature class, it should be awarded only one row of information in the table, not seven. A multipart feature accomplishes this task. Any type of feature, including points, lines or polygons, may be stored as multipart features if needed.

"Feature class" is a generic term that encompasses many different file formats associated with different GIS programs over the years. The specifics of data storage vary from format to format, but nearly all use this structure of spatial features linked to information in a table. We will discuss several formats used by the ArcGIS software later in this chapter. Some data formats permit (or require) feature classes to be combined into a related functional group, known as a **feature dataset**, which helps organize similar data or serves special functions.

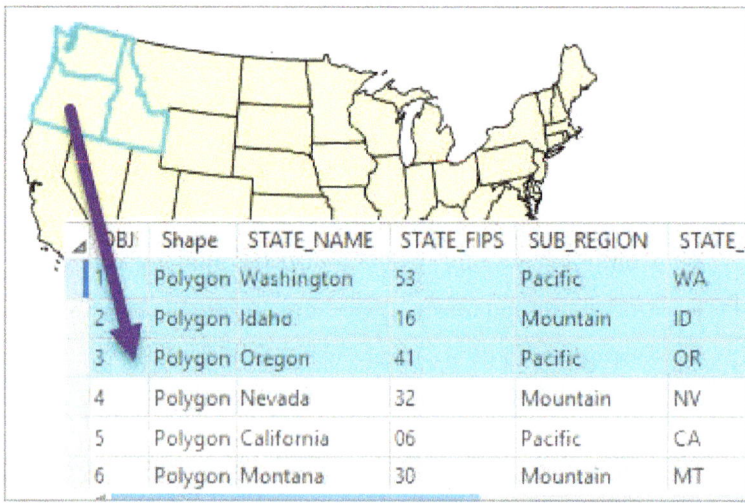

Fig. 5.2. Each state is represented by a spatial feature (polygon), which is linked to the attributes in a table.

Source: Esri

The benefits of the vector data model are many. First, it can store the location and shape of individual features, such as roads and parcels, with a high degree of precision. Second, the linked attribute table provides great flexibility in the number and types of attributes that can be stored about each feature. Third, the vector model is ideally suited to mapmaking because of the high precision and detail of features that can be obtained. The vector model is also a compact way of storing data, typically requiring a tenth of the space of a raster with similar information. Finally, the vector model is ideally suited to certain types of analysis problems, such as determining perimeters and areas, detecting whether features overlap, and modeling flow through networks.

However, the vector model has some drawbacks. First, it is poorly adapted to storing continuous surfaces, such as elevation or precipitation. Contour lines (as on topographic maps) can be used to represent surfaces, but calculating derived information from contours, such as slope, flow direction, and aspect, is difficult. Finally, some types of analysis are more time-consuming to perform with vectors than they are with rasters.

Modeling feature behavior with topology

Two types of vector models exist: the **spaghetti model** and the **topological model**. A spaghetti model stores features as independent objects, unrelated to each other. Two adjacent states may share the same boundary, but a spaghetti model stores them as separate polygons, saving the shared boundary twice without recognizing the relationship between the two states. Simple and straightforward, this type of model is found in many types of applications that store spatial data. It is also commonly used to transfer vector features from one GIS system to another.

A topological data model stores features, but it also contains information about how the features are spatially related to each other: whether two parcels share a common boundary (**adjacency**), whether two water lines are attached to each other (**connectivity**), whether a company sprayed pesticide over the same area on two different occasions (**overlap**), or whether a highway connects to a crossroad or has an overpass (**intersection**). Although computer algorithms can determine whether these spatial relationships exist between features in a spaghetti model, storing topology can save time if the relationships must be used repeatedly. A topology can also define rules for how features can be related to each other, determine whether the rules have been broken, and assist the user in finding and fixing these errors.

Parcels, for example, should have identical shared boundaries. A topology can test for this condition once it is known which parcels are adjacent to each other. Navigation applications that find directions from one place to another must know which streets connect to others, which ones are one-way and which ones are two-way. This type of information can be encoded in a network topology. Other applications of networks include modeling water flow in streams, traffic along roads, flights in and out of airline hubs, and utilities through pipes or electrical systems.

Characteristics of GIS data

Data files stored for use in a GIS differ in many ways from data used in other applications, such as word processing, spreadsheets, videos, or music. Some important characteristics of GIS data must be understood in order to use and manage them effectively.

GIS data files may be very large. Some data sets, such as a parcel feature class for a large city, can have hundreds of thousands of features and records. Images and rasters can run to hundreds of megabytes or several gigabytes. Point clouds from LiDAR systems may store billions of x-y-z locations.

GIS data are often shared by multiple users. Many data sets are collected and used across organizations rather than belonging to one individual. A parcel database must be accessed by many clerks and several city departments. Giving each person a copy of the data causes problems. First, it wastes space. Second, such databases are often updated, and ensuring that every person has the most recent version is a difficult task. Thus, many GIS data sets are intended to be shared; one copy is maintained on a central drive or server, and everyone uses it.

GIS data come from many sources, yet they must work together. GIS data may come from satellites, GPS units, scanners, paper maps, tables, and other sources. They can use different coordinate systems and projections and file types. GIS requires ways to label and keep track of how files are stored and which coordinate systems are used. Some GIS systems rely on users to keep track of everything, but more commonly today, GIS applications use headers or other label files to keep track of data set characteristics and to handle the files properly. Users need to make sure that this additional information about the data is present and correct.

GIS data include local files and Internet services. Most people work with data sets that reside on the computer being used. If a PDF is found on the Internet, the user is likely to make a copy of it on the computer for later use. Many GIS data sets work this way also, but GIS data services are becoming more prevalent. A GIS service, such as that provided by ArcGIS Online, allows the data to be used over the Internet, but it often does not allow it to be copied to the local computer.

GIS data formats do not always follow the latest conventions. Computer operating systems have become more permissive in the naming of folders and files over time. Originally, computer file names were restricted to eight alphanumeric characters (letters or numbers), and computers used the space character to signify the end of one file name and the beginning of another. Uppercase letters were not interchangeable with lowercase letters. Nonalphanumeric characters such as $ or % or # had special meanings. Very long names could exceed program file name limits. Today, operating systems are more flexible, but not all software applications evolve at the same rate or follow the same conventions. Even current versions of an application may contain old bits of programming code that can fail when they encounter files or folders named according to the new rules. GIS users must follow more restrictive rules about where to put data and what to name them.

Organizing data files

Today's computers try to do file management for us—they figure out whether we are saving text, music, or videos and drop them into preorganized folders. They open recently visited folders automatically. They recognize what type of file is being opened and select the right application. They even suggest what to name files. These features can be convenient for a casual user, but they cause problems in complex applications. A GIS professional must take responsibility for learning some precepts of sound file management. These precepts are common to many computing fields and are useful guidelines to follow at any time, not just for GIS.

Be mindful when saving files. Don't let the computer save files with default names. A default name may not reflect what is in the file, or it may be too long, or it may not follow the proper naming conventions. Be careful to note where the computer is trying to save a file; in some cases this may be a default location or your last-used folder, rather than the appropriate place. Always put files in an organized place where they belong. The Windows Desktop is a terrible place for GIS data.

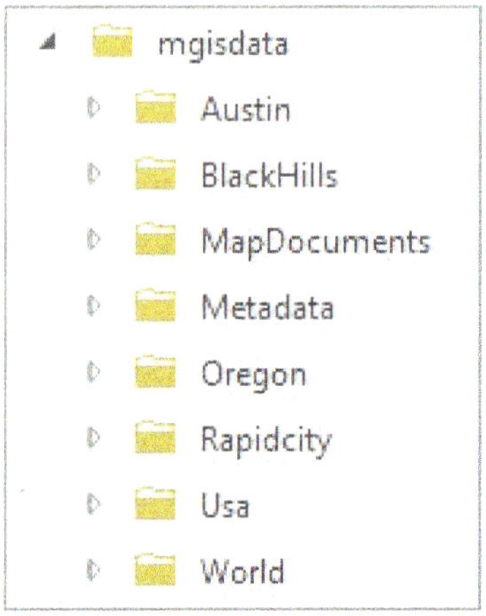

Organize files using folders. Finding files is much easier when data are organized in a logical structure of folders. Imagine how hard it would be to find things if all the data for this class came in one giant folder, rather than organized by geographic area (Fig. 5.3). Take the time to develop a thoughtful system of folders to contain GIS data. There is no one right way to organize a data folder tree, but folder names and the structure should make sense and enable users to find things easily.

Fig. 5.3. Folder organization is very important for GIS data.

Source: Esri

In GIS, it is helpful to separate static data folders—those that contain finished, permanent data—from working folders in which data are being downloaded, files are being added and deleted, and intermediate results clutter the folder. Separating active work folders reduces the chance of accidently deleting a data set or losing track of which data sets are permanent and which are works in progress. Once a project is finished, clean up the folder and move the data to a static location.

Be conservative when naming files and folders. To make file systems compatible with any type of software, keep folder names short and use only letters, numbers, and the underscore character. Avoid spaces, and never use nonalphanumeric symbols. Spaces may be used for file names if the program accepts them, but it is best to stay away from nonalphanumeric characters for all file names.

VERY IMPORTANT TIP: GIS folders and data set names should not contain spaces or special characters, such as #, @, &,*. Use the underscore character _ to create spaces in names if needed.

Store GIS data in top-level drive folders. Windows uses personal Library and Documents folders as default locations for user data, but these locations often cause problems when used for GIS data. In a network environment, the pathnames may contain problematic characters or spaces that may prevent the GIS from working properly in all situations. It can also be difficult to find these locations, and the routes to these folders can be tedious to navigate. The best practice is to create GIS data folders at the topmost level, directly under the drive letter designation, such as C:\ gisdata or C:\mydata. From that initial point, a user can create as many subfolders as needed.

Pay attention to file extensions. Most files on a computer have an extension, a three- or four-letter code that indicates the type of file. A Word document has a .doc or .docx extension, and a picture might have a .jpg or .png extension. Computers use this extension to help determine which application should be used when a user double-clicks a file to open it. By default, the computer hides file extensions from the user, but it is often helpful to see them, especially when downloading files and working with data the computer may not recognize. It is wise to change the default setting in the File Explorer Options in the Control Panel in Windows 10 (Fig. 5.4); Windows 7 has a similar Folder Options setting.

Understanding file locks

When accessing a file, computers distinguish between **read access** and **write access**. Read access allows a program to take information out of a file without changing it. Write access allows changes to be made to the stored file. File systems use access restrictions to protect data. In a network file system, a folder may be designated to have read access for most users in order to protect it from being accidently (or purposely) damaged. You will generally have both read and write access (read-write access) to all files on your own computer.

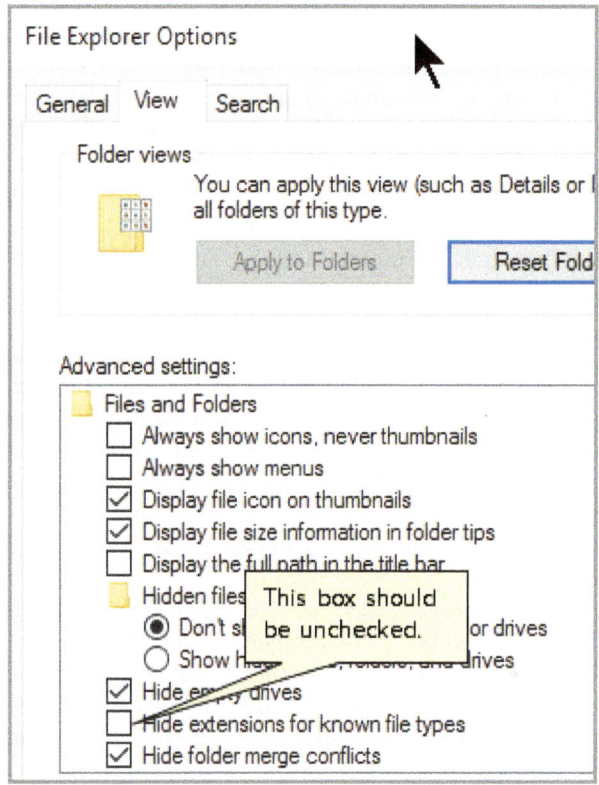

Fig. 5.4. Setting Windows 10 to show file extensions
Source: Microsoft

Access rights are important when many people use the same files. You may have experienced opening a Word document that was already in use and received a message to that effect, with an offer to open a read-only copy. In this case, the already open file had read-write access, but the second copy only had read access. If the second copy is changed, it must be saved using a new name.

Programs keep track of which documents are being used by creating a small **lock file** in the same folder as the document while it is open in an application. When the program releases the file, the lock is deleted, and another program is free to access the file. When a document is opened, the application checks for the file lock. If it is present, the user will get the "in use" message.

Because ArcGIS Pro is set to allow editing of any layers for which the user has write permission, data sets open in Pro will have locks on them. To avoid conflicts, one should avoid running two instances of Pro on the same computer at the same time. However, it is possible that another application is using the file (e.g., the user was editing in ArcMap or has a file open in Excel and tries to add it to Pro). If Pro recognizes a file lock on a data set, the user is allowed to manipulate it in any way except for actions that change the file (editing, adding fields, etc.). Attempts to do so will generate an error message that the file is already in use by another application.

TIP: ArcGIS is prone to glitches in which it does not recognize that a lock has been removed, and it may generate an "in use" error even when no other applications appear to be using the file. The best way to clear the spurious lock is to save the project and restart ArcGIS Pro.

Finding data for a project

Obtaining data for a project is an open-ended and complex activity, especially since the GIS data landscape is continually changing. New data sets emerge; old ones become unavailable or are moved to new locations. New methods for discovering and formatting information mingle with ancient, nearly forgotten file types. The volume, diversity, and complexity of available data are bewildering, but we will begin to develop the skills needed to find what we need. Getting proficient requires flexibility, perseverance, and a willingness to experiment, and it takes time and practice. A high tolerance for frustration and an experienced mentor are valuable assets in the search for data.

Generally, it is more difficult to find preexisting data for large map scales or very specialized projects. General base data for roads, cities, and political boundaries are widely available, even at large scales. Very large-scale data, like utilities, parcels, or building footprints, will generally be found for only small areas at a time, and they may not be available for download by the public because of privacy or security concerns. Specialized data sets, such as land use categories or geologic maps, are most often available either at state or national scales or as detailed maps in isolated areas. Source scale is an important characteristic to review before deciding to use a data set.

If nothing can be found, then either the project must be revised or the data must be created in-house. Original data can be time-consuming and expensive to create, which is why specialized large-scale data are difficult to find. For class projects, it is wise to choose topics and scales for which data may be easily located. Sometimes it is easier to find interesting data first and then develop the topic!

Searching for data can start at an agency web site or at a generic clearinghouse. A list of good places to start is included at the end of this chapter, but the Internet changes so quickly that some of the links may well be obsolete by the time the book goes to press. One can sometimes have good luck with general Internet search engines by including the term "shapefile", "GIS", or "e00" along with a topic or place name; for example, "Wyoming GIS" or "geology shapefile".

ArcGIS Online can be a valuable source of data. However, two things must be remembered. First, many data sets on ArcGIS Online are designed to be used online and do not allow the user to make a local copy. The most common data types that allow downloading are feature layers and layer packages (we will learn more about AGOL data types in Chapter 12). Second, almost anyone can share data on ArcGIS Online. Some of it is provided by agencies and companies and is excellent, but some is of little value. As with anything on the Internet, let the searcher beware.

Data tables can also be good sources of data, such as text files, Excel spreadsheets, or even a copy/paste of tabular data found on a web page and saved in a text file or spreadsheet. If the table has fields with x-y values in them, these rows can usually be converted to point data with attributes. If the table has attributes like county names, state names, or FIPS codes (a federal identifier of political units in the United States), then these tables can often be joined to existing feature classes. Chapter 7 has more information about working with tables.

After downloading some data, the first step is to ascertain the data format because some additional processing is often needed before the GIS can work with, or even recognize, the data. Always start by examining the downloaded files in Windows Explorer, taking note of the file extensions in particular.

Data downloaded from the Internet is typically **zipped**, a process that combines many individual files into a single file and compresses the information to use less space. A single compressed file, known as an **archive**, is easier to download than multiple files, but the archive must be unzipped before the GIS will recognize the data. Zipped files typically have file extensions of .gz, .zip, .tar, .tgz, or .tar.gz. Many different zip utilities are available to "unzip" files, some free, and each one works differently. Ask your instructor what is typically used on campus, or ask a coworker what is used in your organization, and learn how to install and use it.

After unzipping, one must ascertain the data format of the new files. Sometimes data are already in a geodatabase format and can be copied directly to a geodatabase. More commonly, the data need to be converted. The computer terms **import** and **export** usually refer to the process of copying a data file to a new location while simultaneously converting it from one data storage format to a different one. Most GIS programs include conversion tools that allow data sets to be changed from one data format to another, although the number and types of conversions allowed will vary.

Compiling data

One of the first steps of most projects is to assemble the required data. Sometimes this data is accessed from a network drive or GIS services, and no local copy is required. At other times, data are compiled especially for the project, for several reasons: (1) It is more convenient to work with a smaller subset of data. (2) Editing is required, but the source data are read-only. (3) The data come from a non-GIS file type, such as an Excel table. (4) The data must be created, perhaps by editing or collecting them with a GPS. (5) The data must be available when wireless or network connections are unavailable (such as out in the field). In this section, we will look at common ways to compile data for a project from existing sources. We tackle the more complicated process of creating data from scratch in Chapter 8.

Importing and exporting data

An **import** function copies a data set from one location into the geodatabase, converting formats if needed. An **export** function goes the other direction, copying a data set from the geodatabase to another location. Pro uses a variety of tools to accomplish this work. A perusal of the Conversion Tools toolset in the Toolbox is a good way to get familiar with the different data formats that can be converted to ArcGIS files.

Mapping *x-y* coordinates from a table

Tools are available that take x-y coordinates from a table and convert them to locations on a map (Fig. 5.5). The x-y values must be given in a real-world coordinate system, such as latitude-longitude or UTM meters. The coordinate system details must also be known, including the geographic coordinate system/datum and the projection, and it must be explicitly set. Moreover, it should match the x-y values in the file, rather than the coordinate system of the map. For example, if the map is in UTM and the table x-y coordinates are given as longitude-latitude values, the coordinate system must be set to a GCS in the tool.

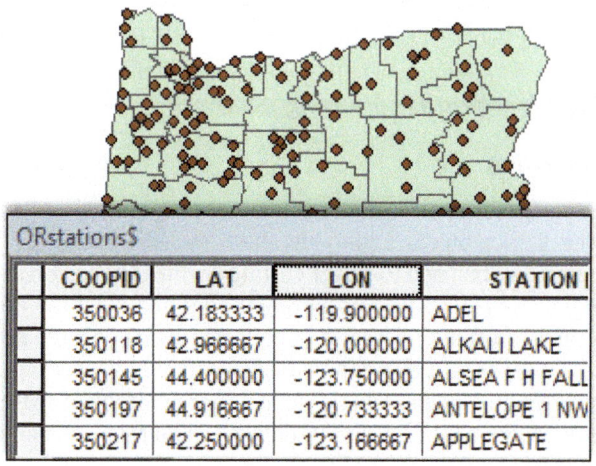

COOPID	LAT	LON	STATION
350036	42.183333	-119.900000	ADEL
350118	42.966667	-120.000000	ALKALI LAKE
350145	44.400000	-123.750000	ALSEA F H FALL
350197	44.916667	-120.733333	ANTELOPE 1 NW
350217	42.250000	-123.166667	APPLEGATE

Fig. 5.5. Climate stations from a spreadsheet displayed as points on a map
Source: Esri

Global positioning systems (GPSs) often provide tables of x-y data. A GPS receiver can calculate its location by triangulating distances and directions from a constellation of orbiting GPS satellites and may produce tables of x-y locations as output. Other examples of x-y data sources include benchmarks, surveyed points, and locations measured on a map. Regardless of the source, the procedure for mapping x-y locations from a table is the same.

Some tools may create a point feature class directly; others may create a temporary spatial layer known as an **event layer**. This terminology comes from considering each entry in the table to be an event, such as an earthquake or a traffic accident. Event layers appear similar to point feature classes, but they remain a table of objects represented by spatial points. Some tools will not accept event layers as input. Usually the event layer will be exported to a shapefile or a geodatabase feature class to make it permanent.

Latitudes and longitude outside GIS systems are often reported in degrees-minutes-seconds and must be converted to decimal degrees before they can be used to create an event layer:

$$X \text{ in decimal degrees} = \text{degrees} + \text{minutes}/60 + \text{seconds}/3600$$

For example, $103°\ 30'\ 15''$ would convert to $103 + 30/60 + 15/3600 = 103.504167$. Keep in mind that degrees are large units, and six or seven decimal places should be kept to ensure precision and accuracy. At the equator, a meter is equal to approximately 10^{-5} degrees. Also, longitudes west of the Prime Meridian in England are negative, but the minus sign is often omitted for convenience. If this longitude is reported by a GPS unit in Nebraska, for example, the true value needed by the GIS is -103.504167.

Geocoding

Geocoding is the process of converting addresses in a table to x-y locations on a map. Geocoding relies on comparing the address to a reference feature class that contains information on streets, including attribute data on their names, address ranges, cities, zip codes, and so on. Such data may be used to create a **locator**. ArcGIS Online has a built-in locator called the ArcGIS World Geocoding Service (or similar; the name seems to change often) that it uses when asked to locate a single address in the Search box. With a subscription account for ArcGIS Online, batches of addresses in a table may be geocoded, although this action will consume service credits. (Geocoding thousands of addresses is one way to exhaust an institution's credits in short order, and the AGOL administrator may restrict the geocoding capability or a user's credit quota to prevent it.) With the right data, one can create custom locators and geocode without using credits.

Modifying and managing data

Extracting

Sometimes a source data set has more features than needed. Imagine compiling data for a project on New Jersey by extracting just the New Jersey features from data sets for the entire United States. A query can be used to select certain features based on a value in an attribute field (e.g., the state name equals New Jersey). Another method is needed if the table does not contain a suitable field. Roads, for example, often cross state boundaries and thus have no field to indicate which state they are within. In these cases, a **clip** can be performed, which uses a polygon feature class to delineate the area of interest and keeps only the features or portions of features inside the boundary. The resulting data set will be identical to the input data set in all respects except for the region of features included.

Merge and append

The **merge** function is used to combine the features of two or more feature classes and place them into a new feature class (Fig. 5.6). The merged layers must have the same feature type

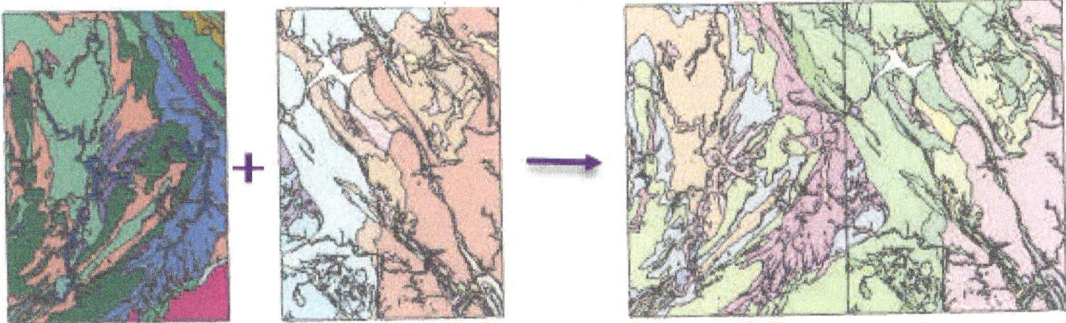

Fig. 5.6. Append or merge can be used to combine adjacent data sets, if the tables match.
Source: South Dakota Geological Survey

(points, lines, or polygons) and must share the same coordinate system. Overlapping of layers is permitted, although commonly the merged data sets are adjacent to each other.

The output data set has a single table. If the input data sets have the same fields with the same names, then the field information is easily copied to the new table. However, it is common for the input data sets to have differently configured tables, in which case the user must specifically define how the new table is structured, a process known as field mapping. The greater the differences in the input tables, the more time-consuming and challenging the field mapping process becomes.

Consider two input data sets containing wells. In the first feature class, WellsEast, the well ID number field is named WellID. In the second data set, WellsWest, the corresponding field is named WellNumber. The user would need to establish a well ID field in the output that draws information from the corresponding field in each input table. This specification would need to be repeated for any fields without matching names and data types.

Append is similar to merge, except that it copies features from one or more input data sets to a target data set that already exists. The output fields are already defined and cannot be modified, although field mapping can still specify where the output fields draw data from the input tables. Append is more often used to add new features and records to an ongoing data set. For example, a national forest might add polygons representing each year's fires to an historical feature class containing all the fires in its jurisdiction. Alternatively, each national forest office might append the fires from its jurisdiction to a national data set containing all fires in the United States.

Simplifying data

Sometimes a data set may be too detailed for effective use. A data set with a large source scale may take forever to draw because there are too many vertices for the scale at which it will be used. It might include dozens of attributes when the user is interested only in one or two. In these situations, simplifying the data may result in better performance for the intended task. Chapter 3 discussed a number of cartographic reasons for simplifying data sets to produce clearer maps.

The **Generalize** tool is used to simplify the spatial complexity of line or polygon features by removing vertices that are closer than a specified tolerance value, such as simplifying this coastline (Fig. 5.7a). The green polygon represents the original coastline, and the black line shows the polygon boundary after generalizing. The output features have fewer vertices and a simpler shape, although the accuracy will be lower than that of the input features.

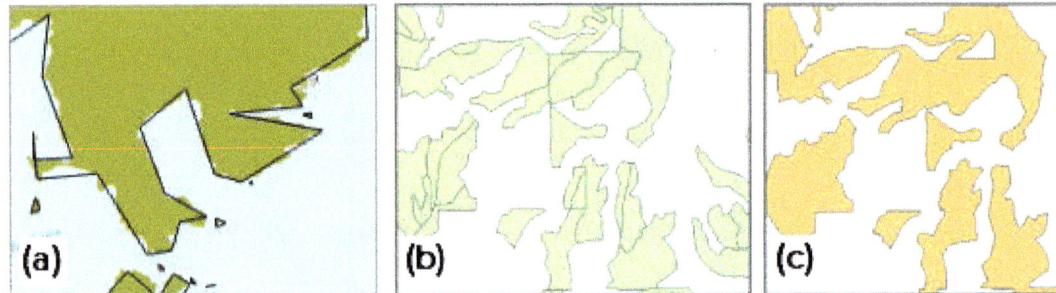

Fig. 5.7. (a) Generalizing a coastline; (b) polygons before dissolving; (c) polygons after dissolving
Source: Esri

The **Dissolve** tool is used to merge features when they share the same attribute value(s). For example, the map in Figure 5.7b shows tree stands with the same cover type (ponderosa pine) but different tree ages. Dissolving on the cover type field removes the age boundaries and yields polygons based on the cover type only (Fig. 5.7c). All other attribute fields other than the dissolve field are eliminated from the output, although the user may request statistical summaries of the fused features, such as the average crown cover of the tree stands combined to make each new feature.

The output from a dissolve operation may produce either simple features or multipart features. In Figure 5.7c, the orange polygons would constitute one large multipart feature by default, with a single record in the output table. If separate features with separate records are desired, the multifeature option must be declined.

Documenting data

Quality geospatial data sets are typically accompanied by **metadata**, or data about the data set. Metadata provide the means to discover GIS data when searching, and they also store valuable information about the provenance and quality of a data set, allowing a user to determine whether the data set is fit for a given purpose (Fig. 5.8). Once created, metadata usually travel with the data set. Data downloaded from ArcGIS Online or the Internet will often have metadata as part of the file (or in a separate file). When downloading,

U.S. States (Generalized)

Type   File Geodatabase Feature Class

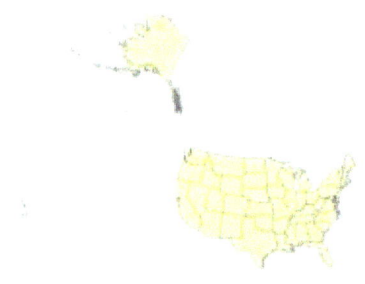

Tags   polygon, area, population, households, demographics, society, boundaries, farming, United States, U.S. States, States, 2002, 2010, 2012, 1992, 2011, 2013, 2014

Summary

U.S. States (Generalized) provides 2010 U.S. Census demographic information and generalized state boundaries to improve draw performance and be used effectively at a national level.

Description

U.S. States (Generalized) represents the 50 states and the District of Columbia of the United States.

Credits

ESRI Data and Maps (ESRI, Inc., Redlands, CA)

Fig. 5.8. The Item Description
Source: Esri

watch for links to metadata and be sure to download them along with the data set. It is a good practice to skim the metadata to learn some basic things about them, such as the originator and the source scale. The metadata may need to be updated if the user makes any changes to the original files.

Compiling data for a project usually requires some work with the metadata. By default, ArcGIS Pro and ArcGIS Online use a simple style called the **Item Description** (Fig. 5.8), which is designed for users to quickly review or enter the most important metadata fields of a data set and includes the following information.

Tags record a few terms that can be used for searching for the data. The *Summary* briefly describes the purpose of the data set. The *Description* typically contains an extended explanation of the contents, such as what the data set contains, how it was developed, what the attributes mean, and information about the data quality. The Description may include a couple of sentences or several paragraphs. The *Credits* record the source(s) of the data, such as a formal citation for a publication or a URL for a downloaded data set. The *Use Limitation* describes restrictions placed on the data, such as whether they can be copied or shared. The *Appropriate Scale Range* suggests the scales at which the data are valid and should be used. The *Bounding Box* is automatically entered by the software from the x-y extent of the features. It is always in degrees and is used by geographic-based search engines.

It is always necessary to develop or update metadata, but the level of effort can vary. Sometimes GIS data are used only by their creators. In this case, the metadata can be minimal, although they are still valuable for recording some information in the Item Description for future reference, such as a citation documenting the source of the data, and any information found on the web site, such as the source scale or the year the data were published. In this chapter, we will learn how to quickly document a data set. More extensive types of metadata will be discussed in Chapter 12.

TIP: Make a practice of quickly documenting any data set that you intend to keep.

Metadata also record access and use constraints. Some GIS data, including most data sets derived from federal agencies, can be freely copied and redistributed. Other data are licensed to the specific users who purchased them. Every user is responsible for understanding the access constraints placed on any data set and for abiding by them. Failure to do so can result in civil and criminal penalties against an individual or the organization.

About ArcGIS

Discrete data formats

ArcGIS can read a variety of GIS file formats (Table 5.1); some are native to ArcGIS, and some originate from other sources. In addition, some may be local, and some may be hosted as online data from ArcGIS Portal.

Geodatabases

The geodatabase is the native storage format for both ArcGIS Desktop and ArcGIS Pro. A geodatabase can contain many different data objects, including feature classes, networks, tables, rasters, and topology. Figure 5.9 shows a geodatabase named oregondata. Feature classes may exist as individual objects in a geodatabase (as do the cities or counties), or they may be grouped into **feature datasets**, a collection of related feature classes (but not rasters) with the same coordinate system, such as the Transportation feature dataset in Figure 5.9. For now, think of a feature dataset as a folder for organizing feature classes (although they have additional capabilities). Feature classes copied or imported to a feature dataset will automatically be converted to the feature dataset's coordinate system.

Table 5.1. Vector model data sources used by ArcGIS

File type	Description
⊡ ⊡ ⊠ Shapefiles	Shapefiles are vector feature classes developed for the early version of ArcView, and they have been carried over into ArcGIS.
⊞ ⊞ ⊞ Coverages	A coverage is the vector data format developed for Arc/Info and is the oldest of the data formats. Not read by ArcGIS Pro.
⊟ ⊞ ⊡ ⊡ ⊠ Geodatabases	Geodatabases are the recommended model for storing spatial information for ArcGIS.
⌐ Database connections	Database connections permit users to log into and use data from any supported relational database.
◇ Layer files	A layer file references a feature class and stores information about its properties, such as how it should be displayed.
▤ ▤ Tables	Tables can exist as separate data objects that are unassociated with a spatial data set.
▣ TINs	TINs are triangulated irregular networks that store 3D surface information, such as elevation, using a set of nodes and triangles.
▦ CAD drawings	Data sets created by CAD programs can be read by ArcGIS, although they cannot be edited or analyzed unless they are converted to shapefiles or geodatabases.

Three types of geodatabases are used by ArcGIS. A **personal geodatabase** is stored in a single Microsoft Access file. This file is limited to 2 GB in size and works only in the Windows operating system. ArcGIS Pro does not recognize this format, and it must be upgraded to a file geodatabase in ArcCatalog before it can be used with Pro. A **file geodatabase** is stored in a system folder, and each file can be up to 1 TB in size. File geodatabases can be accessed by multiple operating systems, including Linux and Unix. An **enterprise geodatabase** stores GIS data within a commercial relational database management system (RDBMS), such as Oracle® or SQL Server®. Enterprise geodatabases are designed to meet security and management needs for large data sets edited by many simultaneous users. In this book, we use file geodatabases.

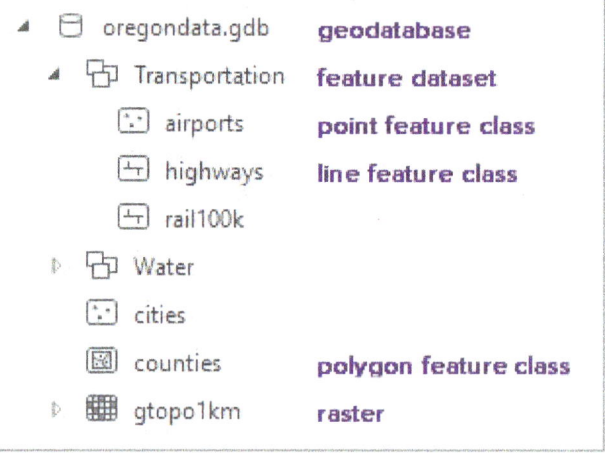

Fig. 5.9. Contents of a geodatabase
Source: Esri

TIP: File geodatabases are folders containing GIS data sets, making it *possible* to store files like spreadsheets in them. However, this practice can damage the geodatabase if a user is not careful. Never use Windows or other applications to save or manage files inside a file geodatabase. Instead, save them in the project folder (above the geodatabase folder).

Shapefiles

A **shapefile** is a spaghetti data format containing a feature class composed of points, lines, or polygons. The table attributes are stored in a **dBase** file. Shapefiles have a common format that is recognized by many GIS and GPS systems, and they are often used to share information between different GIS and remote sensing programs.

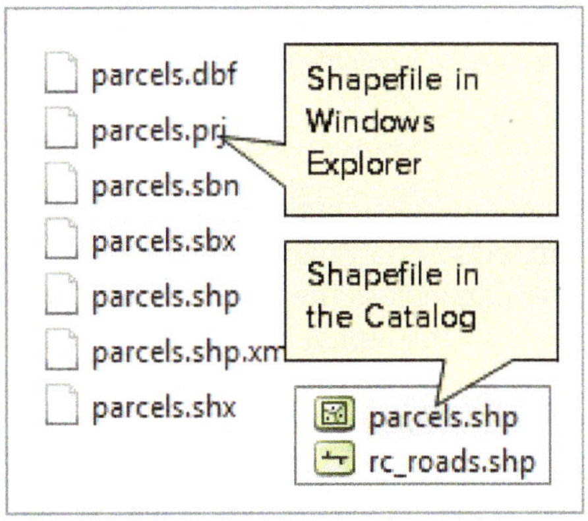

Although a shapefile appears as one icon in ArcGIS, it is actually composed of multiple data files as seen in Windows Explorer (Fig. 5.10). The parcels shapefile consists of seven different files. The .shp file stores coordinate data, the .dbf file stores attribute data, and the .shx file stores a spatial index that speeds drawing and analysis. These first three files are required for every shapefile to function. Additional files may also be present: the .prj file stores projection information, and the .xml file contains metadata. To copy a shapefile, all of these files must be included. Windows Explorer does not do this automatically, so it is a good practice to only use a GIS program to manage GIS data.

In a shapefile attribute table, the first two columns of data are reserved for storing the feature identification code (FID) and the coordinate geometry (Shape) field. These fields are created and maintained by ArcGIS and cannot be modified by the user. All other fields are added by the user.

Fig. 5.10. Shapefiles are groups of files but appear as single entries in the Catalog.
Source: Esri

Coverages

A **coverage** is the oldest vector format, developed for Arc/Info. ArcGIS Desktop has limited functions for managing coverages, and ArcGIS Pro does not recognize them at all, so most users will encounter them simply as an old data format that must be converted to a shapefile or geodatabase using ArcMap or ArcCatalog. Several things are helpful to know in this process.

Coverages are structured as feature datasets that contain multiple feature classes, which may store points, arcs, polygons, and polygon labels. Coverages also store topology, and the tables have several attribute fields reserved for this purpose. Figure 5.11 shows these fields for a coverage called LANDUSE (some fields use the coverage name as part of the field name). It makes sense to delete these fields during or after the conversion, for they serve no useful purpose afterwards.

The old Arc/Info program used a proprietary zipping/conversion format to convert coverages into a single text file, known as an **Interchange file** and bearing the file extension .e00. Files with the .e00 extension must be converted back into coverages using the *Import from E00* tool in the Conversions Tools toolset in ArcToolbox. Afterwards, the coverage can be converted to a geodatabase or a shapefile. ArcMap or ArcCatalog must be used for this process.

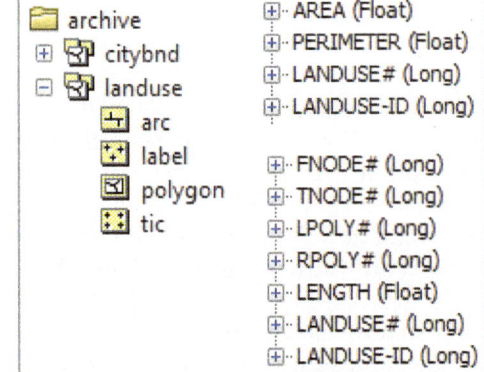

Fig. 5.11. Coverages contain fields of little value once they are converted.
Source: Esri

CAD files

Data sets from CAD programs can be read by ArcGIS, although they cannot be edited or analyzed unless they are converted to shapefiles or geodatabase feature classes. A CAD file may contain multiple feature classes that correspond to layers of the drawing and can be opened separately.

Database connections

A user can connect to a database management system (DBMS) on a network through a database connection. This connection can be open, or it can require a login and password. Once inside, the user can access the tables within the database. Database connections may be available on a local network through a mounted drive, but some can be accessed online as database services.

Because GIS data formats can be complex, it is best to use only Pro, ArcMap, or ArcCatalog to manage them; for example, to copy, rename, delete, and move GIS files on the computer. ArcCatalog and the Catalog pane understand the unique characteristics of GIS data and are better at managing GIS files than Windows Explorer. For example, a single shapefile feature class is composed of multiple files with the same prefix and different extensions (roads.dbf, roads.shp, roads.prj…). Windows Explorer would allow a user to delete one of these files, thereby destroying the feature class and rendering it unusable. Pro, however, knows that these files form a unit. It shows them as a single icon and requires users to save, copy, or rename them as a unit. It is wise to always use the Catalog pane to manage GIS data.

VERY IMPORTANT TIP: Do not use Windows to copy or delete coverages, shapefiles, geodatabases, or project .aprx files. Use caution when using Windows to move project folders.

Projects and shared data

A **project file** (identified with an .aprx extension) organizes all the pieces of the project and keeps track of the data, settings, and resources needed to make the project work. The project file itself does not store any spatial data files. Instead, it stores the name and disk location of the spatial data—its **source data**. When Pro needs the data, to draw some features, for example, it uses the information to find and access the data *from their original location*. In Chapter 2, we learned how layers point to data sets and control how they appear in the map through the layer properties settings. This approach is critical to the ability to share data sets: each user accesses the same source data but can customize how the data appear in different maps (or even within the same map).

The home geodatabase stores GIS data sets unique to the project, but projects may also employ data stored outside the project folder and accessed through folder connections. Storing information separately from the project has several benefits. First, it makes it easy for multiple users to share data. Each user accesses the same data but can use them in multiple projects and customize how they appear and behave. Second, it saves space. Each user need not store a copy of the data. Third, it allows updates of a central database to be easily distributed to every user.

Data sharing has many conveniences, but it also presents certain problems. If someone accidentally damages or deletes a data set, many people are affected. Because the project file only stores the location of the source data, it can lose track of that location if the data are deleted, moved, or renamed. If someone e-mails a project file, the data won't be sent with it. Unless the recipient has access to the same data in the original location (on a network drive, for example), the project will show up empty. Chapter 12 discusses ways of sharing a layer, a map, or even an entire project with someone else.

The sharing of data requires that data managers implement precautions to ensure that an editing mistake does not affect dozens of people. Sometimes the files are write-protected, and only a small number of users are authorized to make changes. In a more complex environment where GIS data must have multiple editors, organizations may use an enterprise geodatabase that is designed to support multiple editors.

A Universal (or Uniform, or Unified) Naming Convention (UNC) path is designed to locate information on a local area network of computers. It takes the form \\computer_name\shared_ directory\, such as \\mybigserver\mypublicdata\. UNC paths are convenient because they are not affected by which letter to which a network resource is mapped. However, they will not work for anyone who does not have access to the local network.

Finally, a Uniform Resource Locator (URL) path specifies the address of any document on the Internet. Most people are familiar with this type of path, as in http://www.esri.com.

Sharing data also raises challenges for small teams working on a joint project. If each person has his or her own data, then updates cannot easily be shared. If all team members access the same data in one location, it becomes possible for two people to make simultaneous or conflicting edits to a feature class, which can potentially corrupt the database. Unless the team has an enterprise geodatabase available, it must devise a strategy for data sharing that eliminates the possibility of two people editing the same feature class at the same time.

Projects and pathnames

Projects keep track of the source files by storing the location of each file as a **path** or **pathname**, which lists the successive folders traversed to a data set, separated by backslashes. Consider the file system represented in Figure 5.12. The blue box is the hard drive C:\, the green boxes are folders, and the gray cylinders are geodatabases. The project file inside the CraterLake folder, CraterLake.aprx, contains a map that uses the counties feature class in the oregondat geodatabase. It could refer to counties using an **absolute pathname** that starts at the drive letter (blue line) and traverses the folders to find the feature class, C:\gisclass\mgisdata\Oregon\ oregondat\counties. Alternatively, it could use a **relative pathname**, which starts at the CraterLake.aprx file and moves up two levels to the gisclass folder (green line) and then down

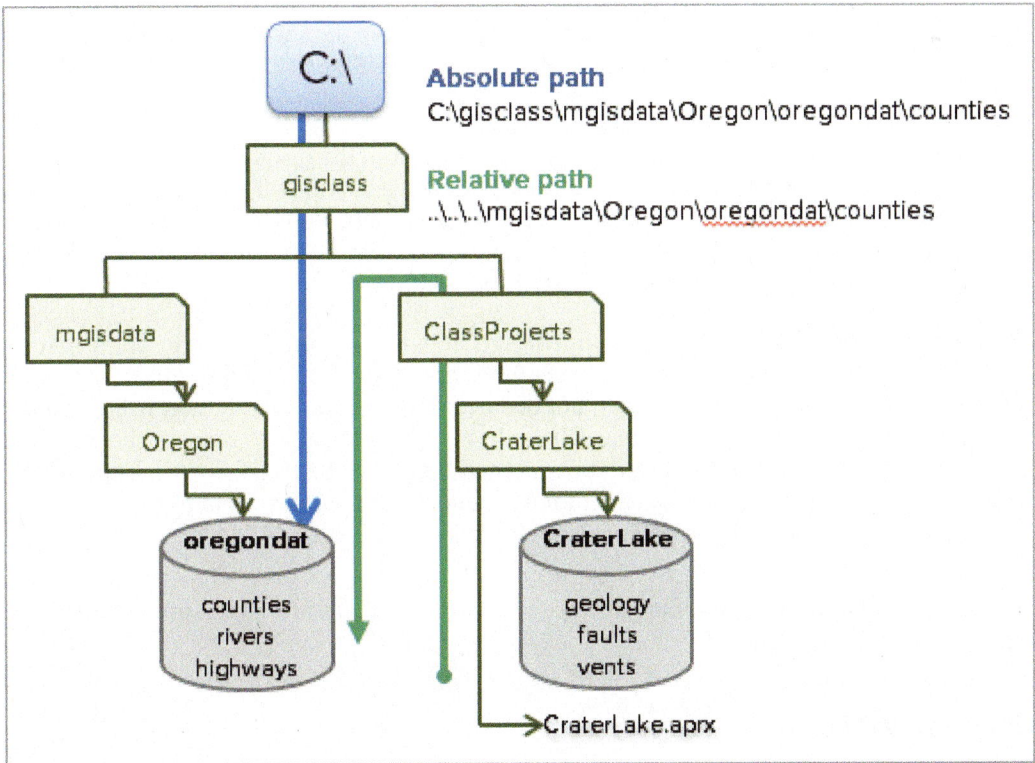

Fig. 5.12. Absolute and relative pathnames

through the mgisdata folder. In a relative pathname, ".." signifies going up one level, so the pathname would be ..\..\mgisdata\Oregon\oregondat\counties.

Projects may use a combination of absolute, relative, or UNC paths depending on the situation. Data on the same hard drive or network location as the project will be accessed using relative pathnames. Items stored on a different hard drive or network location will be referenced using absolute pathnames. This structure works reliably in most cases to keep track of the project resources. For example, the data that come with this book will work fine as long as the gisclass folder is copied from one place to another in its entirety. Since the data were all in the same location to begin with, relative pathnames are used and will continue to work no matter where the folder is placed.

However, if a project in the ClassProjects folder references information on a university's file server, those file server paths will be absolute. If a student copies the gisclass folder to a new computer on the campus network, all the paths will continue to work as long as the new computer has the *same network location* mounted with the same drive letter. But if the student puts it on a thumb drive and gives it to a friend on a different campus, the file server links will be broken because the friend does not have access to the same resources.

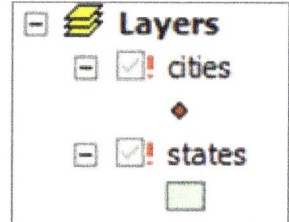

If a project cannot find the data based on the stored path, red exclamation points will appear beside the lost data sets when the map is opened (Fig. 5.13). Any time a folder or file name is changed or the location is moved, the potential exists to break a pathname link and cause the project to lose track of the data. The link can be reestablished manually by opening the layer properties and resetting the path to the correct location (provided the data still exist).

Fig. 5.13. The exclamation point indicates that the pathname link is broken
Source: Esri

Avoid creating projects that reference data on a removable USB drive. The data will be given absolute paths, but thumb drives are not always given the same drive letter each time they are inserted. If the letter changes, access to the data will be lost. If one must store information on a USB drive, create the entire project on the USB drive.

Be aware that using Save As in Pro to save a project in a new location only makes a copy of the .aprx file itself. The paths will be updated to refer to data from the original project folder. If the project is saved in a shared location such as a network drive, it will still work for the creator who has access to the original data on the local computer, but others won't be able to use it. Even worse, using Windows to copy just the project .aprx file to a new location breaks the relative path links for everyone because they won't be updated by Windows.

Pathnames should NEVER contain spaces. ArcGIS will not permit a user to put a space in a GIS data file name, but it is particularly insidious about spaces in folder names. They are permitted and will work for the most part, but occasionally one tool or function will crash if it is used on data with spaces in the pathname. The only way to proceed is to rename the folder, which may cause projects to lose track of the data, and then they must be fixed as well. Experienced GIS users flinch when they see a space in a folder name. Never put spaces in folder names anywhere on a computer, and you will be happier for it.

TIP: What would happen if you removed a space from a folder name when a project already refers to data inside that folder?

Environment settings

The operation of tools can be modified by changing the **Environment settings.** Each tool has an Environments tab that can be used to affect the output of the tool when it is run. Figure 5.14 shows some of the environment settings for the Copy Features tool, which is used when exporting a data set. The first setting, Coordinate System, can be used to modify the output coordinate system without actually having to run the Project tool as a separate step. If the input data are saved in a GCS, but an output data set in UTM is desired, simply modify this setting before running the tool.

ArcGIS Pro has many different Environment settings, and each tool may use only some of them. Only the applicable settings will be shown in the Environments pane for each tool. Once the tool is closed, the setting returns to its default value and will not apply unless it is set again.

Environment settings can also be set for the project, in which case they affect every tool each time it is run (provided the setting is applicable to the tool). These global settings are accessed and set using the **Analysis** ribbon, and they will be saved with the project.

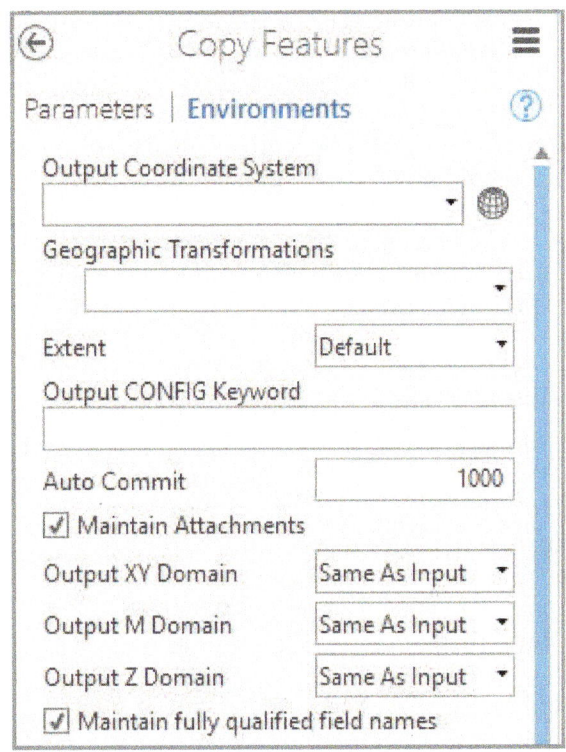

Fig. 5.14. Environments pane of the Copy Features tool
Source: Esri

Summary

▶ GIS data are different from more common data types such as word processing or spreadsheet documents, and they require special considerations.

▶ GIS data have different naming conventions than many other types of data. Folder names and file names should be short and should use only numbers and letters and the underscore character.

▶ GIS users must take responsibility for setting up organized folder trees to store GIS data and making sure that files are saved in the appropriate place with descriptive names.

▶ Data compiled for a project must be taken from data with source scales suitable for the geographic area and intended purpose of the geodatabase.

▶ Finding data for a GIS project requires many general computer skills, such as Internet searching, downloading, unzipping, and converting between file formats. Building these skills through practice, experimentation, and training is important.

▶ Metadata provide information about data sets and are needed for the user to evaluate their fitness for a given purpose. In-house data may use the Item Description for quick documentation.

▶ Map documents store references to feature classes and rasters rather than storing the data. Absolute or relative pathnames may be used, depending on the situation.

▶ Moving, renaming, and deleting GIS data can break links from map documents to the data. Care should be taken during these activities to avoid having to fix broken links.

▶ File locks are used by software to prevent multiple users from editing the same file. Shared GIS data require planning ahead to ensure that multiple users have the access and permissions needed without putting the data at risk.

▶ Compiling data for a project may involve converting formats, geocoding, extracting subsets, simplifying features, or merging data sets.

▶ Environment settings can change the way tools operate and can be useful for streamlining operations or enforcing desired results. The settings can be adjusted for individual tools or for the entire project.

Important Terms

absolute pathname	export	intersection	read access
adjacency	feature class	Item Description	relative pathname
append	feature dataset	locator	shapefile
archive	Feature ID (FID)	lock file	source data
attributes	generalize	merge	spaghetti model
clip	geocoding	metadata	topological model
connectivity	geodatabase,	multipart feature	vertex; vertices
coverage	enterprise	node	write access
dBase	file	Object ID (OID)	zipped
dissolve	personal	overlap	
Environment settings	import	path, pathname	
event layer	Interchange file (.e00)	project file	

Chapter Review Questions

1. What characters should be avoided when naming GIS files, folders, and map documents? Explain why these restrictions are necessary.

2. What is a file extension, and what does the computer use it for?

3. Although ArcGIS Online is a good source of many types of data, what factors may interfere with using it to compile local data for a project?

4. List ways in which users of GIS and other large-scale software systems need to be more careful about working with data.

5. Explain the difference between a feature class and a layer. How does this difference support the ability of users to share the same data sets?

6. What does a red exclamation point next to a map layer mean? How does one fix it?

7. Extracting subsets of a data set is easy; combining different data sets into one feature class is difficult. Explain why.

8. What is the difference between the Summary and Description fields in the Item Description?

9. Explain the difference between the terms *feature, feature class, feature dataset,* and *geodatabase.* Which of them corresponds to a shapefile?

10. Explain what happens when a feature class is generalized or dissolved. What are some reasons why it would be done?

Extend your knowledge

Extend your knowledge by reading these sections of the ArcGIS Pro Help.

> Help > Data > Geodatabases > The geodatabase
>
> Help > Data > Geodatabases > Add datasets to a geodatabase
>
> Help > Data > Data types > Shapefiles
>
> Help > Data > Data types > Geocoding
>
> Help > Metadata > View and edit metadata
>
> Help > Analysis and geoprocessing > Basics > Geoprocessing environment settings

Mastering the Skills

Teaching Tutorial

The following examples provide step-by-step instructions for doing basic tasks and solving basic problems in ArcGIS. The steps you need to do are highlighted with an arrow →; follow them carefully.

TIP: Make sure your computer is set to show file extensions; in Windows this setting is generally found in the Control Panel under Folder Options (see Fig. 5.4).

Understanding types of GIS data

This chapter introduced the variety of forms that GIS data may take, as well as some complications in where they are stored and how they are used. In this first section, we will explore these concepts. Later, we will assemble a geodatabase for New Jersey, so we will start by creating a New Jersey project.

> 1→ Start ArcGIS Pro and choose to create a new blank project. A default name and location are suggested, but wise users mindfully select appropriate names and locations for data.
>
> 1→ Name the project *NewJersey* and specify the gisclass\ClassProjects folder as the location to store it.
>
> 1→ Use **Insert: Project: New Map** to create a new map. Rename the map *NJ Data*.
>
> 1→ Use the Catalog Favorites tab to add the mgisdata folder to the project.

  1. Why is it all right to use a space in a map name but not in a folder or feature class name?

Recall from Chapter 1 that new projects are normally created inside a folder with the same name, which contains a home geodatabase, also with the same name. Non-GIS files like reports or spreadsheets may be saved within the folder or within subfolders created to keep this information organized. GIS outputs will be saved in the home geodatabase by default.

> 2→ Open the Catalog view in the display area (not the Catalog pane). Remember to use the Contents pane, not the Catalog pane, to navigate when in Catalog view.
>
> 2→ Click the Databases entry to see the home geodatabase, NewJersey.
>
> 2→ Click the Folders entry to see the home folder, NewJersey, and the connection to the mgisdata folder just added.
>
> 2→ Double-click the mgisdata folder to see its contents. The data underneath is organized geographically.

2→ Open the Oregon folder and the oregondata geodatabase.

2→ Choose **Catalog: Options: Display Type > Tiles**. Examine the data type beneath each data set name (Fig. 5.15).

2→ Explore the data in the mgisdata folder. Try to find an example of each data type listed in Table 5.1.

2. Which data types in Table 5.1 are NOT found in mgisdata?

3→ In the Contents pane, click the Portal > All Portal entry. Type *New Jersey* in the Search Portal box and press Enter.

3→ Examine the different icons and the data type each represents. Try to find an example of a feature layer and a layer package.

3→ Click an item to view its metadata. Examine the metadata for several data sets, and note which appear to be professionally authored and which appear slapdash. The quality of the metadata is an important clue to the quality of the data set.

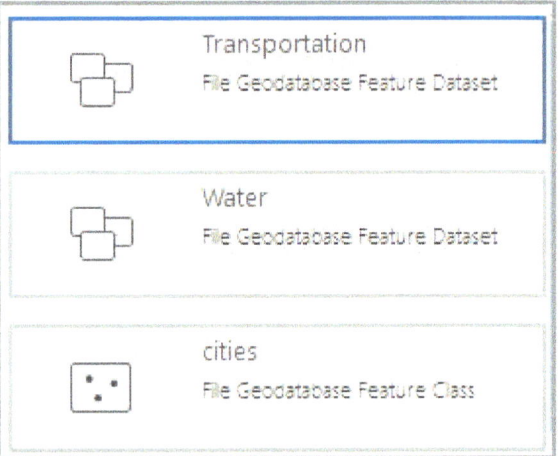

Fig. 5.15. Feature datasets and a feature class in the Catalog view

Source: Esri

Now we will begin compiling data for New Jersey and saving it in the NewJersey home geodatabase. As in Chapter 4, we will project all data sets to a common coordinate system, but we will also learn additional techniques, such as subsetting data to a region of interest, using Environment settings, getting data from Internet sources, and getting data from ArcGIS Online.

Importing and exporting data

Best practices dictate choosing a common, appropriate coordinate system for data within the home geodatabase. We have chosen NAD 1983 New Jersey State Plane as the project's coordinate system, and we will start by setting the map to it.

4→ Switch to the NJ Data map view and open the map properties for the NJ Data map. View the Coordinate Systems settings.

4→ Type *state plane* in the search box and click Enter. Expand the State Plane entry in the Projected coordinate system folder.

4→ IMPORTANT: Scroll down and expand the plain NAD 1983 (meters) entry (Fig. 5.16). Do NOT choose ones labeled CORS96, PA11, HARN or anything else!

4→ Find the NAD 1983 StatePlane New Jersey FIPS 2900 (Meters) coordinate system and click to select it. Also click the star to add it to your Favorites. Click OK.

▷ NAD 1983 (CORS96) (Intl Feet)

▷ NAD 1983 (CORS96) (Meters)

▷ NAD 1983 (CORS96) (US Feet)

▷ NAD 1983 (Intl Feet)

▷ NAD 1983 (Meters)

▷ NAD 1983 (PA11) (Meters)

▷ NAD 1983 (PA11) (US Feet)

▷ NAD 1983 (US Feet)

▷ NAD 1983 HARN (Intl Feet)

Fig. 5.16. The standard NAD 1983 datum is usually the correct choice.

Source: Esri

TIP: For the rest of the tutorial, we will refer to the chosen coordinate system as NJ State Plane instead of using the longer formal name.

We will start by importing some data available in the mgisdata\Usa folder. The data cover the United States, but we only need the New Jersey features, and we can subset them during the export. We will be using both the Catalog pane and the Geoprocessing pane for a while, so it will be convenient to have them both open at once.

TIP: Remember to always save output to the home geodatabase unless directed otherwise.

5→ Arrange the Geoprocessing and Catalog panes side by side. (If both are currently in the same window, click the <u>tab</u> of one to separate it before docking it elsewhere.)

5→ Make sure the Catalog pane shows the Project tab with the local data. Expand the Folders entry to see the contents of the mgisdata\Usa\usdata geodatabase.

6→ In the Geoprocessing pane, open the Feature Class to Feature Class tool (Fig. 5.17).

6→ In the Catalog pane, find the states feature class in the usdata geodatabase. Click and drag it to the Input Features box of the tool.

6→ In the tool, leave the Output Location set to the home geodatabase, NewJersey.

6→ Name the Output Feature Class njbound.

6→ Click Add Clause. Enter the expression: STATE_NAME Is Equal to New Jersey.

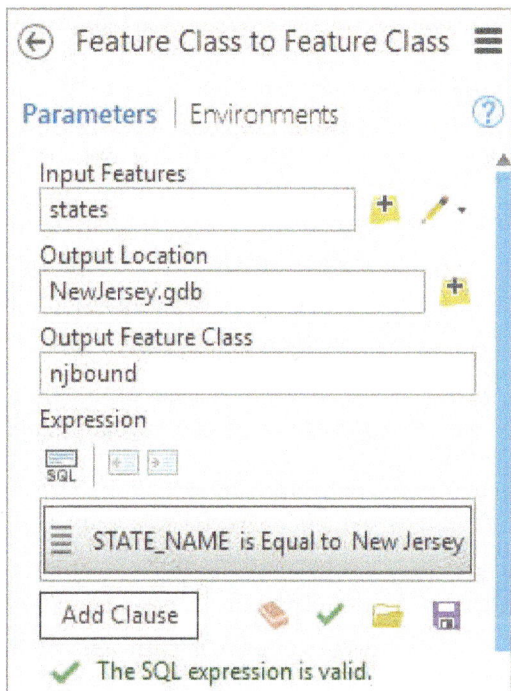

Fig. 5.17. Importing the state boundary
Source: Esri

6→ Click the green check mark under the clause to verify the expression.

The Field Map section of the tool is used to change the fields in the table when the data set is copied. We want to retain all the fields, so don't set anything in this section.

6→ Click Run to run the tool. The state of New Jersey appears on the map.

6→ In Contents, open the layer properties for the njbound layer and examine the Spatial Reference entry in the Source settings.

The source data were stored in a North America Equidistant Conic coordinate system, which was copied to the new data set but is not what we want. We should run the Project tool next. We also notice that the boundaries of the state are generalized, a result of the small source scale.

7→ In the Geoprocessing pane, open the Project tool.

7→ Set the Input Dataset to njbound, and name the output njbound_gen.

7→ Set the Output Coordinate system. Since we set the map CS, we can simply choose Current Map [NJ Data] from the drop-down to set it to NJ State Plane.

7→ Run the tool.

We no longer need the njbound feature class in the wrong projection, so let's take a moment to delete it from the home geodatabase. There is no Undo, so be careful to delete the right one.

8→ In the Catalog pane, expand Folders\NewJersey and then expand the NewJersey home geodatabase.

8→ Right-click the njbound feature class and choose Delete. Click Yes to confirm.

Using Environment settings

Although the process worked, it is tedious to export, project, and delete the intermediate feature class each time. Environment settings can streamline this process by setting the desired output coordinate system within a tool. Let's try it on the next export, the New Jersey cities.

9→ Open the Feature Class to Feature Class tool.

9→ From the Catalog pane, click and drag the cities feature class from the usdata geodatabase to the Input Features box on the tool.

9→ Leave the Output Location as the NewJersey home geodatabase.

9→ Name the Output Feature Class *njcities*.

9→ Click Add Clause and enter the expression: ST Is Equal to NJ.

Note that the expression is different this time. Each table contains different fields and values, so the expression will usually change each time. Now let's set the environment.

9→ Near the top of the tool, click the Environments tab.

9→ Set the Output Coordinate System drop-down to Current Map [NJ Data]. Leave the other settings alone.

9→ Click the Parameters heading to return to the tool parameters. Run the tool.

9→ In Contents, open the layer properties for the new njcities layer and examine the spatial reference in the Source settings.

The njcities have the State Plane coordinate system, and the process was far more efficient. Let's do it the same way for the next feature class, for practice.

10→ The Feature Class to Feature class tool is probably still open. Click the back arrow and reopen the tool to clear the current parameters.

10→ Set the Input Features to counties by dragging them from the Catalog pane.

10→ Name the Output Feature Class *njcounties*.

10→ Enter the expression: STATE_NAME Is Equal to New Jersey.

10→ Click the Environments tab and set the Output Coordinate System to Current Map [NJ Data].

10→ Switch back to the Parameters tab and click Run.

TIP: Sometimes it is better to clear the tool parameters using the back arrow, and other times it is convenient to simply edit the current settings. Feel free to experiment.

The Environments setting is easier than the Project tool, but one can easily forget to set it each time. We can set Environments at the application level so that it applies every time a tool runs.

11→ Click the **Analysis: Geoprocessing: Environments** button.

11→ Under Output Coordinates, set the Output Coordinate System to Current Map [NJ Data]. Leave the other settings unchanged and click OK.

Let's test it on the congressional districts feature class (cd114). It is now unnecessary to set the Environments in the tool.

12→ Clear the Feature Class to Feature Class tool and reopen it. Export the cd114 feature class and save it as *njdistricts*, selecting only the New Jersey districts, and save them as *njdistricts*.

12→ Check the output and make sure that it includes only the New Jersey districts and is stored in the State Plane coordinate system.

Until we change the Environment setting, every new feature class will be saved in the New Jersey State Plane coordinate system.

Understanding layers

As the database was assembled, the new data sets were saved in the home geodatabase and added to the map. Recall the distinction between layers and data sets. Data sets are stored on the hard drive or come through a portal, and they are shown in the Catalog pane. When a data set is added to a map, a layer appears in the Contents pane. A layer stores the location of the source data set and has properties that can be manipulated to change its appearance in the map.

13→ In the Catalog pane, expand the NewJersey geodatabase to see the feature classes.

13→ In the Contents pane, view the layers that belong to the NJ Data map.

13→ From the Catalog pane, drag the njcounties feature class into the map, making two njcounties layers.

13→ To help keep track, rename the new layer njcounties2.

13→ Turn off all layers except njcounties, njcounties2, and the basemap.

Each layer is distinct and can have its own properties.

14→ Change the color of njcounties2. Note that the njcounties symbol stays the same.

14→ Open the layer properties for njcounties and examine the Source database and feature class name. Make a note of it and close the window.

14→ Open the layer properties for njcounties2 and compare the data sources.

14→ In Contents, right-click njcounties2 and choose Label to turn on the labels.

14→ Turn njcounties2 off and on to verify that njcounties has no labels.

Both layers refer to the same location and the same feature class in the home geodatabase. Each layer presents a different view of the same data. However, a change to the source data, such as deleting a field, would affect both layers in the map.

14→ In Contents, right-click the njcounties2 layer and choose Remove.

TIP: Remember: The Catalog pane shows feature classes and other data sets stored on disks and servers. The Contents pane of a map shows layers and controls the display properties. Most changes to layers affect only the layer properties, but some changes, like deleting fields, affect the data set.

Updating metadata

The metadata are automatically copied when a data set is imported or exported. However, in this case, the original data set was modified (by subsetting it to New Jersey) and saved to a new location. Both of these changes should be recorded in the metadata.

15→ Auto-hide the Geoprocessing and Catalog panes to get more space, if needed.

15→ Open the Catalog view in the display area.

15→ Navigate to the NewJersey geodatabase and expose its contents.

15→ Click the njcities feature class to view its metadata.

16→ Click the **Catalog: Metadata: Edit** button. An editing view opens for njcities.

16→ In the Description box, add this statement at the end: *Subset to New Jersey from a national data set by <your name> on <current date>.*

16→ In the Credits box, construct a citation to document the source of the data: *Mastering ArcGIS Pro, Tutorial Data, 1st edition by Maribeth Price (2019) [download]. McGraw-Hill Higher Ed: Dubuque, Iowa.*

16→ Click **Metadata: Manage Metadata: Save** and close the njcities editing view.

17→ To update the thumbnail, click the Preview tab (beneath the metadata).

17→ Click the **Preview** ribbon.

17→ Zoom or pan to adjust the view and click Create Thumbnail.

17→ Return to the Metadata tab. If the new thumbnail does not immediately appear, click a different data set and then return to njcities.

TIP: The Item Description should be updated when a data set is created or modified. Although a formal citation in the Credits box isn't required, it saves time when a citation is needed for a map.

Let's quickly update the metadata for the remaining items in the database.

18→ In the Catalog view, with njcities selected, highlight the citation text and click Ctrl-C to copy it (to save typing).

18→ In the Catalog view window, click the next item, njcounties.

18→ Click the **Catalog: Metadata: Edit** button to open the metadata for editing.

18→ Paste the citation into the Credits box, and edit the Description as before

18→ Save the edits and close the njcounties metadata editing view.

18→ Switch to Preview and update the thumbnail (as in Step 17); then switch back to Metadata.

19→ Edit the metadata for the remaining items in the geodatabase. (You can use Word or Notepad to copy and paste the repeating text rather than typing it each time.)

19→ When finished, leave the Catalog view open, but switch back to the NJ Data map.

19→ Save the project.

Two types of metadata may be edited. The previous steps modified metadata that are stored as part of the data set, which remain with the data set when it is moved, copied, exported, or renamed. Another type of metadata are merely created within the project, the layer metadata.

20→ In the Contents pane, open the layer properties for the njcities layer and examine the Metadata settings.

20→ At the top, click the drop-down box and examine the two choices.

The layer metadata can show the data set metadata (the default), or they can contain separate metadata entered by the user just for this layer, which might be simpler, more user friendly, or more germane to the layer's function in this particular map.

20→ Set the drop-down to *Layer has its own metadata*. All the boxes go blank.

20→ Enter some brief metadata items specific to New Jersey and the educational purpose of this data set. Click OK.

The layer metadata only appear <u>in this project and for this layer</u> (although if the layer were copied to a different map in the same project, the layer metadata would be copied with it). Editing the layer metadata is NOT the same as editing data set metadata in the Catalog view.

TIP: Click *Copy data source's metadata to this layer* to be able to start with and modify the data set metadata instead of starting from scratch. Any existing layer metadata will be replaced.

3. Notice that if the term *layer* is used in the instructions, then you are directed to use the Contents pane. If the term *feature class* is used, then you are directed to the Catalog pane. Explain why.

Clipping data

So far we have used information in the attribute table to select only New Jersey features. This strategy works well when the table has a suitable field to use, but that is not always the case.

21→ If you wish, unhide the Catalog and Geoprocessing panes so they remain open.

21→ In the Catalog pane, expand the Transportation feature dataset in the usdata geodatabase to see the feature classes inside it.

21→ Drag the majroads feature class into the NJ Data map.

21→ In Contents, right-click the majroads layer and choose Attribute Table.

21→ Examine the fields in the table, and close it when finished.

The majroads feature class has no STATE field, so we cannot select the New Jersey roads with an expression. However, the Clip tool extracts features from one data set within the boundary of another.

22→ In the Geoprocessing pane, search for Clip. Two Clip tools appear, Clip (Analysis Tools) and Clip (Data Management Tools).

22→ Hover over each Clip tool and read the pop-up description.

4. What is the key difference between these two tools? Which is the correct one to use in this case, and why?

22→ Open the Clip (Analysis Tools) tool and click on the circled question mark instead of hovering. The tool Help opens. Skim through it and examine the diagrams.

TIP: It is a good practice to consult the Help of any tool before using it for the first time.

23→ In the Clip tool, set the Input Features to majroads and the Clip features to njdistricts (Fig. 5.18).

23→ Name the output *njroads*. Run the tool.

23→ Remove the original majroads layer from the map.

23→ Zoom to the stubby peninsula of southeastern New Jersey and compare the njdistricts and njbound_gen layers to the basemap by turning them on and off.

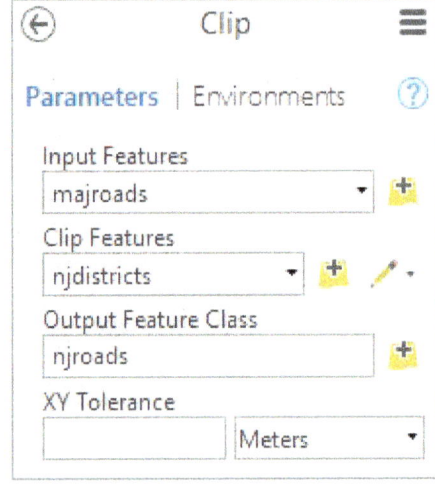

Fig. 5.18. Clipping the roads
Source: Esri

5. Why do the districts and the state boundary layers have different boundaries? Why is the state boundary so different from the basemap boundary for New Jersey?

6. What would happen if you clipped the njdistricts layer with the njbound_gen layer instead of using the Feature Class to Feature Class tool? Why would the result be a problem?

TIP: Clip may modify the original boundaries of a polygon data set, so it is preferable to use the Feature Class to Feature Class tool with a selection expression whenever possible. Reserve Clip for times when the table has no appropriate field.

Exporting from the Contents pane

An alternate method of exporting data to the home geodatabase is to export from the Contents pane instead of opening the Feature Class to Feature Class tool. This method allows the data to be reviewed first. Suppose we wish to add earthquakes to the NewJersey geodatabase but are not sure that the state actually has any. We can check first.

24→ Zoom back to New Jersey. From the Catalog pane, click and drag the usdata\quakehis feature class to the map. A few quakes do appear in the state.

24→ Search for and open the Select Layer By Attribute tool.

24→ Set the Layer Name to quakehis.

24→ Click Add Clause to enter the expression: STATE Is Equal to NJ.

24→ Run the tool. The selected features are highlighted on the map.

25→ In Contents, right-click the quakehis layer and choose Data > Export Features.

25→ Set the Input Features to quakehis and name the Output *njquakes*. Run the tool.

25→ Remove the quakehis layer from the map.

When certain features are selected, the Copy Features tool only acts on the selected ones. Most tools exhibit the same behavior; it is called *honoring the selected set*. As we continue this lesson, use either export method to create the new data sets.

26→ Save the project.

26→ Open the Catalog view and update the metadata for each new data set.

26→ While still in Catalog view, examine the metadata for njbound_gen.

This state boundary comes from a national data set optimized for quick viewing rather than geometric accuracy. The source scale of njbound_gen is not really appropriate for this project.

TIP: Pay attention to source scales when compiling data, and try not to use data sets too far outside their intended scale range.

Using data from ArcGIS Online

Let's look for a more detailed state boundary online.

27→ In the Contents pane (showing the Catalog view), click the All Portal entry.

27→ Type *detailed states* in the search box.

27→ Find the entry USA States, a layer package hosted by Esri™. Read the metadata and write down a citation for the layer package (see Chapter 1 examples).

_____

27→ Right-click the listing for USA States and choose Add to NJ Data.

27→ Switch back to the NJ Data map.

A layer package downloads data to the computer and typically allows features to be exported.

28→ In Contents, expand the USA States group layer, which includes layers from several source scales).

28→ Click **Map: Selection: Select By Attributes**. Set the Layer Name to USA States (below 1:3m) and enter an expression to select New Jersey.

28→ Export the USA States (below 1:3m) layer and save it with the name *njbound_dtl*.

29→ In Catalog view, in Contents, view the NewJersey geodatabase contents.

29→ Update the Item Description for the njbound_dtl feature class to document the subset, and enter this citation in the Credits box: *USA States (2013) [Layer package] Esri on ArcGIS Online, URL: http://ArcGIS.com [<today's date>].*

TIP: Many ArcGIS Online data sets do not allow a user to export them. If the Data > Export Data option does not show when a layer is right-clicked, the layer is not exportable.

Now that we have a state boundary with a suitable scale and coordinate system, we can delete the less detailed version, newjersey_gen.

30→ In the Catalog view, right-click the njbound_gen feature class and choose Delete. Click Yes.

30→ Switch back to the NJ Data map and remove the USA States layer.

TIP: ALWAYS use the Catalog pane or Catalog view to delete, move, or rename GIS data sets.

Getting data from the Internet

Finding and downloading data from the Internet can be challenging, but let's give it a try. First, create a separate folder for downloads, which can be messy and are best kept separate from the permanent project data.

31→ Switch back to the Catalog view and click the Folders entry in Contents.

31→ Right-click the NewJersey Folder Connection and choose New > New Folder.

31→ A new folder appears in the Catalog view, ready to be named. Type the name *Downloads* (Fig. 5.19) and click Enter.

A good way to search for data is to type a search term plus the word *GIS* or *shapefile*.

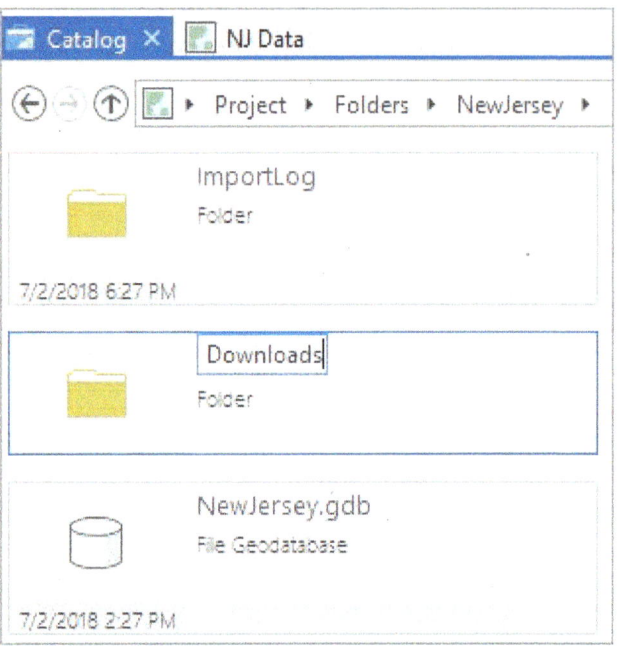

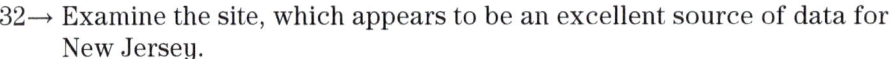

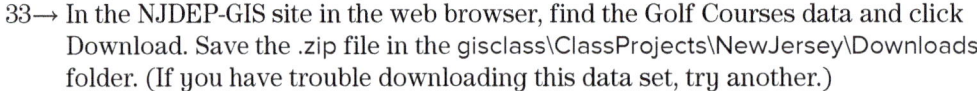

Fig. 5.19. Creating a folder
Source: Esri

32→ Open a web browser and enter the search term *New Jersey shapefile*.

32→ Examine the results. Look for URLs that appear to be state-sponsored rather than commercial, such as the Bureau of GIS – State of New Jersey heading. (URL: http://www.state.nj.us/dep/gis/stateshp.html). Open it.

32→ Examine the site, which appears to be an excellent source of data for New Jersey.

33→ In the NJDEP-GIS site in the web browser, find the Golf Courses data and click Download. Save the .zip file in the gisclass\ClassProjects\NewJersey\Downloads folder. (If you have trouble downloading this data set, try another.)

33→ Minimize (do not close!) the web browser for a moment.

TIP: The data file must be unzipped before ArcGIS can recognize it. The instructions assume that you have a zip utility and can use it. If you need one, download and install 7zip for free.

33→ Open File Explorer and navigate to the NewJersey\Downloads folder.

33→ Use a zip utility to extract the files to the NewJersey\Downloads folder.

Notice the different extracted files with the same root name njgolf and different extensions. Pro can read this shapefile now.

34→ Close File Explorer. In Pro, open the NJ Data map.

34→ Add the njgolf shapefile to the map from the Catalog pane.

34→ In Contents, right-click the njgolf layer and export the features to the NewJersey home geodatabase. Name the output njgolf.

TIP: The Catalog pane does not always refresh when a new data set is created. Right-click the geodatabase or folder and choose Refresh.

The file has not been changed, so the Item Description does not need to be updated. However, it is a good practice to create a citation in the Credits entry to record where it came from.

> **TIP:** Search the web site and/or open the metadata link for the data set to find all the information needed for the citation. That is why we kept the site page open.

35→ Review the data set description on the web site.

35→ Click the Metadata button for the golf courses data set on the web site. A view opens of the formal metadata, in a format known as FGDC.

35→ Prepare a formal citation for this data set: *Golf Courses (1995) [downloaded file] New Jersey Department of Environmental Protection, URL: http://www.state .nj.us/dep/gis/stateshp.html [<today's date>].*

7. Where did the 1995 date for the data set come from?

35→ In Catalog view, click on the njgolf feature class in the NewJersey geodatabase. A message about FGDC metadata appears.

35→ Choose **Catalog: Metadata: Upgrade > FGDC CSDGM Content.**

35→ Edit the metadata and add the citation to the Credits section.

Let's clean up a little.

36→ Switch back to the NJ Data map.

36→ In Contents, switch to the List By Data Sources panel.

36→ Remove any of the original sources that remain, and make sure that all data in the map have their source in the NewJersey home geodatabase.

36→ Open the map properties for NJ Data and examine the Coordinate Systems settings. Expand the Layers folder and verify that all layers except the basemap are in NJ State Plane (Fig. 5.20).

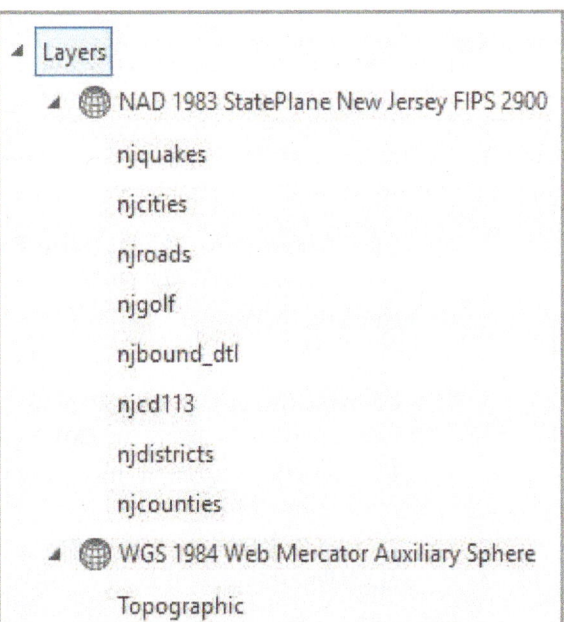

Fig. 5.20. Check the final layers' coordinate systems
Source: Esri

> **TIP:** ArcGIS Desktop comes with an extensive set of GIS data called ESRI Data and Maps, which can be downloaded by a GIS software administrator. Ask your instructor about obtaining access. Some of the data are available on ArcGIS Online, like the USA States layer package used earlier.

Adding data from tables

In some cases, information from a table, such as a text file or Excel spreadsheet, can be converted to locations on a map. One method uses explicit x-y locations in the table to create points.

171

37→ Add the climatestations2005 table from the mgisdata\Usa folder. It contains longitude (x) and latitude (y) locations of climate stations in the continental US.

37→ Open the table and examine the fields.

38→ Use the Table: View: Select By Attributes button to select the New Jersey climate stations using the expression: state is Equal to NJ.

38→ Check to make sure that the 12 NJ stations are selected in the table.

39→ In Contents, right-click the table and choose Display XY Data, which opens the XY Table to Point tool (Fig. 5.21).

39→ Name the Output Feature Class *njclimatestations*.

39→ Keep the X Field set to long_ and the Y Field set to lat.

 39→ Click the *Select coordinate system* button and select the GCS_North_American _1983 coordinate system (it should be in the Favorites). Click Run.

39→ Remove the climatestations2005 table.

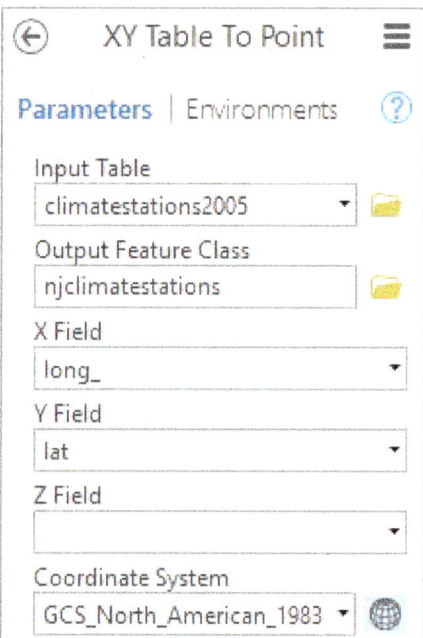

Fig. 5.21. Display *x-y* table values as points

Source: Esri

The second method uses geocoding techniques to convert addresses to locations on a map. This method requires an ArcGIS Online subscription and consumes credits, so we will be careful to geocode only the New Jersey locations of Wal-Mart stores.

40→ Add the store_openings.csv file from the mgisdata\Usa folder.

40→ Open the table and examine it, particularly the address fields.

A csv file, containing comma-delimited text, cannot be queried by Pro directly, so we first must export the file to the home geodatabase as a table.

40→ Click **Standalone Table: Data: Export: Export Table**. Save it in the project geodatabase as *AllWalmarts*.

40→ Open the AllWalmarts table.

41→ Open the Select Layer By Attribute tool and select the New Jersey stores in the AllWalmarts table using the expression: STRSTATE is Equal to NJ.

41→ Examine the bottom of the table to see how many rows were selected; it should read *41 of 3176 selected.*

41→ Choose **Standalone Table: Data: Export: Export table** and save the output as *NJWalmarts*.

42→ Open the NJWalmarts table and confirm that it contains 41 rows and that all of them contain NJ in the STRSTATE field.

42→ Remove the AllWalmarts and storeopenings.csv tables.

43→ Open the Geocode Addresses tool. Set the Input Table to NJWalmarts (Fig. 5.22).

43→ For the Input Address Locator, select the ArcGIS World Geocoding Service.

43→ Set the Address or Place box to STREETADDR, the City field to STRCITY, and the ZIP field to ZIPCODE.

43→ Name the Output Feature Class NJWalmartPts.

43→ Check the box for United States for the country.

43→ Click Run. It may take a few minutes to finish. Wait for it to display the new points on the map.

43→ Save the project.

TIP: If the tool does not recognize the locator, try saving the project, closing Pro, and starting again. If the tool runs and fails, just go on.

Appending feature classes

Data sets are often tiled, or split between regular squares or rectangles. For a study, it is often more convenient to merge the tiles into a single data set. We will illustrate this process using the Black Hills water study data from Chapter 4.

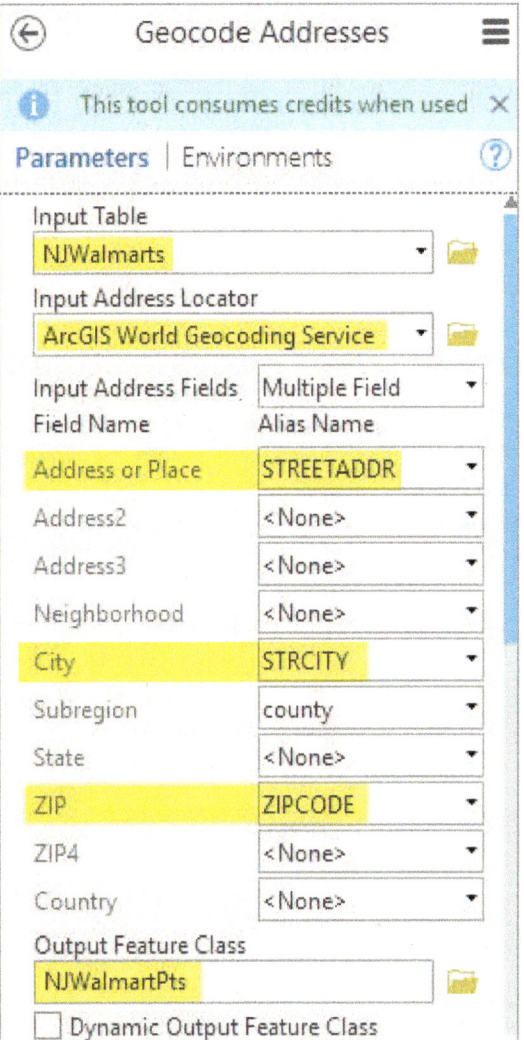

Fig. 5.22. Geocoding the stores
Source: Esri

44→ Choose Project > Open and click your recent project BHWaterStudy to open it.

44→ Open the Study Quads map if it is not already open.

44→ Click **Analysis: Geoprocessing: Environments** and set the Output Coordinate System to match the Current Map (NAD 1983 UTM Zone 13).

A colleague has already downloaded the files from the South Dakota Geological Survey web site.

45→ Turn off all layers except the Topographic basemap.

45→ In the Catalog pane, expose the files in mgisdata\BlackHills\Downloads. Add the HillCityStructure.shp and MtRushmoreStructure.shp shapefiles to the map.

45→ Open the tables for both shapefiles and compare the fields.

The Hill City shapefile has one user-defined field, Category. The Mount Rushmore file has a Name field and a Category field. We will retain both fields, although the Hill City features will be given null values for the Name field since they have no corresponding values in their table.

46→ In the Geoprocessing pane, search for and open the Merge tool (Fig. 5.23).

46→ Click the Input Datasets drop-down box and select MtRushmoreStructure.

46→ Click the second Input Datasets drop-down box and select HillCityStructure.

46→ Name the Output Dataset GeolStructures.

46→ Examine the Field Map.

Merge will attempt to keep all fields. The fields that appear in both input files have a (2) by the field name. The other fields are found only in the Mount Rushmore data set. These fields will contain data for the Mount Rushmore features and null values for the Hill City features.

47→ Run the tool.

47→ Remove the original HillCityStructure and MtRushmoreStructure layers.

47→ Examine the table of the merged GeolStructures feature class.

47→ Close the table and save the project.

Now let's try merging the geologic rock units.

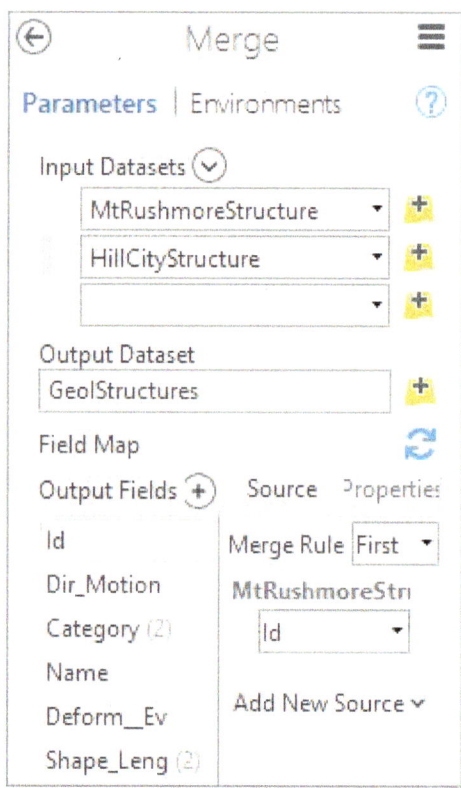

Fig. 5.23. Merging the geologic structures
Source: Esri

48→ Add the HillCityGeology.shp and MtRushmoreGeology.shp shapefiles from the mgisdata\BlackHills\Downloads folder.

48→ Open the two tables and compare the fields.

In this case, none of the field names match, so we must use field mapping to transfer the right data from the old fields to the new fields. We want the output data set to contain a field for the unit name (such as Alluvium) and a field for its shortened geologic unit code (Qal).

48→ On a sheet of paper, write the field names in HillCityGeology that contain these two types of information. Repeat for MtRushmoreGeology.

49→ In the Geoprocessing pane, clear and reopen the Merge tool.

49→ Set the Input Datasets to HillCityGeology and MtRushmoreGeology.

49→ Name the Output Dataset Geology.

The next step is to build a field map that draws information from both data sets into the fields we want to keep. We will keep the better-named MtRushmoreGeology fields Fm_Name and Unit, and match them to the corresponding fields Formation1 and Formation_ in HillCityGeology.

50→ Under Field Map, click the Fm_Name field (Fig. 5.24a).

50→ Leave the Merge Rule set to First. The box underneath indicates that this field gets its value from MtRushmoreGeology.

50→ Click the Add New Source drop-down, which opens a pop-up field selector with the layers on the left and their fields on the right.

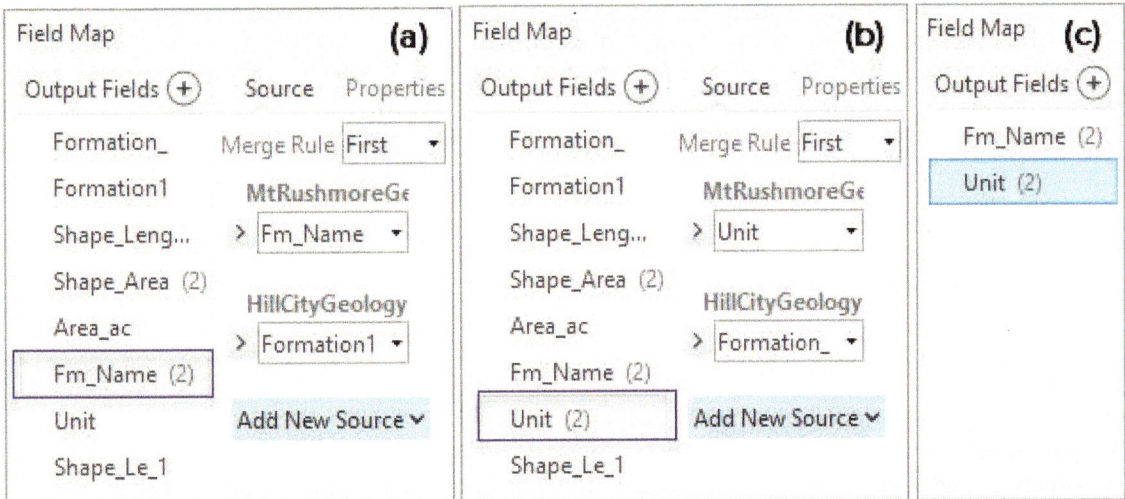

Fig. 5.24. Merging the fields
Source: Esri

50→ In the pop-up, select HillCityGeology, check the Formation1 field, and click Add Selected. The Field map should appear as in Figure 5.24a.

50→ On the left, click the Unit field name (Fig. 5.24b).

50→ On the right, click Add New Source, select HillCityGeology, check the Formation_ field, and click Add Selected.

Examine the list of output fields in Figure 5.24b. Fields followed by (2) have information from both input tables, including the critical information for Fm_Name and Unit. For a cleaner table in the output, we should delete fields that only have one data source.

51→ Under Output Fields, click Formation_ and use the red X to remove it from the list.

51→ Remove all of the output fields except Fm_Name and Unit. (New Shape_Leng and Shape_Area fields will automatically be created for the output data set.)

51→ The final field map will appear as in Figure 5.24c. Run the tool.

52→ Open the table for Geology and examine the fields. Scroll down and make sure that every feature has both a formation name and a unit symbol. Close the table.

52→ Remove the original HillCityGeology and MtRushmoreGeology layers.

53→ Symbolize the Geology layer with a unique values map based on the Unit field.

Notice the vertical line separating the polygons along the merged edge. The Dissolve tool can remove these lines.

54→ In the Geoprocessing pane, open the Dissolve tool.

54→ Set the Input Features to Geology.

54→ Name the Output Features *Geology_Dissolve*.

54→ Set the Dissolve_Fields to Fm_Name and Unit.

54→ Uncheck the box to create multipart features. We want each polygon to remain separate in the output dataset. Run the tool.

55→ Create a unique values map for the Geology_Dissolve layer, based on the Unit field, and examine the boundaries.

Most of the boundary lines disappeared except a few near the top, where the unit name is slightly different for the two data sources, even though the unit symbol was the same. Editing would be required to remove these final lines.

This is the end of the tutorial.

→ Save the project and close ArcGIS Pro.

Practice Exercises

Choose any state except New Jersey.

1. Create a new project for the chosen state, saving it in the gisclass\ClassProjects folder.

2. Open a new map and set the coordinate system to the most appropriate State Plane zone in the State Plane > NAD 1983 (meters) folder. If there is more than one zone for the state, just pick one, preferably the central one.

3. Set the Analysis Environment to match the chosen coordinate system for the map.

4. Export each of the vector feature classes from the usdata geodatabase to the new state geodatabase, taking care to export only the features in the state.

5. Update the Item Description for each data set exported.

6. Find at least two additional feature classes for the state geodatabase, either from ArcGIS Online or a web site, choosing ones that you find interesting. Import them to the geodatabase as well.

7. Create a map of the state layers, appropriately symbolized (omit some layers if the map is too cluttered).

8. Compile these items into a document to submit to your instructor:

   a. Screenshots of the map, including the Contents pane showing the layers
   b. A screenshot of the Layers folder in the map properties (as in Fig. 5.20) to show that all layers are in the same coordinate system
   c. A screenshot of the map showing only the Internet data obtained
   d. A list of citations, one for each data set

Chapter 6. Managing Raster Data

Objectives

▶ Understanding the raster data model and properties of rasters

▶ Managing raster data sets and their coordinate systems

▶ Understanding models for 3D data sets

Mastering the Concepts

GIS Concepts

Discrete data models deal with spatial objects as points, lines, or polygons, but other approaches are needed to store and model continuous surfaces that are based on measured quantities, such as elevation, that vary from place to place on the earth. One of those quantities may be reflectance as measured by satellites or aerial camera systems. Although rasters are among the most common methods of storing surface data, they can also represent discrete objects when needed. Other approaches to storing surfaces include contours, TINs, point clouds, and more. Furthermore, elevation surfaces also provide the means of extending GIS into the third dimension. This chapter focuses on storing and managing raster data, imagery, and 3D geospatial data.

The raster model

The raster model has the benefit of simplicity. The geographic area represented is divided into small squares, called **cells** or **pixels** (Fig. 6.1). Each pixel contains a number indicating a single measured value or attribute, and the raster is stored as an array of numeric values with a fixed number of rows and columns. Each pixel has an "address" indicated by its position in the array, such as row = 3 and column = 6. Rasters can be used to store many kinds of data, including a photograph or a scanned document.

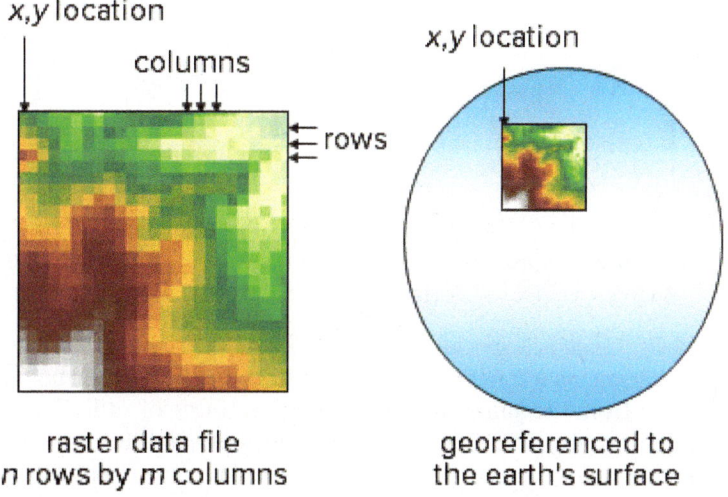

Fig. 6.1. Rasters store arrays of numbers fixed to the earth by a single *x-y* location
Source: USGS

When a raster is used to store geospatial information, it is georeferenced to the earth's surface by declaring an *x-y* location for one corner, a dimension for the cell in real-world units such as degrees or meters, and a real-world coordinate system (a GCS or a projection). Usually the upper-left corner is chosen as the known location, and the *x* and *y* pixel dimensions are the same so that the pixels are square. The location and cell size make it possible to calculate the coordinates of every pixel based on its row and column position. In this sense, the georeferencing of the pixels in a raster data set is implicit—one need not store the x-y location of every pixel. The x-y location may refer to the upper left corner or the center of the first cell, depending on the raster format.

Rasters excel at storing continuous data, which are quantities or variables that change over the earth's surface. A **digital elevation model (DEM)** like the one shown in Figures 6.1 and 6.2a, stores elevation values. Cells are unlikely to have the same elevation as their neighbors, and the values range smoothly into one another, forming a continuous surface (or continuous field). Therefore, they are commonly called **continuous** rasters. Precipitation, population density, temperature, and cloud cover are other examples of variables that can be measured over space and stored in raster format as continuous data.

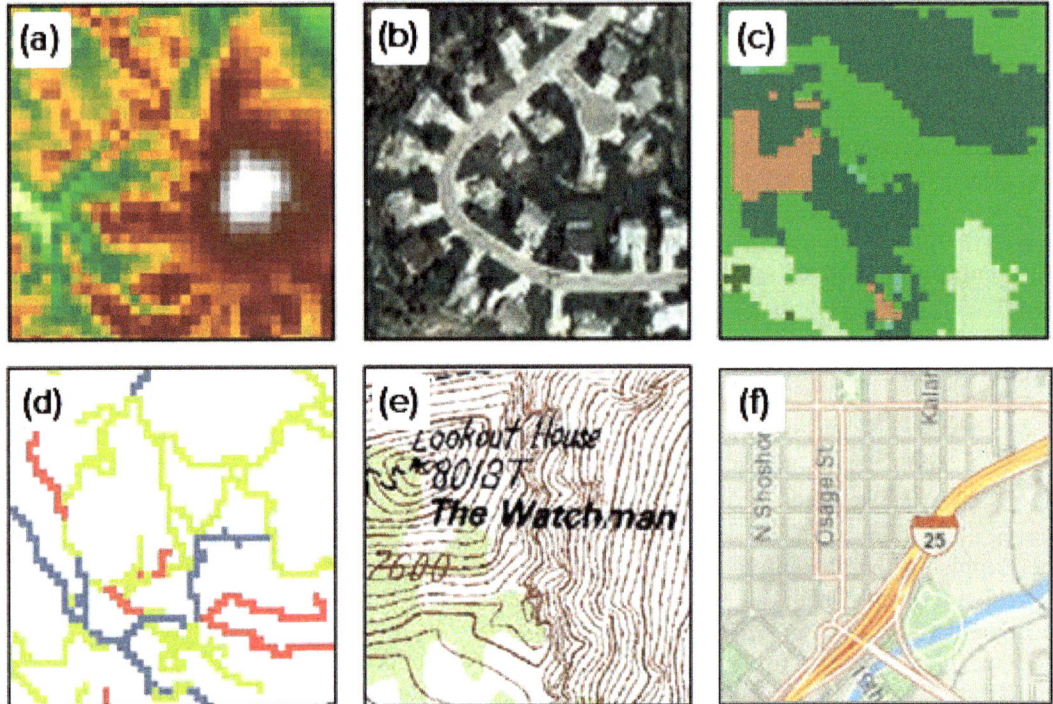

Fig. 6.2. Types of rasters: (a) digital elevation model (DEM); (b) aerial photograph; (c) land use; (d) roads; (e) scanned map; (f) ArcGIS Online basemap tile
Source: (a), (e): USGS; (b), (f): Esri; (c), (d): Black Hills National Forest

Reflectance is a special type of continuous variable that measures the intensity of electromagnetic radiation, as a function of wavelength, bouncing off or being emitted from the earth's surface. Light from the sun is the most common source of radiation and serves as the basis for aerial photography (Fig. 6.2b) and many types of satellite imagery. Other radiation sources may include radar or thermal energy. These data sets constitute a special class of rasters with unique properties and uses, and they are commonly termed **images**. The pixel values in an image are usually stored in the form of a digital number (DN) scaled to the actual reflectance value by a constant. Color images often contain a red band, a green band, and a blue band for each pixel, and the mixture of values in each band defines the color of the pixel, as described in Chapter 2.

Discrete features, such as road lines or land use polygons, can be converted to raster format if needed for analysis or display. Discrete features typically have many attributes stored in a table, but a raster can be created by selecting a single attribute to be stored in the cells. Rasters can only store numbers, not text. If a numeric attribute, such as crown cover percent in a forest stand, is selected, then the percentages can be stored directly in the raster cells. If a text attribute is chosen, such as a land use type, then each unique text value must be assigned an integer code to represent it. In Figure 6.2c, the cells store numeric values representing a land cover type, such as 46 for conifer forest or 23 for hardwood forest. Each value is given a different color for display. The roads shown in Figure 6.2d were originally vector line features with a text attribute indicating a primary, secondary, or primitive road type. When converted to a raster, the number (1, 2, or 3) is used to represent each road that passes through a cell. The cells that don't contain roads are given a null value. Rasters that store vector-type features in a raster format are sometimes called **discrete** rasters.

Raster formats are often used to transfer or display a complex representation of maps based on vector features. Figure 6.2e shows a raster that portrays a US Geological Survey topographic map, obtained by scanning a paper map. The result is known as a **digital raster graphic** (**DRG**). Each cell stores an index code representing a different color, such as 5 for the brown contours and 1 for the white background. Scanning is a common way to convert a paper map to digital form, but it has its limitations. Although the map appears to contain vector features (contours), it is actually stored as pixels, and one cannot select individual lines or analyze them. Their primary use is limited to display as a background map.

Basemaps in ArcGIS Online are another example of using raster formats to portray vector data. Figure 6.2f shows a portion of the National Geographic basemap from ArcGIS Online, which starts as vector features that have been symbolized, labeled, and displayed using vector techniques (often with different layers and properties at different scales). Then the map is divided into equal squares (**tiles**), and a raster snapshot of the map is recorded for each square, like taking a picture of a map using a smartphone. For Internet map services, it is often quicker to send the raster representation of a complex vector map than to send the features and render them on the client's device.

Raster resolution

The x and y dimensions of each pixel define the **resolution** of the raster data. The higher the resolution, the more precisely the data can be represented. Consider the 90-meter resolution roads raster in Figure 6.3. Since the raster cell dimensions are 90 meters, the roads are represented as much wider than they actually are, and they appear blocky. A 10-meter resolution raster could represent the roads more accurately; however, the file size would increase by 9×9, or 81 times.

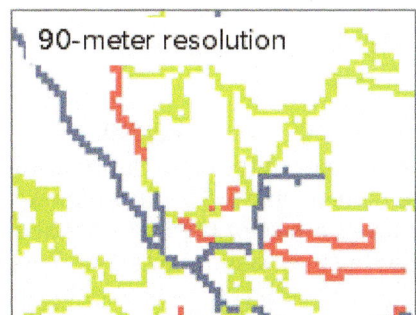
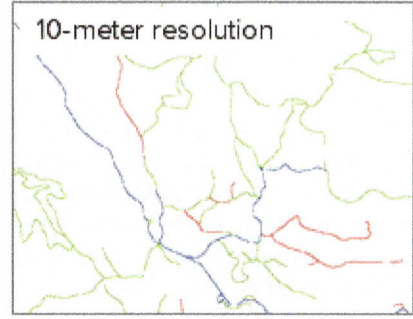

Fig. 6.3. Impact of raster resolution: the 10-meter resolution raster stores the roads more precisely, but it takes 81 times as much storage space.
Source: Black Hills National Forest

The trade-off between raster precision and storage requirements is one of the key limiting factors in the use of raster data. Representing a state with 1-meter resolution aerial photography would require huge volumes of data and overwhelm any attempt to perform analysis for most computers. Making raster data usable requires one of two strategies: limiting the extent or limiting the resolution. A manageable raster of the continental United States would need to have a low resolution with a pixel size on the order of 500 to 1000 meters. However, a small town could easily be represented by a raster with a 1-meter or even 1-foot resolution.

Satellite and aerial photography viewed on Internet servers like ArcGIS Online or Google Maps may seem to defy this precept, but that is because the mechanics of the trade-off are hidden from the casual user. A computer screen only has a few thousand pixels, and it cannot display every pixel of a raster that is millions of pixels wide. Instead, a stack of imagery with different resolutions is usually employed. Imagery of the continental United States is displayed as a simplified raster with a low resolution. As the user zooms in and reduces the extent of the map, the image may be replaced with a higher resolution raster. In Internet services, rasters are typically stored as separate tiles so that each individual file is manageable in size. Once the extent is reduced, the display only needs to pull the raster tiles that show up in the current view extent. The larger the scale, the smaller the extent becomes, and the higher the resolution that each tile can be. These services essentially maintain multiple copies of the data at different resolutions, tiled into manageable chunks, and they use spatial indices to select the right tiles based on the view extent so that the GIS or browser is not overwhelmed with too much information.

Raster pyramids

Single rasters, even those of reasonable size, may still have a higher density of cells than can be displayed onscreen at a given scale. When this happens, the computer must skim through the raster and pick one cell from a larger cluster to display, and this takes time. A common alternative is to build **pyramids** for the raster. A pyramid is constructed by taking a block of adjacent cells, commonly four, merging them to form a single new pixel, and selecting one value to represent the original group (Fig. 6.4). The new, larger cells are used to create a raster with one-quarter the storage space and resolution of the original. The procedure is repeated to create another raster of yet lower resolution, until an entire set is created, appropriate for different scales of display. Resampling can also be based on larger block sizes if desired, such as using a 3×3 block of nine pixels, a 4×4 block of 16 pixels, or a 5×5 block of 25 pixels.

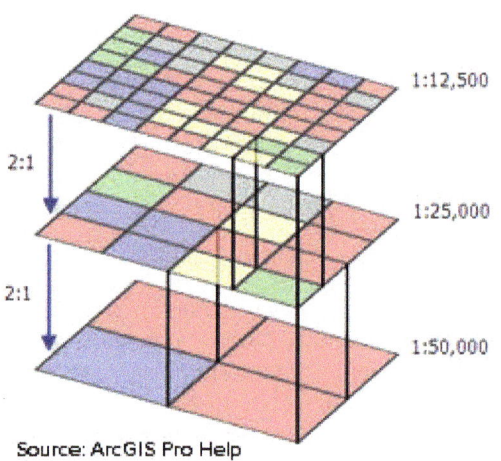

Source: ArcGIS Pro Help

Fig. 6.4. Raster pyramids
Source: Esri

The pyramids speed drawing because the computer only needs to select the right scale of raster to display rather than sorting through it to choose pixels. The disadvantage of pyramids is that storing the additional copies increases the size of the raster by about 50%. Creating pyramids is optional, depending on whether the user is more interested in quick display or smaller file sizes.

Resampling

Building pyramids is only one instance when the cell size of an existing raster must be changed, whether to a lower or a higher resolution. This process is called **resampling**. In the pyramids example, the resampling is performed by grouping existing cells, ensuring that the new cells always contain whole original cells. The procedure is known as **block resampling**. Another type of resampling simply specifies a new cell size without reference to the original cells. Figure 6.5 shows an example of resampling from a 10-meter (colored boxes) to a 12-meter pixel (black lines). In this case, the new cells may contain portions of original cells.

Fig. 6.5. Resampling raster cells using the nearest neighbor method

Any type of resampling requires specifying a rule for determining the value of the new pixel. Examine the four pixels in the left corner in Figure 6.4. At the highest resolution, the four pixels include three red ones and a gray one. The new pixel in the first pyramid is a red pixel. At the next level of pyramid, the grouping of three blue cells and one red cell is converted to a blue pixel. This procedure represents a majority rule for resampling: the new cell contains the value that occurs most frequently in the contributing original cells. (In practice, a backup rule is needed in case no majority occurs, as when each of the four pixels is a different color).

If the cells represent a category like land use, one must choose one of the values from those present in the block. However, if the data are numeric, like elevation or precipitation, then additional options are available, such as averaging the values of the four cells.

When resampling to an explicit cell size, as in Figure 6.5, four common options are available. In **nearest neighbor resampling**, the new cell is given the value of the old cell that falls at or closest to the center of the new cell. This method is required when categories of values are being represented, like land use or soil types. It is also the fastest method because no calculations are required. If the pixels contain numeric data like elevation, additional options come into play. An **area-weighted average** would compute the average of the original cells weighted by their area fraction within the new cell. In **bilinear resampling**, a distance-weighted average is taken from the four nearest cell centers. The **cubic convolution resampling** method determines a new value by a curve fit through the nearest 16 cell centers.

Resampling to a coarser resolution will degrade the precision and accuracy of a data set. In Figure 6.4, notice the disappearance of certain colors and the progressive generalization of the information in the subsequent pyramids. In this case, the original image is retained and the purpose is merely to speed display, so it doesn't matter. However, in Figure 6.4, it is easy to spot the loss of information that can result even from a relatively modest increase in cell size.

Resampling to a finer resolution, conversely, does not actually improve the precision or accuracy of data. In Figure 6.5, consider resampling to a cell size of 2.5 meters, so that each original cell is converted to four new cells. The precision may seem higher, but each of the four new cells merely contains the same information as the larger parent cell. The resolution of the file has increased, but the resolution of the information has not.

Storing rasters

Rasters are used in many different programs and come in a plethora of formats, such as JPEG, TIFF, GEOTIFF, BMP, MrSID, and raw files (BIP, BIL, and BSQ). Rasters may also be stored in a geodatabase as a File Geodatabase Raster (FGDR). Most raster formats consist of the number array plus a header that gives information about the file, such as its number of rows and columns and, in the case of geospatial data, its coordinate system and georeferencing information. The header information may be stored in a separate file or as the first part of the array. In any raster, the definition of the pixel properties is paramount.

A **byte** is the basic unit of computer storage. It is composed of a string of eight digits (bits), which may be zeros or ones. These zeros and ones represent a number in base 2; such numbers are called **binary** numbers. One million bytes constitute a megabyte (MB), and one billion bytes equal a gigabyte (GB), terms that sound familiar as measurements of data storage capacity.

A single byte can store a binary value from 0 (00000000) to 255 (11111111). The largest number that can be stored in a given set of k bits is $2^k - 1$; for one byte, $2^8 - 1 = 255$. The more bytes allotted to a value, the larger the number that can be stored. Two bytes can store values up to $2^{16} - 1$, or 65,535. However, if negative numbers are required, then the top bit must be used to designate whether the number is positive or negative, in which case the capacity of a single byte is reduced to $2^7 - 1 = \pm127$, and a two-byte number is limited to $2^{15} - 1 = \pm32,767$.

In a raster, the number of bytes allotted to store each pixel can be varied by the user to meet the requirements of the data. This property is known as the **pixel depth**. In a black-and-white photograph or document scan, for example, the values are typically stored using one byte as DN values, taking on values from 0 to 255. In a DEM, however, elevation values on earth could be greater than 7000 meters or 21,000 feet, and they would require two bytes to store the largest numbers. Pixel depth may be specified in bits or in bytes; for example, an 8-bit pixel is the same as a 1-byte pixel. Generally, increments of whole bytes are used, although some remote sensing instruments may have less common values, such as 11-bit pixels. The pixel depth directly affects the storage requirements of a raster: a raster with 2-byte pixels will require twice as much storage space as a raster with the same number of 1-byte pixels.

Bytes are designed to store integers. If a raster needs to store decimal values, the most efficient approach is to store the numbers as integers but to specify a **scale** in the header that will convert to the desired values. For example, if one decimal place is needed to store precipitation values for the desired precision, then one multiplies each value by 10, converting 3.7 inches to 37 so that it can be stored as an integer. However, the user is then responsible for ensuring that the scale is applied when interpreting or analyzing the data.

The other approach is to store the values as exponential or floating-point data. This notation stores values composed of a **mantissa** (the decimal part of the number) and an exponent. For example, the number 123456789 stored as an exponential number would consist of a mantissa of 1.23456789 and an exponent of 8, yielding 1.23456789×10^8 (often written as 1.23456780e08). Floating-point data types are more flexible than binary types because they can store either very large or very small values using the same field. However, at least four bytes are required to store a floating-point value with eight significant digits (and more are required for greater precision).

Bands

It is possible to combine rasters with identical extents and cell sizes together as a single file unit. Each of the rasters is known as a **band**, and the combination may be called a **multiband raster** or a **band stack**. Multiband images are most commonly encountered when dealing with imagery and satellite data.

Color images, including geospatial data and photographs, typically use a red-green-blue (RGB) color model to specify colors, as discussed in Chapter 2. A color digital camera measures brightness separately for three wavelengths of light (red, green, and blue) and stores three brightness values for each pixel in three separate bands. Each combination of color values produces a unique color. RGB images are sometimes called true-color images because they allow more than 256^3 colors to be represented (over 16 million).

Satellites often collect electromagnetic reflectance over different ranges of wavelengths: including the common visible spectrum of red, green, and blue as well as nonvisible wavelengths such as near-infrared or thermal. The Landsat family of satellites is one of the oldest and most used sources of satellite imagery; Landsat 5 included seven different bands, Landsat 7 had eight, and Landsat 8 has 11 bands. Figure 6.6 shows seven bands of a Landsat 5 image over Crater Lake; each band looks slightly different because it is recording a different wavelength of light. Notice that the Band 1 image (top left) is the blue band and shows the lake more brightly than any other band because of the blue reflectance of the water. For more information about the band designations, see http://gisgeography.com/landsat-program-satellite-imagery-bands/.

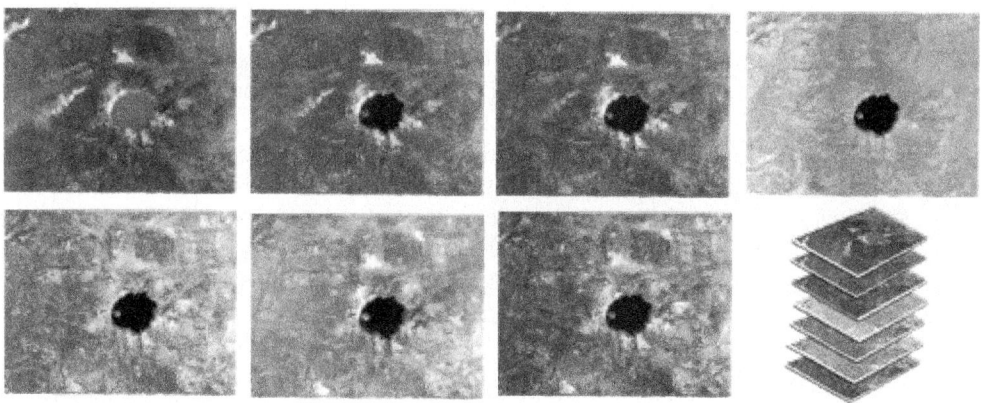

Fig. 6.6. Landsat bands over Crater Lake, numbered 1 to 7 from the top left
Source: USGS

Compression

Rasters may become very large files, so many raster formats employ some type of **compression** to reduce the storage requirements. For example, imagine that the top row of a 1-byte raster stores the values (0 1 1 1 1 1 3 3 3 3 3 3 5 5 5 5 5 5 7 8), which is equivalent to 20 bytes of information (one byte for each value). When values tend to repeat this way, one might instead store pairs of bytes that encode a value and how many times it occurs, turning the example row into (01 15 36 56 71 81), which takes only 12 bytes (the spaces are included only for visual clarity). This scheme, known as run-length encoding, would only be effective for rasters that had many adjacent cells with the same value (note that it doubles how much space is required to store the nonrepeating cells with values of 7 and 8). Different approaches would be needed for other types of rasters.

The plethora of raster formats is largely a function of the variety of compression techniques; many formats employ different and sometimes proprietary schemes. A discussion of different compression techniques is beyond the scope of this text, but one aspect of compression is important to understand, which is whether or not the compression is reversible. Consider the preceding example; if the compression rule is known, one can unpack the compressed version to re-create the original string of numbers. In contrast, imagine a compression rule that truncates decimal cell values to integers (1.2 3.6 4.2 9.8) to (1 3 4 9). This rule is irreversible; there is no way to determine the original values from the new data set alone. An irreversible compression scheme is often called **lossy compression** because the compression causes the loss of information.

Lossy compression is more effective at reducing file sizes, and the result is adequate for many purposes. Human perception is usually incapable of distinguishing the difference between photographs compressed with lossless and lossy techniques (although professional photographers assert otherwise and prefer raw image formats with no compression, and it is true that the differences may become apparent if the photo is enlarged). However, viewing a photograph is less demanding than analyzing raster values, when preserving the original data may be important.

Care must be taken when choosing raster formats or when converting rasters from one format to another because the conversion may not be reversible. A TIFF format uses a reversible (or lossless) compression, but the JPEG format uses an irreversible (lossy) one. Converting a TIFF to a JPEG file drops information, and if the resulting file is converted back to a TIFF, it will not contain the same data as the original.

Data rasters versus picture rasters

By now, we have seen the raster data format play multiple roles, but there is one distinction that is important to make when working with a raster, which is whether the raster contains analyzable data or whether it is simply a picture of data. A DEM represents a **data raster**; it contains elevation values that can be used in analysis tools, such as those used to calculate slope (see Chapter 11). DEM rasters can also be symbolized in different ways. However, imagine using a computer screenshot program to snap a picture of the DEM and save it in JPEG format. Now it is no longer elevation data but a picture of elevation data, or a **picture raster**. The JPEG does not contain elevation values; it stores RGB combinations of colors that were displayed on the screen. It may look exactly like the symbolized DEM layer, but in fact it is quite different: one cannot change its symbols, calculate slope, or examine the elevation value in a cell.

This distinction may seem even more puzzling for satellite imagery. Landsat 8 satellite imagery has brightness values in 11 different bands, which are critical in image processing applications such as mapping land cover types. However, if the primary goal is to display the land surface, all the information increases the file size and slows the display, especially for an Internet service. Thus it is common to convert the file to a compressed format, such as JPEG, which stores the color combinations and is thus much smaller. It may still look like the same satellite image, and it will display more quickly in a service, but one can no longer perform image analysis; the original bands of reflectance are gone.

Raster coordinate systems

Rasters have coordinate systems and projections just as vector feature classes do. Figure 6.7 shows a South Dakota elevation raster in a geographic coordinate system (GCS) and in a conic projection. The shape of the state is different, and so is the ratio of columns to rows. Like vectors, rasters can be projected from one coordinate system to another, although the algorithms are more complicated. The GCS version clearly has more pixel columns than the State Plane version, so the projection algorithm must modify the extent, spacing, and aspect ratio of the pixels. Put simply, projecting rasters requires resampling. Users must specify an output cell size and a resampling method when projecting a raster.

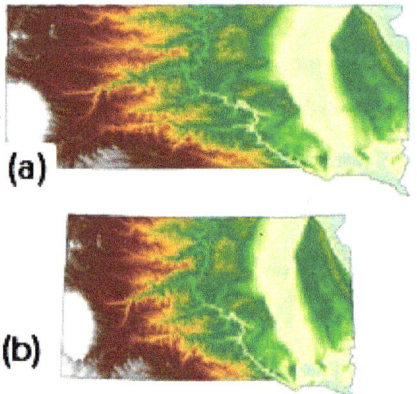

Fig. 6.7. An elevation raster in (a) GCS and (b) State Plane coordinate systems
Source: USGS

For this reason, projecting rasters takes more memory and processing power than projecting vectors. This fact has some practical implications. First, the resampling has the potential to introduce artifacts and errors, so one should minimize the number of times a raster is projected. Second, when displaying vector and raster data with different coordinate systems in the same map, relying on on the fly projection, it is faster and more efficient to set the map to match the raster coordinate system. This technique is especially useful when using basemaps and other GIS raster services; display performance is usually improved if the map is set to the basemap coordinate system (usually WGS 1984 Web Mercator Auxiliary Sphere).

Spatial analysis depends on correct areas and distances, so most rasters come in a projected coordinate system. However, DEM data distributed by the US Geological Survey are usually provided in GCS format to allow the user to select the projection. The downloaded files will have cell size units of degrees: the standard resolutions include 1 arc second (~30 m), 1/3 arc second (10 m), and 1/9 arc second (3 m). When projecting, the user must specify the desired output cell size in projected units (meters), as well as a resampling method. Bilinear interpolation is usually the best method for DEMs, to ensure that the output does not contain artifacts or discontinuities.

Assigning coordinate systems to rasters

Many rasters are provided with a coordinate system at the outset (they are georeferenced). However, sometimes a user must assign the coordinate system. Two cases commonly occur.

Case I involves a raster for which the coordinate system is known, but the information has not been stored in accessible form. Some image formats store georeferencing information inside a header file or in the image itself. Other formats use a **world file**, which is a six-line text file containing the parameters needed to georeference the raster. If the georeferencing information is missing, the user can create a world file with the appropriate parameters. A description of the contents of the world file can be found in the ArcGIS Help.

TIP: A world file has the first and third characters of the image file extension plus a "w" on the end. A .tif file would have a .tfw world file. A .bil file would have a .blw world file. Other formats may use an auxiliary (.aux or .aux.xml) file.

Case II involves a raster without a specified coordinate system, such as saved pictures from the Internet or a scanned map. The coordinate system is in pixel units from the screen or the scanner rather than real-world coordinates in feet or meters. To georeference this raster, the user must create **ground control points** by matching locations on the image to the same locations on another image or feature class that has a real-world coordinate system. Road intersections or other distinctive features are typically used. A minimum of three points are required, but additional pairs typically give more accurate results. A mathematical transformation is calculated from the matching pairs, which can then be used to convert the raster from the arbitrary to the real-world coordinate system.

Different types of transformations can be calculated and applied (Fig. 6.8). A first-order (**affine**) transformation can translate, rotate, and skew the image to match up the points. For a scanned map with a known projection, the affine transformation is usually sufficient because the shape of the map is faithfully reproduced; it just needs its coordinates converted to a new set of x-y values.

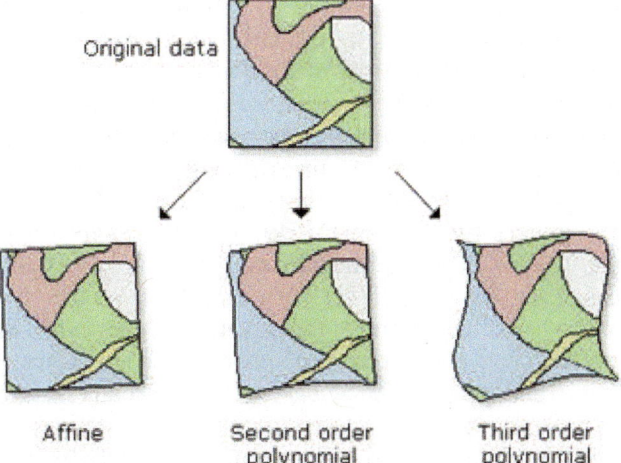

Fig. 6.8. Transformation methods for a raster
Source: Esri

185

An air photo or satellite image may have geometric distortions due to topography, focal effects, or lens distortions. A map with an unknown projection may not exactly match the shape of the coordinate system to which it is being georeferenced. In these cases, a second-order or third-order transformation allows the image to stretch or contract locally for a better fit, known as **rubbersheeting**, which requires additional control points to complete the calculation of the transformation equations. Generally one should use the lowest order of transformation that produces an acceptable result. Using a higher order than needed can actually introduce new distortions.

When the transformation equations are applied to the ground control points, an offset is usually found between the specified locations and the actual locations, due to slight errors in locating the points or to differences in shape between the old and new coordinate systems. This offset is evaluated by calculating a **root mean square error** (**RMSE** or RMS error), which serves as a measure of the accuracy of the transformation. The RMS error is reported in map units of the new coordinate system and should be reported in the image metadata.

To finalize the georeferencing, the user may use one of two methods. (1) The GCPs and transformation parameters are saved to a world or auxiliary file; this method retains the original image unchanged and simply projects it on the fly using the parameters in the file. (2) The parameters are used to **rectify** the original image and save it as a new output file. Rectification requires resampling, so the user must specify an output cell size and a resampling method. With RGB imagery, rectification may disrupt the colors; this change can often be reversed by copying the color map of the original image to the new one.

3D data

Rasters are one method of storing surfaces such as elevation. A conventional map reduces the earth to a flat surface, but at times, it is advantageous (and fun) to view geographic features in three dimensions. A variety of methods are used to store and visualize data in 3D, which is based on three coordinate axes, x and y in the horizontal plane and z to indicate height. Both raster and vector data formats exist for 3D data.

Raster data sets, as already described, are one way to provide 3D information. In a digital elevation model (DEM), the pixel arrangement defines the x-y axes and the elevation value in the raster provides the height (z). Such a data construct is known as a surface, field, or grid. Another raster approach to 3D data storage involves assigning each pixel a fixed height in addition to its x and y cell dimensions. The cubic or rectangular volume thus created is known as a **voxel**, and its stored value represents some quantity defined for that volume, such as the transmissivity of an aquifer or the concentration of a plume from a contaminant spill.

Rasters can be directly visualized in three dimensions, and they are also useful for assigning elevation values to vector features. Consider a line feature class containing roads, with x-y pairs defining the vertices. Each vertex can be matched to the corresponding x-y location on a DEM and assigned a z value from the raster, so that each vertex is now represented by a 3D (x,y,z) triplet. The z-values generated this way are known as **base heights**. Points and polygons can also be treated in this way. The 2D feature class has been converted to a 3D feature class and can be visualized as sitting on top of the raster, a process called **draping**. Other rasters may also be draped on a DEM, such as when an aerial photo is used to portray a realistic-looking landscape. Figure 6.9 shows an aerial photographic basemap and a roads feature class draped on a DEM of Crater Lake.

Elevation values may also come from an attribute field in a feature class. A polygon feature class of buildings may contain a field recording the height of the building, or the number of floors. Once a base height is established, the shape of the building can be projected above its base to the appropriate height, a process known as **extrusion**. A similar process can be applied to points representing wells, using negative heights to project them below the ground surface. These techniques provide a way to visualize the buildings or wells in three dimensions.

A point data set with (x,y,z) values is known as a **point cloud**, and it is the basis of many types of 3D modeling, including LiDAR, a commonly used technique for collecting detailed elevation data. LiDAR (Light Detection and Ranging) is an active remote sensing instrument that emits light at a known frequency and measures the time it

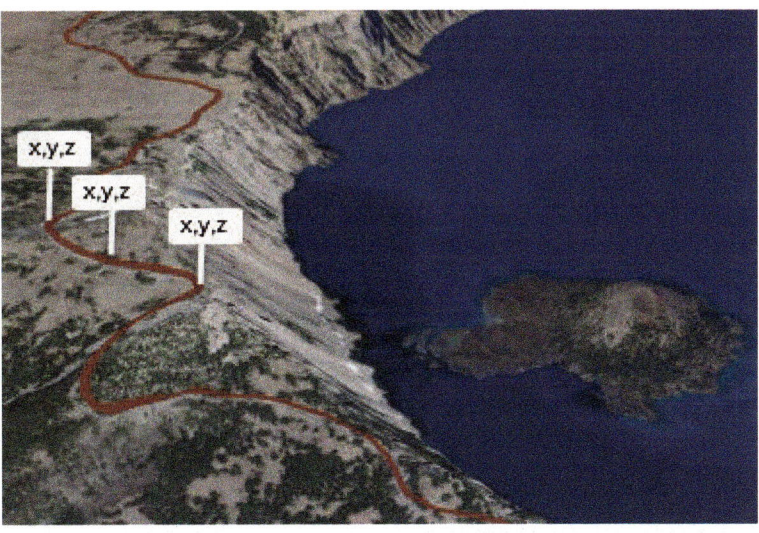

Fig. 6.9. Roads and aerial photography draped on a DEM
Source: Esri

takes to reflect from a remote target and return to the sensor, effectively measuring the distance between the sensor and target. LiDAR instruments are often mounted on planes or helicopters to collect ground surface elevations. Since the light may also reflect from trees, buildings, and other objects, these features must be subtracted out if it is the ground surface elevation that is desired, but information on building heights or canopy heights are often of interest and can be retained as well. Figure 6.10 shows a 3D view of a point cloud of Manhattan and the Ed Koch-Queensboro Bridge, with the points colored according to height (blue low, red high). Ground-based LiDAR units place the instrument on a tripod and send the light sideways in order to image the shapes of things, such as the façade of a building, the details of a statue, or the volume of a tailings dump.

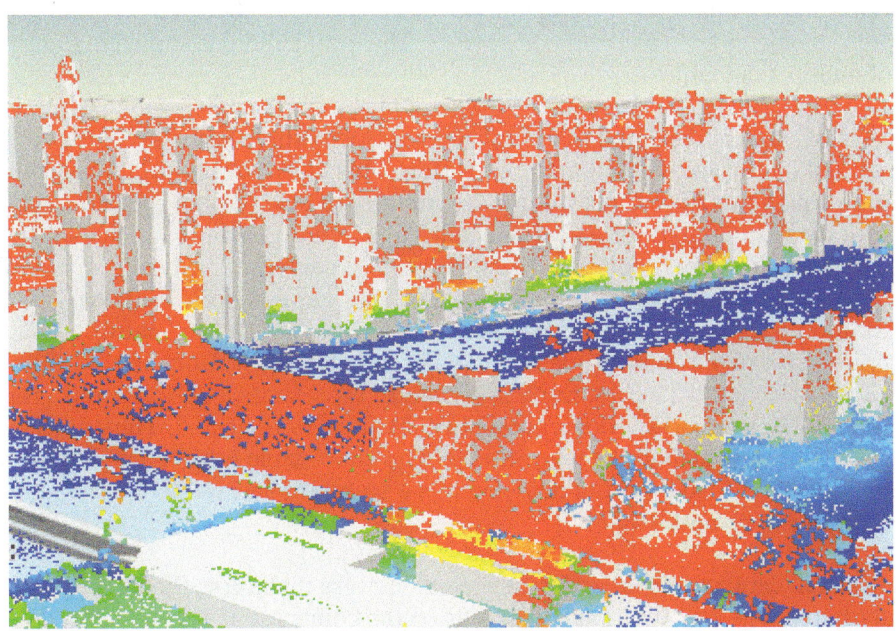

Fig. 6.10. Point cloud of Manhattan viewed in a 3D scene with multipatch buildings
Source: Esri

Finally, 3D surfaces can be represented by a vector data format known as a **TIN (triangular irregular network)**. Any flat 3D surface (plane) can be uniquely defined by three (x,y,z) triplets; if connected by lines, these points form a triangle with a particular orientation in space. Each triangle may be connected to an adjacent triangle representing another plane. The shared triangle boundaries are called edges, and the locations where the triangle corners connect are termed nodes. These networks of triangles form

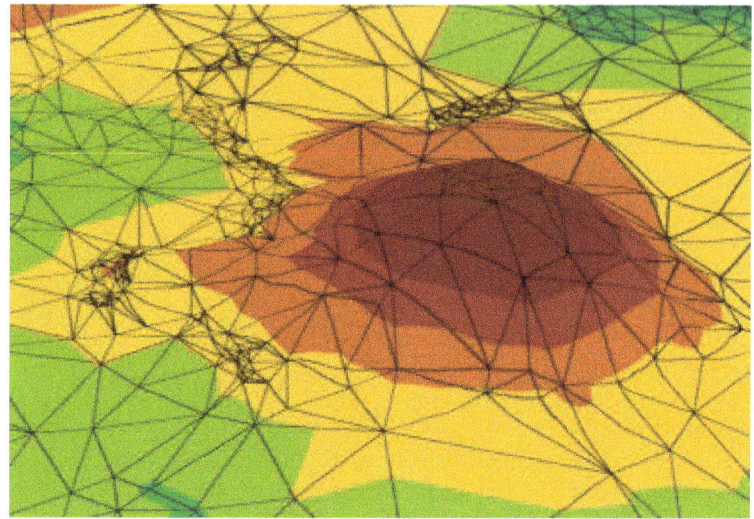

Fig. 6.11. A TIN representation of Wizard Island in Crater Lake
Source: USGS

a faceted representation of a surface or object in three dimensions. Figure 6.11 shows a TIN representation of Wizard Island in Crater Lake in 3D perspective view, draped over color shades representing elevation contour classes. TINs can be very efficient because, unlike rasters, the resolution of nodes can be varied according to the terrain. Flat areas can be represented with fewer nodes, and mountainous terrain can be given a great density of nodes, even within the same TIN. Furthermore, the nodes and edges can be placed on the precise inflection points of the surface, such as the very peak of the mountain or the center axis of a stream valley, so it can capture the true highs and lows of the surface better than a raster.

A data type related to TINs is known as **multipatch features**. Multipatch features are enclosed volumes defined by triangular or rectangular patches combined with three-dimensional rings, used to represent objects such as buildings or trees. The gray buildings shown in Figure 6.10 are examples of multipatch features.

A final type of 3D data is designed to visualize quantities other than elevation, such as time represented on the z-axis instead of elevation. This construct is called a space-time cube. Imagine a map of crime density for a certain month placed on a regular grid, with the number of crimes counted for each grid cell. Then imagine a series of such maps depicting crime for an entire year. Stack them on top of each other, giving each map a designated thickness (as for a voxel), and a **space-time cube** is created. With a 3D viewer we can examine patterns in space as well as through time. Figure 6.12 shows a space-time cube constructed for an 8.8 magnitude earthquake near Chile on February 27, 2010, and its aftershocks over the next 72 hours. Each map square represents a 30-km square region, and each cube in the vertical stacks represents a six-hour time period. The purple colors indicate the sum of the earthquake magnitudes. The space-time cube is used for many types of analysis, not just GIS, and storing one employs a standard format called the network Common Data Form, or **netCDF**.

Three-dimensional data are visualized with special viewers that allow the data to be scaled, moved, and rotated. The computational demands on the computer are great, so trade-offs in resolution or area are often required for reasonable performance.

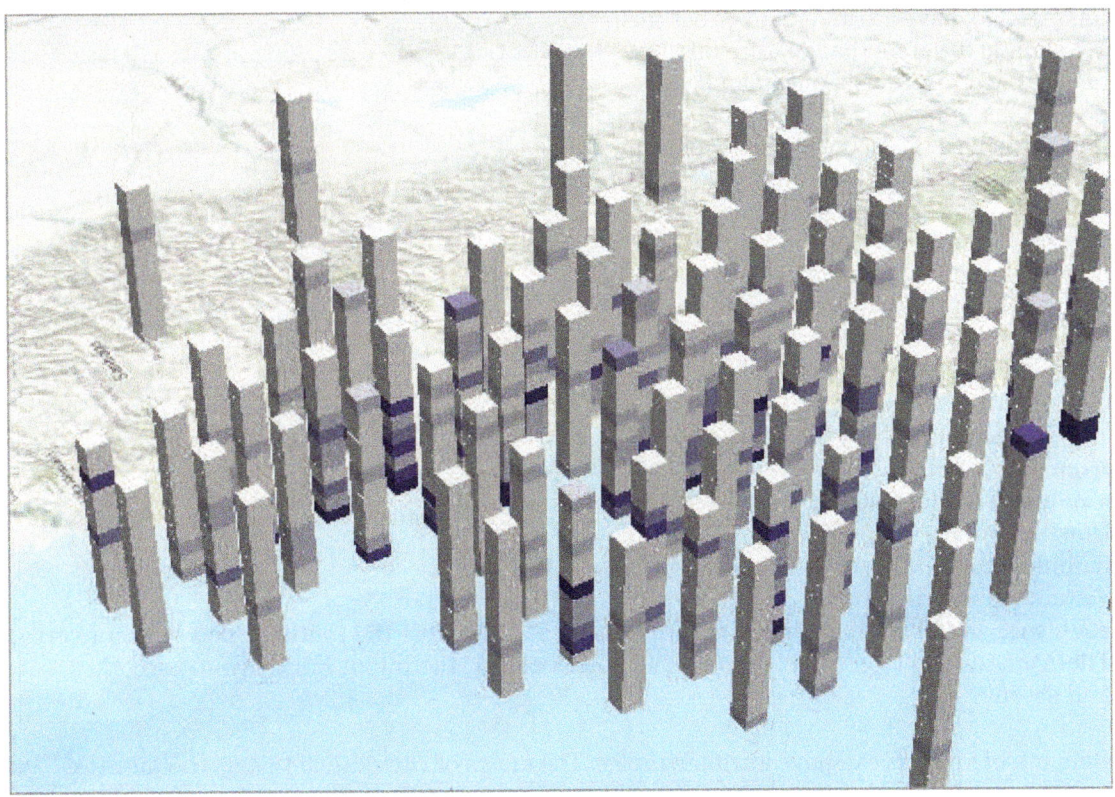

Fig. 6.12. Space-time cube visualization of aftershocks from a Chilean earthquake in 2010
Source: USGS (quakes); Esri (basemap)

About ArcGIS

Storing rasters

ArcGIS Pro can read and display most common raster data formats (see the Help for a complete list). For spatial analysis, however, only four different formats are supported. Prior to ArcGIS Version 10, the software used a format unique to ArcGIS, called a **grid**. At version 10, three additional formats were added: TIFF, ERDAS, and FGDBR (file geodatabase raster). In this book, we will work almost exclusively with geodatabase rasters. This section describes some specific characteristics of ArcGIS-supported rasters that may not be found in other formats.

> **TIP: NEVER copy, move, rename, or delete grids using File Explorer.** Grids have complex storage formats, and Windows cannot manage them properly. Always use Pro tools to manage grids, and be careful when using Explorer to manage other raster formats.

ArcGIS supports single and multiband rasters; the pixel types may include unsigned integers (only positive values), signed integers (values may be negative), and floating-point values. Many raster formats use a special value, often zero, to indicate missing or null data. Because zero is sometimes a legitimate measured value (0.0 in. of precipitation last week) and can skew image statistics, ArcGIS formats use a special **NoData** value, which is ignored during processing. NoData values are typically displayed with black or transparent color.

Rasters may have attribute tables, but they are slightly different than feature attribute tables. Each record lists one unique value present in the raster cells. The table has three standard fields: the ObjectID field, containing a unique ID for the table rows; the Value field, showing each unique cell value; and the Count field, indicating how many cells contain that value (Fig. 6.13). In this table, the Inyan Kara Group covers the largest area because more cells (43,450) have that value than any other.

⊞ geolraster ✕			
OBJECTID	Value	Count	UnitName
1	1	29190	Alluvium
2	2	43450	Inyan Kara Group
3	3	25934	Mowry Shale
4	4	11962	Terrace deposits
5	5	31422	Morrison Fm
6	6	3129	Belle Fourche Shale
7	7	26767	Spearfish Fm

Fig. 6.13. The attribute table of a geology raster
Source: Esri

The attribute table in Figure 6.13 comes from a raster that was converted from a polygon feature class using the UnitName field, so the UnitName is included in the table. The Value field contains numeric codes representing each rock unit, which were arbitrarily assigned at the time the feature class was converted. The Alluvium unit is represented by the code value 1, the Inyan Kara Group by code 2, and so on.

Only integer rasters can have attribute tables. The tables of categorical rasters are the most useful because they document a fairly small set of discrete values. However, an attribute table is automatically created for any raster with fewer than 500 unique values.

TIP: Rasters should not be stored in folders that have pathnames containing spaces because some functions will not work. GIS users are advised to avoid spaces in ALL folder names.

TIP: Raster names must contain only letters, numbers, or an underscore. Grids have even stricter naming conventions and are limited to 13 characters. Consult ArcGIS Help for more information on supported formats and naming rules.

Summary

▶ The raster data format is optimized to store continuous surfaces, such as elevation and imagery from photographic or satellite sensors. A raster contains an array of numbers representing map values for cells or pixels.

▶ The cell or pixel dimensions of a raster are known as its resolution. Resolution is the primary factor affecting the precision of mapping afforded by raster data. Rasters typically store large areas with low resolution or small areas with higher resolution.

▶ Resampling a raster means converting from one pixel resolution to another. Resampling to a coarser resolution nearly always results in degradation of the precision and accuracy of the data. Resampling may be used to derive a stack of progressively coarser representations of a raster in order to speed display.

▶ Raster pixels are stored using binary numeric values. The number of bytes allotted to store a pixel is known as the pixel depth, and more bytes allow larger numbers to be stored.

▶ Compression schemes are commonly used to reduce the disk space needed to store rasters. Compression is characterized as reversible or irreversible depending on whether the uncompressed version can be perfectly reconstructed from the compressed version.

▶ Rasters with identical cell meshes may be grouped into a multiband raster or a band stack, used to store different measurements over the same area. Color and satellite images often use this technique to store the reflectance of different wavelengths of light.

▶ Rasters have assigned coordinate systems just as vector feature classes do. Projecting rasters is a more complex process than projecting vectors; it requires resampling and is not perfectly reversible.

▶ Discrete rasters typically include a raster attribute table that lists each unique value in the raster and the number of cells that contain the value. Text information can be tied to each unique value to document what it means.

▶ Rasters are commonly used to represent surfaces in three dimensions, and they can also be used to assign base heights to vector features to visualize them in 3D.

▶ Other methods to represent 3D data include TINs, point clouds, voxels, and multipatch features. Space-time cubes can be used to impart time to a 2D data set.

Important Terms

affine transformation	digital raster graphic (DRG)	NoData	resolution
band	discrete	picture raster	root mean square error (RMSE)
band stack	draping	pixel depth	rubbersheeting
base heights	extrusion	pixel	scale
binary	grid	point cloud	space-time cube
byte	ground control points	pyramids	tile
cell	image	rectify	triangular irregular network (TIN)
compression	lossy compression	resampling,	
continuous	mantissa	area-weighted average	voxel
data raster	multiband raster	bilinear	world file
digital elevation model (DEM)	multipatch feature	block	
	netCDF	cubic convolution	
		nearest neighbor	

Chapter Review Questions

1. Compare and contrast the relative advantages and disadvantages of the vector and raster data models.

2. Consider creating rasters for these data sets, each of which will be stored using integers. What pixel depth, in bytes, would be the most efficient choice for each type of data: (a) elevation in meters; (b) crown canopy percent; (c) temperature in degrees for a thermal camera used by firefighters; (d) median county income in dollars? Explain your choice for each.

3. Describe the difference between a thematic raster, an image raster, and a picture raster.

4. What extra step is performed when projecting rasters that is not needed when projecting vector data? What happens during this step?

5. Choose whether nearest neighbor or bilinear resampling would be the best choice for these data sets, whether they would require signed integers, and explain why: (a) land use codes; (b) precipitation; (c) Landsat satellite imagery; (d) basemap tiles.

6. What is a ground control point, and what is its purpose? What types of locations make the best ones?

7. Professional photographers advocate capturing and storing photographs using a raw or TIFF format rather than JPEG. Explain why. What is the trade-off?

8. Contrast the difference in the ways that rasters and TINs store an elevation surface.

9. Discuss the statement, "ArcGIS Pro supports more than 10 different raster formats." In what sense is it true? In what sense is it not true?

10. Discuss why having a distinct NoData value in a raster format is an advantage over using a zero value, as many image formats do.

Extend your knowledge

Read the following sections of the Help to learn more about raster and 3D data

> Help> Data> Data types> Imagery and raster
>
> Help> Data> Data types> Triangulated irregular network (TIN)
>
> Help> Data> Data types> LiDAR and LAS dataset
>
> Help> Data> Data types> Multidimensional data

Mastering the Skills

Teaching Tutorial

The following examples provide step-by-step instructions for doing basic tasks and solving basic problems in ArcGIS. The steps you need to do are highlighted with an arrow →; follow them carefully. Click on the video number in the VideoIndex to view a demonstration of the steps.

> → Start ArcGIS Pro and open the ClassProjects\CraterLake project. Remember to save often.

Exploring raster format and function

To begin, let's explore some different raster data types and work on understanding their properties.

> 1→ Close all open maps.
> 1→ Choose **Insert: Project: New Map**. Name the map *Rasters*.
> 1→ In the Catalog pane, expand the Databases entry and CraterLake.
> 1→ Click and drag the dem30m raster into the map (click OK if you get a transformation warning).

We should set a custom extent for this map to easily return to the area of interest.

> 2→ In Contents, double-click the Rasters icon to open the map properties.
> 2→ Click the Extent section and fill the button for Custom Extent. Set it to *Calculate from the current visible extent* and click OK.

This digital elevation model (DEM) stores elevation values in meters. It is a thematic raster, storing a measured property of the earth's surface. The shades represent different elevations.

3→ Click the **Map: Navigate: Explore** button and click on one of the cells.

3→ Examine the Pixel Value containing the elevation. Is it an integer?

3→ Examine the Stretched value; it will be a number between 0 and 255.

3→ Click on one of the lowest (black) cells and examine the Stretched Value.

3→ Click on one of the highest (white) cells and examine its value.

3→ In Contents, examine the range of elevations in the legend for dem30m.

The elevation values range from 1532 to 2720 meters, a difference of over 1200 meters. However, a color (or in this case, grayscale) ramp has only 256 values, from 0 to 255. To display the raster, the lowest elevation is assigned to the lowest symbol in the ramp (0–black), the highest elevation is assigned to the top color (255–white), and the intervening elevations are stretched into the remaining colors. Hence this display is known as a stretched raster display.

4→ Close the pop-up window. In Contents, click to select the dem30m layer and open the **Raster Layer: Appearance** ribbon.

4→ In the **Rendering** group, click the **Symbology** drop-down and examine the choices, noting that **Stretch** is the first option. It is the appropriate option for a continuous raster and was automatically selected when the raster was added.

4→ Click the **Symbology** drop-down again and select **Stretch** to open the Symbology pane. This pane looks different for rasters.

4→ Choose several different color schemes and examine each one. Then set it to the Elevation #1 ramp (Fig. 6.14). (Hold the cursor over a ramp to see its name.)

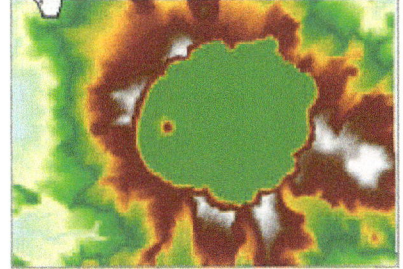

Fig. 6.14. Crater Lake DEM
Source: USGS

5→ In Contents, open the layer properties for dem30m. Examine the General section, and then skim the information in the Metadata section.

5→ Click the Source section and examine the information under Data Source.

5→ Expand the Raster Information heading and examine the information carefully.

This section contains a wealth of information. This raster is stored in the file geodatabase raster format (FGDBR) and has 685 rows and 470 columns, one band, and a cell size of 30. It is stored as floating-point values with a 32-bit (or 4-byte) pixel depth. It has three pyramids created with nearest neighbor resampling and no colormap. To determine the units of the cell size, examine the spatial reference information.

5→ Expand the Spatial Reference heading and find the Linear Unit entry.

1. What are the coordinate system and *x*-*y* units for this raster?

5→ Expand the Statistics and Extent headings and examine them.

5→ Close the layer properties for dem30m.

6→ From the Catalog pane, click and drag the hillshade10m raster to add it to the map.

A hillshade raster is derived from a DEM by calculating the illumination of a neutral surface based on elevation data and a specified illumination source. It is an effective way to visualize terrain. The brightness is stored as digital numbers (DN) from 0 to 255.

6→ Set the **Raster Layer: Appearance: Rendering > Stretch Type** setting to None.

6→ Click on light and dark values of the raster and examine the pixel values.

With the stretch turned off, the DN pixel values are matched exactly to the 0–255 grayscale color ramp. The Count entry in the Identify window indicates that this raster has a table.

6→ In Contents, right-click the hillshade10m layer and choose Attribute Table.

6→ The Value field shows each unique value stored in the raster. Scroll through the table to see the lowest (0) and highest (254) values.

6→ The Count field shows how many pixels in the raster have those values.

2. Sort the Count field in descending order. Which Value occurs most frequently (by a factor of 10)? Why this value is so abundant?

7→ Close the table.

7→ Choose a location on the crater wall and zoom in until the individual pixels of the hillshade10m layer can be distinguished.

7→ Turn hillshade10m off and on a few times to compare the pixel size of the hillshade10m and dem30m rasters.

7→ Open the layer properties for the hillshade10m raster to learn its cell size (if you haven't guessed by now). Examine its other properties and close the window.

Notice the pixel type (unsigned char) and the pixel depth (8-bit). Storing DN from 0 to 255 allows a single byte to be allotted for each pixel. Thus, even though the hillshade has nine times as many pixels as the DEM, it only uses about twice the storage space. (It was generated from a DEM with 10 m resolution, not from dem30m).

Some Landsat satellite imagery is stored in the CraterLake project folder.

8→ Return to the full extent of the map.

8→ In the Catalog pane, expand the Folders entry and the CraterLake folder. Find the LT05_sep11_2011_P45R30clip.tif image and expand it.

8→ Drag the image into the map. Rename it Landsat.

This is a multiband image with seven different bands corresponding to different wavelengths of light. For display, one band each is assigned to the red, green, and blue color channels on the computer monitor. In this image, the lake is red because Band 1(blue light) was assigned to the red gun by default. Let's explore the display settings.

9→ In Contents, select the Landsat image.

9→ In the **Raster Layer: Appearance: Rendering** group, click the **Band Combination** drop-down and select Custom.

9→ Name the custom setting Landsat Natural Color, and make sure the bands are set to Band 3 for red, Band 2 for green, and Band 1 for blue. Click Add and examine the image (Fig. 6.15a).

Band 4 records near-infrared (NIR) light, not visible to humans but reflected strongly by vegetation. Before satellites, special NIR-sensitive film was used in aerial cameras to create "false color" images. The Landsat band combination 4-3-1 appears similar (Fig. 6.15b).

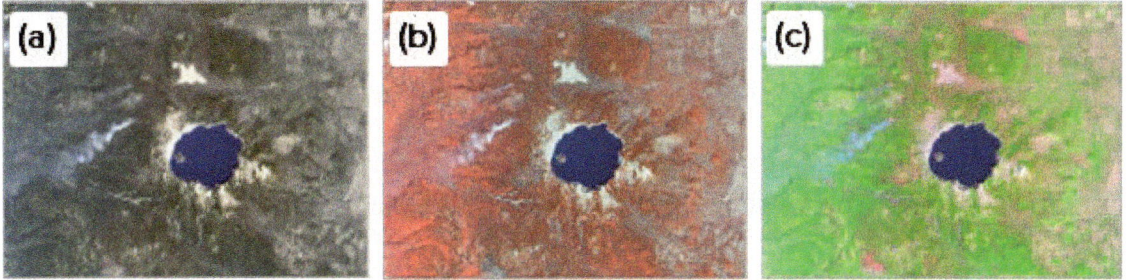

Fig. 6.15. Different band combinations for Landsat imagery of Crater Lake
Source: USGS

10→ In the **Raster Layer: Appearance: Rendering** group, click the **Band Combination** drop-down and select Custom.

10→ Name the custom setting *Landsat False Color* and make sure the bands are set to Band 4 for red, Band 3 for green, and Band 1 for blue. Click Add.

TIP: To delete a custom setting, right-click it in the Band Combination menu and choose Remove.

The band combination can be modified without creating a special setting. Another common Landsat band combination, 7-4-1, shows vegetation as green, bare ground as pink or brown, and urban areas as purple.

10→ In Contents, select the Landsat layer and click the **Raster Layer: Appearance: Rendering: Symbology** button.

10→ In the Symbology pane, set Red to Band_7, Green to Band_4, and Blue to Band_1. Examine the result (Fig. 6.15c)

Multiband imagery has more uses than we can explore here, but one technique performs calculations on bands to highlight different aspects of the terrain. The Normalized Difference Vegetation Index (NDVI) is a common calculation used to show green vegetation density. Pro can calculate indices on the fly, creating a layer displayed according to a specified formula.

11→ Click the **Imagery: Tools: Indices** drop-down and select **NDVI**.

11→ Set the Near Infrared Band Index to 4 and the Red Band Index to 3. Click OK.

NDVI ranges from about 0 to 1 and is primarily sensitive to chlorophyll. In this display, black areas (like the lake) have no green vegetation and bright areas have thick, healthy vegetation. Notice that the smoke from some fires in the northwest is black in this index.

11→ Remove the NDVI layer and save the project.

So far we have explored a continuous data raster and imagery. Let's examine a discrete raster. In fact, let's create one by converting the geology polygons to raster form. The polygons have multiple attributes, but a raster can only store one; we will use the RockType field.

TIP: Remember to always save output to the home geodatabase unless directed otherwise.

12→ In the Geoprocessing pane, open the Polygon to Raster tool.

12→ For Input Features, click the Browse button, navigate to the CraterLake home geodatabase, and select geologyunits. Click OK.

12→ Set the Value field to RockType.

12→ Name Output Raster rocktype.

12→ Hover on the information icon for the *Cell assignment type* and read through the options. Select the *Maximum combined area* option.

12→ Leave the Priority field set to none. Set the cell size to 30, to match the DEM.

12→ Switch to the Environments tab and set the Snap Raster to dem30m, to ensure that the cells of the two rasters align. Run the tool.

12→ In Contents, examine the legend of the rocktype layer.

A discrete raster typically has a restricted set of unique values and uses a different display type than a continuous raster. The correct display type, Unique Values, was automatically selected.

13→ In Contents, select the rocktype layer and click **Raster Layer: Appearance: Rendering: Symbology** to open the Symbology pane.

13→ Change the Value field to Value and examine the legend again.

Remember, a raster can store only numbers. When converting vector features based on a text field, an arbitrary integer is assigned to each value, to be stored in the cell, and the text field is included in the raster table to identify which numeric value corresponds with each text value.

13→ In Contents, right-click the rocktype layer and choose Attribute Table.

3. Which rock unit (give both number and name) is most abundant at Crater Lake?

14→ In Contents, double-click the rocktype layer to open its properties.

14→ View the Source section and expand the Raster Information heading.

This raster has the same cell size and extent as dem30m, yet the file size is much smaller. With only 14 different values, not even a full byte is required to store them, and the conversion used an unsigned char pixel type and a 4-bit pixel depth (compared to the 32-bit depth of the DEM).

14→ Close the layer properties and the table. In the Symbology pane, change the Value field back to RockType.

14→ Save the project.

Exploring 3D data

Now let's examine how DEM data can be used to generate 3D scenes.

15→ Click the **Insert: Project: New Map** drop-down and select New Global Scene. In Contents, rename the Scene title My3D.

15→ From the Catalog pane, drag the rocktype raster from the CraterLake home geodatabase to the map.

15→ In Contents, select rocktype and use the **Raster Layer: Appearance: Effects: Transparency** slider to set the transparency to about 50%.

15→ Hover over the **Map: Navigate: Explore** button to review the pop-up and get reacquainted with 3D navigation, if needed. Explore the scene in 3D.

The rocktype raster does not contain any 3D information. In a scene, a GIS elevation service is used to assign elevations to 2D data sets, or an alternate elevation source can be assigned.

16→ Scroll to the bottom of the Contents pane to examine the Elevation Surfaces entry, showing the current Ground elevation surface, WorldElevation3D/Terrain3D

16→ Right-click the Ground entry and choose Add Elevation Source. Navigate to the home CraterLake geodatabase and select dem30m. Click OK.

16→ Notice that dem30m appears above the Terrain3D service. It takes priority where it exists, but the Terrain3D service will continue to be used elsewhere.

16→ Explore the scene.

The pale Topographic basemap is not a particularly good background for visualizing the terrain. We have a better one, the hillshade raster.

17→ Add the hillshade10m raster to the map from the CraterLake geodatabase.

17→ In Contents, move the rocktype layer above the hillshade10m layer.

17→ In Contents, right-click and remove the Topographic basemap layer.

17→ Add the CraterLakeGeologyGroup.lyrx group layer file from the CraterLake home folder (not the geodatabase).

17→ For faster 3D rendering, if needed, turn off the complex Classes and Labels layer.

Notice that the geology layers drape onto the surface of the lake rather than on the crater floor where they belong. Both elevation sources in this scene assume that the lake surface is the ground. We can try to visualize the crater using contours, a vector method used to store surfaces.

18→ Add the bathymetry vector feature class from the CraterLake geodatabase.

18→ In the Contents window, right-click the bathymetry layer and choose Attribute Table. Examine the fields and find the contour value fields (in feet and meters).

The contours help, but the geology is still flat. We need an elevation source for the crater, and fortunately we have one. The contours from the land surface and the bathymetry were combined to create a TIN, a collection of triangular facets in various orientations. A TIN is built from point elevations and the lines, or edges, connecting them.

19→ Close the table and turn off all layers.

19→ Find the craterlaketin2 TIN data set in the CraterLake home folder and add it to the scene.

19→ Make sure that cratertin2 is selected in Contents. Click the **Tin Layer: Appearance: Drawing: Symbology** drop-down. Select *Symbolize your layer using lines: edges.*

TINs are efficient. A raster must store elevation at the same resolution everywhere, but TINs use more points in rugged areas like the crater walls and fewer points in flat areas. To visualize the geology on the crater floor, however, we must add the TIN as an elevation source.

19→ Remove the TIN and turn the Crater Lake Geology group layer back on.

19→ In Contents, right-click the Ground entry and add the TIN as an elevation source.

19→ Explore the scene.

Extrusion modifies the display of 2D features by adding a thickness, either a constant thickness as specified here, or a thickness from a field. The base height is the elevation of the feature along the ground, in this case.

20→ From the Catalog pane, add the faults and vents feature classes.

20→ In Contents, select the faults layer and open the **Feature Layer: Appearance** ribbon.

20→ In the **Extrusion** group, set the **Type** to Base Height. Set the **Unit** to Meters.

20→ Click the Extrusion Expression button. In the Expression Builder that opens, type *500* in the Expression box and click OK.

21→ Extrude the vents to a constant base height of *250* meters to better visualize them.

21→ When finished exploring, close the My3D scene. The Rasters map should still be open; if not, open it now. Save the project.

Exploring image and map services

ArcGIS Online is a marvelous source of raster data. Let's examine the basemaps, designed for use anywhere in the world at nearly any scale. A basemap is a picture of a map based on vector features and/or rasters.

22→ Use Ctrl-click to turn off all layers and collapse them.

22→ Click the **Map: Layer: Basemap** drop-down and select Imagery.

22→ Click the Full Extent button to return to the default map extent.

22→ Use the mouse wheel to quickly zoom in to a small area. Individual tiles will likely appear in groups as the higher resolution data are transmitted.

22→ In Contents, select the World Imagery layer and access the **Layer: Appearance** ribbon. The Symbology button is inactive. Tile layer symbology can't be modified.

Although ArcGIS Online is a fantastic data source, many GIS servers exist outside it. Some are run by companies and are secured to certain users, requiring a login. Others are public. To use one, you need to know the service URL.

23→ In the Catalog pane, switch to the Portal tab and choose the All Portal icon.

23→ Type *tile* in the search box and click Enter. A collection of hosted Tile Layers appears. Hover over a Portal item to see the pop-up description and look at the bottom of the pop-up for the Location URL.

One disadvantage of the imagery basemap is not knowing when the imagery was collected. The National Agriculture Imagery Program (NAIP) provides annual imagery as a GIS service.

24→ Click the **Insert: Project: Connections** drop-down and select New ArcGIS Server.

24→ Type *https://gis.apfo.usda.gov/arcgis/services* into the Server URL box.

24→ This is a public server, so no login or password is required. Just click OK. Wait a moment for the connection to initialize.

24→ In the Catalog pane, switch back to the Project tab. Find the Servers entry and expand it. The new server link is displayed.

24→ Expand the apfo.usda.gov server link and the NAIP folder. Find the Oregon imagery and drag it to the map.

24→ Save the project.

Server connections are saved with the project, so this NAIP imagery will be available in future sessions. The NAIP service is an image layer that serves actual pixel values. Let's explore some ways that an image service differs from a basemap.

25→ In Contents, open the layer properties for the NAIP layer and examine the Source section. Expand the Spatial Reference section and examine the coordinate system.

25→ Expand the Raster Information section and examine it carefully.

This is a gigantic 2.25-terabyte (TB) raster (just for Oregon!). It is also a four-band raster with a 1-meter cell size and 8-bit pixel depth (per band). When using image services, better performance can usually be achieved by setting the map coordinate system to match the service, so on the fly projection is not necessary.

25→ Use the Rasters map properties to set the map coordinate system to WGS 1984 Web Mercator Auxiliary Sphere to match the NAIP layer.

25→ Zoom to Wizard Island. Turn the NAIP imagery off and on a few times to compare it with the World Imagery basemap. Then turn off World Imagery.

An image service will have most of the display capabilities of a local multiband image. In the legend, notice that this raster has red in band 1, green in band 2, and blue in band 3, which is typical for aerial imagery (but not for Landsat, which reverses the order of the visible bands).

26→ In Contents, select the NAIP layer and click the **Image Service Layer: Appearance: Band Combination** drop-down and select Color Infrared.

26→ Click the **Imagery: Tools: Indices** drop-down and select NDVI. Enter 4 for the Near Infrared Band Index and 1 for the Red Band Index. Click OK and wait.

26→ After examining the result, remove the NDVI layer.

If the service provider allows it, a portion of the image can be exported and saved locally. Exporting a raster is more complicated than exporting features. For best results and to avoid resampling, we should keep it as close as possible to the original image.

27→ In Contents, right-click the NAIP layer and choose Data > Export Raster.

27→ The Output Raster defaults to the CraterLake home folder, which is fine. Change the name to *WizardNAIP*, keeping the .tif extension (Fig. 6.16).

27→ Make sure that the Output Format (near the bottom of the tool) is set to TIFF.

27→ Keep the same coordinate system and don't specify a transformation.

27→ If needed, make the Export Raster pane a floating window so it doesn't diminish the display extent. Set the Clipping Geometry to Current Display Extent.

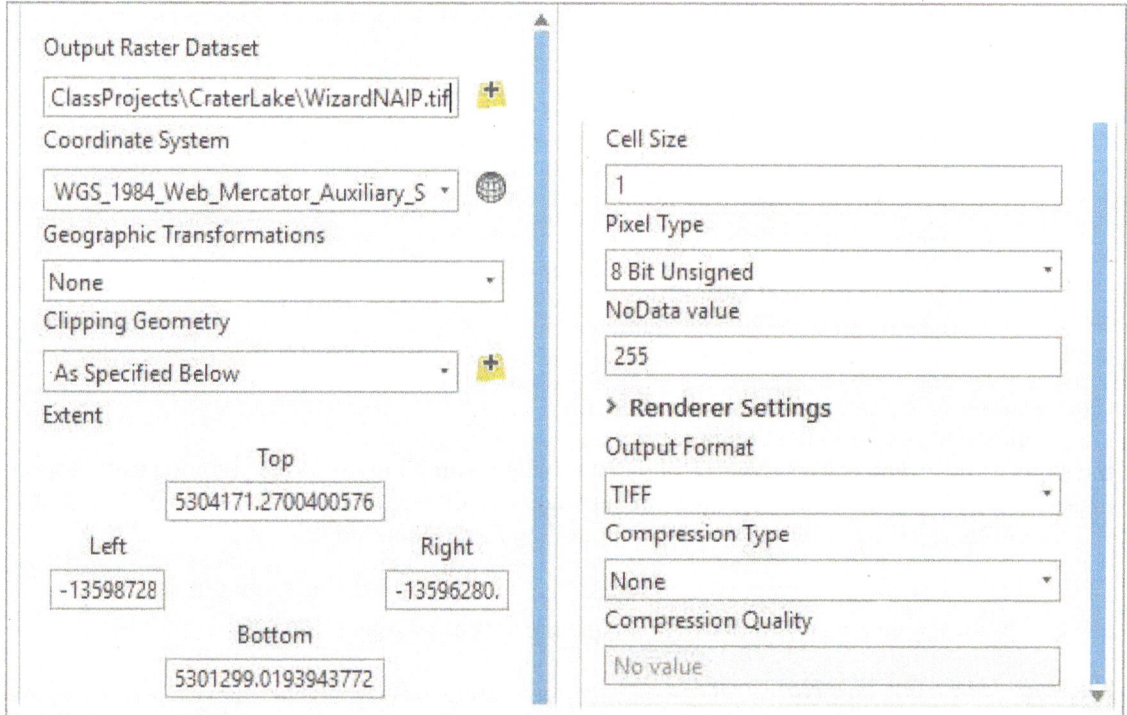

Fig. 6.16. Exporting NAIP imagery for Wizard Island
Source: Esri

27→ Examine the other settings, but keep the defaults. Click Export.

27→ Close the Export Raster window and remove the original NAIP image service.

The new image looks a little different because the display statistics are being calculated from the local image instead of being optimized on the server side.

Even this tiny piece of imagery is over 26 MB! Next, we should project this image raster to match our other data. Again, projecting rasters is more complicated than projecting features because it requires resampling. For multiband spectral images, a nearest neighbor sampling technique works best to preserve the original spectral measurements and to avoid strange color artifacts.

28→ In the Geoprocessing pane, search for and open the Project Raster tool.

28→ Set the Input Raster to WizardNAIP.tif (Fig. 6.17).

28→ For the Output Raster Dataset, click the Browse button. Navigate to the CraterLake home folder and name the output *WizardNAIPutm.tif.* (The .tif extension is required to specify the output file format, so be sure to include it.) Click Save.

28→ Click the Output Coordinate System drop-down and select dem30m to match the existing data sets. Use the suggested Geographic Transformation from WGS_1984 to NAD 1983.

28→ Use the Nearest neighbor resampling technique and set the X and Y Output Cell Size to 1. Leave the registration point blank. Run the tool.

29→ Zoom to the west side of the island and turn the new image off and on a few times.

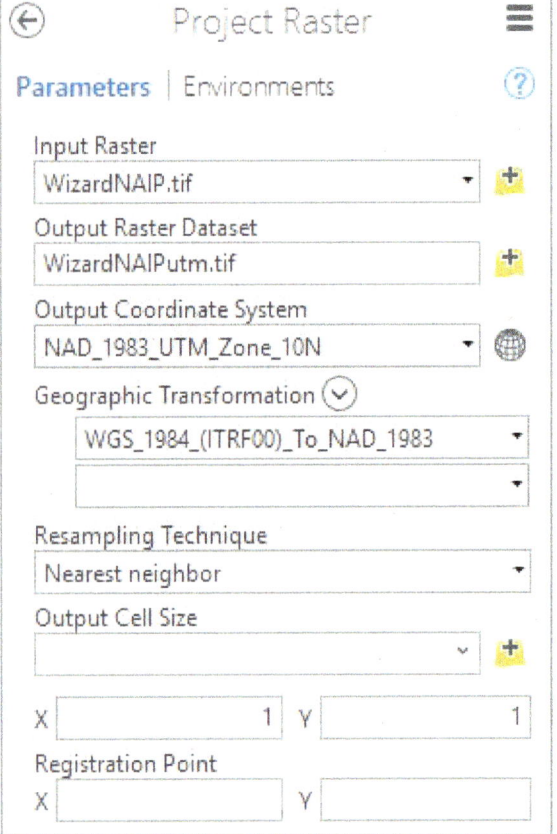

Fig. 6.17. Projecting the Wizard Island image
Source: Esri

Close examination reveals that the projected image, although similar in quality, is not quite the same as the original. This example illustrates why resampling should be avoided except when necessary. If one needed to project this image again to a different coordinate system, the original WizardNAIP.tif should be used, to avoid compounding the resampling effects.

29→ Use the map properties to set the map coordinate system to match dem30m.

29→ Remove the WizardNAIP.tif image and save the project.

Getting data from The National Map

The US Geological Survey maintains a portal that provides a variety of GIS data and services, called The National Map (TNM). It is often a good source when looking for consistent base data like elevation, hydrology, and land cover, and it has the benefit that it can be downloaded.

TIP: Before using data from The National Map, be sure to review the extensive information provided about the data sources on The National Map web sites.

30→ In a web browser, open The National Map: http://nationalmap.gov.

30→ Click the heading on the right: Find Data + View & Download.

30→ Under GIS Data, click the yellow-highlighted Download GIS Data.

30→ Both a data list and a map should appear; widen the computer window, if needed, to see both.

Internet information is volatile and can go out of date quickly. If the viewer platform no longer matches the instructions in this tutorial, use the How To link on the page for latest information.

31→ Above the map, type Crater Lake, Oregon, in the search box (Fig. 6.18) and click Go (1).

31→ Above the map, make sure the Use Map box is checked and fill the Box/Point button (2). Click and drag to draw a rectangle that encompasses the area of Crater Lake that we have been working with (3). It doesn't need to be exact.

31→ On the left, on the Datasets tab, check the box for the National Land Cover Database (NLCD) product (4). Keep the (NLCD) - 2011 subcategory checked.

31→ Keep the default 3 × 3 degree extent option and click the Find Products button (5).

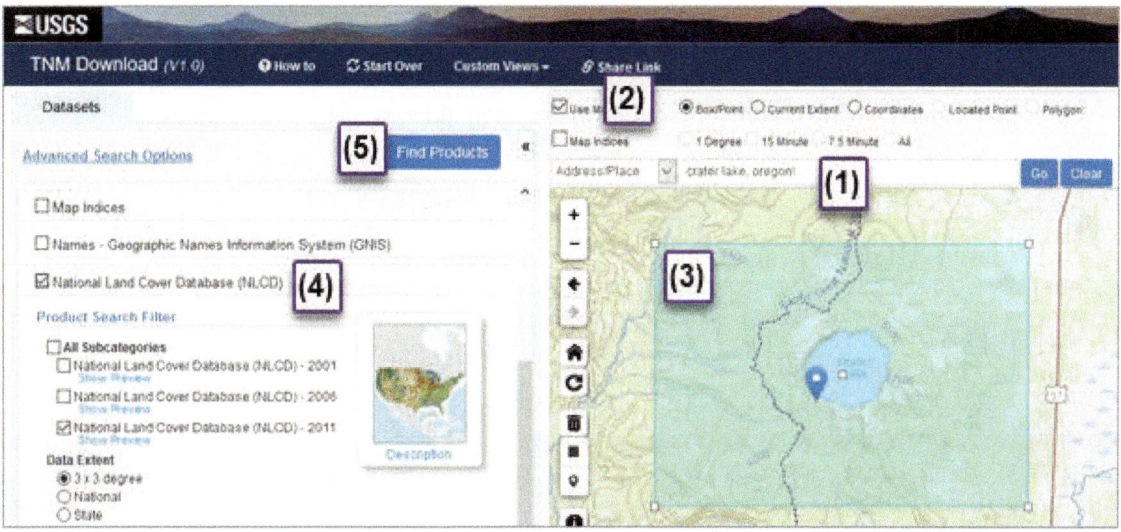

Fig. 6.18. Downloading land cover data from The National Map
Source: USGS

The web site searches for matching products that cross the rectangle and finds three. We will download the land cover.

 31→ Click the Add to Cart button for the NLCD 2011 Land Cover.

TIP: For more organized GIS work, set the web browser to save files to a location chosen each time, rather than automatically saving to the default Downloads folder.

32→ At the top, click the View Cart button.

32→ Click the Download link for the product. Navigate to the gisclass\ClassProjects\ CraterLake folder. Click New Folder to create a new folder and name it *MyDownloads*.

32→ Navigate inside the MyDownloads folder and click Save. Wait for the download to complete.

32→ Minimize (don't close) The National Map web site for now.

The downloaded file must be unzipped before Pro can recognize it as GIS data.

32→ Open File Explorer and navigate to the CraterLake\MyDownloads folder. Unzip the file and close Explorer.

33→ In Pro, in the Catalog pane, expand Folders if necessary. Right-click the CraterLake home folder and choose Refresh to make sure it is up to date.

33→ Expand the CraterLake and MyDownloads folders. Drag the NLCD2011_LC_N42W. tif image (not the jpeg) into the map. Zoom to its extent.

33→ Open the layer properties for the NLCD layer and examine the Source section.

33→ Expand and examine the Spatial Reference and Raster Information headings.

The 3 × 3 degree tile is much larger than we need. It is also rotated because of the map projection being used. We need to both clip and project this raster as we add it to the CraterLake geodatabase. Using Environment settings, we can do both at once. Since resampling will be required, we should use the nearest neighbor method, since this data stores land use categories.

34→ In the Geoprocessing pane, search for clip. The Clip (Analysis) tool is for vector data; the Clip (Data Management) tool works on rasters.

34→ Open the Clip (Data Management) tool (Fig. 6.19a).

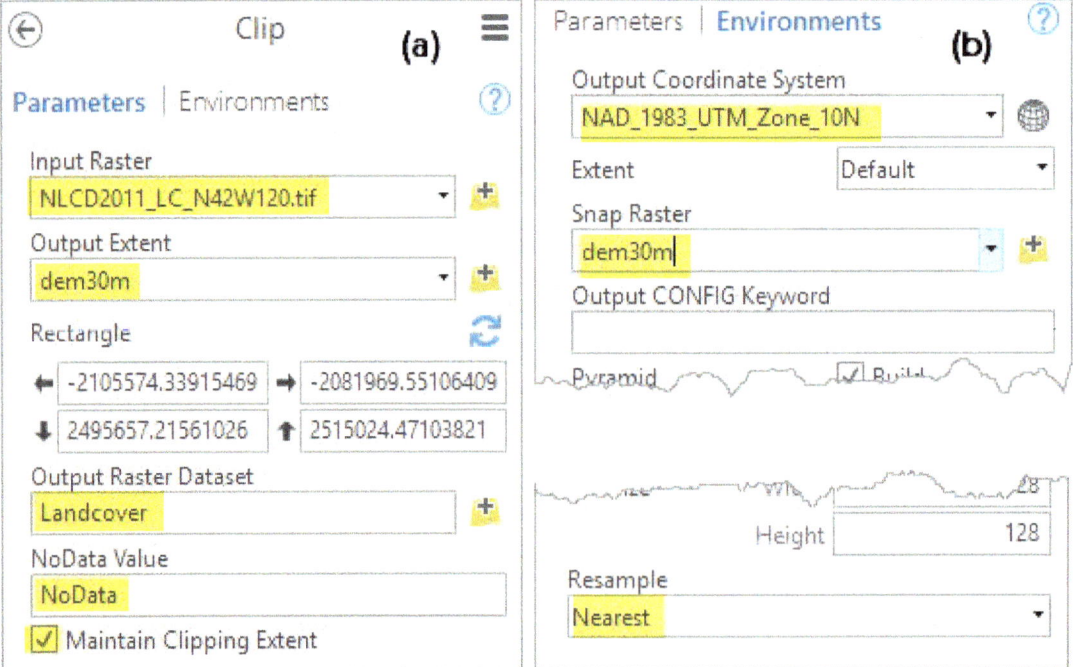

Fig. 6.19. Clipping and projecting the land cover raster
Source: Esri

34→ Set the Input Raster to the NLCD layer. Set the Output Extent to dem30m.

34→ Name the Output Raster *Landcover*.

34→ Set the NoData value to NoData and check the box to Maintain Clipping Extent.

34→ Click the Environments heading on the tool (Fig. 6.19b).

34→ Set the Output Coordinate System to dem30m so they match.

34→ Set the Snap Raster to dem30m also so that the cells align.

34→ At the bottom, set the Resample option to Nearest. Run the tool.

35→ Remove the original NLCD layer and zoom to the new raster.

35→ Open the Landcover attribute table to see the land cover categories.

35→ Close the table. Save the project.

> **TIP:** Don't forget to use the Catalog view to create an Item Description for the new data.

Working with elevation data

The standard worldwide elevation product is a 30-meter resolution DEM, available everywhere. An ArcGIS Online image service is the simplest way to obtain this data. Suppose one needed elevation data for the entire park, not just the small piece we have in the database. We can clip this data set using a park boundary polygon.

36→ In the Catalog pane, find the ParkBoundary feature class in the CraterLake geodatabase and add it to the map.

36→ In the Catalog pane, click on the Portal tab and choose the All Portal icon.

36→ Search for *ground surface*. Locate the Ground Surface Elevation – 30m imagery layer and add it to the map.

36→ Examine the image source properties to learn its coordinate system, pixel type, and depth.

This time we will save the raster inside the geodatabase instead of as a TIFF in the folder.

37→ In Contents, right-click the Ground Surface Elevation layer and choose Data > Export Raster.

37→ For the Output Raster, click the Browse button and navigate into the CraterLake home geodatabase. Name the output raster *parkdem30m*.

37→ Set the coordinate system to match the ParkBoundary layer.

37→ Set the Clipping Geometry to ParkBoundary, and check the box for Use Input Features for Clipping Geometry. Leave the Clipping Type set to Outside.

37→ Round the X and Y Cell Sizes to 30. Set the NoData value to *NoData*.

37→ Switch to the Settings tab and change the Resample method to Bilinear (the top one, not the Pyramid one). Click Export.

37→ Close the Export Raster window when it finishes and turn off ParkBoundary.

37→ Remove the Ground Surface Elevation imagery layer and save the project.

Higher-resolution DEM data are sometimes available. For one potential source, we return to The National Map. However, the process requires multiple steps, complicated by large file sizes.

TIP: Elevation data are an important resource. Find some time to read the USGS information on the 3DEP (3D Elevation Program) to learn about the different elevation products and how they are developed. http://nationalmap.gov/3DEP/3dep_prodserv.html

38→ Reopen The National Map browser window that was minimized earlier.

38→ Click the Datasets tab at the top left. Uncheck the NLCD product and check the Elevation Products (3DEP).

38→ Uncheck the 1/3 arc second product (30 m) and check the 1/9 arc second product.

38→ Note that the data extent tiles are 15-minute quadrangles and the file format is IMG.

38→ Click Find Products. Eight of them will usually appear.

For Crater Lake, the 1/9 arc second data are only available inside the park boundaries. The thumbnails in the results show that large areas of the tiles are black and have no data. We will download the lake area plus one tile to each side to illustrate the method.

39→ Use the thumbnail to find the product with the lake. Click the Thumbnail link to show it on the map. Add it to the Cart.

39→ Use the thumbnails to identify the quadrangles east and west of the lake. Check by displaying the thumbnail on the map. (To remove a wrong thumbnail from the map, click the same thumbnail link again.) Add the right ones to the Cart also.

39→ When done, the map should look similar to Figure 6.20, and the Cart should contain three products.

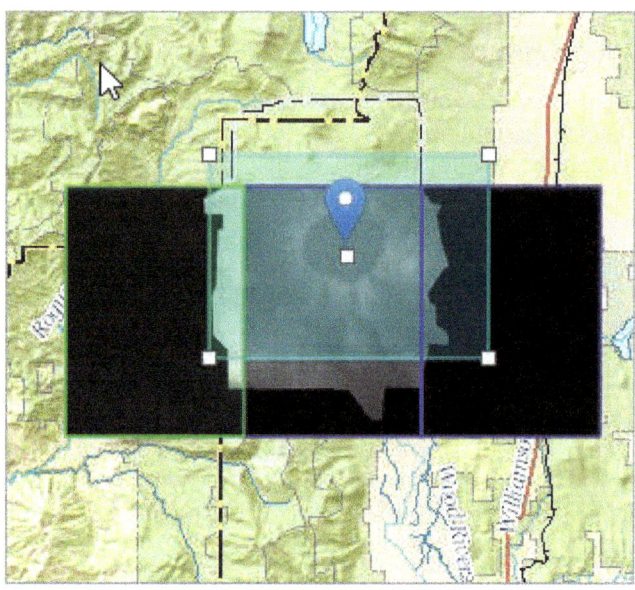

Fig. 6.20. The 1/9 arc second products for Crater Lake
Source: USGS

40→ Click View Cart. Download the products to the CraterLake \MyDownloads folder. Be sure to note the names of the files as they are saved. Wait for each download.

40→ Minimize the browser. Use File Explorer to unzip all three downloaded files. Click Yes to All when asked to replace the NED_DataDictionary.url file; some similar metadata files are downloaded with each piece.

TIP: Notice how many files the unzipping creates. Using the MyDownloads folder to store this mass of files keeps the project's permanent data folder uncluttered, and the MyDownloads folder is easy to delete once the compilation work is finished.

41→ In Pro, in the Catalog pane, in the Project tab, refresh the MyDownloads folder.

41→ Turn off all layers in the map.

In this case, because there are three files and the file names are long, it is more convenient to use the Add Data button instead of the Catalog pane to add the files.

41→ Click **Map: Layer: Add Data** and navigate to the MyDownloads folder.

41→ Hold the Ctrl key down and click each of the three DEM rasters to select them (they all start with ned and have an .img extension; don't add the .jpg files by mistake).

41→ Figure out which is which and rename the layers nedwest, nedlake, and nedeast.

41→ Open the layer properties for nedlake and examine the Spatial Reference and the Raster Information sections.

The seams between the files occur because the display statistics are calculated for each file separately, so the color ramps do not match. The seam will disappear once the files are merged.

The nedlake raster is saved in IMAGINE Image format. It comprises 250 MB, with a cell size of 0.00003 degrees and a 32-bit floating-point pixel type. The coordinate system is GCS North American NAD 1983. We need to clip the files to the area of interest, merge the pieces, and project them. In theory this could be accomplished in one step, but the large files require a long processing time and might cause program failure from insufficient memory. The reliable approach to combine large rasters is to clip first, then project, then mosaic. A boundary polygon for the area of interest has been prepared already. Its coordinate system matches the NED files, to ensure that no automatic projection will be attempted.

42→ Save the project (in case we encounter an error).

42→ Add the topocliparea polygon feature class from CraterLake. It is in the same coordinate system as the rasters, GCS NAD 1983.

42→ In the Geoprocessing pane, clear and reopen the Clip (Data Management) tool.

42→ Set the Input Raster to nedwest and the Output Extent to topocliparea. Name the output dataset *nedwestclip*.

42→ Change the NoData Value to NoData.

42→ Read the information pop-up on the Maintain Clipping Extent options. We want to avoid resampling right now, so leave it unchecked.

42→ Run the tool.

TIP: Clip works best when the clip polygon is the same coordinate system as the raster data.

43→ In the still open Clip tool, change the input to nedlake and the output to *nedlake-clip*. Leave the other settings and run the tool again.

43→ In the Clip tool, change the input to nedeast and the output to *nedeastclip* and run.

Now that the file sizes have been reduced by clipping, it is time to project each piece. Although the data resolution supports resampling to 3 meters, we will save space and resample to 10 meters instead.

44→ Turn off topocliparea and open the Project Raster tool.

44→ Set the Input Raster to nedwestclip and the Output Raster to *nedwestprj*.

44→ Set the Output Coordinate System to match dem30m. No transformation is needed.

44→ Set the Resampling Technique to Bilinear interpolation and the Output Cell Size to 10 for X and Y. Run the tool.

45→ Rerun the Project Raster tool for nedlake-clip and nedeastclip, naming the output *nedlakeprj* and *nedeastprj*, using the same settings as before.

Finally, we will mosaic the rasters into a single file. Projecting might have rotated and shifted the pixels, and the raster edges may no long be aligned. Overlapping areas should be minimal but should be blended where they occur. If there are offsets across the edges, we wish them to be resampled, not shifted, so we will force resampling and choose the bilinear method to ensure that it is done smoothly.

46→ Open the Mosaic tool. Read the Help and the information pop-ups.

46→ Set the Input Rasters to nedeastprj and nedwestprj (Fig. 6.21).

46→ Click the Browse button for the Target Raster and set it to nedlakeprj.

46→ For Mosaic Operator, use the Blend option to mix any overlapping areas.

46→ There is no colormap, so leave the Mosaic Colormap Mode to the default.

46→ Set the NoData value to NoData.

46→ Keep the Mosaicking Tolerance set to 0 to force resampling rather than shifting.

46→ Click the Environments heading. Set the Resample method (at the bottom, not the Pyramid one) to Bilinear.

46→ Run the tool.

46→ Turn off all layers except nedlakeprj, the final product.

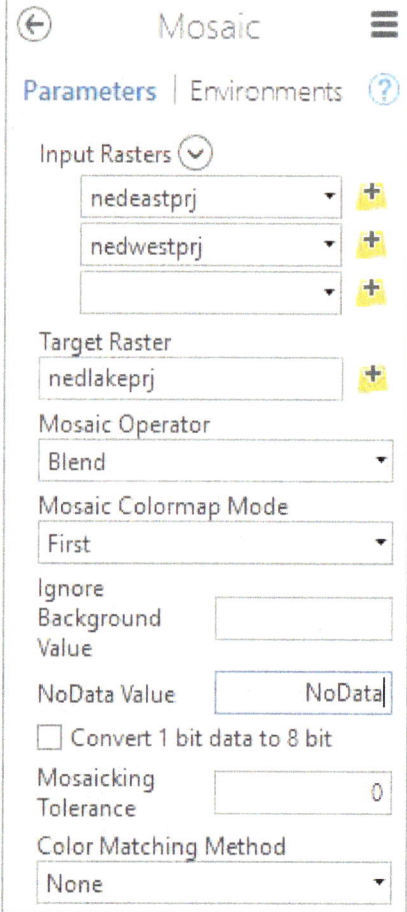

Fig. 6.21. Mosaicking the elevation rasters

Source: Esri

TIP: The Mosaic to New Raster tool can mosaic and project in a single step, but this elevation data set is too large for it.

This long process produced several intermediate rasters that are large in size. We should clean out the database of these intermediate files. Always use Pro tools to delete raster files.

47→ Open the Catalog view and expand the contents of the CraterLake geodatabase.

47→ Right-click the nedeastclip raster and choose Delete. Click Yes.

47→ Delete all the other ned rasters except the final product, nedlakeprj.

48→ Take a minute to fill out the Item Description for nedlakeprj. Include a citation and briefly outline the processing steps taken to create it.

48→ Save the project.

Georeferencing a raster

Georeferencing creates a real-world coordinate system for an image that lacks one, usually because it is a map picture or scan. A georeferencing session starts by adding the data with the coordinate system, the reference layer, first. Then add the image.

49→ Close the Rasters map. Use **Insert: Project: New Map** to create a new map. Name the map *Georef*.

49→ Add the ParkBoundary feature class from CraterLake. Symbolize it with a hollow symbol with a 2-pt-thick line that is easily seen.

49→ Add the crlamap1.jpg image from the CraterLake home <u>folder</u>.

The image appears in the legend but not in the map. It lacks a real-world coordinate system, so it is being displayed somewhere else dictated by its arbitrary coordinate system. The first step is to bring it closer, which facilitates finding the control points.

50→ Click **Imagery: Alignment: Georeference** to open the **Georeference** ribbon.

50→ Click the **Georeference: Prepare: Fit to Display** button to bring the image into the current view.

At this point, the park boundary should both be visible as the polygon and on the image, but not aligned. We will add control points matching the locations of one boundary to the other. Always add the image control point first, then the reference layer point. Zooming in makes it easier to enter precise points. Use the mouse wheel to zoom and wheel-click/drag to pan while entering points.

50→ Click **Georeference: Adjust: Add Control Points**.

50→ Mentally match two corresponding corners of the northwest park boundary. Zoom in but keep both versions of the corner visible.

50→ Carefully click the corner on the image first and then click the corresponding corner on the ParkBoundary polygon. Pro will help by snapping to the vertex of the polygon.

50→ Notice that the map updated after the first control point was added.

51→ Zoom to the southeast corner of the park, using the mouse wheel to avoid adding another control point.

51→ Click the location on the image first and then click the corresponding location on the ParkBoundary polygon.

The match will already look much better, but at least three points are required, and more are usually best. Be careful not to snap to a ParkBoundary vertex when adding an image point.

52→ Zoom or pan to the southwest corner of the park. Click the Snapping button in the lower left part of the display to turn off snapping while entering the image point.

52→ Turn snapping back on and enter the polygon boundary point.

53→ Add a fourth pair of points at the northeast corner.

53→ Add two additional points on the east and west boundaries if offsets between the two boundaries are still visible.

Some maps require additional control points in misaligned areas until no improvement is seen when new points are added. This map is looking good, even when zoomed in close.

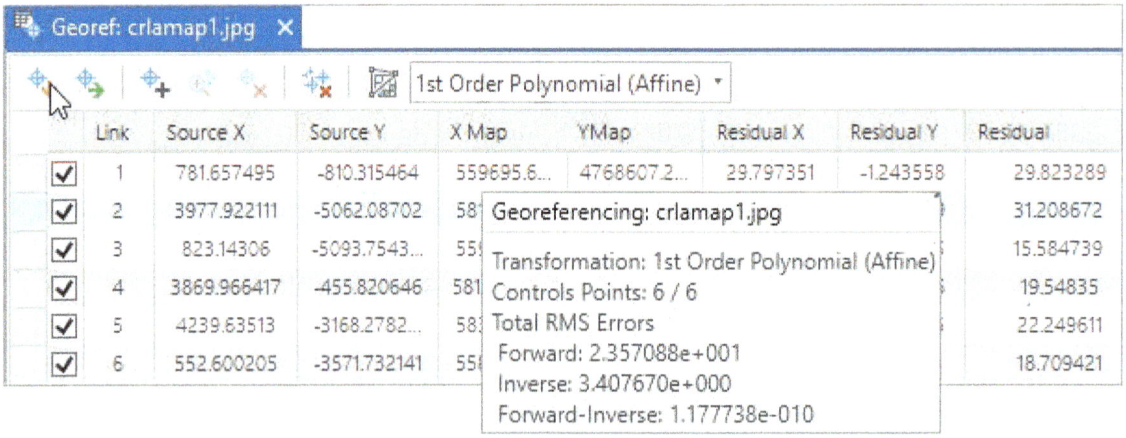

Fig. 6.22. The Control Point Table shows the points added.
Source: Esri

54→ Click the **Georeference: Review: Control Point Table** button to see the control points.

54→ Examine the Georeferencing: crlamap1.jpg information box (Fig. 6.22).

The points are listed in the order added, and yours will differ from the figure. The X and Y Source fields show the coordinates of each control point in the original image units (pixels). The X Map and Y Map columns show the point coordinates in map units (meters). The 1st Order Polynomial (Affine) Transformation method is being used.

Note the residuals for each point and the total RMS error in the Georeferencing information box, reported in map units. In Figure 6.22, the residuals range between 15 and 31 meters, and the RMS error is about 24 meters. Your values should be similar.

TIP: If the RMS error is large and does not improve when more points are added, try a second-order transformation. Second-order transformations require more control points. See the ArcGIS Pro Help for a complete discussion of transformation methods.

54→ Close the Control Point table. If the RMS error is less than 20 meters, you did well. Write it here in order to add it to the metadata later: _____

54→ To save the transformation parameters with the image, click **Save**.

54→ Click **Close Georeference** to close the ribbon.

55→ Switch to Catalog view and update the Item Description of crlamap1.jpg (in the CraterLake folder) with the RMSE. Include the units. Put it in the Description box.

At this point the image is georeferenced, and the control point information has been saved with it. The next time it is added to a map, it will appear in the correct location.

56→ For fun, add the crlamap1.jpg to the My3D scene and explore the map in 3D.

56→ Close the scene and save the project when done.

TIP: Save as New will create a copy of the image. It may require resampling and may affect the image colormap, but the georeferencing will be a permanent part of the image, rather than information stored in an auxiliary file.

Visualizing space-time data

To explore the capabilities of using time as the third dimension, we will explore the space-time distribution of aftershocks from an 8.8 magnitude earthquake off the coast of Chile in 2010.

57→ Open the gisclass\ClassProjects \ChileQuake project.

57→ Open the attribute table for the Quakes_ 36hrs layer and examine the fields.

The time field we will use, TimeReformat, is at the end of the table. It had to be reformatted from the original time field so that ArcGIS could interpret the date and time correctly. The first step is to create a space-time cube stored in netCDF format. To produce the cube, we will sum the magnitude of earthquakes occurring in 30 × 30 km cells over six-hour periods.

58→ In the Geoprocessing pane, open the Create Space Time Cube By Aggregating Points tool (Fig. 6.23).

58→ Set the Input Features to Quakes_36hrs.

58→ Click the Output Space Time Cube browse button and navigate to the ChileQuake folder. Name the output *quakes30km.nc*. The *.nc* extension is required.

58→ Set the Time Field to TimeReformat. Leave the Template Cube blank.

58→ Set the Time Step Interval to 6 Hours and the Time Step Alignment to Start time.

58→ Keep the Fishnet grid and set the Distance Interval to 30 Kilometers.

58→ Click the plus symbol to add a Summary field. Choose the mag field, enter the Sum statistic, and select Zeros for the value to fill empty bins. Click Tab to make sure the field specification is complete.

58→ Run the tool.

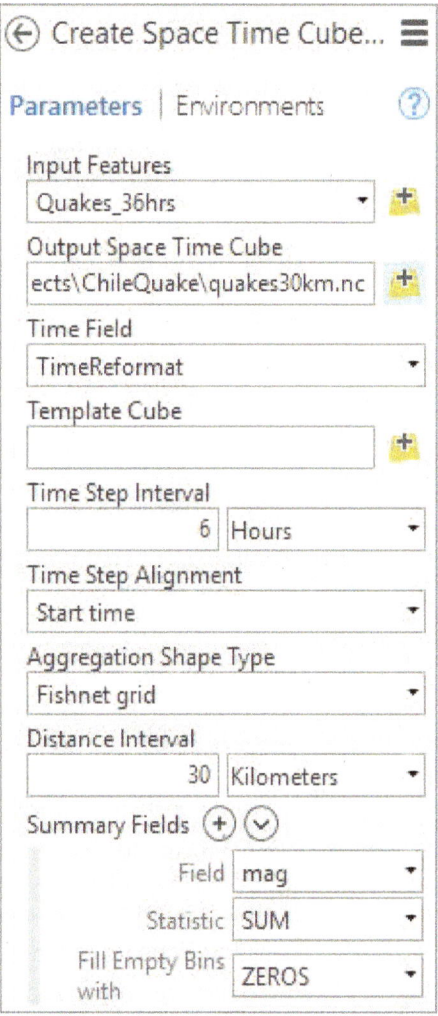

Fig. 6.23. Creating the space-time cube
Source: Esri

Now the data cube is created. Another tool prepares the data cube for 3D visualization.

59→ Use the **Insert: Project: New Map** drop-down to open a New Global Scene.

59→ In the Geoprocessing pane, open the Visualize Space Time Cube in 3D script.

59→ Click the Browse button and select quakes30km.nc as the Input Space Time Cube.

59→ Set the Cube Variable to MAG_SUM_ZEROS and the Display Theme to Value.

59→ Run the tool.

Nothing shows up at first because the scale range is not set appropriately.

60→ Click the **Feature Layer: Appearance** ribbon and set the Visibility Range setting to Out Beyond 10,000,000 feet.

60→ Copy the Main Shock layer from the map to the scene and enlarge the symbol to 36 pt. so that it is visible below its stack.

60→ Explore the cube. It should appear similar to Figure 6.12.

Each stack represents a 30-km square, and each cube in the stack represents a six-hour time period. The purple cubes show where and when the aftershocks occurred and the summed magnitudes of aftershocks.

This is the end of the tutorial.

→ Save the project and close Pro.

Practice Exercises

1. Choose a national park other than Crater Lake.

2. Select an appropriate coordinate system for the park.

3. Create a project for the park and save it in the gisclass\ClassProjects folder.

4. Find at least three vector data sets at a suitable scale for the park. Use any source of data other than the gisclass folder. Project them all and clip them to the park's boundary.

5. Find at least three raster data sets at a suitable scale for the park. Use any source of data other than the gisclass folder.

6. Find at least one map image of the park and georeference it. Record the RMS error.

7. Prepare the following items:
   a. A layout with two maps or scenes showcasing the data and including a full list of complete, formal citations for each data set
   b. A screenshot of the project database expanded to show the data sets
   c. A screenshot of the map properties Layers folder to show that the data sets are all in the matching appropriate coordinate system

Chapter 7. Attribute Data

Objectives

- ▶ Understanding how tabular data are stored and queried
- ▶ Understanding joins and cardinality concepts
- ▶ Exploring data in tables using statistics and summarized statistics
- ▶ Defining appropriate field properties
- ▶ Editing and calculating fields in tables
- ▶ Generating point layers from x-y coordinates in tables

Mastering the Concepts

GIS Concepts

Tables are a fundamental data structure associated with both vector and raster data, and they are similar in many ways to tables found in commercial database software. This chapter covers methods for managing tables and analyzing the information found within them.

Overview of tables

A **table** is a data structure for storing multiple attributes about a location or an object. It is composed of rows, called **records**, and columns, called **fields** or **attribute fields**. Figure 7.1 shows an example of a table containing attributes of counties.

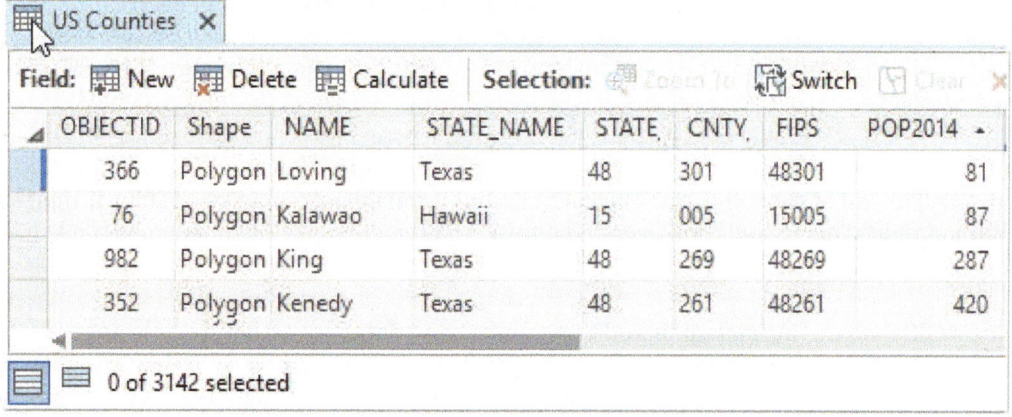

Fig. 7.1. A table with information about counties in the United States
Source: Esri

In GIS, tabular data files fall into two main categories: **attribute tables** and **standalone tables**. An attribute table, such as the one shown in Figure 7.1, contains information about features in a geographic data set. An attribute table always contains one, and only one, row of information for each spatial feature. In a shapefile, the row is linked to the spatial feature in a separate file using a unique integer ID number called a **Feature ID**, or **FID**. In a geodatabase, the file stores both the attributes and the x-y coordinates in the same data file, although the coordinates are not

visible in the table, and it uses an **Object ID**, or **OID**, instead of an FID (Fig. 7.1). In contrast, a standalone table simply contains information about one or more objects in tabular format instead of having information about map features. A standalone table might come from a text file, an Excel® spreadsheet, a global positioning system data file, or a commercial database. Standalone tables exist independently of a geographic data set and may be only incidentally related to map features. They also have an OID rather than an FID. Both the FID and OID essentially count the records in the table in the order in which they are stored. In Figure 7.1, the table was sorted on a different field, so the OBJECTID field appears out of order.

Database management systems

GIS tables share a history and many current properties of the database programs that are routinely used in commercial, governmental, and academic settings. Most vector-based GIS programs use an underlying database to store data, such as the ancient INFO database used for coverages, the dBase table structure used for shapefiles, the Microsoft Access engine used for personal geodatabases, or a large-scale relational database management system (RDBMS), such as SQLServer, used for enterprise geodatabases. These programs are designed to store, manipulate, analyze, and protect tabular data of all kinds. Governments use them to store information about citizens, parcels, taxes, and more. Companies use them to store information about customers. Universities use them to manage information about students, classes, and faculty. Three types of databases have traditionally been used.

A **flat file database** stores rows of information in a text or binary file. Finding information requires parsing the table and selecting the records of interest. They are simple but not efficient.

A **hierarchical database** has multiple files, each of which contains different records and fields. Parent tables can be linked to child tables through a specified field called a key. For example, a table of college classes might be linked to a table of students in each of the classes through a course ID number. Relationships between tables are fixed, which makes looking up information quick. However, the relationships are inflexible, designed to permit a predefined set of operations.

A **relational database** also has multiple tables stored as files. However, the relationships are not defined ahead of time. Instead, the user can temporarily associate two tables if they share a common field. This association is called a **join**, and the common field becomes the key. This database model is extremely flexible and is the preferred choice for GIS systems.

Because GIS data are intimately linked to an underlying database structure, it is not a stretch to incorporate tables from database programs into GIS analysis. ArcGIS Pro can connect to and work with RDBMS data files directly, expanding the type of data available. If a county keeps its parcel tax records in an RDBMS such as Oracle or SQL Server, for example, a GIS analyst can bring the database tables into ArcGIS Pro as standalone tables.

Joining and relating tables

In an RDBMS and in a GIS, tables are commonly combined using a join in order to bring different sets of information together. The tables are combined using a common field called a **key**. The key field must be of the same data type in both tables. When a join is performed, the two separate tables become one and contain the information from both tables (Fig. 7.2). The join is a temporary relationship and may be removed when it is no longer needed.

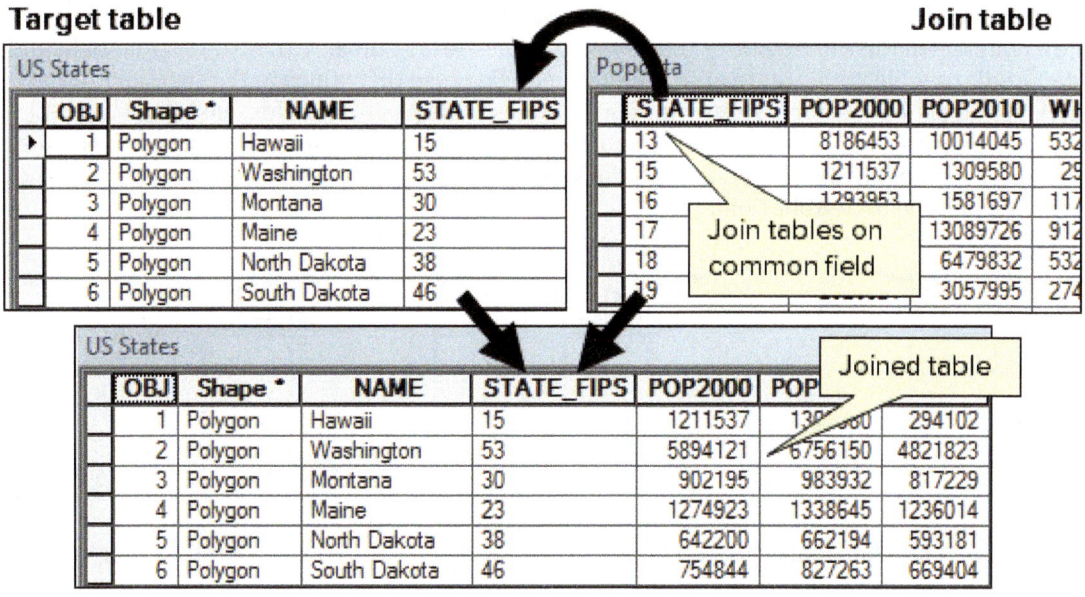

Fig. 7.2. Joining on a common field. These two tables are joined into one using the common field STATE_FIPS. The data from the join table are placed into the target table.
Source: Esri

TIP: Joining Census data is often accomplished using FIPS codes, which stands for Federal Information Processing Standard. Each state, county, city, block, and so on is assigned an identifying number, and these are combined to produce a unique identification for each census unit. For example, the state of Arizona is assigned '04' as its state FIPS code. Santa Cruz County is assigned the county FIPS code '019'. The county name is not unique (California also has a Santa Cruz County), but the combination of the state and county FIPS codes is. The code '04019' refers to Santa Cruz in Arizona, and '06087' refers to the one in California. A join using county names would not work because they are not unique, but a join on the FIPS code does work. The quotes are used in this discussion to show that FIPS codes are stored as text values; the leading zero on some codes would be dropped if they were stored as numbers.

Joins have a direction. The table containing the information to be appended is called the **join table**. The table that receives the appended information is called the **target table**. (Some GIS programs, including previous versions of ArcGIS, refer to the join table as the **source table** and the target table as the **destination table**.) In Figure 7.2, the target table, US States, is a feature attribute table. The join table containing demographic data is a standalone table. When the two tables are joined, the demographic data are appended to the US States attribute table. Often joins are used to bring data from standalone tables into an attribute table for mapping or analysis.

Join direction matters. After the demographics are joined to the table of the states shapefile, then the information could be used to make a map of the population data. If, however, the join had been performed in the opposite direction, with the demographics as the target and the states as the join table, the resulting table would be a standalone table. In this case, a map showing the demographics could not be made because the demographics table is not a feature class.

Multiple joins can combine three or more tables. In ArcGIS, the target table can have multiple join tables joined to it; each join is initiated with the same target table but a different source. Source fields in the first join can be used as key fields in the next join. However, a table cannot simultaneously be a source for one join and a target for another.

Imagine that Sally is comparing the average teacher salary in counties to the state average (Fig. 7.3) for all counties in the United States. She has a counties salary table with a state name field and a states salary table with a state abbreviation field. Because the values in the two fields don't match, a join cannot be done. However, a third States table has both fields, enabling a multiple join. The counties salary table is the target both times. First Sally joins the states to the counties salary table using the StateName field as the key; then she joins the teacher salary table to the counties using the StateAbbr field.

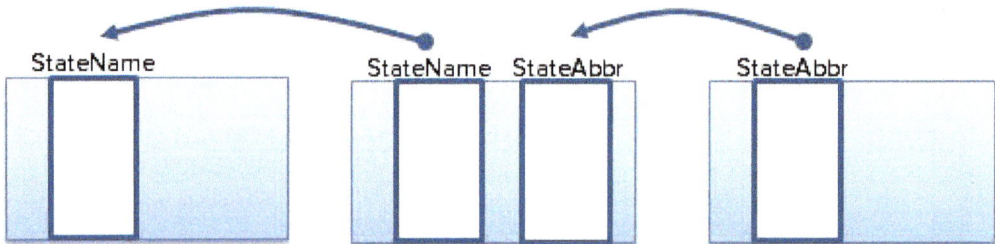

Fig. 7.3. A multiple join used to compare county teacher salaries with the state average.

Cardinality

When joining, the **cardinality** of the relationship between the tables must be considered. Cardinality is the numeric relationship between the objects in one table and their matches in the other. The simplest kind of relationship is *one-to-one*, in which each record in the target table matches one record in the join table. In Figure 7.2, each state has one corresponding record of demographic data. In a *one-to-many* relationship, each record in the target table could match more than one record in the join table. For example, a state typically has many school districts. In a *many-to-one* relationship, many records in the target table would match a single record in the join table, such as many cities falling within one state. Finally, a *many-to-many* relationship indicates that multiple records can appear in both tables. For example, a student may take more than one class, and most classes have more than one student.

The direction of the join must be taken into account when ascertaining the cardinality of a relationship. The target table is the point of reference and comes first; that is, the relationship cardinality is reported as {target} to {join}. Imagine two tables containing *states* and *counties*. If one performs a join with *states* as the target table and *counties* as the join table, the cardinality is one-to-many because each state contains many counties (and a county may belong to only one state). If the join is reversed and *counties* is the target table, then the cardinality becomes many-to-one because there are many counties in one state.

TIP: Putting the target table first when stating cardinality is not a universal convention and may be found reversed in other data management applications. It is used here to match the convention adopted by Esri in its publications and Help documents. It may help to always imagine the target table on the left, as in Figure 7.4.

The cardinality of a relationship dictates whether the tables can be joined. ***The Rule of Joining stipulates that there must be one and only one record in the join table for each record in the target table.*** Consider the four tables shown in Figure 7.4.

One-to-one cardinality. In Figure 7.4a, a table containing the number of earthquakes and total damage in each state (join table) is being joined to a states attribute table (target) using the common field provided by the state abbreviation. Each state occurs once in each table, so there is no ambiguity about matching the join records to the target records.

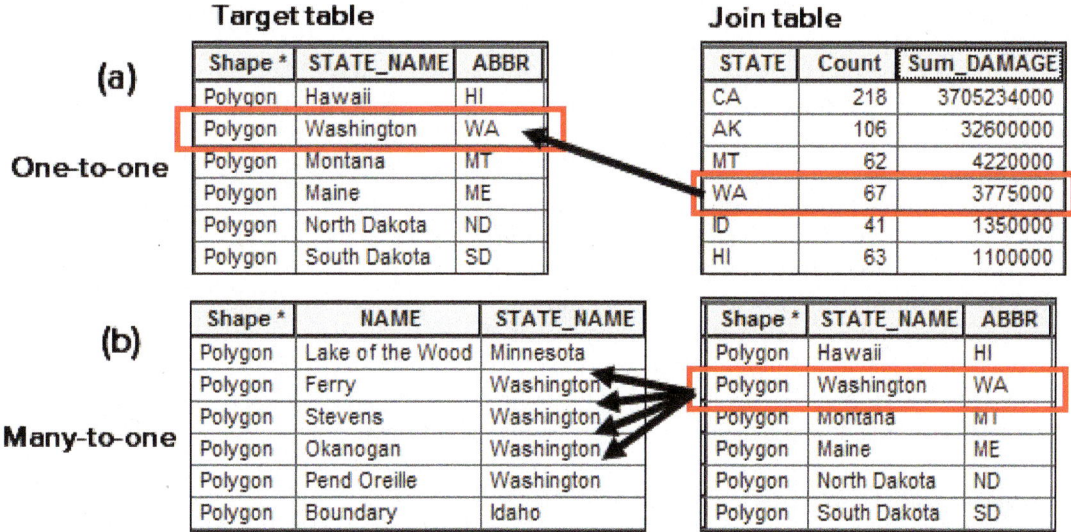

Fig. 7.4. A cardinality of one-to-one or many-to-one permits tables to be combined without violating the Rule of Joining. (a) One-to-one cardinality. (b) Many-to-one cardinality.
Source: Esri

Many-to-one cardinality. In Figure 7.4b, a table containing state information (join table) is being joined to a table of counties (target) using the common field provided by the state name. Although there are many counties in each state, there is only one record for each state in the join table and no ambiguity in linking the join records to the target records. The state record does get used more than once, but it still does not violate the Rule of Joining.

One-to-many cardinality. Figure 7.5 shows the reverse of the join in Figure 7.4b. Now states are the target table and counties are the join table. Many county records in the join table match each state record in the target table. The Rule of Joining is violated, and it becomes ambiguous which county record should be matched to the state. One cannot perform a join if a one-to-many relationship is present. Instead, a different operation is used, called a **relate**.

Target table

states

	Shape *	STATE_NAME	STATE_FIPS
	Polygon	Hawaii	15
	Polygon	Washington	53
	Polygon	Montana	30
	Polygon	Maine	23
	Polygon	North Dakota	38
	Polygon	South Dakota	46
	Polygon	Wyoming	56

Join table

counties

	Shape *	NAME	STATE_NAME	FIPS
	Polygon	Lake of the Wood	Minnesota	27077
	Polygon	Ferry	Washington	53019
	Polygon	Stevens	Washington	53065
	Polygon	Okanogan	Washington	53047
	Polygon	Pend Oreille	Washington	53051
	Polygon	Boundary	Idaho	16021
	Polygon	Lincoln	Montana	30053

Fig. 7.5. A one-to-many relationship violates the Rule of Joining because more than one record in the source table matches a record in the destination table.
Source: Esri

In a relate, the two tables are still associated by a common field, but the records are not joined together. The two tables remain separate. However, if one or more records are selected in one table, then the associated records can be selected in the other table. For example, selecting the state of Washington in the states table allows the selection of all the counties in Washington in the related table, as shown by the red boxes in Figure 7.5. Many relationships in the world have cardinalities of one-to-many, such as a school to its classes or a well to its yearly water quality tests, and relates provide valuable support in dealing with those features and their attributes.

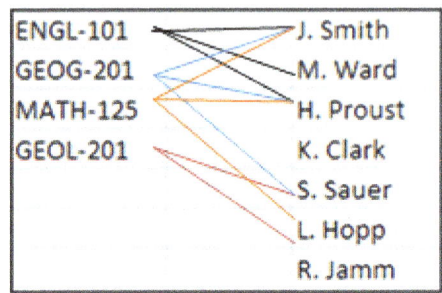

Fig. 7.6. A many-to-many relationship

Many-to-many cardinality. Many-to-many relationships exist but are difficult to deal with in a relational database. Consider a university. Each class has many students in it. However, most students are taking more than one class (Fig. 7.6). In practice, one would have to use multiple relates to represent the relationships and work with them one at a time. A one-to-many relate could be set up from the classes table to the students to allow all the students in each class to be found. Another one-to-many relate would allow the list of classes for each student to be generated.

Sometimes a join is performed in which most records have matches, but some do not. Occasionally, a user makes a fundamental error in crafting the join, and none of the records have matches. In these cases, the join will still be performed, but the records without matches will be given zero or <Null> values in the attribute fields from the join table. In Figure 7.7, an earthquakes table (highlighted) was joined to a states table (white) based on the state abbreviation and a many-to-one cardinality. Earthquakes in Puerto Rico had no

Earthquakes table fields		States table fields		
STA	LOCATION	STATE_	STATE	POP201
PR	Puerto Rico Region	<Null>	<Null>	<Null>
PR	Puerto Rico Region	<Null>	<Null>	<Null>
PR	Northwest of Puerto Ri	<Null>	<Null>	<Null>
AL	Irondale, Alabama	Alabama	AL	473559
AL	Near Birmingham, Alab	Alabama	AL	473559
AL	Near Birmingham, Alab	Alabama	AL	473559

Fig. 7.7. Result of no match for the destination records (highlighted) in the source table (white)
Source: Esri

match in the states table, and the joined fields are given <Null> values. It is always a good idea to view the table after a join and confirm that it behaved as intended. If missing values are encountered, a plan can be made to handle them. In this case, one might choose to exclude the earthquakes outside the 50 states from further consideration or to find a different states feature class that includes territories.

Statistics on tables

Exploring data through statistics is a fundamental first step in data analysis because it is important to understand the structure and properties of the data before applying more advanced techniques. A standard approach begins with calculating statistical measures such as the minimum, maximum, mean, median, and standard deviation, and viewing a histogram to evaluate whether it is potentially a normal distribution. Careful interpretation of this information can yield some important insights, especially when measured against one's expert knowledge.

Consider Figure 7.8, showing basic statistics and a frequency diagram (histogram) of earthquake magnitudes for a feature class of historical earthquakes in the United States. A geologist would recognize two odd things about these data. First, over 300 earthquakes have a magnitude of 0, which makes little sense for this data set and might indicate missing data. The missing values affect the current statistics, which must be recalculated without the missing values. The second oddity is more subtle. Without the zero values, the histogram resembles a slightly skewed bell curve. However, a geologist knows that earthquakes follow an inverse power law, with many more small earthquakes than large ones. In this data set, the small quakes were purposely excluded. This choice makes sense for the objective of the data set (to portray historic earthquakes), but it does render statistics like the minimum and average quake magnitudes less meaningful.

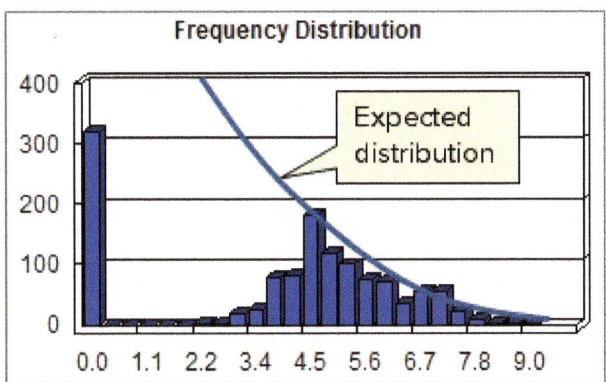

Count: 1264
Minimum: 0
Maximum: 9.23
Sum: 5006.1
Mean: 3.95
Std. Deviation 2.5

Fig. 7.8. Statistics and frequency diagram (histogram) for earthquakes
Source: Esri

This example illustrates the important step of exploring data and thinking about them critically before blindly calculating and accepting a result. Not only does the practice help prevent serious mistakes in inferring things about data sets, it can raise interesting questions. Why are there so many missing data values? Why are any magnitude 2 or magnitude 3 earthquakes included in this data set at all, since they typically do little damage and are rarely even felt?

A data set that passes the initial exploration can become the subject of more sophisticated statistical tests that are often provided with some GIS programs; one can also export the table for analysis in standard statistical software packages. GIS professionals can only improve their expertise by becoming more knowledgeable about statistics in all its forms.

Although calculating statistics for an entire table, or a set of selected records, is valuable, sometimes it is beneficial to group the records in a table and calculate statistics separately for each group. One can specify a single field, or a combination of fields, known as the **case field**, used to delineate the groups. Normally the case field will contain categorical data. For example, one might group a table of earthquakes according to which state the earthquake is in and generate earthquake statistics for each state separately.

Figure 7.9a contains a table showing major historical earthquakes in the United States. Attributes for each quake include the state in which it occurred, the number of deaths it caused, the total damage caused, and the Richter scale intensity (MAG). Sam would like to know which states have suffered the most from historical earthquakes. In particular, he wants to know the total deaths, total damage, and maximum magnitude for each state. He calculates statistics on the table but chooses STATE as the case field so that separate statistics are calculated for each state.

quakehis

STATE	DEATHS	DAMAG	MAG	LOCATION
MO	7	0	7.88	New Madrid, Missouri
SN	51	0	7.36	Northern Sonora, Mexico
AK	0	0	8.15	Yakutat Bay, Alaska
AK	0	0	8.26	Southeast Alaska
AR	7	0	7.68	Northeast Arkansas
(a)	3000	52400000	7.8	Near San Francisco, Calif
	12	6000000	7.48	South of Bakersfield, Calif

quakesum

STATE	Count_	Sum_DEATH	Sum_DAMAG
CA	218	3777	3705234000
AK	106	125	32600000
MT	62	32	4220000
WA	67	15	3775000
(b)	41	2	1350000
	63	79	1100000

Fig. 7.9. Summarizing: (a) earthquakes table and (b) summarized earthquakes table by state
Source: Esri

The quakesum output table, shown in Figure 7.9b, has one record for each state. It also has a Count field indicating the number of earthquakes in each state and one field for each of the statistics we requested. To better interpret the data, Sam sorted the table by the damage field in descending order to find the most grievously affected states.

The next step might be to join the quakesum table to a US States layer and create a map showing the total damage or total deaths for each state (Fig. 7.10). Using Summarize, Sam collapsed a one-to-many relationship between states and earthquakes into a one-to-one relationship between states and earthquake statistics, allowing the data to be mapped.

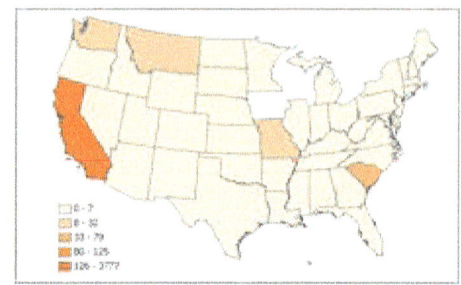

Fig. 7.10. Earthquake deaths by state
Source: Esri

Field types

Creating database tables always begins with an analysis of the fields that the table will contain. Unlike a spreadsheet, in which any cell can contain any type of data, a database field (column) must contain only one type of data, perhaps text, perhaps integers, but never both. Each field must be defined, and the type of its contents must be specified, before any data are entered. Furthermore, once a **field definition** is set, it cannot be changed. The most common field types are numbers, strings, and dates. Each field may also have parameters that further specify what it may contain, such as the maximum number of characters allowed in a text field.

A field STREETNAME would be a text field because street names contain letters. The **field length** defines how many characters can be stored in the name. If the STREETNAME field is given a length of 10, then any name longer than 10 characters will be truncated to fit. Thus, "Elm Street" fits perfectly (including the space), but "Maple Street" would be truncated to "Maple Stre" to fit in the field. It is important to ensure enough space in fields to hold the longest likely value. When designing a database, try to use the smallest field widths that will store every possible value because extra spaces increase the size of the database.

Numeric fields have different parameters. When defining a numeric field, the user may designate a storage width (**precision**) and the number of decimal places (**scale**). For example, a field with a precision of 5 and a scale of 0 could store a number between –9999 and 99999 (the minus sign takes one space). A field with a precision of 5 and a scale of 2 could store a number between –9.99 and 99.99 (the decimal place also takes one space).

Numbers can be stored in a variety of ways. Most databases offer several formats for storing numeric values, and these formats can differ from database to database. The next section describes the common options in general terms.

ASCII versus binary

As discussed in Chapter 6, a byte is the basic unit of computer storage. Numbers are usually stored in base 2, or binary format. The number *16* would be stored in binary as *00010000* in a single byte of information. A single byte can store an integer from 0 to 255; more bytes are needed to store larger numbers.

Text is also stored as bytes, but as codes. Every number, every letter, and every symbol (such as $) is assigned a single-byte code defined by a standard known as **ASCII**. To store the word *cat*, the computer stores the code for *c*, the code for *a*, and the code for *t*. Thus, it takes three bytes to store *cat*. The word *horse* requires five bytes. Numbers may also be stored using the ASCII one-byte code for each numeral. This method requires three bytes of data to store *147* and five bytes to store *147.6*. Text files and HTML files, among others, are stored in ASCII.

Although numbers may be stored in binary or ASCII, binary is more efficient. The number *14456* would require five bytes in ASCII but only two bytes in binary. It is also faster to compute with binary values because computers perform all calculations in base 2. If the number is already stored in binary, the computer does not need to convert it before calculating. Many types of files use a binary encoding scheme, including spreadsheets, word processing documents, and shapefiles. Raster and image data are also stored as binary data.

Very large numbers, such as 1,000,000,000,000, require many bytes to store, as do very small numbers, such as 0.0000000000001. A floating-point format, with a mantissa and exponent, is typically used when such values are needed. The computer truncates the mantissa at a certain level of precision—when storing values in the trillions, differences of tens or hundreds are rarely of interest. In ArcGIS, a **single-precision** floating-point field stores eight significant digits of information in the mantissa. A **double-precision** field stores 16 significant digits. Floating-point data types are more flexible than numeric or binary types because they can store either very large or very small values using the same field.

Every attribute must be defined before use; that is, the field type must be specified (text or numeric) and the field properties set. Once a field is defined, the definition cannot be changed. If a mistake is made defining the field, one must delete the field and redefine it. Different GIS systems may have different allowable data types, although most use the basic categories of text, integers, and floating-point values. Table 7.1 shows the specific data types used in ArcGIS. Geodatabases also have a Raster and GUID type for advanced users.

Table 7.1. Field data types available for feature classes (*geodatabases only)

Field type	Explanation	Examples
Short	Integers stored as 2-byte binary numbers *Range of values −32,000 to +32,000*	255 12001
Long	Integers stored as 10-byte binary numbers *Range of values −2.14 billion to +2.14 billion*	156000 457890
Float	Floating-point values with eight significant digits in the mantissa	1.289385e12 1.5647894e−02
Double	Double-precision floating-point values with 16 significant digits in the mantissa	1.12114118119141e13
Text	Alphanumeric strings	'Maple St' 'John H. Smith'
Date	Date/time format for calendar dates and times	07/12/2008 10/17/1963 13:24:06
BLOB*	Binary large object; any complex binary data, including images, documents, etc.	

Domains

Many databases allow users to specify lists of allowed values for a field, rather than letting a user type anything in it. Imagine a field in a trees feature class designed to contain "Deciduous" or "Coniferous". Allowing a user to type the value each time is tedious and may result in spelling mistakes or inconsistent upper- or lowercase usage, like "Deceduous" or "deciduous" or "dec". Any of these alternates present problems in a query because, for most databases, a Conifer is not the same as a conifer.

These issues can be avoided by creating a **domain**. A domain is a rule about the allowable contents of a field. A **coded domain** provides a list of values that the user must pick from to enter data in the field (coded domains are sometimes also called pick lists). Figure 7.11 shows a coded domain with two values for specifying a tree type. A **range domain** specifies a minimum and maximum value for a field. A Percent field, for example, might be assigned a range of 0 to 100. Coded domains can be defined for text or integer fields. Range domains can be assigned to any numeric field.

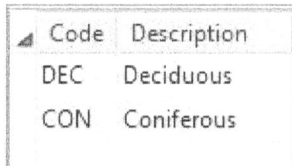

Fig. 7.11. A tree type coded domain
Source: Esri

Most databases allow domains, although they differ as to whether they are created for particular fields or for the database as a whole, which makes them reusable. A coded domain serves several functions. (1) It speeds data entry because the value can be chosen from a drop-down menu instead of being typed. (2) It allows a simple code to be stored for efficiency while presenting a more understandable term to the user. (3) It helps protect the database from inconsistent data entries. A range domain can be used to test for incorrect or nonsensical values in a numeric field.

The fields defined for a table, along with their data types, properties, domains, and other characteristics, are known as the **schema** of the table. The schema can be used to make a copy of a table that is identical in every way except that it contains no records of information.

About ArcGIS

Tables in ArcGIS

Tables may come from any of the underlying database programs supported by ArcGIS. The tables may be in different formats for storage, but the table itself always looks the same and has the same functions so that users don't need to learn different commands for working with different file types. However, not all commands and functions are available for every file type. In ArcGIS Pro, tables are examined and manipulated in a **table view** (Fig. 7.12). The view contains an Options menu and several tools at the top. More commands are accessible from a context menu that appears if a field is right-clicked.

When a feature class is added to a map, its attribute table is considered part of the layer, and certain modifications to the table, such as sorting or choosing which fields to display, can be made without affecting the source data. These table properties are analogous to the Symbology properties applied to the layer. The table view has the same relationship to its source standalone table that a layer has to its feature class. However, some changes made in a table view, such as sorting or changing the displayed field width, only apply to the view and do not affect the stored table. Other changes, such as adding a field or calculating values, constitute permanent changes to the source table. It is important to understand which type of change is made by a given action.

In a shapefile, the table is stored in a dBase format file and includes a unique feature identification number (FID) that links the spatial and the attribute data. In a geodatabase and in any standalone table, each record contains an Object ID (OID) analogous to the row number in a spreadsheet. Both shapefiles and geodatabase feature classes contain a Shape field that

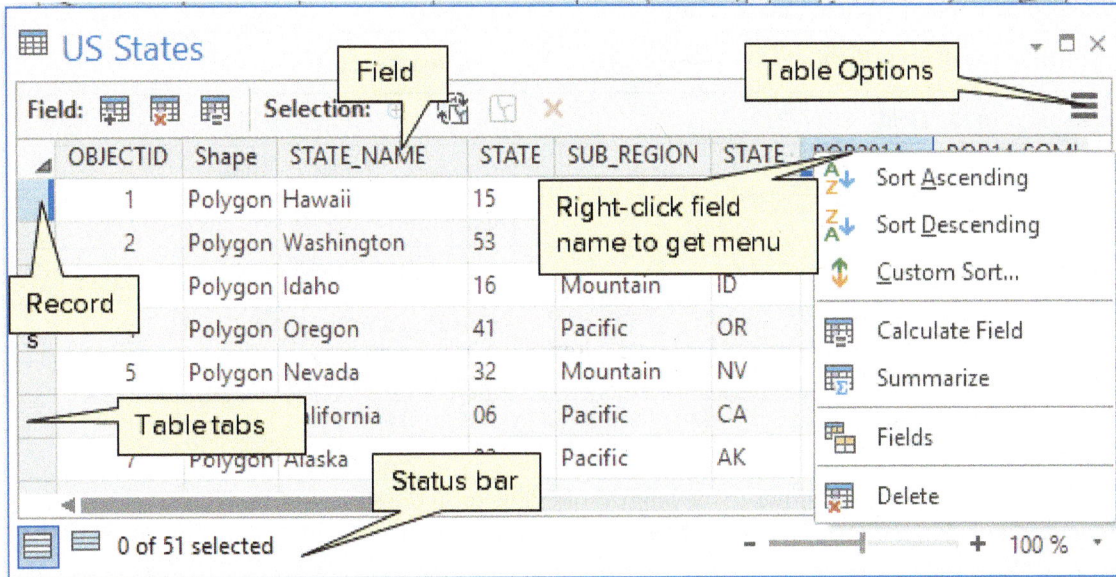

Fig. 7.12. Parts and terminology of a table
Source: Esri

represents the *x-y* coordinate values of the feature. Geodatabase tables for line and polygon feature classes also have Shape_Area and/or Shape_Length fields, which keep track of the areas and lengths of features.

Figure 7.12 shows some terminology associated with tables. A table consists of rows and columns. A row is called a record, and it contains information about a single object or feature. A column is called a field, and it stores one type of information. Each field has a name shown in the top row. Field names must contain 13 or fewer characters and should contain only letters, numbers, and underscores. An alternative name, called an **alias**, can be set to give a field a more descriptive name that does not have to follow the naming rules. For example, the somewhat perplexing field MEDREN could be given a more understandable alias, such as MedianRent.

Fields view

The **Fields view** is used to view and modify properties of tables. In some cases only the layer properties are modified, meaning that the changes only apply to the view of the table shown in the map layer (or table view for a standalone layer). Changing the width of a field, turning fields on or off, modifying the alias, or changing the number format are examples of settings that only apply to the layer. However, Fields view can also make permanent modifications to the stored data file, such as adding, deleting, or renaming a field name. In other words, the schema of the table is being modified.

When the Fields view is open, a drop-down menu specifies whether edits are being made to the layer/table view (US States in Fig. 7.13) or to the underlying data set (Data Source). Buttons and commands that affect only the layer are not shown when the Fields view is set to edit the Data Source. However, some edits in layer mode, such as adding a field, will still modify the actual data set. It is important, when working in Fields view, to understand when changes affect only the layer and when they will change the data (Table 7.2).

Fig. 7.13. Managing types of edits in Fields view
Source: Esri

Table 7.2. Scope of edits in Fields view

Edits affecting only the table view	Edits affecting the source data
Displaying the field	Adding/deleting a field
Setting read-only status	Changing the field name
Changing the field width	Changing the field width
Formatting numeric options	Creating domains
Highlighting a field	Creating subtypes
Changing the alias	Changing the alias

The alias can be modified either in the layer/table view or in the source data. If the Fields view is set to the layer, the alias is modified only within that table view. If the Fields view is set to the Data Source, then the alias change is permanently stored with the data set and will be visible in any view of the table.

TIP: Make a habit of saving schema edits in the current view before moving to another ribbon. Unsaved table edits or schema changes in one ribbon will prevent saving changes in another ribbon.

In an attribute table, some fields are required, and they are created and updated by the program that creates the geographic data set. For example, a shapefile has a Feature ID (FID) field and a Shape field. The geodatabase feature class has an Object ID and a Shape field. *These fields should never be altered by the user.* Likewise, users of shapefiles must make sure that they never delete records in an attribute table unless the accompanying features in the spatial data set are also deleted. Editing features only within an editing session in Pro will meet this criterion, as does deleting rows in a geodatabase feature class.

As we learned in Chapter 5, different GIS data formats use different underlying databases. The coverage model uses a database called INFO. Shapefiles use a dBase format file. Personal geodatabases use the database underlying Microsoft Access. File geodatabases use their own format. Enterprise geodatabases use a large-scale commercial RDBMS, such as Oracle or SQL Server. ArcGIS can read comma-delimited text files (fields separated by commas) and tab-delimited text files (fields separated by tabs) as well as Excel™ spreadsheets if they are correctly formatted.

Editing and calculating fields

ArcGIS offers two ways to change the values in a table: by typing the information directly into the fields or by calculating the value of a field. The Calculate Field tool is used to generate arithmetic expressions using fields as variables. For example, the percentage of Hispanics in each state could be calculated from the two fields containing the total state population and the number of Hispanics (Fig. 7.14). Expressions may use operators $(*, /, \&, +, -, =)$, functions like sin() or cos(), and values typed in the expression box.

TIP: Calculating destroys the original data in the field, so usually calculations are performed in an empty field added to the table. Be especially careful when performing a calculation on a field that already contains data.

The Add Geometry Attributes tool (Fig. 7.15) offers another type of field calculation. This tool calculates one or more geometry attributes of a feature class, such as the areas or perimeters of polygons, the lengths of lines, the centroid locations of polygons, etc. Although areas and lengths are stored in the geodatabase Shape_Length and Shape_Area fields, the units are always in the coordinate system units. This tool calculates more than length or area, works on shapefiles as well as feature classes, and permits the units to be chosen by the user. The tool adds specific fields with predetermined names to the data set. If run on a data set that already contains those fields, the fields will be updated. Creating an alias for these fields that incorporates the units of the geometry calculation is a wise procedure.

Importing tables

Many GIS projects encounter situations when external tabular data must be imported for use in ArcGIS. Several different formats may be used, but the layout of the records and fields must meet certain requirements. Field names must be present in the first row and must follow the ArcGIS naming conventions. Columns that are supposed to hold numbers must hold only numbers and avoid NoData or xx or Null or N/A to indicate missing values. Blank rows, merged cells, and formulas are not allowed. Most import formats are read-only within ArcGIS. They can be viewed but not

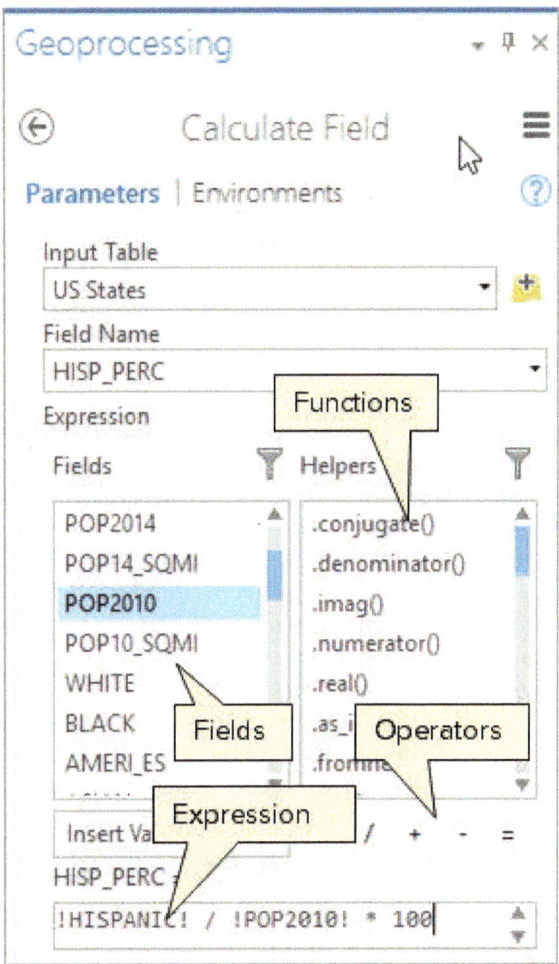

Fig. 7.14. The Calculate Field tool
Source: Esri

edited or changed, and some tools and functions may not work on them. Usually, the import file is exported to a dbf file or geodatabase table if changes to it are needed.

Dbf files, or dBASE files, are commonly produced by database programs. These files can be directly imported to and edited within ArcGIS.

Text files, such as those shown in Figure 7.16, come from a variety of sources. They might come from text copied from a web site, data copied from a word processing document, or values produced from a statistical program. Nearly any program that handles tabular data has the option to save them as text, and text is one of the most basic and standard formats for transferring data between programs. All text files are stored using ASCII characters. Three primary formats are found. Figure 7.16 shows the same tabular information stored in all three formats.

A tab-delimited file has the field values separated by the tab character (Fig. 7.16a). A comma-delimited file has the field values separated by commas (Fig. 7.16b). The fixed-column format (also known as a formatted or space-delimited file) is most often encountered as the output of a computer program, such as FORTRAN. It appears similar to a tab-delimited file, but no tabs are present. Instead, each row is composed of a string of characters, and each field starts and ends at a specified character location (column). In Figure 7.16c, the first field includes columns 1 to 5; the station field goes from columns 6 to 30, and so on. (The red lines are not actually present in the file but have been added to the figure to help visualize where each field starts and stops.)

TIP: ArcGIS can read tab-delimited or comma-delimited text files, but it cannot read fixed-column text files. (However, Excel™ can read fixed-column files and save them as comma-delimited CSV files.)

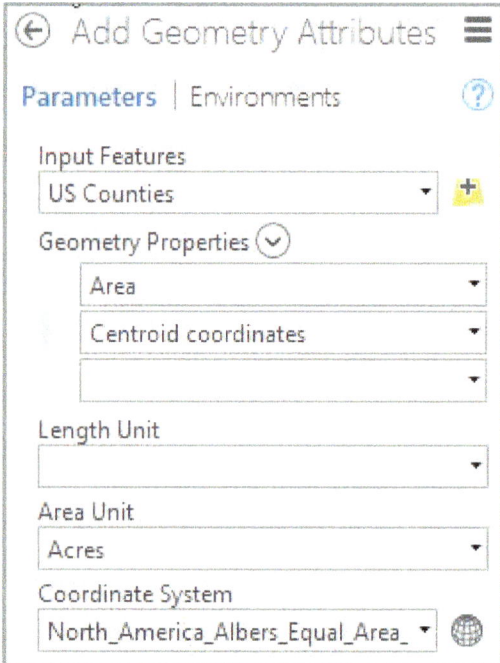

Fig. 7.15. The Add Geometry Attributes tool
Source: Esri

Pro tries to interpret the field type by examining the first row of data, but it does not always choose the desired type. A short text field, such as a two-character state abbreviation, will usually be given a field length of 255. A zip code will often be interpreted as a number, which will cause a leading zero to be removed: 08577 becomes 8557. Putting quotes around the zip code values will usually fix this issue. Sometimes it is helpful to create a first row of dummy values that will force Pro to select the right data type. In the example CSV file that follows, the "ddd" entries force a text interpretation for the zip codes, the 999 entry will encourage an integer field to be created for NUM, and the 99.9 value will require a floating-point field for PERCENT.

> NAME, NUM, STREET, CITY, STATE, ZIP, PERCENT
>
> ddd, 999, ddd, ddd, ddd, ddd, 99.99
>
> Smith, 527, W. Main St., Peoria, IL, 45908, 38.3
>
> Jones, 6904, Jackson Blvd., Nashua, NH, 07958, 46

Text files may also contain formatting problems, such as illegal field names with spaces, or an "xx" in a numeric column to indicate missing data. Such issues may need to be fixed before the import can be successful. Some issues will not cause problems when read by Pro, but they will prevent the file from being exported to a geodatabase format.

Simple editing of text files can be done using system tools, such as the Notepad program. However, more advanced and powerful editing is available using the Microsoft Excel™ program. It can open and save a variety of text formats and can be used to translate one type to another, such as converting a fixed-column text file to a tab-delimited file. (The files shown in Fig. 7.16 were all saved from an Excel™ worksheet.)

Preparing Excel data

ArcGIS Pro can read the Excel format directly, but not all spreadsheets are suitable. They must be simply formatted and should follow the rules commonly imposed on database tables, such as having legal field names and storing only one type of data in a column. Using Excel data requires careful preparation. For best results, make sure that the spreadsheet meets the following requirements.

- ▸ The first row of the worksheet contains the field headings.

- ▸ Field headings must start with a letter, must have no more than 13 characters, and must not contain spaces or shift characters such as %, $, #, and so on.

- No blank lines occur within the data-containing rows.

- The spreadsheet contains no formulas and no merged or split cells.

- The bottom of the spreadsheet has no extraneous data, such as column totals.

- Every column contains consistent data types. Numeric columns must have only numbers.

- Missing values in a numeric column must be flagged with an impossible numeric value, such as –99 in the population field, rather than with text characters like *xx* or *na*.

- Text fields contain no commas unless all the text is enclosed in quotes.

An Excel workbook may have multiple worksheets inside (Fig. 7.17), each of which must be opened separately in Pro. If they have not been named, they will appear with the default names Sheet1$, Sheet2$, and Sheet3$; usually Sheet1$ will contain the data.

TIP: The spreadsheet or CSV file in Excel must be closed before it is opened in Pro.

TIP: Pro cannot read the older .xls format; convert to .xlsx first. If a spreadsheet will not open correctly in Pro, try saving it as a CSV file instead. In general, Pro seems to work better with CSV files than with Excel files.

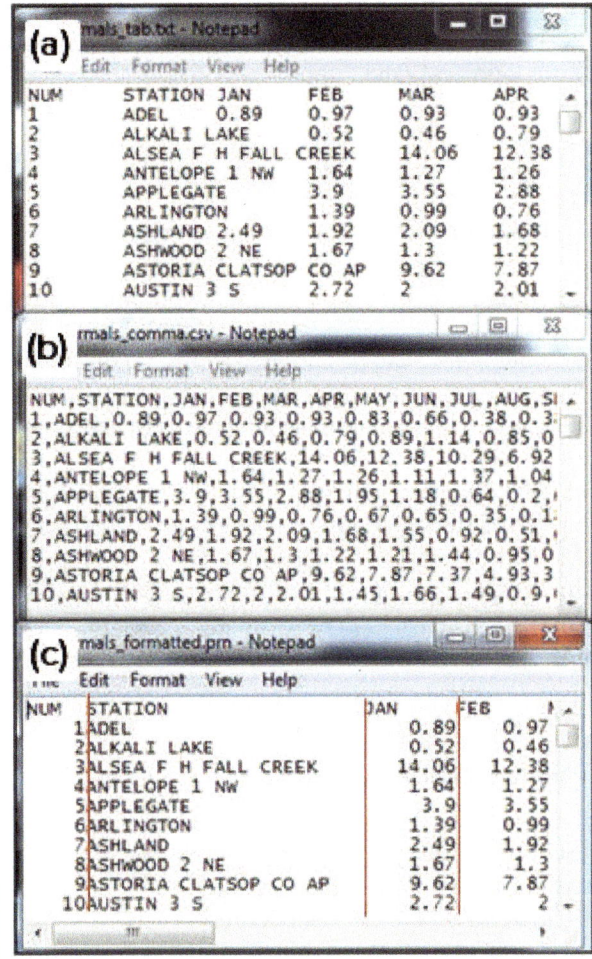

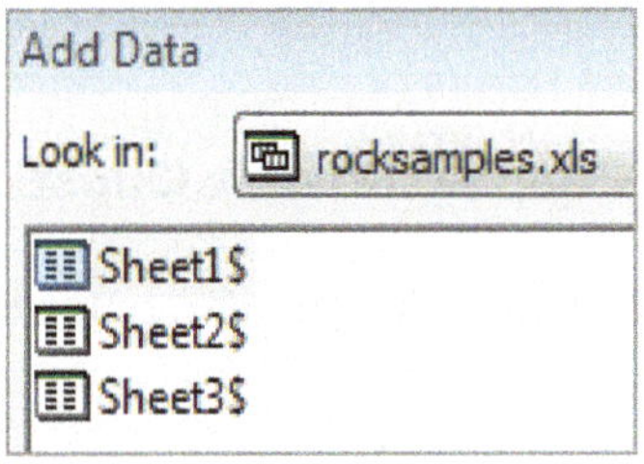

Fig. 7.16. Text file formats: (a) tab-delimited; (b) comma-delimited; (c) fixed-column
Source: Microsoft, National Climatic Data Center

Fig. 7.17. An Excel workbook with multiple worksheets
Source: Esri

Summary

- Tables consist of rows and columns of information. A row is associated with one feature and includes columns of information called fields.

- Tables associated with spatial data sets are called attribute tables and contain records, one for each feature in the data set. Standalone tables are not associated with map features.

- Relational database management systems construct temporary links between data tables and are the preferred model for GIS software.

- Tables may be joined or related on a common field in order to access information in one table from another. Joins may be performed on tables with one-to-one or many-to-one cardinality. Relates must be used on tables with one-to-many or many-to-many cardinality.

▶ The Statistics function calculates basic statistical values for all selected records in a table. The Summarize command generates statistics about groups of features defined by a categorical field.

▶ Fields must be defined to contain a specific type of data, such as text, integers, dates, or floating-point values. Once defined, the field type cannot change.

▶ Domains provide rules about the values placed in attribute fields to help avoid errors and foster consistent data entry. Coded domains include lists of values to pick from. Range domains specify the maximum and minimum allowed values for numeric fields.

▶ New fields may be added to tables by defining names and field types. New tables may have data entered by typing in values or by using the Calculate function. The Calculate Geometry function can put areas and lengths in attribute tables for the corresponding features.

▶ ArcGIS tables can display and manipulate tabular data from a variety of sources, including dBase files, INFO files, geodatabases, SQL queries, or comma-delimited text files.

TIP: Field names must have 13 or fewer characters; may include letters, numbers, and the underscore character (_); and should not contain spaces or special characters, such as @, #, !, $, or %. Field names must also start with a letter, not a number.

Important Terms

alias	double-precision	join table	scale
ASCII	Feature ID (FID)	key	schema
attribute field	field	Object ID (OID)	single-precision
attribute table	field definition	precision	source table
cardinality	field length	range domain	standalone table
case field	Fields view	record	table
coded domain	flat file database	relate	table view
destination table	hierarchical database	relational database	target table
domain	join	Rule of Joining	

Chapter Review Questions

1. Describe the difference between an attribute table and a standalone table.

2. Find a table of ASCII codes on the Internet and decipher this message written in decimal numbers: 54 104 101 32 83 116 46 32 66 101 114 110 61 114 100 32 115 113 97 116 116 101 100 32 111 110 32 97 110 100 32 83 78 69 101 90 69 68 33

3. Choose the best field type for each of the following types of data in a geodatabase and explain your choice:

student GPA	zip codes
state populations	number of students in a class
state names	birthdays

4. What is the primary purpose of domains? Explain the difference between a coded domain and a range domain.

5. What is the cardinality of each of the following relationships? Assume that the destination is on the left.

cities to counties   parcels to zoning codes

capitals to states   watersheds to wells

6. You have a table of cities and a table of counties, both with a state abbreviation field. Can you join them if cities is the target table? If counties is the target table? Explain your answer.

7. Describe the difference between a join and a relate. When is a relate the only option?

8. What is the function of a case field? Make up an example table with seven records, and use it to illustrate your explanation.

9. For each of the following changes made in the Fields view, state whether it will modify the table view only, will modify the schema, or may do one or the other depending on the Source setting:

deleting a field   turning field visibility off

changing a field name   making the field read-only

changing a field alias

10. List at least four ways that a spreadsheet may not be suitable for importing into Pro.

Extend your knowledge

Read these sections of the Help for more information about tables

Help> Data > Data types > Tables

Help> Data> Geodatabases > The geodatabase > Table basics

Help> Data> Geodatabases > The geodatabase > Data design

Mastering the Skills

Teaching Tutorial

The following examples provide step-by-step instructions for doing basic tasks and solving basic problems in ArcGIS. The steps you need to do are highlighted with an arrow →; follow them carefully.

→ Start Pro and open the TablesPractice project in the mgisdata\ClassProjects folder.

→ Remember to save the project often.

Viewing tables

Let's begin by exploring some tables and learning some basic skills for working with them. Many table settings affect only how the tables appear and do not change the underlying data.

1→ In Contents, right-click the US States layer and choose Attribute Table to open a table view.

1→ Open the **Table: View** ribbon and examine its buttons.

The table view has several display settings that change how it appears.

1→ In the table view, narrow the STATE_NAME field to a more suitable width by clicking and dragging the right border of the field.

1→ Right-click on the POP2014 heading and choose Sort Ascending from the menu.

1→ In the lower right corner, slide the bar or use the drop-down to change the zoom level.

1→ Examine the text at the bottom that says (0 out of 51 Selected), indicating that there are 51 records in this table (50 states plus the District of Columbia).

TIP: Right-click a field name and choose Custom Sort to sort on more than one field. Sorting only changes the view of the data, not the data source.

2→ In Contents, right-click the US Counties layer and choose Attribute Table. It opens as a new table view, most likely in the same window as the US States table.

2→ Experiment with different ways to arrange the tables. Leave them with the map on top and the two tables side by side underneath.

2→ Click on the title tab of each table. The currently active table will have a blue tab, and commands selected from a ribbon will affect only that table.

Using queries and statistics on tables

Tables are commonly used to create and work with subsets of the layer, created using a query based on an expression.

3→ Click the US States title tab to make it the active table.

3→ Click **Table: View: Selection: Select By Attributes** to open the Select Layer By Attributes tool in the Geoprocessing pane.

3→ Make sure the Layer is set to US States and enter the expression: POP2014 is Greater Than 5000000.

 3→ Click the green check mark to verify that the expression is correctly formatted. Run the tool.

The selected states are highlighted in the map and in the Table window. Also, the text at the bottom now reads *22 out of 51 Selected*, indicating that 22 states have more than 5 million people.

4→ Close the US Counties table to better see the US States table.

 4→ Click the *Show selected records* button at the bottom of the US States table to view only the selected records.

4→ Click the *Show all records* button to show them all again.

The Switch Selection button switches the selected with the unselected records.

 4→ Click the Switch button in the Table window.

 4→ Clear the selected set by clicking the Clear Selection button in the Table window.

Summarizing large amounts of data in a table is often the first step in a critical analysis. Suppose we would like to know the total number of whites and minorities for the United States.

TIP: Save all new tables to the home geodatabase unless directed otherwise. Give them descriptive names to facilitate telling them apart later.

5→ In the Geoprocessing pane, open the Summary Statistics tool (Fig. 7.18).

5→ Set the Input Table to US States and name the Output Table *US MinorityTotals*.

Fields may be selected individually using the drop-down Field box, but a faster method can be used when statistics for many fields are desired.

5→ Click the Add Many button next to Field to show a list of field names. Check the following: AMER_ES, ASIAN, BLACK, HAWN_PI, HISPANIC, MULTRACE, and WHITE.

5→ Click Add when finished checking the boxes. Accept Sum as the statistic for each field.

5→ Leave the Case field blank. Click Run. Leave the tool open when it finishes.

5→ Open the USMinorityTotals table and examine the statistics.

Notice that the new table appears under a Standalone Tables heading in the Contents pane. This entry is called a **table view**, and it has properties than can be set independently of the statistics table stored in the project geodatabase. The relationship of a table view to its source data is similar to the relationship of a layer to its feature class.

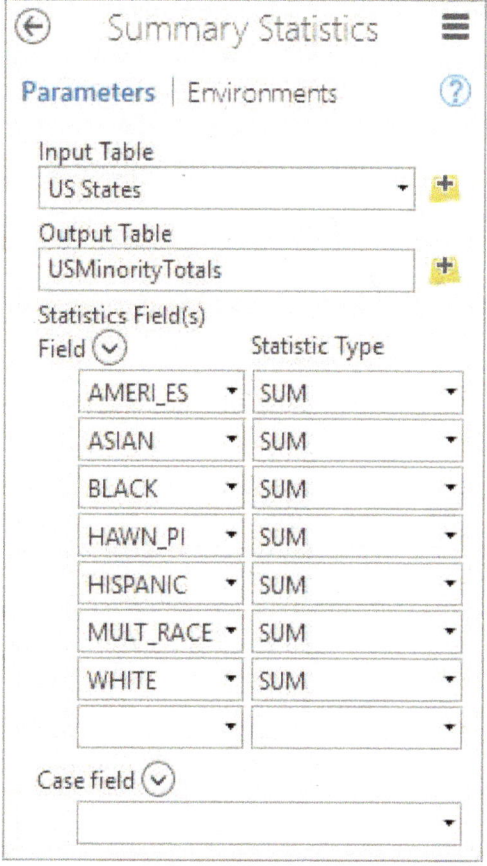

Fig. 7.18. The Summary Statistics tool
Source: Esri

6→ In Contents, right-click the USMinorityTotals table and choose Properties.

6→ Examine the different settings of the table view, then close the Properties window.

1. How many Hispanics and Blacks live in the United States? (Be sure to report the answers using an appropriate number of significant figures.)

The Summary Statistics tool uses all records, unless a subset of records has been selected, in which case it uses only the selected subset to calculate the statistics. Let's determine the same statistics for the Pacific states.

7→ Click **Map: Selection: Select**. In the map, hold down the Shift key and click on Washington, Oregon, and California, so that all three are highlighted (selected).

7→ Return to the Geoprocessing pane. The Summary Statistics tool should still be open with the same fields requested previously.

7→ Change the name of the Output Table to *PacificMinorityTotals*. (If the Fields went blank, specify them again.) Run the tool and leave it open.

7→ Open the PacificMinorityTotals table and compare the statistics to the US totals.

A case field is used to divide the records in the table into groups and calculate statistics separately for each group. The US States table has a SUB_REGION field that assigns the states to a regional group (Pacific, New England, etc.) We can use the Case field to calculate separate statistics for each subregion.

8→ Click the US States table to view it. Click Clear to clear the selected records.

8→ Return to the Summary Statistics tool in the Geoprocessing pane.

8→ Change the name of the Output Table to *SubregionTotals*.

8→ Set the Case field to SUB_REGION. Run the tool.

8→ Open the SubregionTotals table and examine the results.

2. Which subregion has the most Blacks? Which one has the most Hispanics?

Take a moment to appreciate the power of a case field. In Step 7, we selected the Pacific states and calculated statistics for <u>one</u> subregion. Using SUB_REGION as the case field did the same thing for nine subregions simultaneously and put the results in a single table to facilitate comparisons.

9→ Close all of the open tables.

9→ Turn off US States in the Contents pane and turn on 114th Congress.

9→ Open the table for the 114th Congress layer.

9→ Examine the fields, especially noting the field named PARTY.

How many Democratic districts are there? We can find out using a query.

10→ Choose **Table: View: Selection: Select By Attributes**.

10→ Set the Layer Name to 114th Congress and click Add Clause.

10→ Enter the expression: PARTY is Equal to Democrat. Run the tool.

10→ Examine the number of selected records at the bottom of the table.

3. How many Democratic districts are there? _____

Finding the number of Democratic districts using a single query is fine. But suppose one wanted to know the number of districts for all three categories: Republican, Democrat, and Vacant? The Frequency tool is helpful for counting objects in categories defined by a case field.

11→ Click the Clear button in the 114th Congress table to clear the selected records.

11→ In the Geoprocessing pane, open the Frequency tool.

11→ Set the Input Table to 114th Congress and the Output Table to *CountDistricts114*.

11→ Set the Frequency field (i.e., the case field) to PARTY.

11→ The Summary field is optional; it will sum any field for each category. Set it to the SQMI field in order to determine the total area controlled by each party. Run the tool.

11→ Open the CountDistricts114 table to examine it.

See that all three categories are listed, with the FREQUENCY field showing the number of districts in each category and the SQMI field showing the summed area. Some states have only one district; others have more. Suppose that we want to know many districts each state has.

12→ Clear and reopen the Frequency tool in the Geoprocessing pane.

12→ Set the Input Table to 114th Congress. Name the output table *DistsPerState*.

12→ Set the Frequency field to STATE_ABBR and run the tool.

4. Is this new table an attribute table or a standalone table? _____

12→ Open the new DistsPerState table.

12→ Right-click the FREQUENCY field and choose Sort Descending.

12→ Close all the tables and save the project.

Using the Fields view

Tables have a way to view and modify their properties, known as the **Fields view**.

13→ Open the SubregionTotals table.

13→ Click **Table: View: Field: Fields** to open the Fields view.

13→ Examine the rows of fields and the columns of field properties. Examine the different types of data in the Data Type field and compare to Table 7.1.

13→ Examine the Current Layer drop-down in the Fields view, SubregionTotals.

When the Fields view is set to a layer or table view (SubregionTotals in Fig. 7.19), most of the changes only apply to that view, rather than to the underlying data set. Suppose one wishes to compare only Hispanic and White populations. Reviewing the data is facilitated by showing only the fields of interest and formatting the values with thousands separators.

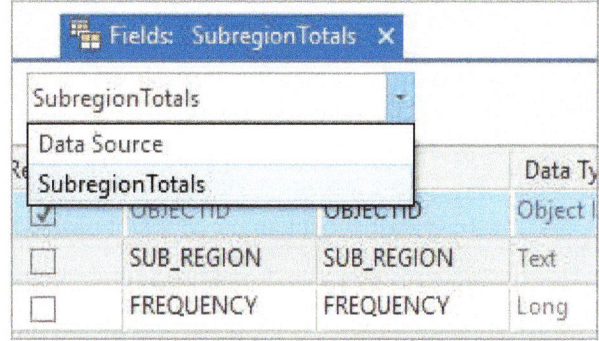

Fig. 7.19. The Fields view can be set to make cosmetic changes to the table view (SubregionTotals) or to make permanent changes to the Data Source.
Source: Esri

14→ In the Fields: Subregion-Totals view, uncheck the Visible box at the top to turn all fields off. Check the boxes for SUB_REGION, SUM_HISPANIC, and SUM_WHITE to turn them on.

14→ To make the table more understandable, replace the Alias (not the Field Name!) for each of the three fields with *Subregion*, *Hispanics*, and *Whites*.

14→ Click **Fields: Changes: Save**.

14→ Click the SubregionTotals table view tab to view the changes.

The changes made to the table are only cosmetic. The other fields are still there, just invisible. We can also format the numeric values in the table to make them easier to compare.

15→ Switch back to the Fields: SubregionTotals view.

15→ On the Hispanics row, click in the Numeric box twice to display the number format button (…). Click the button to open the Number Format window.

15→ Change the Rounding to Significant digits and set it to three.

15→ Check the box that says *Show thousands separators* and click OK.

15→ Repeat the format change for the Whites row.

15→ Click **Fields: Changes: Save**.

15→ Switch back to the SubregionTotals table and examine it.

We have created a nicely formatted view of this table for a layout or a report. The changes only affect the table view for SubregionTotals, not the source data. However, the Fields view can also make permanent changes to tables.

16→ Switch back to the Fields: SubregionTotals view.

16→ Change the Current Layer drop-down to Data Source.

Notice that the view changes, and only settings that permanently modify the data set are shown Fig. 7.20). The name, alias, and length of text fields can be modified (for geodatabases only, not for shapefiles). Fields can be added or deleted in this view, and in some versions of Pro, the order can be changed by clicking and dragging the row to a new location.

16→ In the Field Name column, remove the SUM_ prefix from each name (Fig. 7.20).

16→ In the Alias column, enter a proper name for each field: *Subregion, Native American, Asian, Black, Pacific Island, Hispanic, Multirace,* and *White.*

16→ Click **Fields: Changes: Save**. It takes longer to save changes to the data source.

16→ Switch back to the SubregionTotals table.

The SubregionTotals table does not change (including the plural form of the Hispanics and Whites alias) because those properties were set for the table view, but the recent changes applied to the data source. We can see the difference by adding another view of the table to the map.

Fields: SubregionMinorityTotals	
Current Layer	Data Source

Field Name	Alias
OBJECTID	
SUB_REGION	Subregion
FREQUENCY	Frequency
AMERI_ES	Native American
ASIAN	Asian
BLACK	Black
HAWN_PI	Pacific Island
HISPANIC	Hispanic
MULT_RACE	Multiracial
WHITE	White

Fig. 7.20. Editing the properties of a source table
Source: Esri

17→ In Contents, rename SubregionTotals to *SubregionTotals-First*. (Notice that the table view name changes, but the Fields view does not.)

17→ From the Catalog pane, add the SubregionTotals table from the home geodatabase.

17→ Open the newly added table view. All fields are shown, and the alias changes made to all of the source table fields are visible.

17→ Switch back and forth between the SubregionTotals-First and SubregionTotals tables and compare them.

The earlier cosmetic changes (showing fewer fields and formatting the numbers) apply only to the first table view. The newly added table has its own default properties that are independent of the first view and reflect the changes made to the source table, including the renamed field aliases. We can see the actual field name changes as well.

18→ Click **Table: View: Field: Aliases** to switch between the aliases and actual names. Leave it set to the Aliases when finished.

18→ Close the Fields: SubregionTotals view and close all of the tables. Save the project.

TIP: Understanding the difference between the table view and its data source is challenging, but it is a crucial distinction. Be sure to review these concepts and get help if it is still not clear.

Notice that we practiced using the Fields view in a new table rather than the tutorial data in the mgisdata folder, so that mistakes would not affect subsequent lessons. Modifying data sources must be done mindfully and carefully, and it is wise to have a backup available, or a means to easily re-create the table. Cultivate awareness of when changes in the Fields view are temporary and when they are permanent (see Table 7.2).

TIP: The effect of alias changes depends on the Current Layer setting in the Fields view. If the layer or table view is selected (e.g., SubregionTotals), an alias change only affects the current layer or view. If Data Source is chosen, new aliases become a permanent part of the data set.

Joining tables

Let's find out which districts may have changed their party affiliations since the 113th Congress. Comparing the party in two different tables is facilitated using a join, which requires a key field.

19→ Open the tables for the 113ᵗʰ and 114ᵗʰ Congress and arrange them side by side.

19→ Examine the fields and identify the appropriate key field to join the tables. The key must be unique to each record and have the same value in both tables.

5. What is the best potential key field in this table? _____

We want 114th Congress to be the target table.

20→ In Contents, click the 114th Congress layer to select it.

20→ In the **Table: View: Relationship** group, click the **Joins** drop-down and select Add Join. The Add Join tool appears in the Geoprocessing pane.

20→ Make sure the Layer Name is 114th Congress (target table) (Fig. 7.21).

20→ Choose DISTRICTID for the Input Join Field.

20→ Choose 113th Congress as the Join Table.

20→ Choose DISTRICTID as the Output Join Field.

20→ Run the tool.

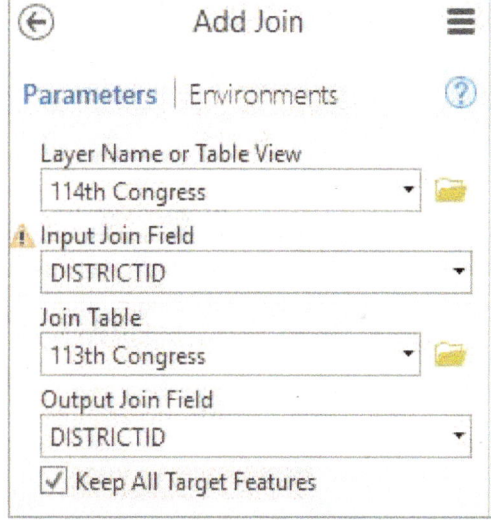

Fig. 7.21. Joining the tables
Source: Esri

TIP: The warning next to the Input Join Field says that the table is not indexed. Indexing speeds access on large joined tables with thousands of records but is not important for smaller tables.

20→ Close the 113th Congress table and examine the 114th Congress table. Scroll to the right, noticing the additional fields now in the table.

It can get confusing to remember which fields came from which table. Pro places a prefix with the table name (the data source name, not the title in the Contents) in front of the field name. Hence, the 114th Congress table fields are prefixed with the name cd114, and the cd113 fields are prefixed with cd113 (e.g., cd114.DISTRICTID and cd113.DISTRICTID). The prefixes only appear when the field names, not the aliases, are being shown.

20→ If the field names don't show prefixes, click **Table: View: Field: Aliases**.

TIP: Occasionally a tool does not work properly on a joined table. If this occurs, export the joined layer or table, and the joined fields will be saved in the output.

Now we will create a layer showing the districts that changed parties. The Make Feature Layer tool is similar to Select Layer By Attributes, but it creates a new layer from the selected features.

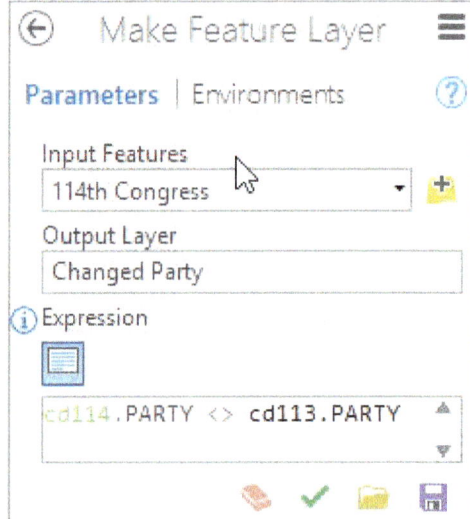

21→ In the Geoprocessing pane, open the Make Feature Layer tool (Fig. 7.22).

21→ Set the Input Features to 114th Congress and name the Output Layer *Changed Party*.

21→ Click Add Clause and enter the first part of the expression: PARTY does Not Equal.

21→ Click the Fields button and select the other PARTY field from the bottom of the list. Click Add.

The prefixes may be missing from the fields shown in the drop-down list, but we presume the second PARTY field came from the cd113 table. We can confirm it by viewing the expression in SQL, the actual language used to execute the expression.

Fig. 7.22. Selecting districts that changed parties
Source: Esri

21→ In the Make Feature Layer tool, click the SQL button under Expression and examine the expression (Fig. 7.22).

21→ Run the tool. Open and examine the Changed Party table.

22→ Create a unique values map showing the *Changed Party* districts using the cd114. PARTY field. Outline the newly Republican districts in red, the newly Democratic districts in blue, and the vacant ones in black.

22→ Display the Changed Party map over the 113th Congress layer (Fig. 7.23).

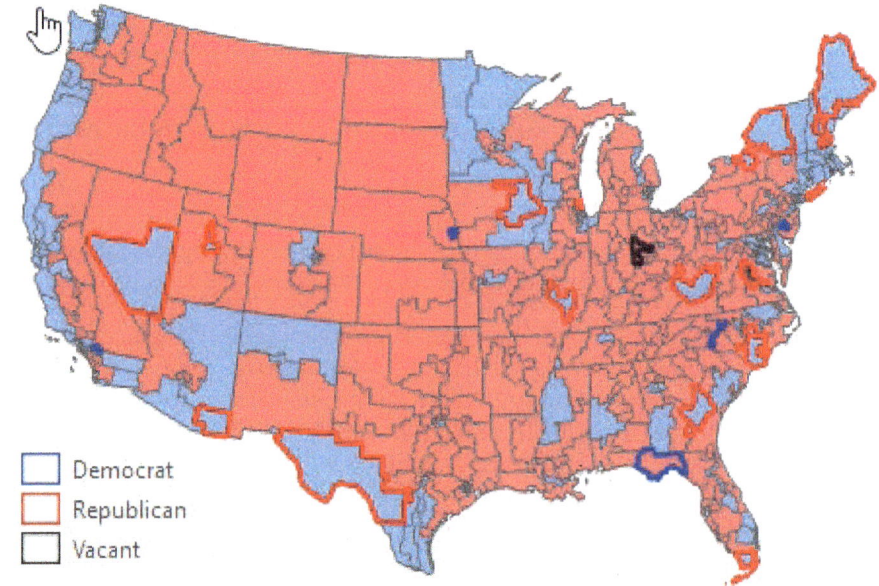

Fig. 7.23. Map of districts that changed parties
Source: Esri

Now that we have the Changed Party layer, we no longer need the join on 114th Congress. The Changed Party layer is a separate layer with its own properties, and it will keep its join.

> 22→ In Contents, right-click the 114th Congress layer and choose Joins and Relates > Remove All Joins.

Relating tables

It would be interesting to group the representatives according to subregion, but the SUB_REGION field only exists in the US States table. Setting up a relate between the US States table and the 114th Congress table will allow us to select a subregion and obtain a list of representatives from it. The common key between the tables is the state abbreviation.

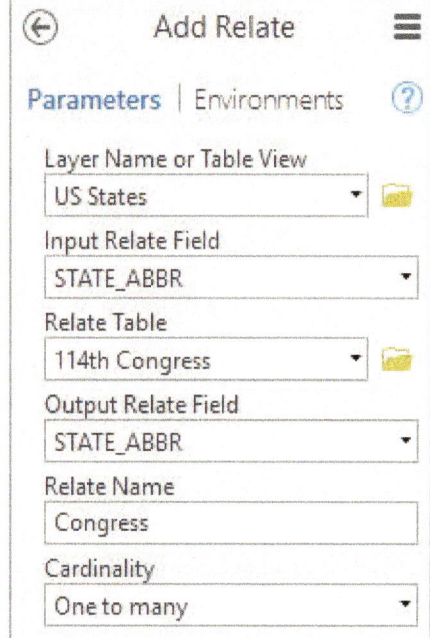

Fig. 7.24. Adding a relate
Source: Esri

6. What is the cardinality between subregions and districts? Which is the join table, and which is the target table?

> 23→ In Contents, highlight the US States layer. Click the **Feature Layer: Data: Relationship: Relates** drop-down and choose Add Relate (Fig. 7.24).

> 23→ Set the Layer Name to US States and the Input Relate field to STATE_ABBR.

> 23→ Set the Relate Table to 114th Congress and the Output Relate Field to STATE_ABBR.

> 23→ Enter Congress for the name of the relate and *One to many* as the cardinality.

> 23→ Run the tool.

Now let's use the relate to find out which representatives are from New England states.

> 24→ Open the US States table and the 114th Congress table and arrange them side by side.

> 24→ Turn off all layers except US States and make sure it is highlighted in Contents.

> 25→ Choose **Table: View: Selection: Select By Attributes**.

> 25→ Set the Layer Name to US States and enter the expression: SUB_REGION is Equal to New England. Run the tool.

> 25→ In the US States table, click the Show Selected button.

Now that the New England states are selected and highlighted, the relate can access the representatives from these states.

> 25→ In the **Table: View: Relationship** group, click the **Related Data** drop-down and select Congress: cd114.

> 25→ The 114th Congress table selects and highlights the related records.

7. How many representatives come from New England states? _____

The selected records from the representatives table will provide a new file that can be saved permanently and given to interested constituents.

> 26→ Click the bar of the 114th Congress table to make sure it is active.

26→ Click **Table: View: Export Table**. It will automatically export only the selected records.

26→ Make sure the Input Rows says 114th Congress and the Output Table is named NewEngReps. Run the tool.

26→ Open the new NewEngReps table and verify that it contains the correct representatives.

Currently, two tables have selections. You could select each one and use the **Table: View** ribbon to clear them, but the **Map** ribbon has a clear button that clears all selections.

 27→ Click **Map: Selection: Clear** to clear the selection in all tables.

27→ Click the US States table tab to make it active. Click the Show All Records button (without a selection, the Show Selected Records view is blank).

 27→ In Contents, right-click the US States layer and choose Joins and Relates > Remove All Relates. (This command is also found in the table options menu.)

27→ Close all tables and save the project.

Editing values in tables

Sometimes fields in a table must be created or updated manually by editing the table. We will characterize each subregion in the SubregionTotals table as East or West.

TIP: Creating a new field makes permanent changes to the stored table, but the changes will not take effect until the edits in the Fields View are saved.

28→ Open the SubregionTotals table.

 28→ In the table, click the Add Field button to open the Fields: SubregionTotals view.

28→ In the bottom row of the Fields view, type a new field name, Section, in the outlined box and in the Alias box.

28→ Click in the Data Type box and select the Text data type.

28→ Scroll all the way right and set the Length to 4.

28→ For a better view when entering data, turn off all fields except SUB_REGION and Section. Click **Fields: Changes: Save**.

28→ Switch to the SubregionTotals table view.

TIP: Decide how many characters or decimal places are needed before creating a field.

Tables can always be edited, although the changes will not be permanent until saved. When working with a table, be careful not to change anything except when intended. It is a good idea to switch to the Edit ribbon before starting to edit.

29→ Click **Edit** to open the ribbon.

29→ In the table, click in the Section field in the East North Central row and type East.

29→ Click in the next row and type East or West as appropriate. Continue until all subregions have been designated as East or West.

29→ Click **Edit: Manage Edits: Save** and answer Yes.

30→ Switch to the Fields: SubregionTotals view and turn all the fields back on. Choose **Fields: Changes: Save** and close the Fields view.

30→ Close and reopen the SubregionTotals table, if needed, to see the fields back on. Close the table when finished examining it.

Creating domains

Domains can speed data entry and prevent typing mistakes. Suppose we wish to characterize each state as primarily Democrat or Republican or Mixed and enter that tag into the table. We can create a list of values to select from, called a coded domain. First we create the new field.

31→ Open the DistsPerState table and click the Add Field button.

31→ In the Fields view, add a new field named PRIMARILY. Set its data type to Text and give it a length of 3.

31→ Click **Fields: Changes: Save**.

Next we create the domain and enter the codes. The domain must have the same field type as the field we will assign it to, Text. Domains are actually created for geodatabases, not individual tables or fields, so that they can be reused in multiple tables. The domain will be saved as part of the home geodatabase, TablesPractice.

32→ Click **Standalone Table: Data: Design: Domains**. The Domains: TablesPractice view opens (Fig. 7.25).

32→ In the Domain Name box, enter PartyStatus.

32→ In the Description box, enter *Code designating the primary party*.

32→ In Field Type, choose Text. The Domain Type will default to a Coded Value Domain and a Codes section will appear on the right.

33→ In the first row of the Codes box, type the Code *DEM* and the Description *Mostly Democratic*. (Use the Tab key to move to the next box.)

33→ In the next row, type the Code *RPB* and the Description *Mostly Republican*.

33→ In the next row, type the Code *MIX* and *Mixed Party*. Do NOT press Tab after entering the last Description—it creates a blank row that precludes saving the domains. Instead, click the blank rectangle next to one of the existing codes.

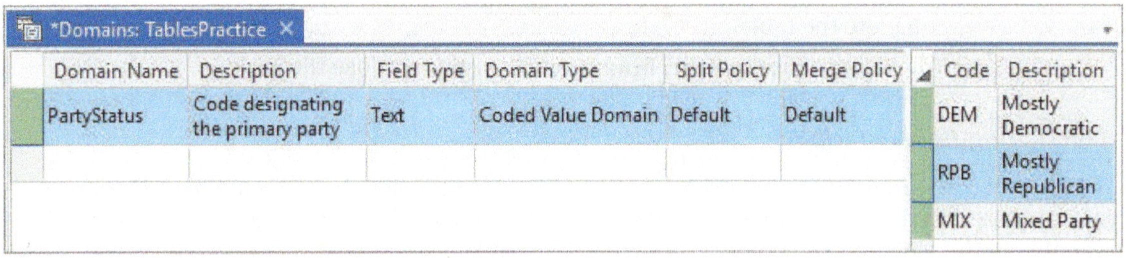

Domain Name	Description	Field Type	Domain Type	Split Policy	Merge Policy	Code	Description
PartyStatus	Code designating the primary party	Text	Coded Value Domain	Default	Default	DEM	Mostly Democratic
						RPB	Mostly Republican
						MIX	Mixed Party

Fig. 7.25. Creating a coded domain for the TablesPractice geodatabase
Source: Esri

33→ When done, choose **Domains: Changes: Save**. Close the Domains view.

TIP: The green boxes in Figure 7.25 indicate that the rows can be saved. If one or more boxes turn red, then an error exists, and the changes can't be saved until it is fixed.

Once the domain has been created, it can be assigned to the field or fields that will use it.

34→ In the Fields: DistsPerState view, in the PRIMARILY row, click in the Domain box (several times if necessary) and select the PartyStatus domain.

34→ Click **Fields: Changes: Save**. Close the Fields view.

Now set up the map so we can enter the designations.

35→ Turn layers on/off so that the 114th Congress layer can be seen.

35→ Symbolize the US States layer as a single symbol map with a 2-pt. hollow shade with a black outline. Place it above the districts (Fig. 7.26).

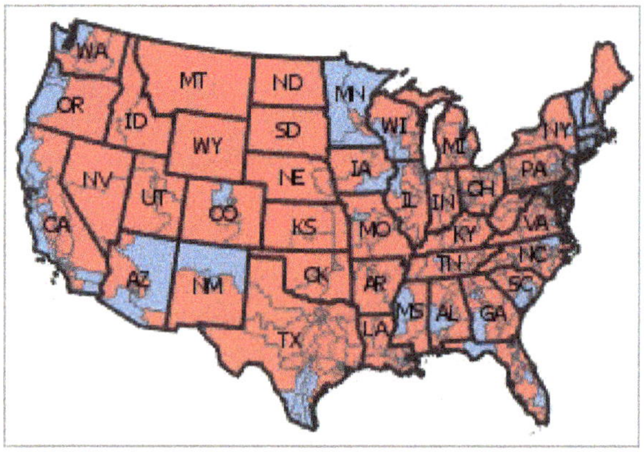

Fig. 7.26. The 114th Congress
Source: Esri

35→ Click the **Feature Layer: Labeling: Layer: Label** button to turn the labels on, and set the Field to STATE_ABBR.

35→ In the DistsPerState table, right-click the FREQUENCY field and choose Sort Descending.

Now we are ready to characterize the states using the domains.

36→ In the DistsPerState table, click the PRIMARILY field box for CA and select the Mixed Party value.

36→ In the row for TX, choose Mostly Republican.

TIP: If an In Use error appears when choosing the value for the next row, it is a bug. Save the project, close Pro, and restart it to clear the spurious file lock. Continue editing.

36→ Assign a few more values in the table based on what the map shows. You needn't complete the table.

36→ When done, choose **Edit: Manage Edits: Save**. Close the table.

Calculating fields in tables

Imagine that we are writing a report and we need a table showing the percentage of Hispanics in each state. The US States table has only the number of Hispanics. Creating a new field and calculating the percentage of Hispanics solves the problem.

37→ Open the US States attribute table and click its Add Field button.

37→ Name the field HISP_PERC and select Float as its type. Click **Fields: Changes: Save**. Close the Fields view.

TIP: If the New Field button is dimmed, the table may have unsaved edits. Choose **Edit: Changes: Save** to examine the edits and save them or discard them.

Now calculate the percentage of Hispanics. The minority population fields in this table represent 2010 census data, not 2014, so we will use the POP2010 field for calculating percentages.

TIP: Be VERY careful when calculating; you may overwrite existing data, and it can't be undone. Don't calculate a field with existing data unless you intend to update the values.

38→ Right-click on the heading of the new, empty HISP_PERC field (located to the far right of the table), and choose Calculate Field.

38→ In the Geoprocessing pane, double-click the field names (the window will add ! around them) and the operator buttons to enter the expression: HISPANIC / POP2010 * 100. (Click in the expression box to type the 100 value.)

38→ Check the settings (Fig. 7.27) and run the tool.

TIP: Do many records equal null? You might have accidentally selected a single record by clicking on the table. Check the table status bar to see if any records are selected. If so, clear the selection and try the calculation again.

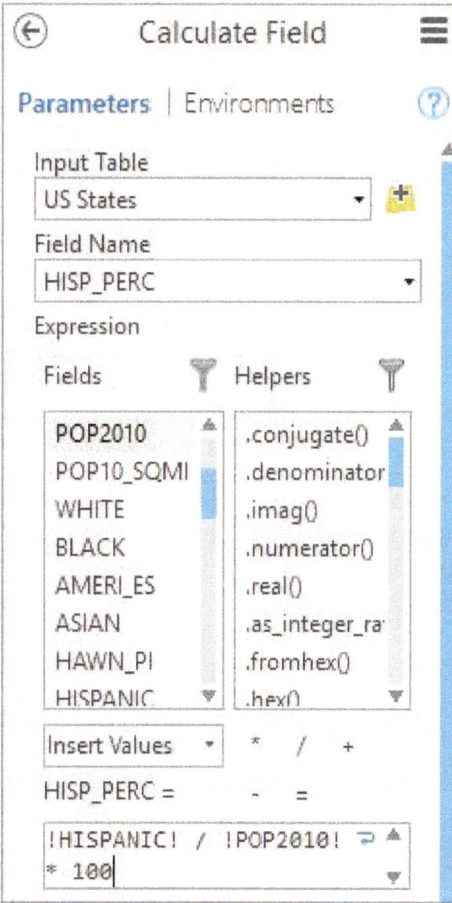

Fig. 7.27. Calculating a field
Source: Esri

Finally, we need to determine the density of Hispanics per square mile in the counties for a report. This goal requires creating and calculating another new field, the Hispanic density.

39→ Close the US States table.

39→ Open the US Counties table and click the Add Field button.

39→ Name the field *HISP_SQMI* and make it a Float type.

39→ Click **Fields: Changes: Save** and close the Fields view.

Now we will calculate the new field.

40→ Right-click the empty HISP_SQMI field and choose Calculate Field. Enter the expression HISPANIC / SQMI and run the tool.

40→ Close the US Counties table and turn off all the other layers in the map.

41→ Symbolize US Counties with a graduated colors map using the HISP_SQMI field. Use the Geometric Interval classification method (Fig. 7.28).

The population density of Hispanics has a strongly skewed distribution due to the presence of densely populated areas like Los Angeles. The Jenks classification method does not show the distribution of Hispanics well, but the geometric interval method does. Try both to see the difference.

Exploring data with charts

Mapping is one part of data exploration, but graphing data is also a good way to investigate patterns. ArcGIS Pro has some charting tools that integrate with maps.

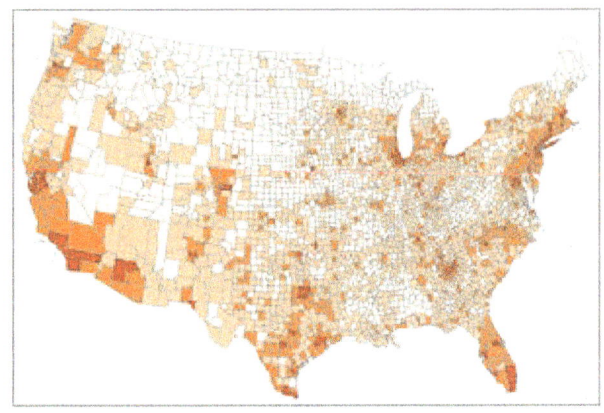

Fig. 7.28. Density of Hispanics
Source: Esri

42→ Add the climatestations2005. dbf file from the mgisdata\Usa folder.

42→ In Contents, right-click the climatestations2005 table and choose Display XY Data. Run the tool to display the climate stations on the map.

One common question about a variable is what kind of distribution the values have (normal, skewed, etc.) A histogram helps answer this question.

43→ In Contents, click the icon to open the List By Charts panel.

43→ Right-click the climatestations2005_XYTableT layer and choose Create Chart > Histogram. A Chart view opens.

43→ In the Chart view, click the Properties button to open the Chart Properties pane.

43→ In the Chart Properties pane, set the Variable to Ave_annual. Set the number of bins to 20.

44→ Experiment with the other settings on the chart properties to see their effects.

44→ In the Chart pane, click the General tab. Title the chart *Average Temperature* and name the X-axis *Degrees F*. Click Tab each time to implement the change.

The chart is connected with the map to facilitate analyzing the spatial distribution of plot values.

45→ Arrange the display to show both the map and the histogram.

45→ Click one of the bars in the histogram. The corresponding stations are highlighted on the map.

8. Notice the spatial distribution of stations with values in the high 40's. Suggest an explanation for why the western values scatter so far from the latitude band.

TIP: To quickly create a histogram for a field in a table, right-click the field heading and choose Statistics.

Another common data exploration technique is to look for correlations between variables. For example, one might expect that the annual temperature is correlated to the latitude of the climate station. Test this hypothesis with a scatter plot.

46→ In Contents, right-click the climatestations2005 XYTableT layer (not the table) and choose Create Chart > Scatter Plot.

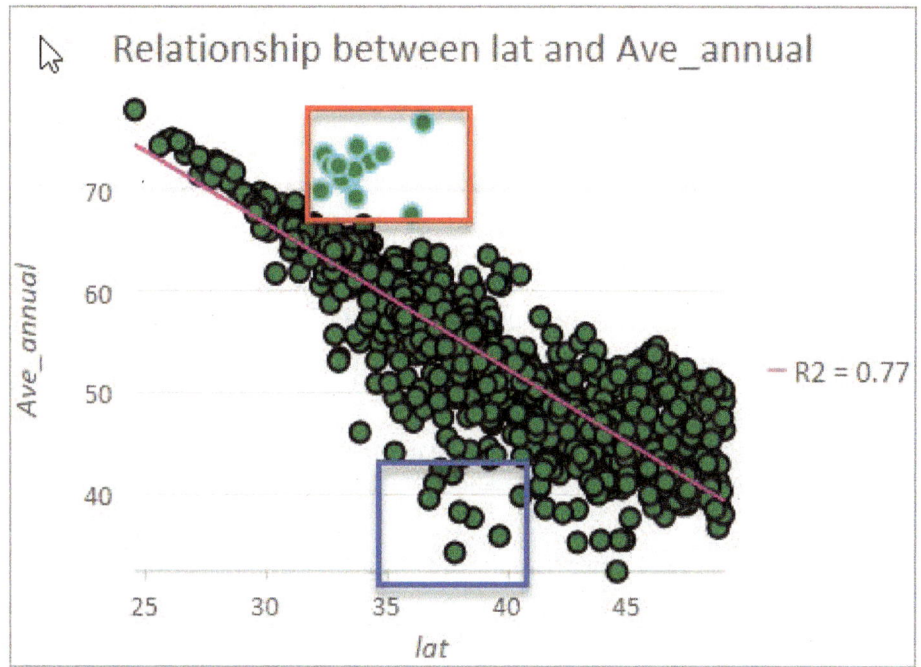

Fig. 7.29. Analyzing data using a scatter plot
Source: Esri, National Climatic Data Center

46→ In the Chart Properties pane, choose lat for the X-Axis and Ave_annual for the Y-Axis.

46→ In the scatter plot, find the group of outliers above the line, shown in a red box in Figure 7.29 (locations hotter than expected given their latitudes).

46→ Click and drag a rectangle around this group to select them. The matching stations are highlighted on the map.

46→ Examine where the cold outliers occur on the map (blue box in Fig. 7.29).

One can also select map features to see where they plot on the chart.

47→ Click the **Map: Selection: Select** tool.

47→ On the map, click and drag a rectangle around the Washington (state) stations to select them. The corresponding stations are highlighted in the chart.

48→ Close all charts and return to the List By Draw Order panel in the Contents pane.

48→ Clear the selection and save the project.

TIP: These charts are intended to explore data. If more control settings are needed to create charts for reports, export the data to dedicated graphing software.

Working with Excel data

Spreadsheets can be a valuable source of data for a project. This example, using precipitation data for climate stations in Oregon, includes opening Excel files, displaying table data as points on a map, joining tables, and exporting to create a final feature class.

49→ In the Catalog pane, open the Oregon Precip map.

TIP: An Excel .xls or .xlsx file is a workbook that may contain multiple worksheets. Only single worksheets may be opened in ArcGIS Pro.

49→ In the Catalog pane, open the mgisdata\Oregon folder and expand the ORstations. xlsx workbook to view the worksheets inside. Drag the ORstations$ worksheet to the map.

49→ Open the ORstations$ table and verify that the fields contain valid data.

49→ In Contents, highlight the ORstations$ table and click **Standalone Table: Data: Design: Fields** to examine the table in Fields view.

49→ Verify that the fields are listed appropriately as numeric or text data.

49→ Close the Fields view and the table.

TIP: Always examine an Excel table after adding it to make sure that it was interpreted correctly.

9. Why won't we use any of the other fields with LAT or LON in their names?

50→ In Contents, right-click the ORstations$ table name and choose Display XY Data.

50→ In the XY Table To Point tool, set the X Field to LON and the Y Field to LAT.

50→ Verify that the Spatial Reference defaulted to a GCS, the appropriate choice for latitude-longitude coordinates. Run the tool.

The precipitation values are stored in a different worksheet. We will add it and join it to the point feature class so that we can map the precipitation.

51→ Add the migsdata\Oregon\precipnormals.xls\Sheet1$ worksheet.

51→ Open the Sheet1$ table and examine the precipitation values for each station.

51→ Open the ORstations$_XYTableToPoint table and compare it with the Sheet1$ table.

10. What is the common field in these two tables? _____

52→ In Contents, right-click the ORstations$_XYTableToPoint layer (not the table!) and choose Joins and Relates > Add Join.

52→ In the Add Join tool, choose STATION NAME as the Input Join Field.

52→ Choose Sheet1$ as the Join Table, and STATION as the Output Join Field. Run.

52→ Examine the ORstations$_XYTableToPoint table to make sure it contains the precip- itation values. Close all tables.

The ORstations$_XYTableToPoint feature class is stored in a GCS. We will export it to the oregondata geodatabase and use Environment settings to ensure that the coordinate system matches the other data in the geodatabase. As an additional benefit, the joined fields will automatically be copied to the new data set.

53→ In Contents, right-click the ORstations$_XYTableToPoint layer and choose Data > Export Features.

53→ In the Copy Features tool, enter *precip* for the name of the output feature class.

53→ In the Geoprocessing pane, click the Environments tab. Click the Output Coordinate System drop-down and set it to gtoposhd. Run the tool.

54→ In Contents, remove the ORstations$_XYTableToPoint layer.

54→ Symbolize the new precip layer with a graduated symbol map based on the ANN field (Fig. 7.30).

This is the end of the tutorial.

→ Save the project and close Pro.

Practice Exercises

Use the store_openings.csv file in the mgisdata\Usa folder to perform the following tasks and answer the questions.

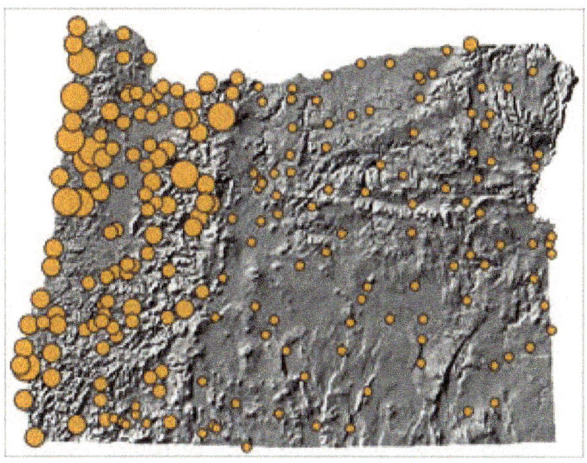

Fig. 7.30. Oregon precipitation
Source: National Climatic Data Center; USGS

1. Export the store_openings table from a .csv file to a table in the TablesPractice geodatabase.

2. How many stores are there? Where did the first one open? Where did the latest one open?

3. Examine the table and the metadata for the table. Is the latest date reported in the previous question actually the youngest store in the United States? Why or why not?

4. Create a table that lists each state and the number of stores that it contains. Which state in this table contains the most stores? How many does it have?

5. Determine the number of stores per 100,000 people in each state, using the POP2014 field. Which state has the most stores per capita? Which has the least? (**Hint**: In a joined table, only the target table fields may be calculated.)

6. **Capture** a map showing the number of stores in each state; include a legend in the map. Also **capture** a map showing the number of stores per capita in each state. Compare the two maps and discuss what they could mean.

7. Use a chart to examine the correlation between POP2010 and the number of stores in a state. **Capture** the chart and refer to it in a discussion that covers these points: How good is the correlation? Which state has a much higher number of stores than one would expect? Which two have much lower numbers?

8. Create a table that contains the minimum and average opening date of stores in each state. Which state has the latest average opening date? What is the average date?

9. Read the Help section in Help> Maps > Time and learn how to set up and operate the time properties and slider. Use it to examine the spread of stores over time based on the earliest opening date in each state. Discuss what you learn. **Capture** a map showing the states with stores by the beginning of 1975 and again at the start of 1985.

10. Find a table on the Internet that includes x-y coordinates or a field that allows it to be joined to a spatial data set, such as counties. Format the table so that it can be read in ArcGIS, and create a map showing data from the table. **Capture** the map with a brief explanation of the data set and the conclusions that can be drawn from it. Include a citation for the table.

Chapter 8. Editing

Objectives

► Understanding topological errors

► Using snapping to ensure topological integrity of features

► Adding features to map layers using basic editing functions

► Using the sketching tools and context menus to precisely position features

► Entering and editing attribute data

Mastering the Concepts

GIS Concepts

Editing can update existing feature classes or create new ones. If a housing subdivision is constructed, the new roads, parcels, sewer lines, and other infrastructure need to be added to the city database. Entirely new feature classes can be created, for example, if the city planning department decides to create a map of garbage collection zones where none existed before. Parcel ownership attributes must be updated when parcels are sold. All of these tasks are done by editing feature classes and tables.

Editing consists of several basic tasks: (1) creating new features, (2) modifying the shape or location of existing features, and (3) entering or modifying attributes in a table. Throughout this process, care must be taken to create and protect the proper relationships between features that are adjacent or connected to each other.

Editing and topology

Topology refers to the spatial relationships between features in terms of adjacency, connectivity, intersection, or overlap. When editing, care must be taken to create and maintain topological integrity between features so that the relationship in the database matches the relationship in the real world. If two parcels share a common boundary, then the boundaries should match exactly. Line features, such as streets or water lines, should connect to each other in the feature class. Lines that cross each other should intersect at a node. Lines and polygon boundaries should not cross over themselves. Polygons should not overlap each other so that an area is part of more than one polygon. These basic rules must be observed to ensure the **logical consistency** of features so that they properly represent the relationships of their real-world counterparts.

Topological data models permit the user to test, locate, and fix topological errors. However, topology is even more important when editing spaghetti models because the user must manage the spatial relationships without help. This chapter introduces two capabilities that aid in creating and maintaining topological integrity: snapping and coincident boundary creation.

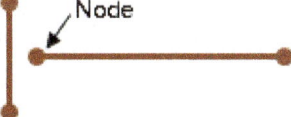

A dangle—the two lines fail to connect.

Correct topology—the horizontal line intersects the vertical one, creating three lines.

Snapping features

When creating line features that connect to each other, such as roads, one must take care that the features connect properly (Fig. 8.1). A

Fig. 8.1. Topological relationships between lines

line that fails to connect is called a **dangle**. Although one may not be able to see the gap between the lines in a map, the gap will exist unless the endpoints of each line (nodes) have exactly the same coordinate values. Even though the map may look right, certain functions, such as tracing networks or locating intersections, will not work properly. It is impossible to intersect lines correctly by simple digitizing.

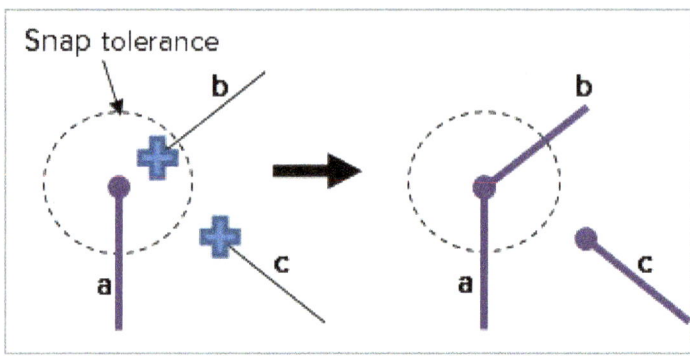

Fig. 8.2. Snapping. Line *b* snapped to *a* because the cursor was inside the snap tolerance. Line *c* did not.

Snapping ensures that the nodes of lines and the vertices of polygons match. It can also be used to match points on the edges or vertices of existing lines or polygons. When snapping is turned on, it affects features being added or modified. When the cursor is placed near an existing feature (Fig. 8.2), the cursor is drawn like a magnet to the node or vertex, and the new feature is snapped to the existing one—that is, the coordinates are matched at that location. The **snap tolerance** specifies how near the cursor must approach to be drawn to the existing feature.

Care must be taken when specifying the snap tolerance. If it is too small, then features won't snap. If it is too large, then features may snap when it isn't appropriate. Snap tolerances are set in screen pixel units or in the map units. A snap tolerance of 10 pixels indicates that features will be snapped if the cursor moves within 10 screen pixels of an existing feature. When set in pixels, the snap tolerance remains the same as the user zooms in and out, which is convenient. Setting the snap tolerance in map units is useful if the user is trying to maintain a particular level of geometric accuracy independent of zoom scale. Four types of snapping can be used.

Point snapping matches a new vertex to an existing point in a point feature class.

End snapping matches a new vertex to the endpoints of an existing line. End snapping can ensure that new streams connect to the ends of existing streams, and only to their ends. End snapping only applies to line features.

Vertex snapping matches a new vertex to any vertex in an existing line or polygon. It can ensure that adjacent parcels connect only at existing corners.

Edge snapping constrains the feature being added to meet the edge of an existing line or polygon feature. In this case, the vertex being added could be placed anywhere on or between vertices. Edge snapping can ensure that a street ends exactly on another street or that a stream gage lies exactly on the stream it measures.

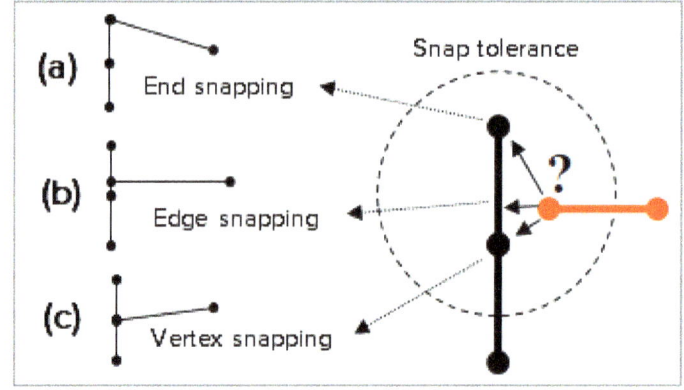

Figure 8.3 illustrates how snapping changes depending on the snapping type. The black vertical line already exists in the layer, and the orange horizontal line is being added. The

Fig. 8.3. Where will it snap? The snap type dictates where the new horizontal line will end.

dashed circle shows the snap tolerance. If end snapping is turned on, then the new line will snap to the endpoint of the existing line (Fig. 8.3a). If edge snapping is turned on, then the line can be snapped anywhere between the end and the vertex (Fig. 8.3b). If vertex snapping is set, then the new line will connect to the closest vertex or end (Fig. 8.3c).

If more than one kind of snapping is turned on, then the most inclusive one takes precedence. To ensure snapping only to ends, therefore, both edge and vertex snapping must be turned off.

Coincident boundaries

Another type of topology to consider is the relationship between two adjacent polygons. A polygon is stored as a series of x-y vertices that completely enclose a space. If two polygons share a boundary, then the boundary gets stored twice. If the shared edge contains exactly the same x-y pairs for both polygons, it is called a **coincident boundary**. If the pairs do not match exactly, then there will be gaps where the polygons fail to meet, overlaps where they cross each other, or both (Fig. 8.4).

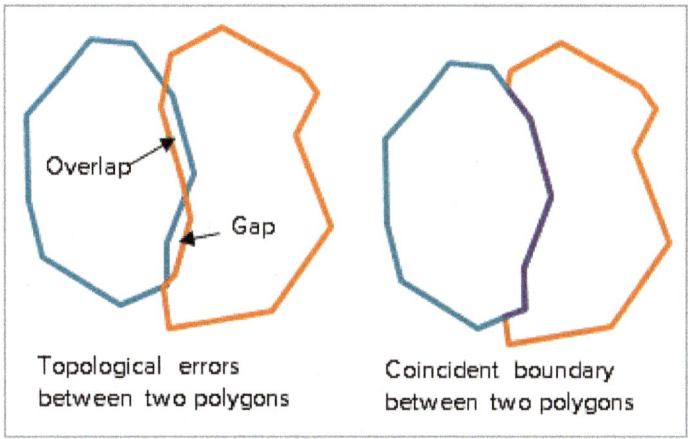

Fig. 8.4. Topological relationships between polygons

Both types of relationships are possible in the real world. If the two polygons represented fires that burned in two different years, then it is possible that the second fire might have overlapped the first, or that some space existed between them, in which case the presence of gaps and overlaps is real and necessary. However, many polygons should have coincident boundaries—soil units, land use zones, school districts, counties, and states all represent features for which gaps and overlaps are errors, usually caused by careless data entry. Imagine the legal headaches that would ensue if the two polygons represented parcels; both overlaps and gaps would likely be in dispute as to who owned what. When editing polygons, it is important to ensure that the right relationship is incorporated into the database and that adjacent polygons have coincident boundaries, except for the instances when gaps or overlaps are justified.

About ArcGIS

The ArcGIS Pro editing environment supports many different environments and tools for editing. This chapter presents the most basic elements of creating and modifying features and attributes. The Pro Help contains information on more advanced editing techniques.

What can be edited?

In general, users can edit any vector data sets for which they have read-write access, including shapefiles and geodatabase feature classes. Certain types of layer sources, such as AutoCAD files cannot be edited because they are not native data formats for ArcGIS. Rasters cannot be edited.

Two types of changes to data sets can be made. **Schema changes** are those that modify the structure of the data set itself, such as adding or deleting fields, changing field names or aliases, or modifying domains. Schema changes are sometimes also called design changes. Changes that affect only features within a feature class, such as adding a new road or changing the owner of a parcel, are called editing. In ArcGIS Pro, editing is always available. The Edit ribbon provides access to the editing commands, tools, and menus. Because editing is always possible, users must take care not to edit features or tables inadvertently. The program tracks when edits have been made to one or more data sets, and it will insist that edits be saved or discarded before you close the project.

Once editing is begun on a layer (and therefore unsaved edits exist), a file lock is placed on the source data set to prevent other users or programs from modifying the data. Within ArcGIS Pro, certain operations that change the data set schema, such as adding new fields to a table, are also inaccessible whenever unsaved edits exist. For example, if Lucy has created several new parcels for a data set and then she decides to add a new field to store the owners' middle names, she won't be able to add the field until the pending edits have been saved or cleared. If you are having trouble performing a function that modifies the underlying structure of a feature class or table, check the Edit ribbon for unsaved edits.

However, spurious file locks can occasionally occur, when changes have been saved but the program still believes the file to be locked. If a user has cleared all edits but is still receiving a "locked by user" error message, the lock may be spurious. Saving the project (and any pending edits) and restarting the program will usually clear these troublesome locks.

Editing and coordinate systems

Recall that a map may contain data sets stored in multiple different coordinate systems, each of which is being projected on the fly to match the map. ArcGIS Pro can edit layers with coordinate systems that differ from the map coordinate system. For example, you may be editing a shapefile of roads stored in a State Plane (feet) projection within a map that is set to match the Streets basemap coordinate system, WGS 1984 Web Mercator Auxiliary Sphere (meters). Although the editing takes place in the Web Mercator coordinate system and uses default units of meters, the edits will be converted to State Plane coordinates in feet as they are saved in the shapefile.

Performance and reliability

Although ArcGIS Pro has many useful capabilities, such as editing across coordinate systems and editing multiple files simultaneously, one must be cautious in taking advantage of these techniques. If an editing session is placing a high demand on system resources by having many files open or by doing many projection transformations, system performance and reliability may suffer. Moreover, because spatial data files are complex constructions, the price of an editing glitch may be the loss of data integrity. An ill-timed system glitch might not only lose recent changes but also corrupt an entire data set. Follow these recommendations to help ensure reliable and responsive editing:

▶ *Always have a backup copy of the file being edited*, stored elsewhere on the disk or on a different medium. This precaution is especially important when working on data sets that would be expensive or impossible to replace should something go wrong.

▶ For extensive editing sessions, create simple map views that use only the layers needed for editing, and close all other maps and layouts in the project.

▶ Save edits often in case of a system glitch. By default, ArcGIS Pro does not automatically save edits, although it can be set to save after a certain number of minutes or a certain number of operations. However, remember that saving clears the Undo memory of edits. After edits are saved, they can no longer be undone.

ArcGIS Pro allows edits to all data sets by default, which is convenient but can also lead to unintended changes. Editing can be turned off for one or more layers in a map. Accident-prone users might wish to turn off editing to prevent accidental changes.

Editing tools

Editing is a process integrated into the ArcGIS Pro framework through a variety of panes, menus, and tools. Like other processes, it is context-sensitive, so the interface is updated dynamically as the user selects certain layers, features, ribbons, and panes. This section briefly reviews the main editing tools and access points used during an editing session.

The Edit ribbon

The Edit ribbon provides access to the editing functions (Fig. 8.5). It includes a menu to control snapping, sets of tools for creating and modifying features, and tools for managing features selected for editing.

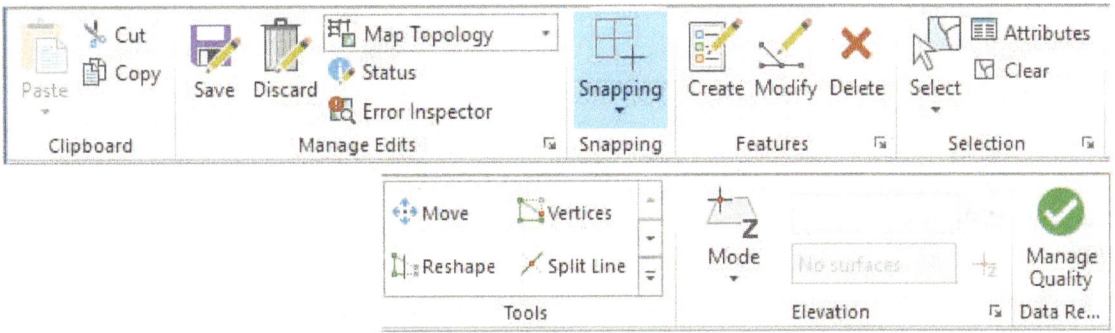

Fig. 8.5. The Edit ribbon contains many editing tools, menus, and commands.
Source: Esri

The Status button in the Manage Edits group allows the user to view the editing status of the layers in the map.

The Create button opens the Create Features pane, used to add new features to a data set. The Modify button opens the Modify Features pane, which provides many tools for modifying the shape of an existing feature, such as moving some of its vertices or reentering a section of a line or polygon. Some of the more commonly used Modify tools are also found in the Tools group on the Edit ribbon. These panes and many other editing functions are discussed in the next sections.

The Contents pane

The Contents pane has several icons to display the Contents as panels that are especially useful when editing (Fig. 8.6).

The View by Selection panel, discussed more fully in Chapter 9, controls which layers can be selected with the map's selection tools. Editing is often focused on one layer at a time, so it is useful to

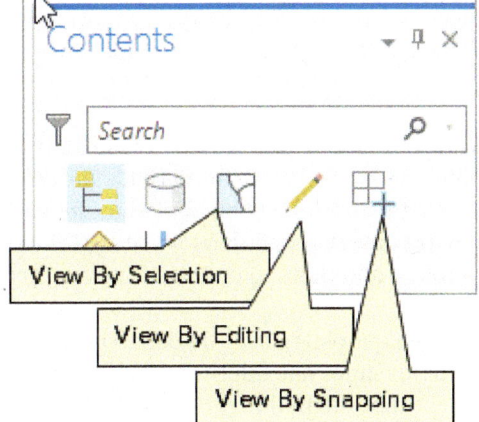

Fig. 8.6. Icons in the Contents pane used for editing
Source: Esri

make that layer the only one selectable in order to prevent accidental changes to other layers. The panel also reports how many features are currently selected in each layer.

The View By Editing panel controls editing permissions for individual layers. Editing can be turned off for one or more layers to protect them from accidental changes.

The View By Snapping panel lets the user turn snapping on and off for each layer. If snapping is off, the cursor will not snap to features in that layer. However, the type of snapping currently in use (point, node, vertex, or edge) is controlled dynamically by the Snapping group on the Edit ribbon (or by a corresponding easy-access pop-up menu below the display window).

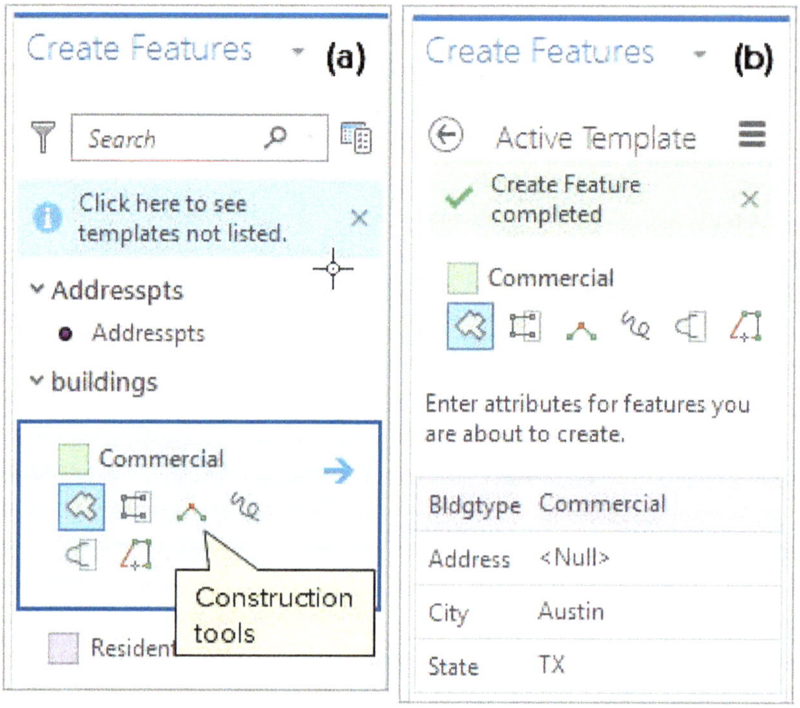

The Create Features pane

The Create Features pane is used to add new features to a data set. When opened (Fig. 8.7a), it shows at least one **feature template** for

Fig. 8.7. (a) The main Create Features pane, showing the available editing templates; (b) the Active Template pane
Source: Esri

every editable layer in the map. A feature template contains construction tools used to create new features and their attributes. The template selected in Figure 8.7 is named Commercial and saves features into a polygon layer called buildings. New features are added by selecting the template, choosing one of the construction tools, and entering the feature in the map. Each geometry class (points, lines, polygons) has its own set of construction tools. Templates are created automatically, but the user can modify existing templates and create new ones if needed.

When a template is selected, a blue arrow grants access to the Active Template pane (Fig. 8.7b), which includes the attribute fields for the feature class. If the user enters information in these boxes, the values will be copied to the next feature created, making data entry more efficient. If you are entering buildings in a city, for example, the city and state names can be placed in the template and thereby entered for every feature created (Fig. 8.7b).

Multiple templates may be constructed for each feature class. As shown in Figure 8.7a, the buildings layer has another template called Residential. It is identical to the Commercial template except that the building type field, Bldgtype, contains "Residential". If the user is digitizing a new commercial building, she clicks on the Commercial template. When the building is added, the default attributes "Commercial", "Austin", and "TX" are automatically placed in the attribute fields. If she wishes to add a residential building, she clicks the Residential template, digitizes the building, and the attributes are entered. Using templates to enter attributes can save a great deal of typing, speeding data entry and minimizing the chance of typing errors. Only the address field, which is unique to each feature, must be entered manually.

TIP: If a template is not visible for a layer, check the List By Editing view in the Contents pane to make sure that editing is turned on for the layer.

The Create Features pane only shows templates for layers that are currently editable. The symbol associated with a template is based on the layer symbology and will update if symbol changes are made. If a layer uses a unique values symbolization, a template will be created for each unique value, with the field defining the unique value already entered in the template. This software procedure provides a quick way to generate multiple templates for a layer.

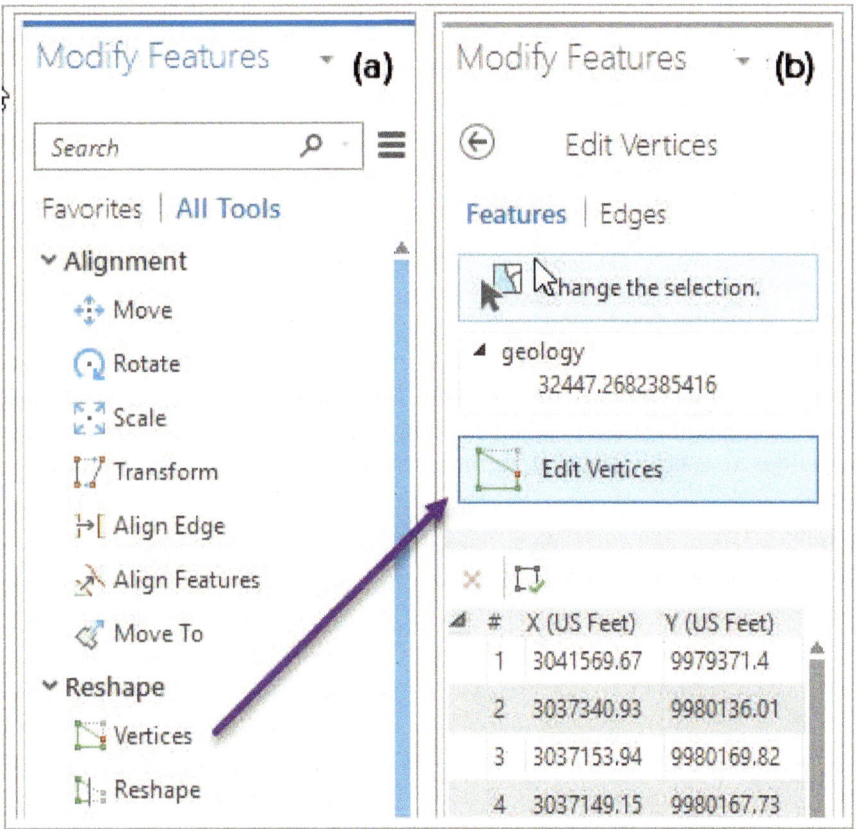

Fig. 8.8. (a) The main Modify Features pane, showing the available editing tools; (b) the Edit Vertices pane

Source: Esri

The Modify Features pane

The Modify Features pane (Fig 8.8a) organizes many tools for making changes to features that already exist in a layer, such as moving individual vertices, splitting a line or polygon, replacing a section of a line, and merging features together. The most commonly used tools are also accessible from the Tools group on the Edit menu.

Clicking one of the tools opens a new pane with settings, options, or entries needed for using the tool, as shown for the Vertices tool in Figure 8.8b. Most of the tools require that one or more features be selected before the tool can be used, so the pane includes a menu to create or change the current selection.

Sketches

When a line or polygon feature is being created or modified, a provisional object known as a **sketch** is shown, analogous to a pencil sketch an artist makes before she inks in the final drawing. The sketch shows each *x-y* vertex and the **segments** connecting them. The green outline in Figure 8.9 shows a polygon sketch with green vertices separated by straight-line segments. The most recent

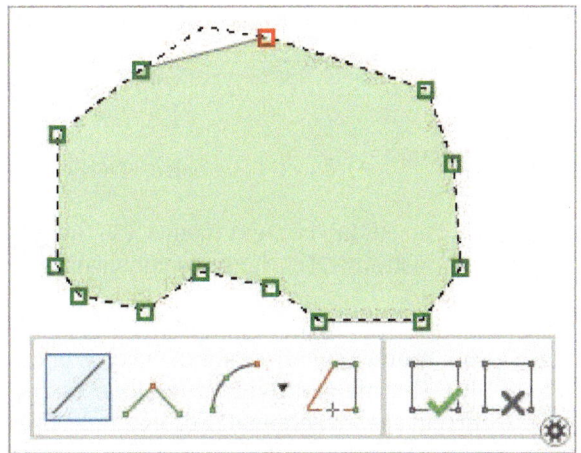

Fig. 8.9. A polygon sketch with the Construction toolbar

Source: Esri

vertex is shown in red. New vertices are added to a sketch using one of the construction tools. This sketch is not added to the data set until it is "finished" by double-clicking the last vertex or choosing Finish from one of the editing menus.

Different types of segments can be used to connect adjacent vertices. The default segment is a straight line, but several curved segment types are available. While the user is sketching, a small toolbar known as the Construction toolbar appears near the bottom of the screen (Fig. 8.9). It provides quick access to segment types, including a trace tool that allows one to trace segments of existing features. It also contains buttons that let one finish or cancel the current sketch.

Context menus and toolbars

During editing, three context menus aid in completing certain tasks or changing edit settings (Fig. 8.10). The General Editing menu (Fig. 8.10a) appears when the Select button is active on the Edit ribbon and the user right-clicks the map, whether or not any features have been selected. It provides access to selection, repositioning, and modification tools for the selected feature(s).

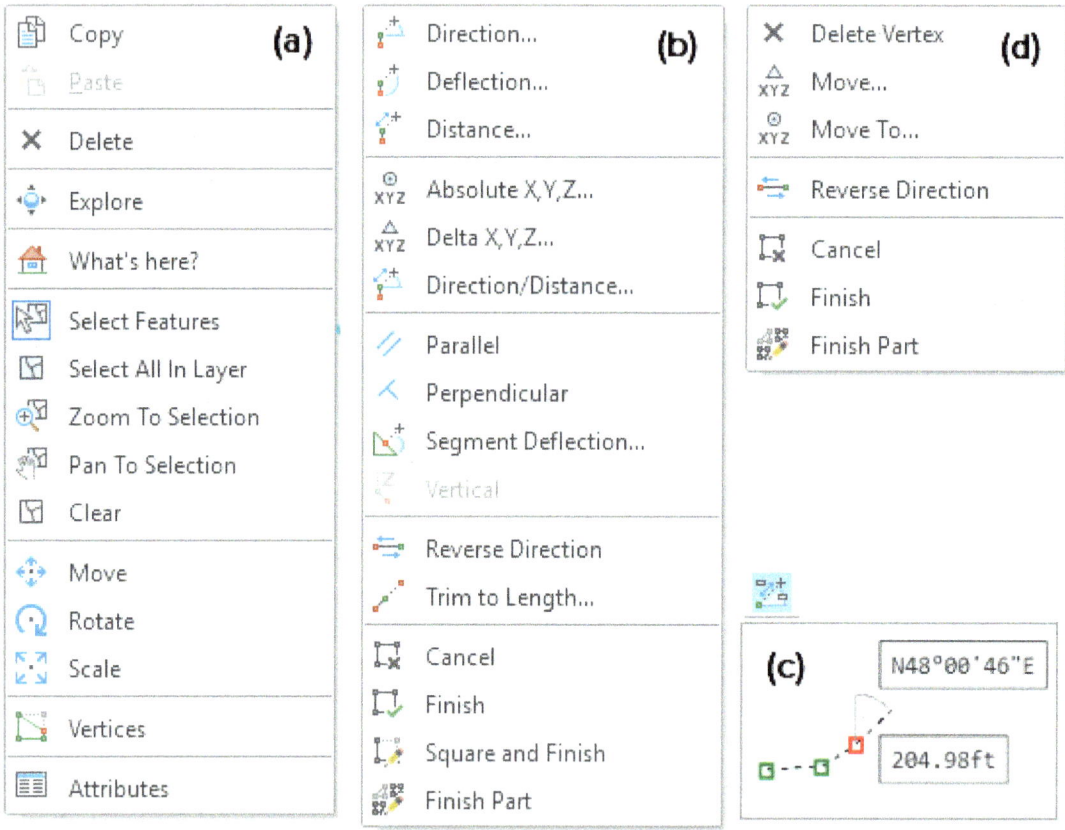

Fig. 8.10. The editing context menus: (a) the General Editing menu, (b) the Sketch menu with segment constraints, (c) dynamic constraints, and (d) the Vertex menu
Source: Esri

The **Sketch menu** appears when the user is engaged in sketching and right-clicks on the map (Fig. 8.10b). This menu provides functions for specifying exact angles, lengths, or distances for the current sketch segment. For example, the user might wish to make a segment parallel to an existing feature or give it a specific bearing and length (common when entering survey data). These menu items are known as **segment constraints** because they constrain the current segment to follow a specified rule. If one uses segment constraints often, the **dynamic constraints** editing mode can be used, which continuously shows the angle and distance of the current segment as a guideline (Fig. 8.10c). The constraints operate the same way as those in the Sketch menu but do not require the menu to be opened.

The Vertex menu appears when the user right-clicks on a vertex or segment of the sketch (Fig. 8.10d). This menu provides functions for adding, deleting, or moving vertices.

These context menus are associated with contextual toolbars that may appear when certain menu items are selected, similar to the Construction toolbar shown in Figure 8.9. For example, selecting Move, Rotate, or Resize from the General Editing menu causes the Reposition toolbar to appear (Fig. 8.11a). Selecting Vertices opens the Vertices toolbar (Fig. 8.11b). These toolbars put the relevant tools within easy reach for efficient editing.

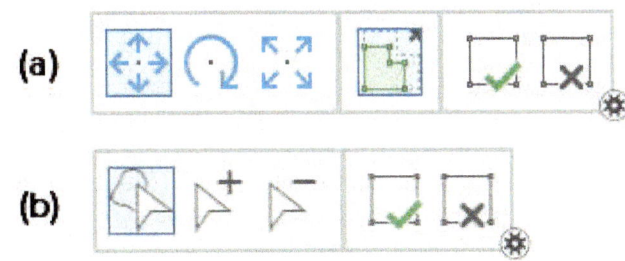

Fig. 8.11. More context toolbars: (a) the Reposition toolbar, and (b) the Vertices toolbar

Source: Esri

The Attributes pane

Editing features often includes editing their attributes. The user can modify attributes in two ways. In the first way, edits are typed directly into an attribute table, as discussed in Chapter 7. The second method uses the Attributes pane to edit the values of single or multiple features.

The Attributes pane (Fig. 8.12) is accessed from a button on the Edit ribbon. The upper panel shows the currently selected records; if no features or records are selected, then both areas will be blank. The lower panel shows the attributes of the highlighted feature in the upper panel. When the feature is clicked, it flashes on the screen, and the attributes in the lower panel may be edited.

The pane also allows editing of attributes for multiple features. If the top row containing the feature class (buildings) is clicked, then updates to the attributes are applied to all of the selected features.

Tools to create and maintain topology

In topological editing, features are not considered to be separate, independent objects; instead they have spatial relationships with each other. When adjacent polygons share a boundary, edits must be performed on both polygons to avoid generating gaps or overlaps. Lines that connect with each other should have their connections maintained. Two aspects of topological editing require attention: (1) creating new features with correct topology and (2) maintaining correct topology when existing features are edited.

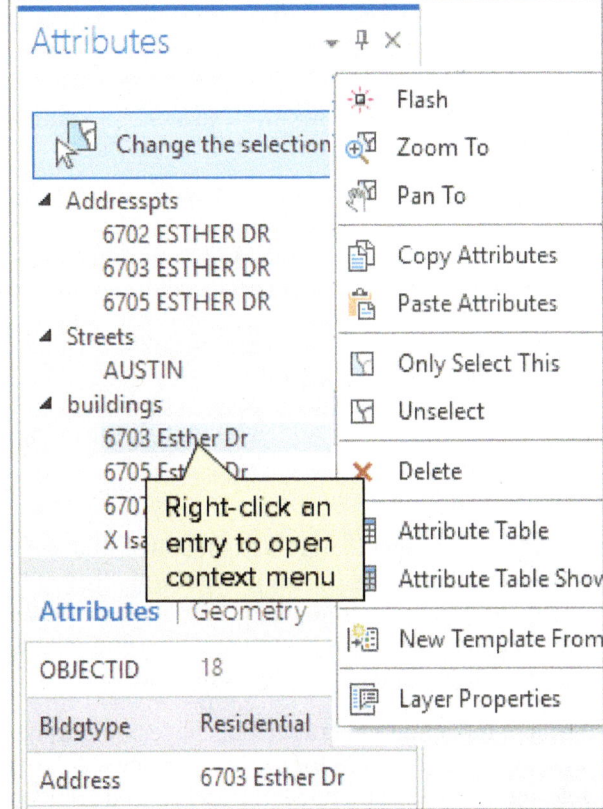

Fig. 8.12. The Attributes pane can display and edit many records at once.

Source: Esri

Several tools assist with creating correct topology in the first place. Snapping ensures that features that are supposed to intersect or touch each other actually do. Polygons with adjacent boundaries are created using the Autocomplete Polygon tool to ensure that shared boundaries are identical.

The Snapping toolbar

A small toolbar allows easy access to snapping tools when sketching. The four basic kinds of snapping (point, end, vertex, and edge, shown left to right in Fig. 8.13) are controlled with the click of a button. Additional icons include intersection snapping (where two lines intersect at a common node), midpoint snapping at the midpoint of a line, and tangent snapping to a curve. This toolbar can be opened from the Edit ribbon, but it is used so often that a pop-up button for it is also available at the bottom of the map view.

Fig. 8.13. The Snapping options
Source: Esri

The snap tolerance can be controlled in this menu using the Snapping Settings. In addition, snapping can be turned on or off for individual layers using the View By Snapping panel in the Contents pane.

Creating adjacent polygons

A special construction tool is used to add adjacent polygons when they must have coincident boundaries (Fig. 8.14). After the first polygon has been created, the Autocomplete Polygon tool is selected and used to digitize only the new part of the

Draw the first polygon with the Polygon tool.

Add a new polygon with the Autocomplete Polygon tool.

Two polygons with coincident boundaries

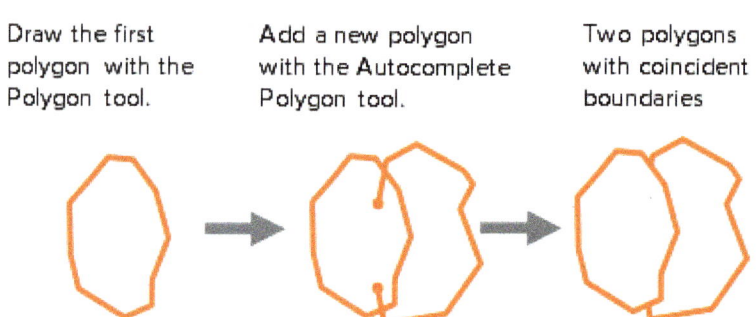

Fig. 8.14. Creating a coincident boundary between two adjacent polygons with the Autocomplete Polygon construction tool

adjacent polygon, ensuring that the polygons share a coincident boundary and are free from gaps and overlaps. The Split tool provides another way to add adjacent polygons. With this method, an outer polygon is constructed around an area encompassing two or more polygons, and then the internal boundaries are added by cutting the polygons along the intended coincident boundary.

Map topology

ArcGIS Pro automatically tracks features that are adjacent or connected to each other using a construct known as the **map topology**. Shared boundaries between polygons are known as **edges**, and shared endpoints of lines are termed **nodes**. Edges can be modified for both polygons simultaneously in order to prevent gaps and overlaps from being introduced. Nodes that are moved will retain their connections to all lines that share the node.

Many of the Modify Features tools have two different tabs, one for Features and one for Edges, as shown in Figure 8.15 for the Edit Vertices tool. In the Features pane (Fig. 8.15a), features are selected and edited individually, and shared polygons are ignored. Thus, changing the vertices opens a gap between the polygon and its neighbor. In the Edges pane (Fig 8.15b), the edge between the polygons is selected and edited. When the change is complete, both polygons are modified so that the edge remains coincident and no gaps or overlaps are created. The Edges pane uses purple instead of cyan as the selection color to help distinguish which mode is being used.

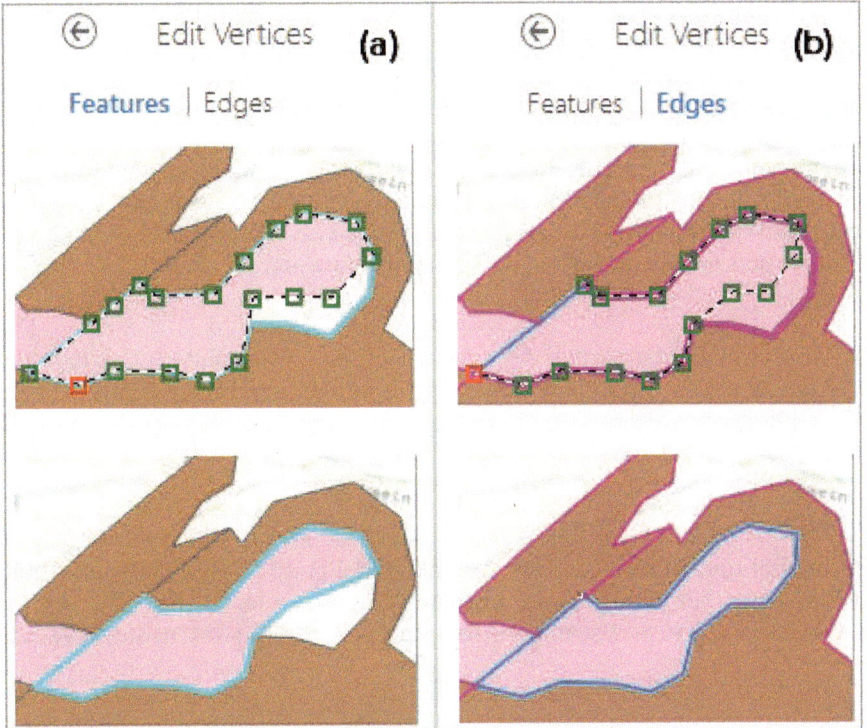

Fig. 8.15. (a) Modifying a feature versus (b) modifying an edge

The map topology allows one to simultaneously modify connected lines. In Figure 8.16, four street segments meet in a node (pink dot) that has been selected in the Move tool. When the node is moved, the streets are moved together to maintain the connection. By default, the shapes of the connected features are adjusted proportionally to create a smooth edit (Fig. 8.16a). This **stretch geometry** option may be turned off so that only the section from the common node to the previous vertex is modified (Fig. 8.16b).

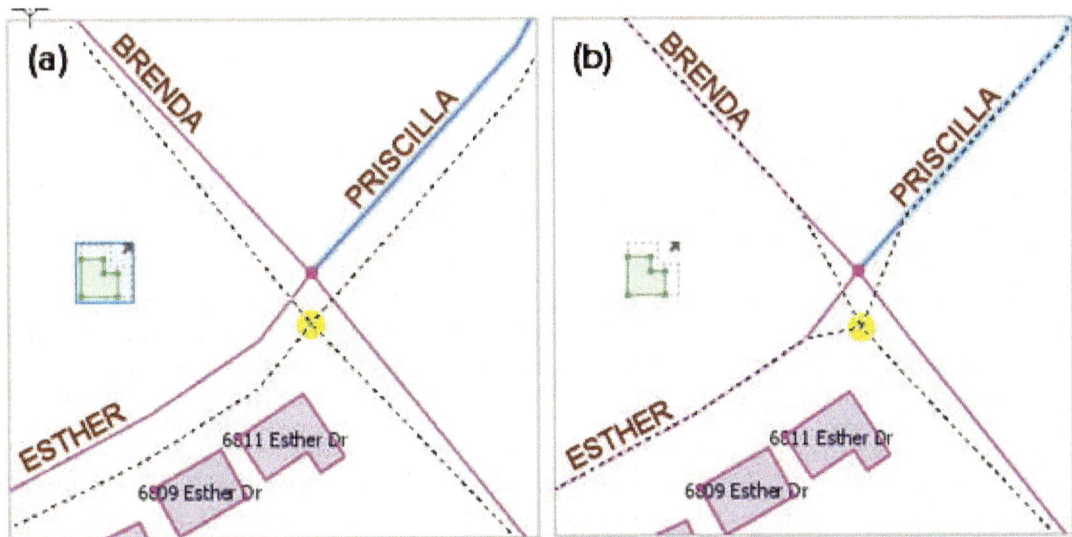

Fig. 8.16. Moving a node and its connected streets with: (a) Stretch Geometry option turned on, (b) Stretch Geometry option turned off
Source: City of Austin

The Stretch Geometry icon (shown in Fig. 8.16) appears on the Reposition toolbar (Fig. 8.11a) and can be used to toggle the behavior as needed.

Topological editing uses a **cluster tolerance** to help enforce snapping and coincident boundaries. When two vertices of features being edited fall closer together than the cluster tolerance, the vertices are made coincident. ArcGIS Pro will automatically calculate a default tolerance that is designed to preserve the accuracy of features. Although the user can set a custom value, it requires some knowledge and experience to choose an appropriate value. A poorly chosen cluster tolerance can distort features and reduce the spatial accuracy of a data set.

TIP: If you have an ArcGIS Standard or Advanced license, more sophisticated topological editing capabilities, including specifying topological rules and automatically locating errors, can be achieved using a **planar topology**. See the Help.

Editing annotation

Annotation is another type of feature class, like a line or a point feature class, except that it stores labels. Like any other feature class, it has an attribute table, and the table stores properties of the annotation, such as the text string, the font, the size, and the formatting (Fig. 8.17).

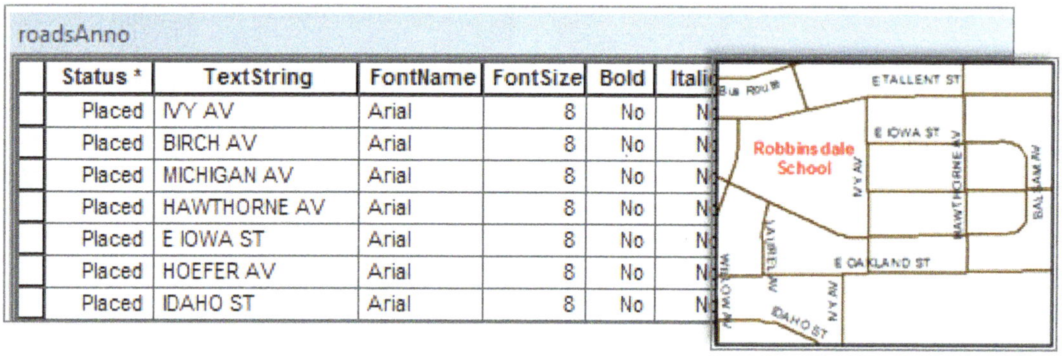

	Status *	TextString	FontName	FontSize	Bold	Itali
	Placed	IVY AV	Arial	8	No	N
	Placed	BIRCH AV	Arial	8	No	N
	Placed	MICHIGAN AV	Arial	8	No	N
	Placed	HAWTHORNE AV	Arial	8	No	N
	Placed	E IOWA ST	Arial	8	No	N
	Placed	HOEFER AV	Arial	8	No	N
	Placed	IDAHO ST	Arial	8	No	N

Fig. 8.17. Annotation feature class in a geodatabase
Source: Esri

Annotation may be created as standard annotation, where each text string is a standalone feature, or it may be created as feature-linked annotation, in which the text string is linked to the feature that it belongs to. With feature-linked annotation, deleting the feature deletes the annotation and vice versa. Feature-linked annotation requires an ArcGIS Standard Advanced license, but standard annotation can be edited with ArcGIS Basic.

Every piece of annotation can be created individually, or they can be created en masse by converting dynamic labels. The initial properties of the dynamic labels (font, size, color, etc.) will be used as the default style, so it is important to set them before the labels are converted. After the labels are created, editing can modify the location, style, and font attributes for each label.

Annotation is always given a **reference scale**, which is the scale at which the labels appear at their assigned size. The user specifies the reference scale when the annotation is created. If the reference scale is 1:24,000, for example, and the annotation size is 10 points, then it only appears as 10-pt. text when the map is shown at 1:24,000. If the user zooms in, the labels will increase in size; if the user zooms out, the labels will decrease in size. Before creating the annotation, users should view the map at the desired reference scale to confirm that the font and size chosen for the labels are acceptable.

TIP: A single annotation feature class may contain different classes of annotation, such as large labels for big cities and small labels for towns, each of which can be given its own default properties and reference scale. See the Help for more information.

Summary

▶ Editing creates and maintains features in a feature class. Care must be taken to ensure topological integrity when editing.

▶ Snapping ensures that features connect properly with each other by automatically matching the coordinates of features within a certain distance of each other (the snap tolerance). Features may be snapped to points, endpoints, vertices, or edges.

▶ When adding adjacent polygons, ensure that they have coincident boundaries by using snapping, the Autocomplete Polygon tool, or the Cut Polygons tool.

▶ Editing sessions are initiated and controlled using the Editor toolbar. You may edit multiple layers of data, but only one geodatabase or folder may be edited at a time.

▶ Edits are stored in the same coordinate system as the layer, regardless of the coordinate system of the data frame. However, editing is streamlined when the layer and frame coordinate systems match so that no reprojection need take place.

▶ Basic editing uses the Edit tool, feature templates, and the construction tools. The Edit tool selects features to edit.

▶ Feature templates define all the properties needed to create new features, including the feature class to place them in, the default construction tool to use, and the attributes with any default values in them.

▶ Multiple templates may be created for one feature class, enabling different attribute values to be given to different classes of features.

▶ The construction tools create preliminary figures called sketches. Finishing the sketch creates the new feature in the feature class.

▶ Different sketching tools provide a variety of functions for placing vertices. Context menus provide additional tools for specialized functions.

Important Terms

cluster tolerance	map topology	sketch	snap tolerance
coincident boundary	node	Sketch menu	stretch geometry
dangle	planar topology	snapping,	topology
dynamic constraints	reference scale	edge	Vertex menu
edge	schema changes	end	
feature template	segment constraints	point	
logical consistency	segments	vertex	

Chapter Review Questions

1. List the main types of topological errors.

2. Explain the difference between feature editing and schema changes.

3. Which type of snapping (end, vertex, or edge) would work best in the following situations? (a) digitizing streams, (b) digitizing parcels, (c) digitizing streets, (d) digitizing traffic lights at intersections, (e) digitizing stream gages on streams

4. Describe the three panels of the Contents pane that are helpful when editing.

5. If a data layer is in North America Equidistant Conic coordinates and the data frame is set to UTM coordinates, in which coordinate system will the edits be stored?

6. How are dangles prevented when digitizing new lines?

7. How does one prevent gaps or overlaps when digitizing new polygons?

8. Explain the purpose of a map topology.

9. Explain the difference between the terms *map scale, display scale,* and *reference scale.*

10. What two ways can annotation features be created?

Expanding your knowledge

Expand your knowledge by reading these sections of the ArcGIS Pro Help.

Help > Data > Edit geographic data (all sections)

Mastering the Skills

Teaching Tutorial

The following examples provide step-by-step instructions for doing basic tasks and solving basic problems in ArcGIS. The steps are highlighted with an arrow →; follow them carefully.

→ Start ArcGIS Pro and open the ClassProjects \EdwardsAquifer project.

→ Remember to save the edits and the project often.

Creating new feature classes

A project for Austin is studying an important aquifer, the Edwards Limestone. A USGS map, already scanned and georeferenced, provides the source for digitizing the geologic features. The first step includes creating empty feature classes to contain the new data.

1→ In the Geoprocessing pane, open the Create Feature Class tool (Fig. 8.18).

1→ Make sure that the Feature Class Location is set to the home geodatabase, EdwardsAquifer.

1→ For the Feature Class Name, enter *karst-features.*

1→ Set the Geometry Type to Point.

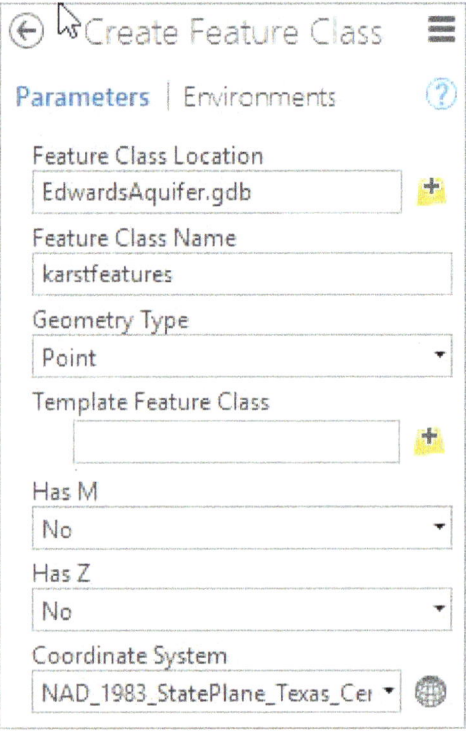

Fig. 8.18. Creating a feature class
Source: Esri

1→ Click the Coordinate System drop-down and set it to match the scanned map, EdwardsCrop.png.

1→ Run the tool. Keep the tool open.

1. What coordinate system is being used for this project? _____

The new feature class is automatically added to the map. Next, we create two more feature classes for faults and geology. Having left the tool open, we need only change the output feature class name and the geometry type.

2→ In the Geoprocessing pane, change the Feature Class Name to *faults* and set the feature geometry to Polyline. Run the tool.

2→ In the Geoprocessing pane, change the Feature Class Name to *geology* and set the feature geometry to Polygon. Run the tool.

These feature classes need fields to store the attributes, including coded domains for the fault type and the karst features. Domains are created for the geodatabase, so the field does not need to exist for a domain to be added. We will create the domains first and then associate them with fields.

3→ Open the Catalog view and expose the EdwardsAquifer geodatabase. Right-click it and choose Domains.

3→ Click in the first Domain Name box and enter *KarstType*.

3→ Enter a Description such as *Type of karst feature*.

3→ Set the Field Type to Text and let the Domain Type default to Coded Value Domain.

3→ Click in the first Code box and enter *spring* for both the Code and Description.

3→ Enter *cave/sink* for both in the next row. Don't hit the Tab key! Instead, select the previous code row to make sure the last value is accepted.

TIP: Pressing the Tab key starts a new code row, and the domain cannot be saved until values are entered. If you press Tab by mistake, enter fake values, then select the row and delete it.

4→ Click in the next empty box and enter a new Domain Name, *SymbolType*.

4→ Enter the description *Type of line used to display feature*.

4→ Set the Field Type to Text. The Domain Type will default to Coded Value Domain.

4→ Click in the first Code box and enter *solid* for both the code and description. Type *dashed* in the second row for both, and do NOT hit the Tab key.

4→ The domains should appear as in Figure 8.19. Click **Domains: Changes: Save** and close the Domains: EdwardsAquifer view.

Next add the fields to the faults and geology feature classes, taking care to assign the domain.

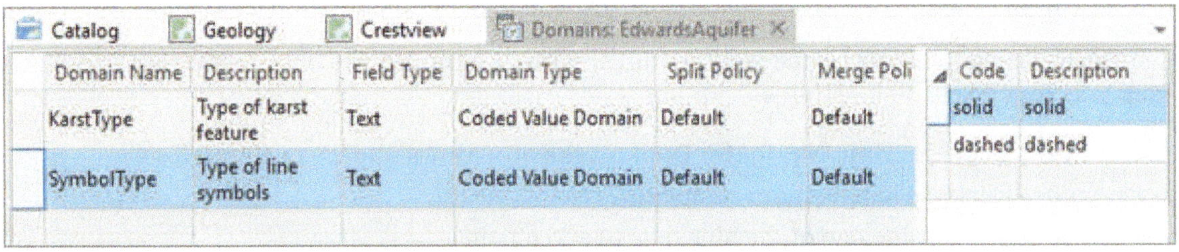

Catalog	Geology	Crestview	Domains: EdwardsAquifer X					
Domain Name	Description	Field Type	Domain Type	Split Policy	Merge Poli		Code	Description
KarstType	Type of karst feature	Text	Coded Value Domain	Default	Default		solid	solid
							dashed	dashed
SymbolType	Type of line symbols	Text	Coded Value Domain	Default	Default			

Fig. 8.19. Creating domains for the EdwardsAquifer geodatabase
Source: Esri

5→ Open the Geology map. In Contents, click on faults to select it and choose **Feature Layer: Data: Design: Fields** to open the Fields view.

5→ Click to add a new field and type *Linetype* for the Field Name.

5→ Click inside the Data Type box and choose Text. Set the Length to 10.

5→ Click in the Domain box and select the SymbolType domain.

5→ Click **Fields: Changes: Save** and close the Fields: faults view.

6→ In Contents, click on the karstfeatures layer and open its Fields view.

6→ Add a new field named *Type*, making it a Text field with a length of 15.

6→ Click in the Domain box and select the KarstType domain.

6→ Click **Fields: Changes: Save** and close the Fields: karstfeatures view.

7→ Open the Fields view for the geology layer and add a text field named *Unit* with a length of 6. No domain is needed for this field.

7→ Click **Fields: Changes: Save** and close the Fields: geology view.

2. Creating fields and domains is making what type of change?

Digitizing points and lines

We will digitize features by tracing them from the EdwardCrop.png image—sometimes called "heads up" digitizing. First, however, take a quick tour of the editing tools and settings.

8→ Click the **Edit** ribbon and examine the groups and tools.

8→ Click the **Options** button in the **Manage Edits** group and examine the Editing settings, but cancel without changing anything.

8→ The **Options** button in the **Features** group opens the Manage Templates pane. Examine and then close it.

8→ In Contents, click the icon to open the List By Editing panel, which controls whether layers can be edited. All layers should be editable (checked) except the raster image and basemap.

8→ In Contents, click the icon to open the List By Snapping panel, which controls snapping for each layer.

The Create Features pane contains a **feature template** for each editable feature class. These templates are used to construct new features, and they contain settings for managing how it occurs. It is convenient to keep this pane open throughout the editing session.

9→ Click **Edit: Features: Create** to open the Create Features pane.

9→ Click the karstfeatures template. Hover over the icons to see the name of each construction tool.

9→ Click the geology template. A larger selection of construction tools appears.

First we are going to digitize some karst features, shown on the map as blue dots for springs or black dots for cave or sink features.

10→ Click the **Map: Navigate: Bookmarks** drop-down and choose First Karst.

10→ Click the karstfeatures template.

10→ Click inside one of the blue dots to add a point.

10→ Add two more points.

Notice that the last point clicked is highlighted, indicating that it is selected.

10→ Click **Edit: Features: Delete** (or press the Delete key). It will delete the highlighted point, the last one entered.

10→ Click on the spring again to reenter the point.

Next, enter attributes for the springs.

11→ In Contents, right-click the karstfeatures layer and choose Attribute Table.

11→ Click inside the highlighted box for the Type field and select *spring*.

11→ Select *spring* for the other two points as well.

It gets tedious assigning the attribute each time. The template provides an easier option.

12→ In the Create Features pane, click the karstfeatures template and click the blue arrow to open the Active Template pane.

12→ Click in the Type box and select *spring*. Keep this pane open.

12→ Click another spring on the map to create a point. When the new feature appears in the table, the spring type is automatically assigned.

12→ Create several more points for springs.

What if we want to create points for cave/sink karst? We can update the attribute in the Active Template pane.

13→ In the Active Template pane, change the Type attribute to *cave/sink*.

13→ Click on several black points in the map to create cave/sink points.

13→ Continue entering points for the springs and cave/sink features that are visible in the current map area, changing the attribute field as necessary.

13→ Close the karstfeatures table.

13→ **Click Edit: Selection: Clear** to clear all selections.

13→ **Click Edit: Manage Edits: Save** to save the edits so far. Click Yes.

14→ In Contents, right-click the EdwardsCrop.png raster and open its properties. Examine the Metadata section.

3. What is the source scale of the EdwardsCrop raster? _____Use the Measure tool to find the diameter of the cave or spring point markers in ground units. _____

It is not easy to get the points in the exact center of the circles, and a circle occupies a large area on the ground. How does the source scale of this map affect the geometric accuracy of the springs? Would it be easy or difficult to find the springs using a GPS?

With templates, it is easy to switch between layers when digitizing.

15→ Choose **Map: Navigate: Bookmarks > First Faults**. Find the fault that travels diagonally across the map, a solid line with two dashed ends.

15→ In the Active Template pane, click the back arrow to return to the main Create Features pane.

15→ Click the faults template and click the blue arrow.

15→ Select *dashed* from the Linetype box and leave the pane open.

15→ Click to place a vertex at the lower end of the lower dashed section.

15→ Click another point or two on the line, staying in the dashed portion.

15→ Double-click to add a final vertex at the top of the dashed section and complete the sketch. A highlighted line feature appears.

TIP: If you are not satisfied with the fault (while it is still highlighted), chose **Edit: Features Delete** (or click the Delete key) to get rid of it and try again. If it is no longer highlighted, use the **Edit: Selection: Select** tool to select it.

16→ In the still-open Active Template pane, set the Linetype to *solid*.

16→ Start entering vertices for the solid section and double-click at the end.

16→ In the Active Template pane, set the Linetype back to *dashed* and create the dashed section at the upper end.

The digitized faults are hard to see because the thin symbol does not show up well. Let's use the Linetype field to symbolize the faults more effectively.

17→ Symbolize the faults using a unique values map based on the Linetype field. Use a 3-pt. thick green solid and dashed line for the two types.

17→ In the Active Template pane, click the back arrow, if needed, to return to the main pane.

TIP: The faults template in the Create Features pane should update to show the new line symbols as separate templates.

18→ Return to the First Karst bookmark.

18→ Digitize several more faults of each type. Use the solid template for solid faults and the dashed template for dashed faults.

18→ Click **Edit: Manage Edits: Save**. Also save the project.

TIP: Saving the project does NOT save the edits, and vice versa.

Working with snapping

Snapping ensures that features connect to each other and helps prevent topological errors. In Figure 8.20, assume that the solid faults will be digitized before the dashed ones. At the green arrows, the end of a new fault should be snapped to the end of an existing one. At the purple arrows, the dashed faults should be snapped to the edge of the solid fault.

19→ Click on the **Edit: Snapping** drop-down to open the Snapping menu.

19→ Make sure Snapping is on. Hover over each icon to see the types of snapping and make sure that only end snapping is on (Fig. 8.21).

19→ Move the cursor near the end of a fault that is already digitized and watch a Snap Tip appear.

19→ Digitize several of the solid faults in this view, being sure to connect them from end to end. Include the solid faults that pass through the intersections marked by purple arrows in Figure 8.20.

The Snap Tips show what the next vertex will snap to. "Faults: Endpoint" appears at the ends of the existing faults.

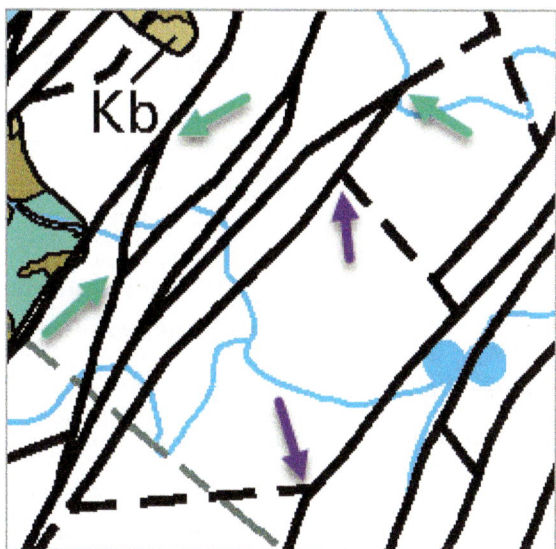

Fig. 8.20. Snapping for faults
Source: USGS

20→ Choose **Edit: Snapping** and turn on edge snapping.

20→ Digitize several of the short dashed cross-faults, making sure they snap to the edge of the existing solid faults.

TIP: One can also access the Snapping menu by hovering over the snap icon in the lower left corner of the map view. Snapping can be turned on or off even during a sketch.

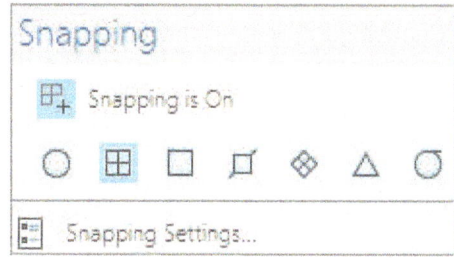

Fig. 8.21. Using only end snapping
Source: Esri

20→ Continue digitizing faults in this view. Whenever a fault gets close to the end or edge of an existing fault, snap them to ensure that no dangles are created.

20→ When you are confident that you understand how snapping works, stop digitizing the faults.

Zooming and panning can be performed in the middle of a sketch.

21→ Zoom to the Long Fault bookmark.

21→ Look above the "Kprd" text in the blue polygon and find the fault that starts dashed at the upper end and changes to solid at the lower end (A in Fig. 8.22).

21→ Click the dashed template and digitize the dashed upper end, moving from NE to SW. Stop where the dash ends, and double-click to finish the sketch.

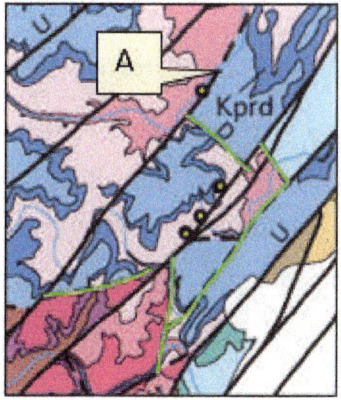

Fig. 8.22. Digitizing faults
Source: USGS

22→ Click the solid template and start the next solid section of the fault.

22→ When you run against the edge of the map view, don't finish the sketch yet.

22→ Click the mouse wheel and drag to pan to the rest of the fault without disturbing the sketch.

22→ Roll the mouse wheel to zoom in or out without interrupting the sketch.

22→ Continue digitizing to the end of the fault at the map edge. Finish the sketch.

23→ Click on the edge of the image to start the next fault below the first.

23→ Alternate between sketching and panning to digitize the entire solid part of this fault, ending where it appears to split into two segments by the caves.

23→ Start another fault on the end, taking the upper branch and digitizing up to where the fault splits again (near the text "Buda").

23→ Choose **Edit: Manage Edits: Save**. Clear any selected features.

Using the sketch context menus

Let's look at ways in which the Sketch menu enhances editing, such as correcting mistakes.

24→ Zoom to the Practice bookmark.

24→ Click on the faults solid template and enter some vertices along a random line. Notice that the last vertex is red and the other vertices are green.

24→ Right-click on top of the last red vertex to open the Vertex menu.

24→ Choose Finish.

Choosing Finish does the same thing as double-clicking at the end of the sketch.

25→ Start another line and add some vertices.

25→ Right-click the last red vertex to open the Vertex menu, and choose Delete Vertex.

The last vertex was deleted, and the second-to-last one is now red. This menu can delete any of the previous vertices.

25→ Right-click on any of the green vertices and choose Delete Vertex.

25→ Right-click another vertex and examine the menu again.

25→ To close the menu without choosing anything, click anywhere not on the map, such as the blank gray area above the ribbon names.

Notice that all this time we are still in the middle of creating this sketch.

26→ Right-click on the sketch *between* two vertices.

26→ Choose Add Vertex from the Vertex menu.

A straight line segment is the default between vertices, but curves can also be created using the Construction toolbar.

26→ Find the Construction toolbar near the bottom of the map window and click the drop-down arrow on the Curves tool to open a menu with a choice of segments.

26→ Choose one of the curves and enter additional vertices to produce curved segments.

26→ Choose a different curve tool and experiment. Try them all.

26→ When done experimenting, finish the sketch.

The Trace tool is useful for following an existing feature, either exactly or using an offset.

27→ Click the Trace tool on the Construction toolbar.

27→ Click the end of an existing line feature to start tracing, and move the mouse along the line. A dashed line indicates the new feature being created.

27→ If the cursor strays too far, the trace line may hop to a different feature. Cancel the sketch using the Construction toolbar button and try again.

27→ Finish the sketch at the end of the feature being traced.

28→ Start tracing another line by clicking on its end.

28→ Type O on the keyboard to open the Options menu.

28→ Check the box to *Trace with offset*, set the offset to 500 ft, and click OK. Trace the line.

28→ To place the trace on the other side of the line, type O again and enter a negative value for the offset, −500.

Remember, clicking ON the sketch gives the Vertex menu. A different menu appears when clicking OFF the sketch. Some of the entries, such as Finish and Cancel, are found in both menus. The sketch can be cancelled and restarted at any time, if needed.

29→ On the Construction toolbar, switch back to the Line tool.

29→ Start a new sketch with several vertices.

29→ Right-click off the sketch to open the Sketch menu.

29→ Choose the first entry, Direction.

29→ Type N45E in the Direction box and press Enter.

As the mouse moves, the cursor is constrained along a diagonal line, forcing the next segment to follow the specified angle. This editing option is called a **segment constraint**.

29→ Click along the line to enter a vertex. The new segment follows the specified angle.

29→ Double-click to finish the sketch.

TIP: Search the Help for "direction formats" and read "Direction formats for editing" to learn more about the different styles, options, and units for specifying directions.

Next, let's create a segment of a specific length.

30→ Start a sketch and add several vertices.

30→ Right-click off the sketch and choose Distance from the Sketch menu.

30→ Type 5000 into the box, accept the default units (ft), and press Enter.

30→ Move the cursor around, noting how the last segment remains the same length.

30→ Click somewhere to enter the vertex.

The Sketch menu can combine length and direction constraints to enter a line with a particular bearing and direction. We will use this feature to enter the next segment of the same sketch.

31→ Right-click off the sketch and choose Direction/Distance.

31→ Type N45E for the direction and 5000 for the length. Click Enter when both boxes are filled.

31→ Note that the vertex was added automatically. Add a few more vertices and double-click to finish the sketch.

Digitizing polygons

Now we will learn how to create polygons, but first we should delete these practice lines.

32→ Choose **Edit: Selection: Select** and draw a box that touches all the practice lines.

32→ Choose **Edit: Features: Delete**.

33→ In the main Create Features pane, click on the geology template. The Polygon construction tool is the default.

33→ Start a polygon by adding several vertices. Notice how the shape remains closed, with the last vertex always connected to the first.

33→ Double-click the last vertex to finish the sketch.

33→ Try adding several more polygons that don't touch each other, for practice.

We will create some adjacent polygons. It is important to use the Autocomplete Polygon tool so that adjacent polygons have a coincident boundary.

34→ Turn on Vertex snapping using the Edit ribbon or the pop-up Snapping button near the map scale readout.

34→ In the Create Features pane, change the construction tool to Autocomplete Polygon.

34→ Click INSIDE an existing polygon to start the sketch (Fig. 8.23a).

34→ Snap to a vertex of the existing polygon on the way out, for neater polygons.

34→ Add more vertices to define the new polygon.

34→ Add the last vertex INSIDE the existing polygon and double-click to finish.

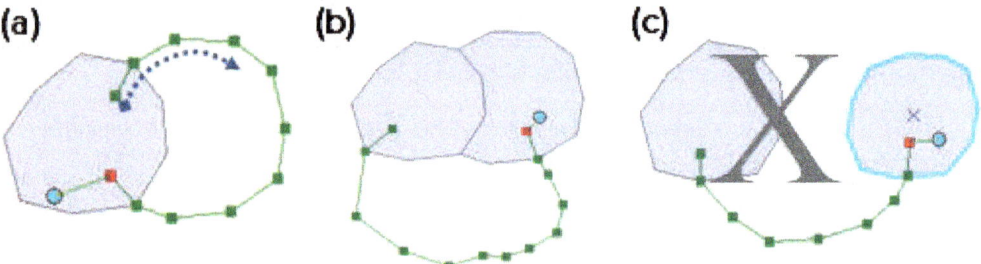

Fig. 8.23. Adding adjacent polygons: (a) digitize only the new part; (b) the new polygon can span multiple existing polygons (c) unless the region cannot be closed.

TIP: The Autocomplete tool will not construct a polygon unless the space defined by the existing polygon(s) and the sketch is completely enclosed (Fig. 8.23b vs. 8.23c).

35→ Begin a new sketch inside one of the polygons, digitize the boundary of a new polygon, and end inside the other polygon (Fig. 8.23b). Finish the sketch.

Splitting provides another way to create adjacent polygons with coincident boundaries.

36→ Click **Edit: Selection: Select** and select one of the polygons if needed (Fig. 8.24a).

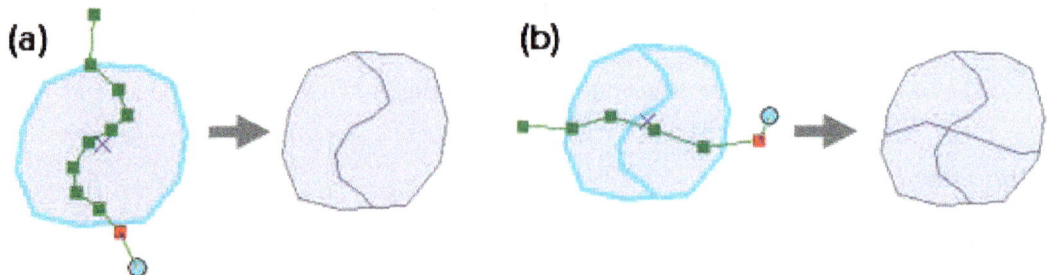

Fig. 8.24. Cutting (a) one polygon or (b) multiple polygons

Split

36→ In the **Edit: Tools** group, scroll or click the scroll drop-down to select the **Split** tool.

36→ Start a sketch OUTSIDE the selected polygon, and add a line that goes through it.

36→ Add the last vertex OUTSIDE the polygon, and double-click to finish the sketch.

Notice that selecting the Split tool opened a Modify Features pane with additional tools and options. We will learn more about this pane later. Splitting works on multiple polygons, too.

36→ Check that the two polygons are still selected. If not, select them (Fig. 8.24b).

36→ Sketch another line through both and finish the sketch.

36→ Close the Modify Features pane.

It's time to do some real polygon editing. First, let's clean up.

37→ Use the Edit ribbon to select and delete the practice polygons.

37→ Zoom to the Polygons AB bookmark.

Now we will digitize some geology polygons on the SW end of the outcrop area. Treat the faults as geological contacts that define boundaries of the polygons.

38→ On the Create Features pane, click the geology template and click the blue arrow. Set the construction tool to Polygon and type *Kkd* in the Unit box.

38→ Start digitizing polygon A, shown in Figure 8.25.

38→ Use the Snapping pop-up button to turn snapping off for now—it tends to interfere with adjacent vertices here.

38→ Take reasonable care digitizing, but the shape doesn't have to be perfect for this exercise. Double-click to finish.

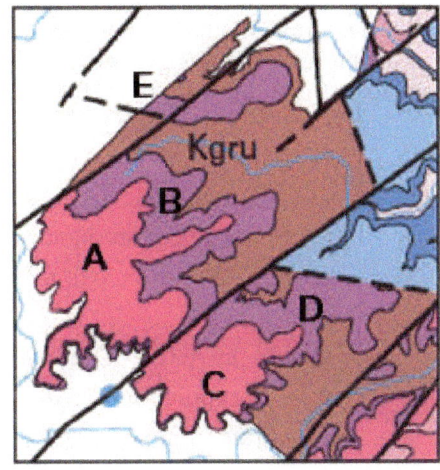

Fig. 8.25. The first polygons
Source: USGS

TIP: For the best results, place vertices at sharp corners. Use fewer vertices on straight lines and more on curves.

It can be challenging to choose a good polygon symbol for editing that does not obscure the background image but allows the completed polygons to be easily seen. A transparent gray symbol with a distinctive outline works well.

39→ Clear the selected polygon.

39→ Open the Symbology pane for geology and set up a Gray 50% color symbol with a 2-pt. bright green outline. Click the Color drop-down for the gray fill color, select Color properties, and slide the Transparency to 50%. Click OK.

39→ Apply the symbol and examine it (Fig. 8.26). Modify the outline color, if needed, before closing the Symbology pane.

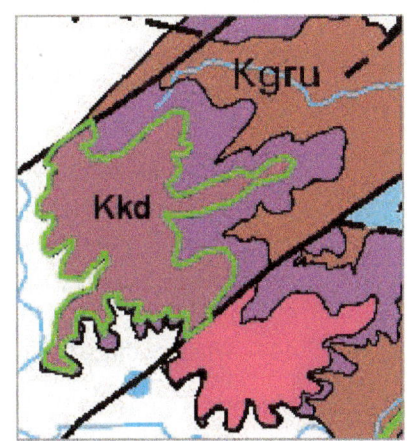

Fig. 8.26. The first polygon, with new symbol and label
Source: USGS

39→ Open the **Labeling** ribbon for the geology layer. Change the **Label Field** to Unit and the font to Arial 12-pt. Bold. Turn on the labels (Fig. 8.26).

The purple polygon B is adjacent to polygon A. To ensure that they have a common boundary, use the Autocomplete Polygon tool to create polygon B.

40→ Turn on snapping and make sure that vertex and edge snapping are on.

40→ In the Active Template pane for geology, click the Autocomplete Polygon construction tool and enter *Kkbn* for the Unit field.

40→ Start digitizing by snapping to the upper-right vertex of polygon A.

40→ Click along the fault until the purple area ends; then follow the geological contact between the purple and brown polygons to the lower fault.

40→ Click along the lower fault. End by snapping to the vertex of polygon A. Double-click to finish the polygon.

Notice that the Autocomplete Polygon tool used the boundary from polygon A, so the two polygons share an identical boundary that did not need digitizing a second time. Once selected, Autocomplete Polygon will remain the active tool, so keep using it until directed otherwise. Next we digitize the pink polygon C below polygon A.

41→ In the Active Template pane, enter *Kkd* for the Unit.

41→ Start at the lower-left corner of the Kkd polygon A and snap to the vertex.

41→ Digitize counterclockwise around pink polygon C.

41→ At the end, snap to an edge or a vertex of polygon B. Double-click to finish.

42→ In the Active Template pane, enter *Kkbn* for the Unit.

42→ Start polygon D in the lower-middle part of polygon C, snapping to its vertex.

42→ Start trying to digitize this narrow slice of polygon D.

Polygon D is close to polygon C in places, and inappropriate snapping may occur. Two things can help: zooming in and reducing the snap tolerance.

43→ Cancel the sketch using the Construction toolbar.

43→ Right-click the mouse and drag to smoothly zoom in until the narrow portion of polygon D fills the screen.

43→ Choose **Edit: Snapping: Snapping Settings**. Change the XY Tolerance from 10 pixels to 5 pixels and click OK.

44→ Start digitizing polygon D again, snapping to the vertex of polygon C as before.

44→ Don't let the sketch cross into polygon C—keep the unit narrow and thin along the existing polygon.

44→ Click the mouse wheel and drag to pan to new areas as needed, or zoom out once past the tricky part.

44→ When ending, be sure to snap to an edge or a vertex of polygon B̲.

Cutting polygons is sometimes easier than using Autocomplete.

45→ Choose the Polygon Slice bookmark and examine the long, thin slice above the fault (below E in Fig. 8.25).

45→ Enter *Kgru* for the Unit field.

45→ Snap to the upper-right corner of polygon B and enter vertices along the fault, then up and around to enclose the entire slice. End by snapping to the edge or vertex of polygon A.

45→ While the slice is highlighted, choose **Edit: Tools: Split**.

45→ Digitize across the slice along the fault, starting and ending outside the slice. The slice is cut in two.

45→ Digitize across the still-selected slice again, following the geological contact between the purple and brown units.

We need to change the Unit attribute of the middle purple slice from Kgru to Kkbn. Instead of opening the table, we will use the Attributes pane.

46→ Choose **Edit: Selection: Attributes** to open the Attributes pane (Fig. 8.27).

46→ In the Attributes pane, click *Change the selection* and click the center purple slice.

46→ Click in the Unit field in the Attributes pane and change Kgru to *Kkbn*. Click Enter.

TIP: It is convenient to arrange the Create Features, Attributes, and Modify Features panes into one window and switch between them as needed (red box in Fig. 8.27). Do it now.

47→ Switch to the Create Features: Active Template pane and enter *Kkbn* for the Unit.

47→ Continue using the Autocomplete tool to digitize the purple polygon below the slice, starting and ending along the slice edge.

48→ Pan or Zoom, if needed, to see the entire large Kgru brown polygon.

48→ Enter *Kgru* in the Unit field and digitize the polygon.

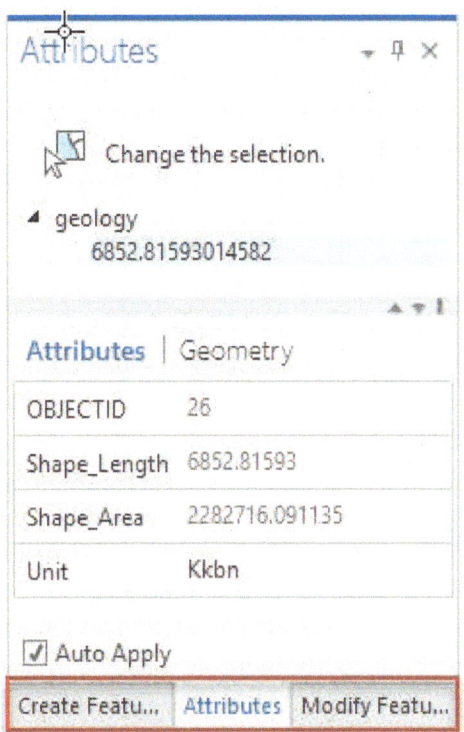

Fig. 8.27. Arranging the editing panes
Source: Esri

The new polygon has completely enclosed the purple Kkbn polygon, and now they overlap each other, a topology error that needs fixing.

49→ Click **Edit: Selection: Select** and click the purple Kkbn polygon.

49→ Click **Edit: Features: Modify** (another way to open the Modify Features pane).

49→ Choose the Clip tool. Choose the Discard option and click the Clip button.

50→ Switch back to the Create Features pane and set the geology unit to *Kkbn*.

50→ Digitize the small purple polygon that remains, snapping it to the polygon above it.

50→ Set the Unit to *Kgru* and close off the last section of brown polygon along the dashed fault.

50→ Zoom to the new polygons (Fig. 8.28).

50→ Save the edits and save the project.

When digitizing adjacent polygons, the order is important. New polygons must enclose a space completely using a single line. This exercise made them in a suitable order, but users must learn to look ahead and plan before committing to digitizing.

Adding buildings

Now we will investigate some additional tools for city editing using a neighborhood named Crestview in northern Austin. We will create a buildings feature class and store it with the other neighborhood data in a geodatabase named Crestview, rather than storing it in the EdwardsAquifer project geodatabase.

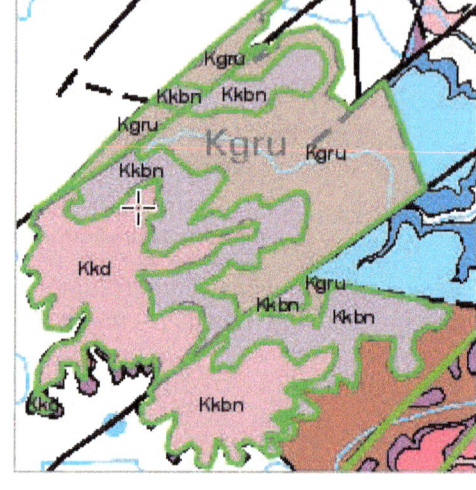

Fig. 8.28. The final polygons
Source: USGS

51→ Switch to the Crestview map view in the project.

51→ In the Geoprocessing pane, open the Create Feature Class tool.

51→ Set the Feature Class Location to the Folders\Austin\Crestview geodatabase.

51→ Enter *buildings* for the Feature Class Name.

51→ Leave the GeometryType set to Polygon.

51→ Click the coordinate system drop-down and set it to match the Streets layer. Run.

52→ In Contents, click the buildings layer to highlight it and choose **Feature Layer: Data: Fields** to open the Fields view.

52→ Click to add a new field. Name it *Bldgtype*. Make it a text field with a length of 5 characters.

52→ Click in the Domain box and select <Add New Domain>.

52→ Name the domain *ParcelType* and enter a suitable description. Create two codes: COM/Commercial and RES/Residential.

52→ Save the domain and close the Domains view.

52→ Click a couple times in the Bldtype Domain box until it becomes a drop-down and the new ParcelType domain is visible. Select it.

52→ In the Fields view, click **Fields: Changes: Save** before adding more fields.

Next we add three more fields for the address, city, and state. We will also define default values for city and state, since all of the parcels are in Austin, TX.

53→ Add another text field called *Address* with a length of 50 characters.

53→ Add another text field called *City* with 25 characters. For this field, click in the Default box and type *Austin*.

53→ Add another text field called *State* with a length of 2 characters. Set the Default value to *TX*.

53→ The fields added to the table schema should appear as in Figure 8.29. Save the changes and close the Fields view.

54→ Zoom to the First Commercial bookmark.

Field Name	Alias	Data Type	✓	Number Format	Domain	Default	Length
OBJECTID	OBJECTIC	Object ID		Numeric			
Bldgtype	Bldgtype	Text			ParcelType		5
Address	Address	Text					50
City	City	Text				Austin	25
State	State	Text				TX	2

Fig. 8.29. The fields added to the buildings table schema (other fields not shown)
Source: Esri

54→ Turn on the Addresspts layer.

54→ In the Create Features pane, click the buildings template and click the blue arrow.

54→ Select *Commercial* for the Bldgtype attribute.

54→ Type *200 W HUNTLAND DR* in the Address box.

The City and State fields already have values, Austin and TX, because defaults were entered for the fields (Fig. 8.30).

54→ Click the Polygon tool and digitize the first large building on the left.

The next building on the right has only right-angle corners, allowing the use of a special construction tool. The address for this building is the same as for the first.

 55→ In the Active Template pane, click the Right Angle Line tool.

55→ Carefully digitize one of the long edges of the building.

55→ Continue entering the other corners. The tool constrains the segments to create right angles.

55→ For the last vertex, right-click on it and choose Square and Finish.

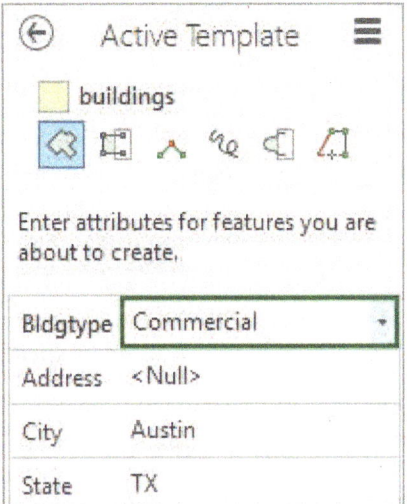

Fig. 8.30. Buildings template
Source: Esri

TIP: For the best results, use the longest, clearest side of the building for the first edge.

55→ Enter the next building's address and digitize it.

55→ Change the address and digitize the next building.

56→ Zoom to the First Residential bookmark.

56→ In the Active Template pane, change the Bldgtype to *Residential*.

56→ Type the address for the first house on the right and digitize the house.

56→ Digitize four or five more houses on this side of the street for practice.

TIP: If you forget to enter the address before digitizing a house, you can switch to the Attributes pane, select the house, and correct the address.

It would be helpful to have two templates, one for residential and one for commercial buildings. The easiest way to create multiple templates is to create a unique values map first.

57→ In the Active Template pane, click the back arrow.

57→ Open the Symbology pane for buildings and create a unique values map based on the Bldgtype field.

57→ In the Create Features pane, the buildings template updates to show both types.

 TIP: If the templates fail to be created or updated, they can be done manually using the Manage Templates button in the Create Features pane.

Using the Attributes pane

Let's try a different workflow to digitize the buildings first and add the address information later, using the Attributes window, to avoid constantly switching between digitizing and typing.

58→ In the Create Features pane, click the Residential template and click the blue arrow.

58→ Select the Right-Angle tool and digitize several more houses in this block.

Now to assign addresses to the buildings. The Attributes pane is useful here (Fig. 8.31).

59→ Switch to the Attributes pane if it is already open. If not, choose **Edit: Selection: Attributes** to open it.

59→ Click the button to *Change the selection*.

59→ Draw a rectangle around the houses just added to select them all.

Examine the top panel of the Attributes pane. It includes selected features from all layers, not just the buildings. Moreover, the buildings are listed by nonsensical numbers. Fortunately, the Attributes pane has context menus to help set up the layers to work better.

59→ In the Attributes pane, right-click the entry for buildings and examine the context menu; then choose Layer Properties.

59→ In the Layer Properties window, examine the Display settings. Set the Display field to Address and click OK.

The Display field setting manages the default field used for labeling (although null values will be listed using the feature ID). The layer name entry in the Attributes pane can be used to modify all of the selected records at once. We will preset the street names to save a little typing.

TIP: Highlight the feature class name at the top of the Attributes pane to make the field edits apply to all of the selected records in that layer.

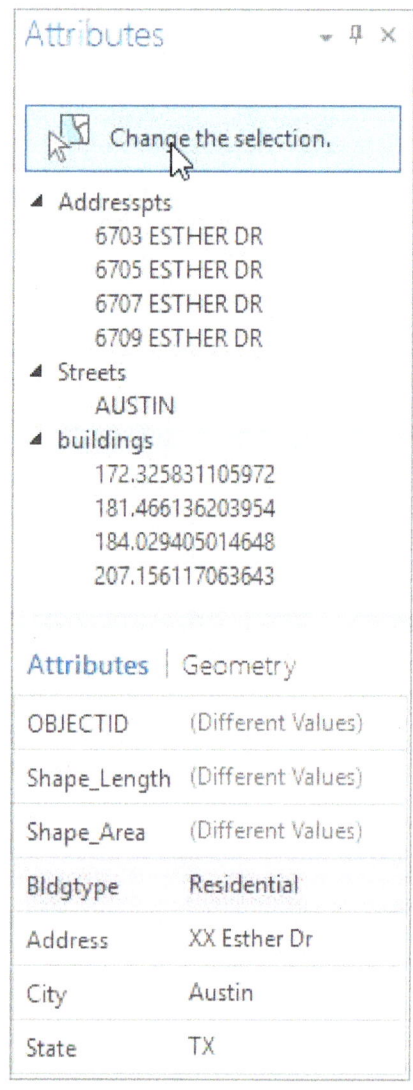

Fig. 8.31. Assigning addresses in the Attributes pane

Source: Esri

60→ Click the buildings layer name in the Attributes pane to highlight it.

60→ Type *XX ESTHER DR* into the Address box and click Tab to enter it.

Now the street name is set, and we only need to edit the house numbers, one at a time.

60→ In the Attributes pane, click on the first buildings feature in the list. The building it represents will flash on the screen.

60→ Click several times, if need be, to see which house number it needs.

60→ Edit the XX in the Address field and replace it with the house number.

60→ Click the next entry and enter its house number. Repeat until all of the addresses are correctly entered.

60→ Save the edits and save the project.

Creating and editing annotation (optional)

Unlike dynamic labels, annotation feature classes provide labels that can be individually edited and reused for many maps. We will create annotation for the Crestview Streets layer.

61→ In Contents, use Ctrl-click on a check box to turn off all layers except Streets.

61→ Make sure that the street name labels are turned on.

61→ Set the map scale to exactly 1:10,000.

This setting will be the **reference scale**, the scale at which the annotation appears at the assigned size. Make sure that the labels appear as desired at this scale before creating the annotation.

61→ In the **Feature Layer: Labeling: Text Symbol** group, change the font size to 8-pt. and the font style to Regular. Confirm that the labels appear acceptable at this scale.

62→ In the Geoprocessing pane, open the Convert Labels to Annotation tool (Fig. 8.32).

62→ Set the Input Map to Crestview.

62→ Set the Conversion Scale to *10000*.

62→ For the Output Geodatabase, click the Browse button and select the Folders\Austin\Crestview geodatabase.

62→ Check the box to *Convert unplaced labels to unplaced annotation*.

62→ Click Run. The annotation will be created and the dynamic labels will be turned off.

First, let's examine the effect of the reference scale.

63→ Zoom in and watch the annotation get larger. Zoom out and watch it get smaller.

63→ Return to 1:10,000. Zoom in a little more to work with the annotation comfortably.

Next we examine how to modify the size, angle, and location of the annotation features.

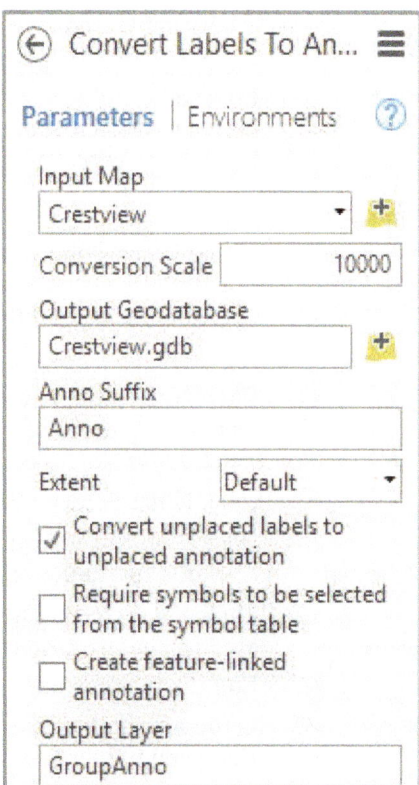

Fig. 8.32. Converting the street labels to annotation

Source: Esri

63→ Choose **Edit: Features: Modify** to open the Modify Features pane.

63→ Click the Annotation tool.

63→ Select one of the annotation features.

63→ An editing template for StreetsAnno appears in the Create Features window.

When annotation is selected, the shape of the cursor changes as the mouse moves over the annotation, indicating the action that will occur.

64→ Place the cursor near one of the square handles on the side or corner until it changes to a two-ended arrow. Click and drag to enlarge or shrink the text.

64→ Click the Finish button on the Construction toolbar to keep the changed size.

Of course, changing the size is probably the least useful thing to do here, since annotation looks better when it matches. You can quickly undo any inadvertent changes.

64→ Click the Undo button on the Quick Access toolbar at the top of the window.

64→ Select the annotation again, if necessary.

64→ Move the cursor until it turns to a four-ended arrow cross. Click and drag to move the annotation to a new location.

64→ Move the cursor to a corner of the annotation until it changes to a curved arrow.

64→ Click and drag to rotate the annotation to a new angle.

64→ Place the cursor near the middle until it shows as an "I" insertion symbol. Click and edit the text to a different street name.

64→ To discard the edits, click the Cancel button on the Construction toolbar.

65→ Find a piece of annotation that could be better placed. Move and/or rotate it to a better position and accept the edits.

65→ Find several more and edit them until you are comfortable with this process.

Unplaced labels are put in the table with the designation "Unplaced" in a Status field. They can be placed manually, but first they must be turned on.

66→ In Contents, expand the GroupAnno layer to see the StreetsAnno feature class.

66→ Right-click the StreetsAnno layer and choose Symbology.

66→ Check the box to *Draw unplaced annotation*, check the box for *Using this color*, and select a distinctive color.

67→ Click **Edit: Selection: Attributes** to open the Attributes pane.

67→ Dock the Attributes pane below the Modify Features pane so both can be seen. Click the Attributes tab in the pane.

67→ Find an unplaced annotation feature. Click the Make/Change a selection on the Modify Features: Annotation pane (not the Attributes pane) and select it.

67→ Edit the location as needed. Click Finish on the Construction toolbar.

67→ In the Attributes pane, click in the Status field and change the drop-down from Unplaced to Placed. Click Tab to enter the change.

TIP: If you see a 0 or 1 in the Status field, save all edits and the project, close Pro, and restart.

If an unplaced label just won't fit, it can be deleted.

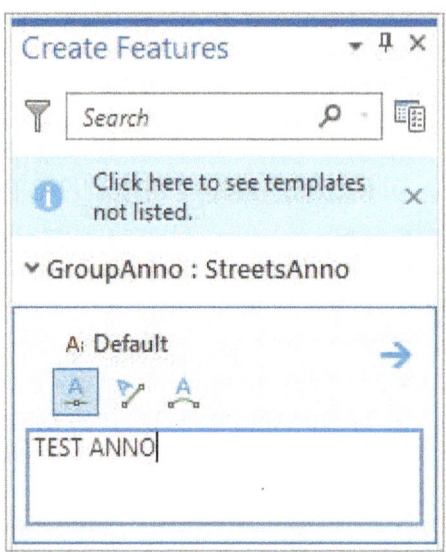

68→ Find the cluster of ANDERSON unplaced labels to the east and select one.

68→ Click **Edit: Features: Delete** to delete it.

If needed, annotations can also be created one at a time.

69→ Use **Edit: Features: Create** to open the Create Features pane.

 69→ Click the Default StreetsAnno template and click the Horizontal Annotation construction tool.

69→ Type a street name in the text box, *TEST ANNO* (Fig. 8.33).

69→ Move the cursor to the desired location and click to place the annotation.

69→ If the position is off, click **Edit: Features: Delete** while the annotation is still selected, and try again.

Fig. 8.33. Creating new annotation
Source: Esri

69→ Click in the text entry box and type a second name, *MORE ANNO*.

69→ Click on the map to place the MORE ANNO annotation.

 70→ Click the Straight Annotation construction tool.

70→ Click two points along a street to define the line and angle. As the second point is entered, the annotation will be placed.

 70→ Click the Curved Annotation construction tool.

70→ Click several points along a curved road. Double-click to end the curve and place the annotation.

Switch tools to stop placing new annotations.

71→ Switch to the Modify Features pane and choose *Change the selection*.

71→ Select some of the new annotation and change its location or angle.

This is the end of the tutorial.

→ Save the edits and save the project. Close Pro.

Practice Exercises

1. Digitize all of the remaining caves/sinks and springs in the Edwards outcrop area map, placing them in the karstfeatures feature class. Remember to assign the attributes.

2. The Travis-Hays County line runs NW-SE across the Edwards outcrop area map, cutting it roughly in half. Digitize all of the faults SW of the county line, placing them in the faults feature class. Remember to assign the attributes.

3. Create a map showing only the Edwards (State Map) layer outline, the karst features, and the faults. Assign different symbols to cave/sink and spring features and to the dashed and solid fault lines. **Capture** the map.

4. Digitize the geological units shown inside the yellow line in Figure 8.34, putting them in the same feature class as the polygons entered in the tutorial. The area is just NE of the polygons digitized in the tutorial. Assign each polygon the appropriate attribute. Be sure to use the Autocomplete Polygon tool or the Cut Polygons tool, and plan ahead.

5. Create a unique values map of the geology polygons based on the Unit field. **Capture** a map showing only these polygons.

6. Digitize the buildings in the full block north of Esther Dr. Also include the houses on that section of Isabelle Dr., as well as the very large building behind them. Finish entering the addresses for all of the buildings.

7. Create a map showing all of the digitized buildings, labeled with their addresses and symbolized by type. **Capture** the map.

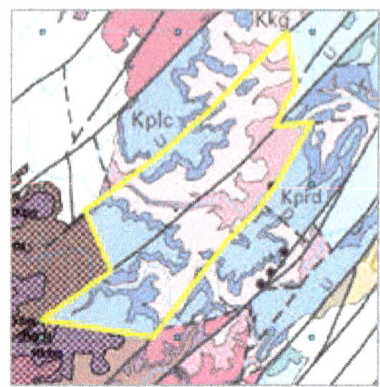

Fig. 8.34. Digitize this area.
Source: USGS

Chapter 9. Queries

Objectives

- ▶ Understanding queries and how they are used
- ▶ Selecting features based on attributes using SQL and Boolean operators
- ▶ Selecting features based on their spatial location with respect to other features
- ▶ Applying selection options, including the selectable layers and the selection type

Mastering the Concepts

GIS Concepts

A **query** is a fundamental tool in GIS (and other types of databases), used in nearly every analysis. Queries serve a critical function in data exploration or in picking one's way through a data set, looking for patterns. They can also be used to extract certain features before performing another operation. This chapter introduces three types of queries: ones for choosing features from a screen or objects from a table, ones based on values in a table, and ones based on spatial relationships between layers.

What queries do

Queries are used to extract a set of features from a map or objects from a table, based on criteria set by the user. The selected features or objects can then become the focus of interest for a subsequent action. Figure 9.1 shows some common types of queries as described below; in each case the selected features are shown in the dark pink color.

- ▶ *Selecting features of interest.* Queries can be used to search a table and to find which features meet certain criteria. Which water wells have iron values above 0.5 mg/L (Fig. 9.1a)? How many for-sale houses fall into my price range? Combined with the statistics function, this tool becomes powerful. What is the average cost of three-bedroom homes in this town?

- ▶ *Exploring spatial patterns.* Creating a map from selected features and examining their spatial distribution can be illuminating. What is the distribution of the mines containing gold (Fig. 9.1b)? Are they widely spaced or clustered together? Do they appear associated with certain faults or fault directions?

- ▶ *Isolating features for more analysis.* Select the high-population-density counties (> 2000 people/sq. mi) and compare the racial diversity index with that of the lower-population-density counties to see whether a difference exists between the two groups (Fig. 9.1c).

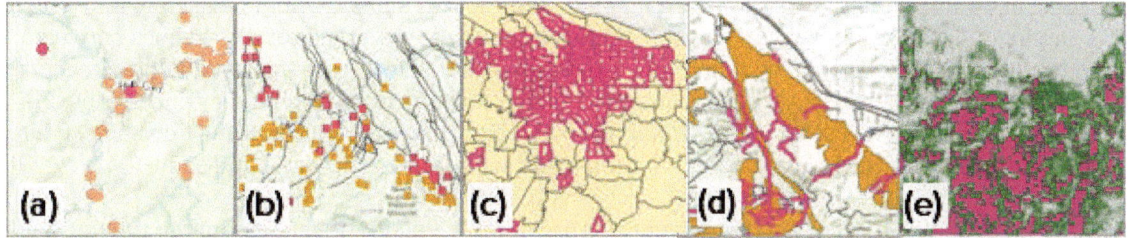

Fig. 9.1. Examples of queries (selected features in pink): (a) high-iron wells; (b) gold mines; (c) high-population-density tracts; (d) roads on limestone; and (e) tree canopy greater than 80%
Source: (a-d): Esri; (a,b,d): South Dakota Geological Survey; (d): Black Hills National Forest; (e): USGS

> ▶ ***Exploring spatial relationships.*** Which roads cross limestone units and pose a risk to aquifers from gasoline or oil spills from car accidents (Fig. 9.1d)? Which parcels lie in the floodplain? Which cities are close to a volcano?

> ▶ ***Creating raster queries.*** Cells of a raster can be queried also. Which cells of a tree canopy raster have canopy cover values greater than 80% (Fig. 9.1e)? Which cells show an increase in vegetative greenness as measured by satellite images taken in 1998 and 1999? Where did land use change between 1970 and 2000?

A query extracts features from a feature class or records from a table and isolates them for further use, such as printing, calculating statistics, editing, graphing, creating new files, or doing more queries. In an **interactive query**, the user looks at a map or a table and uses the mouse to select the desired object(s). An **attribute query** uses values in the attribute table to test a condition, such as finding cities with population greater than 1 million. Attribute queries are sometimes called aspatial queries because they do not require any information about location. A **spatial query** uses information about how features from two different layers are located with respect to one another. A spatial query might choose cities within a certain county, rivers that are within a state, or hospitals within 20 miles of an airport. Attribute queries can be performed on either attribute tables or standalone tables. A spatial query requires two spatial data sets. Spatial and attribute queries may be combined, such as selecting cities with more than 50,000 people who are within 50 miles of an airport. The spatial and attribute queries must be performed separately but can be done in any order.

Attribute queries

In this method, the user specifies a certain condition based on fields in the attribute table and selects the records that meet the criteria, such as choosing all states with a population greater than 5 million. We have already used attribute queries multiple times in this text as a practical tool for exploring tables or exporting subsets of data. This chapter introduces some general concepts about queries and presents more advanced selection options.

Attribute queries test conditions between two or more attributes using **logical operators**, which include the familiar <, >, and = operators. The condition is written as an **expression** in a format called **Structured Query Language (SQL)**, which is used by most major databases to perform selections. SQL expressions can be complex and sophisticated or fairly simple.

SELECT *FROM cities WHERE [POP1990] >= 500000

SELECT *FROM counties WHERE [BEEFCOW_92] < [BEEFCOW_87]

The SELECT *FROM and WHERE key words must be present as dictated by the syntax rules of SQL. The lowercase words (e.g., *cities*) give the name of the table from which to select, and the terms in double quotes or brackets are the field names. This syntax would be used in any SQL-compliant database, so SQL is a transportable language. However, minor differences between database SQL varieties can be found, such as whether field names are enclosed in brackets or quotes or nothing. SQL has specific syntax rules that must be learned to enter expressions. For example, operators must be preceded and followed by a space, text values must be enclosed in single quotes, numbers may not contain thousands separators (commas), and so on.

Care must be taken when performing queries on numeric values, depending on the type of field used to store the data. The = operator can only be used on integers, because a floating-point value is never exactly equal to an integer or to a specific decimal value. The expression NUM = 3 will not work on a floating-point field, and the expression NUM = 3.6 would never work at all. Queries on floating-point fields must use only the < or > operators.

Boolean operators

The term **Boolean** refers to a branch of mathematics in which the variables are true/false statements rather than numbers, and the expressions and operators evaluate conditions based on whether they are true or false. In a logical expression containing two conditions, each condition may be true or false. In some cases, the user is interested in the records that meet both conditions. At other times, the desired records need only meet one of the conditions. **Boolean operators** are used to evaluate the pair of true/false conditions to return a result based on the user's intention.

The Boolean operators include AND, OR, XOR, and NOT. AND requires both input conditions to be true. OR requires either one or both of the conditions to be true. XOR (exclusive or) specifies that one or the other must be true, but not both. NOT requires the first specified condition to be true and the second not to be true. Consider the following queries with two conditions:

SELECT *FROM accounts WHERE Cust = 'COM' AND Balance > 500

SELECT *FROM accounts WHERE Cust = 'COM' OR Cust = 'GOV'

SELECT *FROM accounts WHERE Balance > 500 AND Balance < 1500

The first expression selects records from a customer accounts table that are held by commercial customers and have a balance over $500. The second expression finds the accounts held by commercial and governmental customers. The third expression finds accounts with balances between $500 and $1500.

> **TIP:** SQL requires that a field name appear in every condition of the expression, even if it is the same field. SQL will not correctly evaluate the expression Cust = 'COM' AND 'GOV'.

Figure 9.2 represents Boolean operators graphically using **Venn diagrams**. The large ellipse labeled T represents all the records in the table. Circle A represents the subset of records meeting condition A, circle B represents the subset meeting condition B, and the blue region indicates the records selected by the different Boolean operators.

Examine the first expression in the preceding examples of Boolean operators. The first condition, Cust = 'COM', constitutes condition A; condition B would be Balance > 500. A AND B finds the accounts for which both input conditions are true (commercial customers and balance over $500). If either condition is false, the record is not selected.

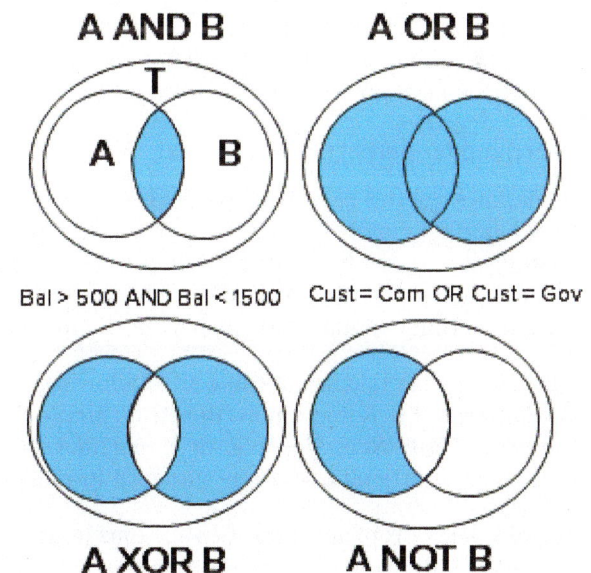

Fig. 9.2. Boolean operators shown as Venn diagrams

The expression A OR B selects the records for which either condition is true, finding all the commercial customers and all the accounts with balances over $500 (commercial or not). A NOT B finds all the commercial accounts but excludes the ones with balances over $500, instead finding the commercial customers with balances under $500. The "exclusive or" operator XOR is not available for attribute queries in ArcGIS but is used in raster analysis. XOR finds instances when one or the other condition is true, but not both.

The AND and OR operators are sometimes confused. In English we would say, "select the commercial and the governmental accounts" if we want both, and it seems natural to use the expression Cust = 'COM' AND Cust = 'GOV'. However, this expression yields no records because a customer account cannot be both commercial and governmental at the same time. The OR operator should be used.

The AND operator can be used to find records within a numeric range. Consider the third expression, Balance > 500 AND Balance < 1500. If a record with Balance = 1000 is being considered, it would test true for both conditions and be selected. Using OR is incorrect; it would return ALL of the values instead of those inside the range. A record with Balance = 300 would test true for the second condition and so be included. A record with Balance = 3000 would test true for the first condition and be included.

The AND, OR, and XOR operators are commutative. For example, A AND B yields the same result as B AND A. They are similar to the arithmetic operators + and ×, where $2 + 3 = 3 + 2$. When selecting commercial customers with accounts greater than \$500, it does not matter which condition is applied first. The NOT operator is not commutative: A NOT B ≠ B NOT A. It is similar to the arithmetic operators – and ÷; for example, $6 ÷ 2 ≠ 2 ÷ 6$. As an example, selecting English majors who are not women yields a completely different result than selecting women who are not English majors.

SQL can accept three or more Boolean conditions. Arithmetic operators have an order of precedence: multiplication is done before addition, such that $2 + 3 * 4 = 14$ and not 20. Boolean operators have no order of preference but are simply executed from left to right. If a different order is desired, parentheses must be used to indicate which conditions are evaluated first. Compare these two expressions for selecting parcels:

(LUCODE = 42 AND VALUE > 50000) OR SIZE > 50

LUCODE = 42 AND (VALUE > 50000 OR SIZE > 50)

In the first expression, all parcels > 50 hectares are chosen regardless of their zoning or value, and additional parcels are added if they are code = 42 and high in value. In the second expression, all of the parcels must have a zoning of 42 and can be either high in value or large in size.

Searching for partial matches in text

Sometimes queries based on text fields need to look for partial matches to a condition rather than a perfect match. Consider a table containing customer names. A user might wish to find all customers with the last name *Smith*. If the first and last names were stored in separate fields it would be easy, but if they were stored together in a single field it would be necessary to search for names that contain *Smith* as part of the full name.

Also, when selecting strings, one must allow for extraneous spaces or inconsistencies in how data are entered. For example, the string *Maple St* with a space following *St* would not match the string *Maple St* or *Maple St.* or *Maple Rd*. Often databases have inconsistencies like this one because it is difficult to enforce standard formats perfectly during data entry.

To solve such problems, most GIS systems (and databases) provide an operator, such as LIKE or CONTAINS, that allows searches for a particular sequence of characters within another string. This operator is helpful in searching for *Smith* within a field containing first and last names, or finding *Maple* in a field of street addresses. The SQL operator LIKE searches for the specified set of characters within the field and returns any record that contains those characters. When using LIKE, a wild card character (* for shapefiles; % for geodatabases) is used to stand for other characters that may appear. Consider selecting cities based on their NAME field. The two queries

NAME LIKE 'New *' *(shapefile)*

NAME LIKE 'New %' *(geodatabase)*

would select all cities beginning with the word New, such as New London and New Haven, but would not include Newcastle because there is no space in the latter. The expression

NAME LIKE '%Smith%' ✎

would return the customers with the name *Smith*. Notice the second % sign used in the expression. One might expect it to be superfluous, and often it would be, but it does ensure that an extraneous space character at the end of the name would not prevent it from being selected.

The selection tools in Pro give the user an easy way to specify searches using the Add Clause method, using the *contains* operator (Fig. 9.3a). However, the actual expression used to select the records from the table is executed in SQL using LIKE (Fig. 9.3b). Either the Add Clause method or the SQL method can be used to enter expressions, although many users will find the Add Clause method more intuitive.

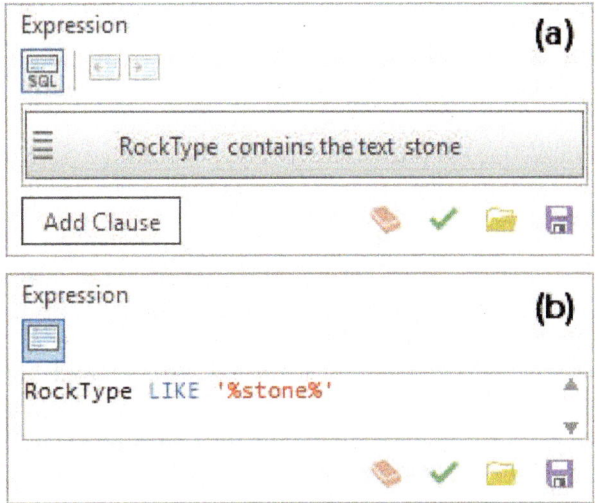

Fig. 9.3. A *contains* expression shown in SQL
Source: Esri

Spatial queries (location.

Spatial queries are a powerful tool unique to GIS because they select based on spatial relationships, such as finding wells within five miles of a river or finding parcels inside a floodplain. A spatial selection uses two layers and one spatial condition. The features of the layer being selected are compared spatially with the features of the second layer to see which ones meet the criteria, and features meeting the criteria are selected.

Panning and zooming techniques, already familiar to the reader, are actually a type of spatial query. Pans and zooms define a rectangle of interest based on x-y coordinates and ask that the software return all features that fall inside it.

Spatial operators

Attribute queries test relationships between attributes using logical operators, such as <, >, or =. Spatial queries test for relationships using **spatial operators**, including *containment*, *intersection*, and *proximity*. Consider two feature classes, A and B. The spatial operators shown in Figure 9.4 test for the relationship of each feature in A to each feature in B and return the features from A that meet the criteria with respect to B.

Many more than the three basic operators are shown in Figure 9.4, but each of the operators falls into one of the three general categories of containment, intersection, or proximity. The containment operators are highlighted in pink, the intersection operators are highlighted in yellow, and the proximity operators are highlighted in teal. The additional operators are special cases of the three basic types.

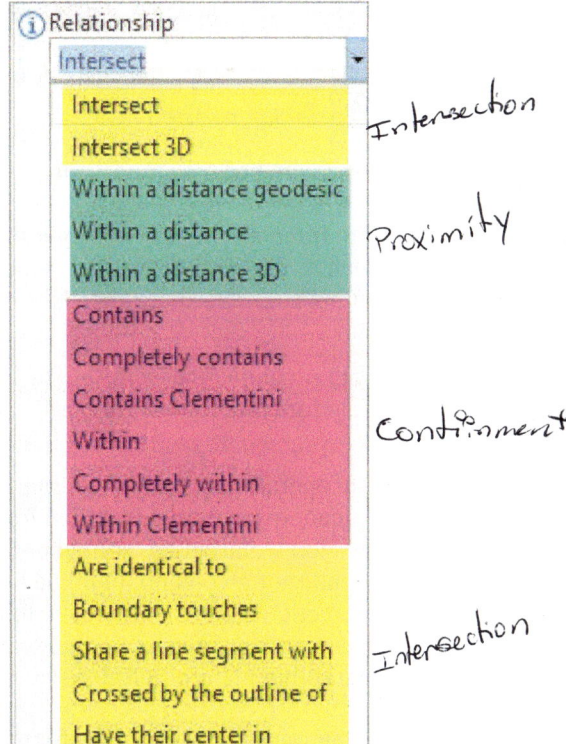

Fig. 9.4. Spatial operators
Source: Esri

The **containment** operators test whether one feature includes another. The test A *Completely contains* B returns all features in A that fully surround the features in B. A *Contains* B if B is inside A but they share a boundary. For example, both the yellow and blue counties in Figure 9.5 are contained by the state of Oregon, but only the blue counties are completely contained by the state. The *Within* and *Completely within* operators are the inverse of *Contain* and *Completely contain*. The *Have their center in* operator tests whether the center of a feature in A lies inside a feature in B.

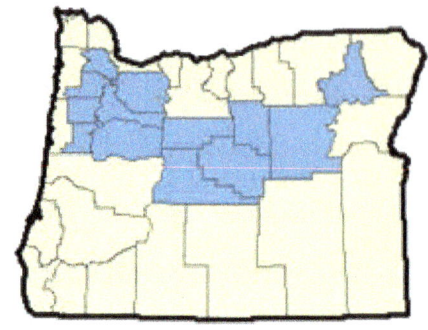

Fig. 9.5. The blue counties are completely contained by Oregon. Source: Esri

Intersection is the most generic operator and returns any feature in A that touches, crosses, or overlaps any part of a feature in B. The *Crossed by the outline of* operator is a special case of intersection that returns features in A that are crossed only by the boundary of features in B, such as the parcels that are crossed by the city limits. The most stringent condition is that the feature from A equals the feature from B exactly (i.e., they have precisely the same geometry). This corresponds to the operator *Are identical to*.

Proximity tests how close features in A are to features in B. The most generic test, *Within a distance*, selects features in A that are within a certain distance of B, such as returning all parcels within two miles of a school. The *Within a distance* operator uses planar Cartesian coordinates based on the map projection to evaluate proximity. The geodesic distance option takes the curvature of the earth into account and is recommended when evaluating distances over large areas such as a country or the world; it can also be used when the data set projection does not preserve distance in order to preserve other properties such as area.

Adjacency is a special case of proximity where the distance goes to zero and the boundaries of the features actually touch each other. These conditions are covered by the *Share a line segment with* and *Boundary touches* operators (which might also be considered to be intersection operators because the features touch).

The *Intersect 3D* and *Within a distance 3D* operators are used for 3D scenes and applications, such as finding intersecting volumes or determining which rooms in a high-rise building are within a specified distance of the fire exits.

TIP: A complete description of the different spatial operators can be found by searching the Pro Help for *Select Layer By Location: graphic examples*.

The operation of spatial queries is best illustrated using examples (Fig. 9.6). Counties that contain volcanoes might be identified for emergency planning (Fig. 9.6a). A boat manufacturer might develop a marketing campaign for counties that intersect major rivers (Fig. 9.6b). Cities within 50 miles of a volcano might be identified for increased risk (Fig. 9.6c). A state environmental protection agency might need to identify the rivers that intersect Texas (Fig. 9.6d). A study of transportation access might identify counties within 50 miles of major airports (Fig. 9.6e). A marketing plan for a Nebraska college might target the surrounding states (Fig. 9.6f).

Sometimes a problem calls for testing spatial relationships using a *subset* of features rather than the entire layer. For example, a user might want to select cities within 50 miles of Interstate 80, rather than cities within 50 miles of any interstate. This procedure begins with an attribute query to isolate the desired interstate, I-80, and then a spatial query to select the cities.

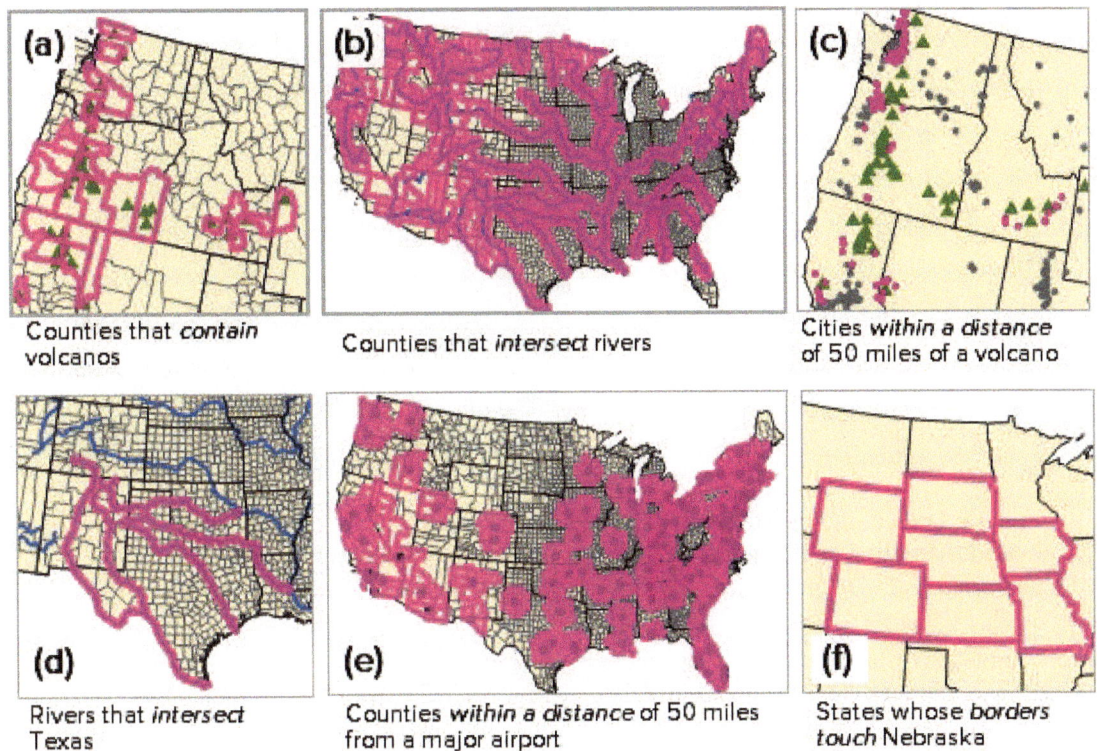

Fig. 9.6. Examples of selecting a layer based on its spatial relationship to another layer
Source: Esri

It must be kept in mind that spatial queries can only select or not select an entire feature. Consider Figure 9.6d, which shows the rivers that intersect Texas. Most of the rivers extend beyond the Texas border. If the within operator is used, then only rivers inside the state are selected (in this case, none of them). A different type of function is needed to extract only the *portions* of rivers that fall inside Texas.

Logical consistency and accuracy

Logical consistency characterizes whether the data representation of features has the same geometric relationships as their counterparts in the real world. If the boundary of a county is not exactly the same as its shared boundary with a state, the county feature will exhibit gaps and/ or overlaps with the state feature (Fig. 9.7a), and the data are not logically consistent. These issues may cause problems during spatial queries, because the expected relationship (the county is inside the state) is not reflected in the database. Furthermore, because feature classes vary in precision and geometric accuracy, it can often happen that two objects which coincide in the real world will not match when the GIS coordinates are compared.

Consider the problem of selecting cities intersecting rivers. A city at a national scale is represented by a point, and a river by a generalized line. Even if the real-life city intersects a river, as Minneapolis–St. Paul is split by the Mississippi River, it would be only luck if the point occurred on the line feature representing the river (Fig. 9.7b).

Using a search radius during a query may be helpful in mitigating logical consistency and accuracy issues. If the state and county boundaries match within 2 kilometers, setting a search radius to that distance when selecting counties within the state would help negate the impact of the logical inconsistency on the spatial query. In the city and river intersection, if one assumes that a typical large city might be represented by a square 10 kilometers on a side, then setting the

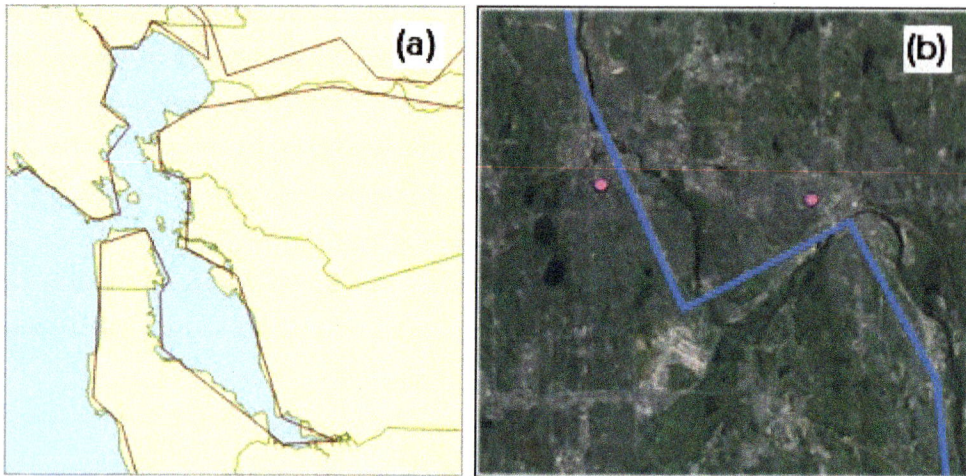

Fig. 9.7. Logical inconsistencies. (a) Gaps and overlaps between county and state boundaries. (b) Generalized river and city features for Minneapolis–St. Paul.
Source: Esri

search radius to 5 kilometers might find most of the cities with the desired real-world relation, although one might still encounter some false positive or false negative results.

Users must always remember the difference between real-world entities and how they are represented by a GIS, and be prepared to mitigate problems that arise because of logical consistency or accuracy issues. If a query does not seem to be producing the anticipated results, a detailed examination of the spatial relationships between the layers is needed.

Buffers

A **buffer** is a type of spatial query used to delineate areas that fall within a certain distance of a set of features. Buffers can be created for points, lines, or polygons (Fig. 9.8). They could be used to find 300-yard drug-free zones around schools or sensitive protected areas within 100 meters of a stream. Negative buffers can be applied to determine setback limits from the edge of a piece

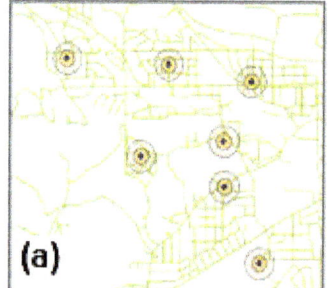
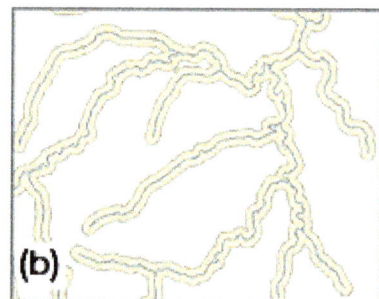

Fig. 9.8. Examples of buffers around (a) wells and (b) roads
Source: (a) City of Rapid City; (b) USGS

of property. Buffers can be created as simple rings or as multiple rings. An attribute can even supply different sizes of buffers for different features—for example, buffering primary roads by 200 meters and secondary roads by 100 meters.

Buffers are created for each individual feature, and they may overlap. This option might be appropriate in analyzing the distribution of soils in buffers around wells. In many cases, however, it is best to dissolve the buffers, which gets rid of the boundaries between them and removes the overlapping areas to create a single region. Figure 9.9 shows the difference in results obtained when the buffers around roads are not dissolved versus when they are. If the areas of the buffers are of interest, then dissolving is always the correct action. If areas were calculated from non-dissolved buffers, they would be grossly overestimated because the overlapping areas would be counted multiple times.

Finally, buffering is an intensive process, and the time involved increases quickly as the number of features increases. Any steps to reduce the number of buffers created at one time will facilitate analysis, for example, by performing queries before buffering rather than afterwards.

Extraction functions

Extraction functions separate features of interest from a larger group. Some consider queries to be one type of extraction. The functions *clip* and *erase* can extract both whole features and portions of features based on a spatial boundary supplied by another feature class and save them in a new feature class. Clip and erase can truncate features when they cross a boundary, for example, a road crossing a state line. This ability differs from the Select Layer By Location function, which must select (or not select) the entire feature.

A **clip** works like a cookie cutter to truncate the features of one file based on the outline of another. In Figure 9.10, the county roads layer (gray lines) has been clipped by the land use layer (beige) to produce a shapefile of the roads that fall inside the city planning boundaries (red). The features being extracted come from the input layer. The feature class providing the boundary is called the clip layer. The clip layer must always be a polygon feature class, but the input layer may be points, lines, or polygons. Only the outside boundary of the clip layer is used for clipping; internal boundaries, if present, have no effect on the output.

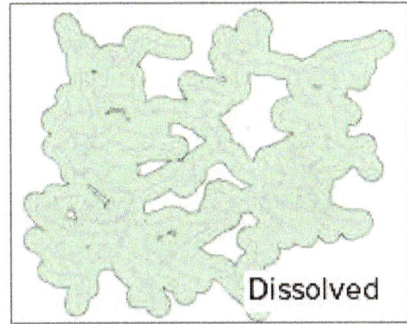

Fig. 9.9. Use the dissolve option to prevent overlapping buffers
Source: Black Hills National Forest

TIP: The map properties menu includes an option to temporarily clip the contents of the map to a specified feature class; it is used for display only and does not affect the source data.

Erase works in the opposite sense of a clip, keeping features that fall outside the erase layer and eliminating those inside. Given the same input and clip layers in Figure 9.10, an erase would keep the roads outside the city (gray) and eliminate the roads inside the city (red). Erase is only available to users with an ArcGIS Advanced license.

Buffers are often combined with clip and erase functions to select or remove areas from

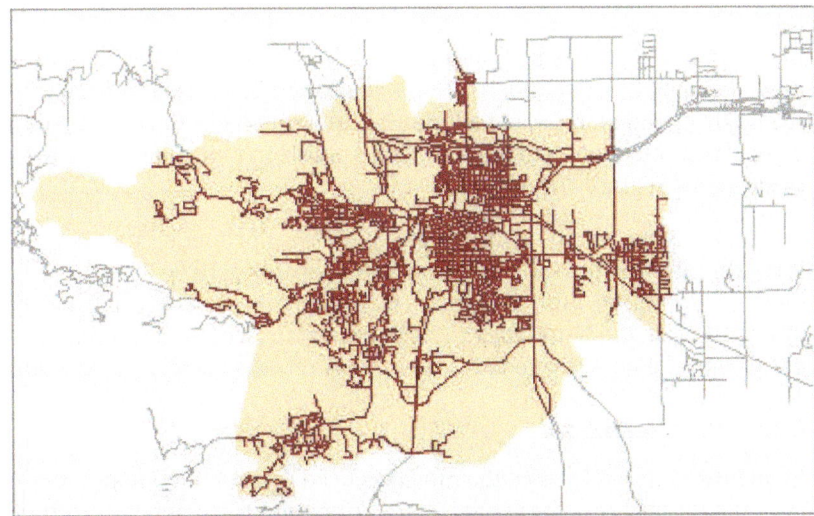

Fig. 9.10. A clip truncates the features in one feature class by the boundary of a polygon feature class.
Source: City of Rapid City

consideration based on distance criteria. For example, buffers around roads could be used to clip potential landfill sites, removing areas that are farther from existing roads than desired. A buffer of streams could be used with erase to eliminate logging sites that are too close to streams.

Attributes from both truncated and nontruncated features in the input layer are preserved intact during a clip or an erase operation. However, attributes from the clip or erase layer are ignored. The result of a clip or erase is a copy of the input layer that occupies a smaller region of interest.

In clip and erase, any features crossing the boundary will be truncated, potentially changing areas, perimeters, or lengths of features. The geodatabase fields Shape_Area and Shape_Length are automatically updated, but user-defined area or length fields are not. A field named MILES or ACRES containing highway miles or tract acres, for example, would contain an incorrect value for any truncated feature. The lengths and areas of user-defined fields must be updated manually using the Calculate Geometry Attributes tool.

Rasters may also be clipped to a smaller area of interest defined by a feature class or another raster. The tools used for raster extraction are different from those designed for feature data, but they work in much the same fashion.

About ArcGIS

ArcGIS Pro offers three ways to select features. **Interactive selection** uses a pointer to select features on the screen. Attribute queries are performed using the Select Layer By Attributes tool or the Make Feature Layer tool. Spatial queries are executed using the Select Layer By Location tool. After a query, the selected features are highlighted in the map and in the table.

Processing layers with selections

ArcGIS follows an important rule when processing features or records in any of its functions or tools: *after a query is performed and a subset of features or records is selected, any subsequent operation on that layer or table honors the selected set.* Previous tutorials have already illustrated this rule, as when the New Jersey counties were selected and exported to create a new data set for a New Jersey project. If someone selected the New England states and calculated statistics, only the New England states would be included. If a user selected parcels in a floodplain and then exported the features to a new feature class, only the parcels in the floodplain would be exported. Once a query has been executed on a layer, the layer behaves as if the unselected features do not exist.

By default, a layer has no query applied. In this state, all of the features are available for processing by a command. If a Statistics command were requested in this state, all of the features would be included in the calculations. Executing a query on a layer pushes it into a new state. The layer has a selection, and any function applied in Pro will restrict the operation of the function to the selected features or records.

Sometimes a user may execute a query that has proper syntax but yields no matching records. For example, it is possible to query for cities above 10,000 feet in elevation, but no cities actually meet the condition, and no features would be returned. When this happens, ArcGIS removes the query and sets the layer back to its default state with all features available. Essentially, it does not permit the user to execute a query that returns no records or features.

Interactive selection

In an interactive selection, the mouse is used to click on one or more objects in the map or table until the desired records are selected. For example, to get all the states beginning with A, one could sort the table by state name and then find and click on each record containing states starting with A. Or, to select the states with a Pacific Ocean coastline, a user could look on the map, hold down the Shift key, and choose California, Oregon, Washington, Hawaii, and Alaska.

Graphic shapes can also be used to select features, such as a rectangle, a polygon, a circle, or a line. Pro includes several tools that select features by constructing a shape on the fly (Fig. 9.11). When selecting using a shape, features only need to touch the shape, not lie within it. Notice that three states are selected in Figure 9.12, even though the selection rectangle, shown in green, does not completely enclose any of them. This default setting can be changed in the Selection options menu.

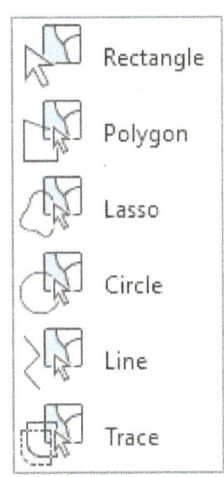

The user can also control which layers may be selected using an interactive query. By default, all layers are selectable, but the Contents pane includes a panel for viewing and setting which layers may be selected (Fig. 9.12a). If all feature layers are checked, a selection rectangle, shown in green, would select all features passing through the rectangle (Fig. 9.12b), including states, counties, rivers, interstate highways, and more. If only the states layer is checked, then the same selection rectangle yields just the three states (Fig. 9.12c). The Selection pane also lists how many of each layer are currently selected.

Fig. 9.11. Interactive selection tools
Source: Esri

TIP: The Selectable Layers option applies only to interactive selection with the mouse. It does not affect the Select Layer By Attribute or Select Layer By Location tools.

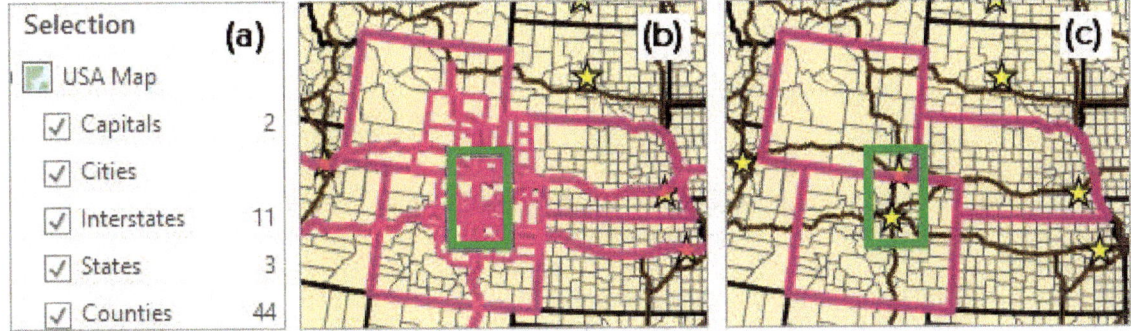

Fig. 9.12. The Selection pane (a) controls which layers may be selected; (b) selecting with all layers selectable; (c) selecting with only states selectable
Source: Esri

Making selections

The Select Layer By Attribute tool (Fig. 9.13a) generates a selection for a layer or a standalone table, highlighting the selected records in the table view and in the map view. The logical expression used to make the selection is entered in the Expression section, in which the user specifies the field, selects an operator, and then enters either a value or another field for comparison (Fig. 9.13b). When the Values box is highlighted, the user may select from the values present in the field or type a value explicitly, as shown in the expression: ST is Equal to AZ. When the Fields box is highlighted, the user can select another field to compare to the first, as in the expression: POP2014 is Greater than POP2010.

The green check mark verifies the expression to make sure it is syntactically correct. Multiple criteria, separated by AND or OR Boolean operators, may be entered. A complicated or often-used expression may be saved and loaded again later. The query builder window includes an SQL viewer, which can be used to view the SQL expression directly. The SQL expression can also be edited, in case the standard one generated by the Add Clause button does not quite generate the

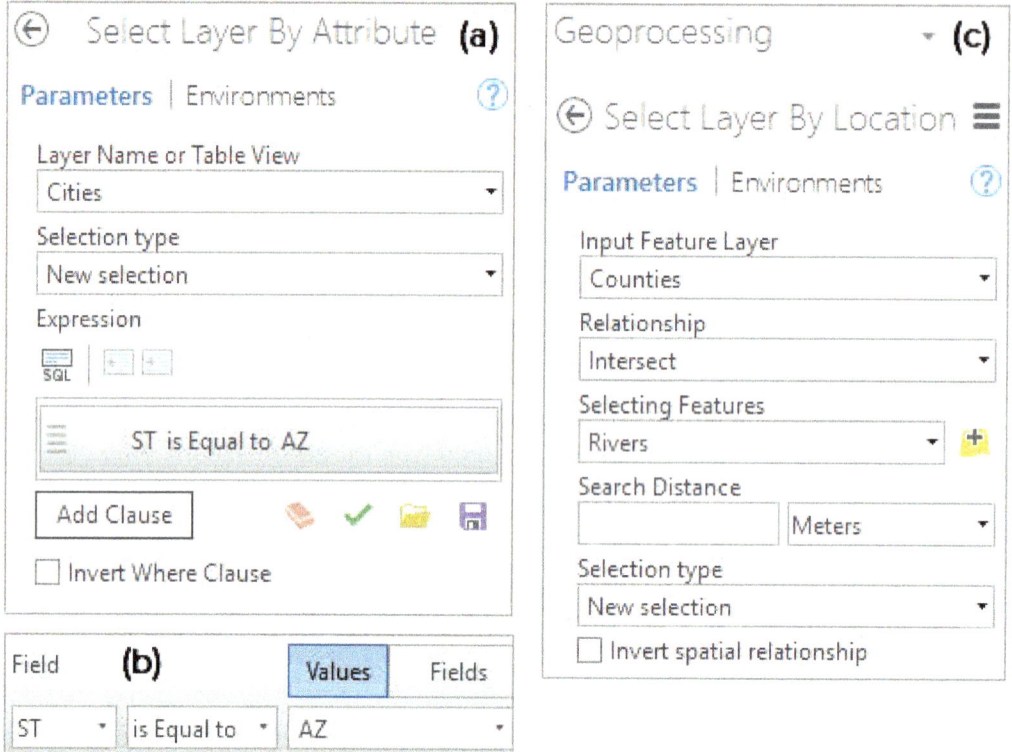

Fig. 9.13. Tools for making selections
Source: Esri

desired result. For example, one might enter an expression with three clauses and then change the Boolean order of operation using parentheses.

The Select Layer By Location tool provides the means to set up a spatial query (Fig. 9.13c). The Input Feature Layer refers to the layer being selected, in this case Counties. They will be the features highlighted after the query. The Relationship identifies the spatial operator used to compare the layers. The Selecting Features identifies the layer to which the input features are compared. The Search Distance option provides a way to manage potential logical consistency issues.

The Select Layer By Location tool will honor any previous selection made on the Selecting Features layer. For example, if the user previously selected the Mississippi River, then only counties touching the Mississippi River would be selected.

Choosing the selection type

ArcGIS Pro has a **selection type** parameter that controls what happens to an existing selection when a new query is performed on a layer (Fig. 9.14). The selection type may be specified in two places. When set from the Selection options menu in the **Map: Selection** ribbon, it affects the interactive selection (only the first four types in Fig. 9.14 are available in the Options menu). In addition, both the Select Layer By Attribute or Select Layer By Location tools include a selection type parameter that affects how the tool does its selection when it runs.

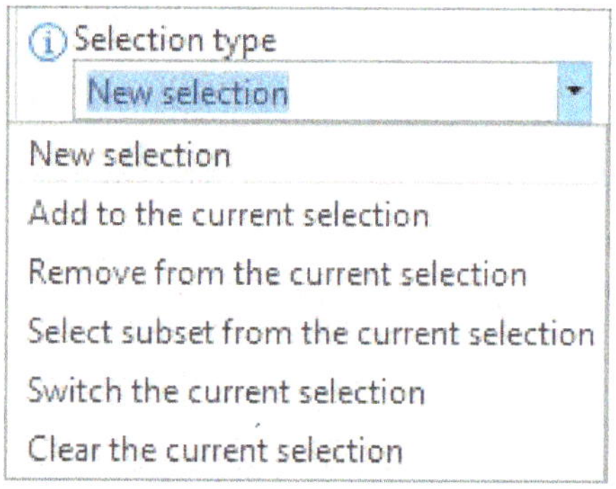

Fig. 9.14. Selection types
Source: Esri

New Selection selects the specified records regardless of what else is already selected, and it replaces any previous selection with the new selected records. For example, if Maria selected green jelly beans and then chose to select red jelly beans, then she would only have red jelly beans selected. This method is the default.

Add to the current selection yields the records that fit the query in addition to any that are currently selected. Thus, if Brandon had green jelly beans already selected and then selected the red ones, then he would have both red and green jelly beans selected. Using this method is equivalent to doing a single selection step with a double expression using the OR operator, such as Color is Equal to Red OR Color is Equal to Green.

Remove from the current selection takes away the records that meet the selection criteria. If Tamara already had red, green, and yellow jelly beans selected and she "selected" green jelly beans, then she would have red and yellow jelly beans left. In this option, the records matching the expression are deleted from the selected set. It is equivalent to the NOT operator.

Select subset from the current selection chooses the records that fit the criteria from the already selected set. If Zach had red, green, and yellow jelly beans selected and chose to select the green ones, then he would have the green ones selected. If he chose to select orange ones, then he would end up with *nothing* because orange jelly beans were not in the original selection. Performing several selections in a row with this method is similar to performing a series of AND expressions.

Switch the current selection replaces the selected set with the unselected records. In other words, if Juan had red, yellow, and green jelly beans available and the red ones were selected, and if he did a switch selection, then he would have green and yellow ones selected.

The ability to set the selection type gives greater flexibility in defining selection sets and enables selection using multiple steps. When performed sequentially, complex expressions involving many criteria are often less confusing than when executed by entering a long expression full of ANDs and ORs and trying to figure out where to put the parentheses.

Imagine selecting earthquakes larger than magnitude 5.0 in Mineral County, Nevada (Fig. 9.15). This problem involves selecting by attribute as well as by location, so it occurs in two steps. First, use Select Layer By Location to select the earthquakes within the county, using *New selection* as the method (the green earthquakes in Fig. 9.15). Then use Select Layer By Attribute to choose earthquakes with MAG > 5. However, using *New selection* again in the second step would yield ALL earthquakes with MAG > 5, not just those in Mineral County. One must specify *Select subset from the current selection* in order to obtain the right earthquakes (circled ones).

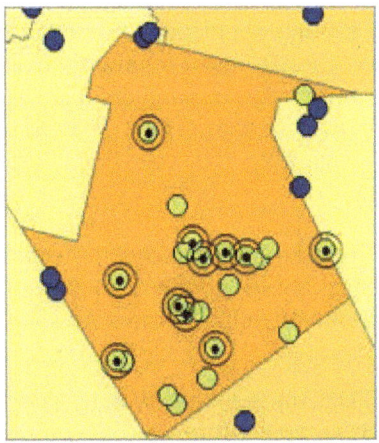

Fig. 9.15. Using selection types on earthquakes
Source: Esri

Managing results from queries

After a query, it is often useful to preserve the selected features for future use. Several options are available. Some affect only the layer properties and leave the source data unchanged; others create new data sets from the selection.

Making a layer

The Make Feature Layer tool is similar to the Select Layer By Attribute tool, except that instead of simply highlighting the selected features in the original layer, it creates a new layer in the map. In Figure 9.16a, the user employed the Make Feature Layer tool to select the South Atlantic subregion

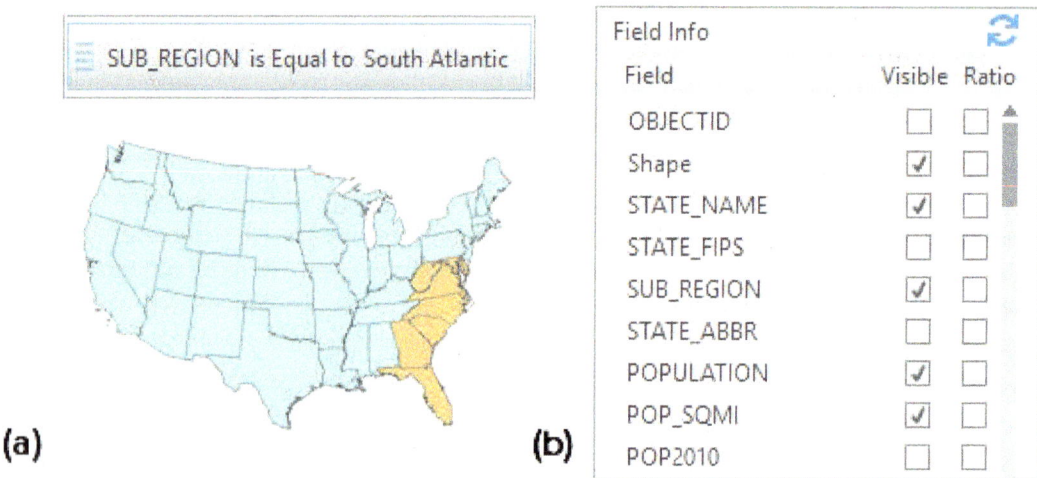

Fig. 9.16. The Make Feature Layer tool creates a new layer showing the selected features and the desired fields

Source: Esri

of the United States using the expression: SUB_REGION is Equal to South Atlantic. The tool created a new layer with only the selected states, shown in orange above the original states layer. The layer still refers to the original states source data. The tool also has a fields section that permits the user to turn the visibility for the output fields off or on (Fig. 9.16b). The new layer will only show the checked fields, although the fields will remain in the source data and can be turned on again if needed.

Creating a selection layer has several advantages. The layer can be given its own symbols and can be displayed separately from the original layer. The layer preserves the selected features for future reference and eliminates the risk of accidentally clearing the selection and having to redo it. Selection layers can be used to input the same set of features to several tools. They are helpful for viewing and recording intermediate results in a complex series of queries. The layer can also be saved as a layer file for use in other maps.

Creating a layer is the best solution when a selected set is intended for use within a single map. It does not create multiple copies of the data, and edits to the source data will be updated automatically. It is the best solution to create a temporary duplicate of certain features.

TIP: Any selection may be saved as a new layer using the context menu item Make Layer From Selected Features.

Definition queries

A **definition query** is a property of a layer. It is similar to a selection layer, except that instead of being created from a selected set, the query is performed in the layer properties window. Instead of showing the selection in a different color than the other features, a definition query causes the layer to appear as if it only contains the selected features (even though they remain available in the stored data set). Like selection layers, definition queries refer back to the original source data, and they share the same advantages and disadvantages.

Exporting data

The **export** function creates a new feature class that contains only the selected features, and it is used when a permanent copy of the selection is needed. Because it creates a physical copy of the data, the link to the original feature class is broken. This can be a disadvantage, such as when an organizational database with frequent updates is the source. Changes made to the database will not be reflected in the exported file, which may eventually become out of date. On the other hand, it can be an advantage if the user needs to make changes but does not have write access to the original data. Exports may also be used to develop a new geodatabase based on a subset of existing data, such as developing a New Jersey geodatabase using feature classes from a United States geodatabase, as we did in Chapter 5.

Summary

▶ Queries extract a subset of records or features from a data set based on a set of one or more conditions.

▶ Attribute queries extract features based on conditions from fields in the attribute table. They are written in SQL and use operators such as =, >, <, AND, and OR. Partial matches to text strings are evaluated with the LIKE operator.

▶ Boolean operators are used to evaluate pairs of conditions for expressions involving multiple criteria. Venn diagrams help us to visualize the results of Boolean operators.

▶ Spatial queries extract features based on criteria of how two layers are spatially related using spatial operators to test containment, intersection, or proximity.

▶ Extraction functions, such as clip and erase, separate features of interest based on a "cookie cutter" provided by a polygon layer and can truncate features at the cutter boundaries. Attributes are carried through without change.

▶ A buffer may be considered a type of spatial query because it finds areas within the specified distance of a set of features. Dissolving of the buffers is typically used to prevent overlap of the polygons.

▶ Data accuracy and scale issues may affect spatial queries, and they must be considered in evaluating the methods and results.

▶ ArcGIS uses only the selected features in a layer when a function, command, or tool is applied to the layer. If no query has been applied, all features are used. If a query has been applied, only the selected features are used.

▶ An interactive query lets the user pick certain features by finding them on the screen and clicking them. Attribute queries are implemented using the Select Layer By Attribute tool, and spatial queries use the Select Layer By Location tool.

▶ The Contents List By Selection panel allows the user to specify which layers can be selected interactively. By default all layers are selectable.

▶ Four selection types add greater flexibility to queries. These types include creating a new selection, adding to the current selection, removing from the current selection, and selecting from the current selection.

▶ Creating a new layer from a set of selected features allows the selection to be stored, displayed, and passed on to tools or commands. Exporting creates a permanent copy.

Important Terms

attribute query	definition query	Interactive selection	spatial operators
Boolean	erase	intersection	spatial query
Boolean operators	export	logical operators	Structured Query
buffer	expression	proximity	Language (SQL)
clip	extraction	query	Venn diagram
containment	interactive query	selection type	

Chapter Review Questions

1. List the six different types of queries discussed in this chapter, and briefly describe each in your own words.

2. Let T be a table containing all students attending a community college in New York. Let A be the subset of students living in New Jersey. Let B be the students with a GPA greater than 3.0. The query A AND B yields 200 records. The query A OR B yields 1100 records. The query A NOT B yields 400 records. Construct a Venn diagram for the sets, labeling each section with the number of students. How many students live in New Jersey? How many students have a GPA greater than 3.0?

3. From the information in Question 2, can the number of students attending the community college be determined? If yes, state how many. If not, explain why.

4. In what ways does the Clip tool differ from the Select Layer By Location tool?

5. Imagine some trail mix composed of peanuts, raisins, almonds, cashews, dried cranberries, and chocolate candies colored red, green, yellow, and orange. Imagine that you apply the following set of "queries" to the trail mix:

> Create new selection all candies
> Add to selection cashews
> Remove from selection red and green candies
> Select from selection all nuts and candies
> What is selected now? _____

6. For each of the following queries, state whether AND or OR is the correct Boolean operator, and explain why.

   a) ZONE = 'Commercial' ___ ZONE = 'Residential' for selecting two types of parcels

   b) COVTYPE = 'SPRUCE' ___ CROWNCOV > 50 for selecting dense spruce stands

   c) POP2000 > 10000 ___ POP2000 < 50000 for selecting medium-size towns

   d) TEMP < 30 ___ TEMP > 80 for selecting towns with extreme average temperatures

7. What is an operator? Describe and give examples of each of the following: arithmetic operators, logical operators, spatial operators, and Boolean operators.

8. Why is logical consistency a potential issue during spatial queries?

9. Nancy and Sarah buffered the same feature class of streams and determined the total area. Nancy got 102 km^2 and Sarah got 59 km^2. Who got the right answer? How do you know?

10. Which one(s) of the following tools or functions create a new data set in a geodatabase?

Select Layer By Location	Make Feature Layer
Buffer	Clip
Definition Query	Export

Expand your knowledge

Expand your knowledge by reading these sections of the ArcGIS Pro Help.

Help > Maps > Interact with maps (the sections on selection and query)

Mastering the Skills

Teaching Tutorial

The following examples provide step-by-step instructions for doing basic tasks and solving basic problems in ArcGIS. The steps you need to do are highlighted with an arrow →; follow them carefully.

→ Start ArcGIS Pro and open the ClassProjects\UnitedStates project. Remember to save the project often.

Using interactive selection

First, let's experiment with selecting features with the mouse. Your selected features will probably be highlighted in cyan, but the figures show the selections in pink for better visibility.

1→ Choose **Map: Navigate: Bookmarks > Texas**.

1→ Click the **Map: Selection: Select** drop-down and choose the Rectangle tool.

1→ Click on the star that represents Austin, Texas (Fig. 9.17). If you don't get the star the first time, try again.

A single click results in one feature being selected. If multiple features overlap at the point clicked, usually the top one is selected.

Fig. 9.17. Austin and Dallas in Texas
Source: Esri

1→ Move away from Austin slightly and select the interstate highway running through it.

1→ Try selecting a different interstate, and then select a river.

1→ Try to select a county.

Instead of a county, the state feature is selected. When overlapping features are clicked, the selection preference is controlled by the order of layers in the Contents View By Selection panel.

2→ In Contents, click the icon to open the List By Selection panel.

2→ Click and drag the Counties layer above the States layer.

2→ Click a county to select it.

2→ Click and drag a box around the Dallas area (Fig. 9.17) and notice that anything in the box or anything that goes through the box is selected.

2→ In Contents, the List By Selection panel shows how many features in each layer are currently selected.

Use the same View By Selection panel to control which layers are selectable.

3→ In Contents, hold the Ctrl key down and uncheck one of the boxes. All of them will become unchecked.

3→ Check the box for the Counties layer.

3→ Draw a box around Dallas again. This time, only the counties are selected.

Selections can be cleared when no longer needed, either all at once or for one layer individually.

4→ Check the boxes to make all layers selectable again, and draw a box around Dallas.

4→ In Contents, right-click the Counties layer and choose Selection > Clear Selection. Only the selection on the Counties layer is cleared.

4→ Click **Map: Selection: Clear**. All selections are cleared when this button is used.

Notice that when a new selection is made, the previous selection disappears. The Shift and Ctrl keys can modify this behavior.

5→ Open the Michigan bookmark.

5→ In Contents, uncheck the boxes for all layers except Counties, making it the only selectable layer.

5→ In Contents, switch to the List By Drawing Order panel.

5→ Uncheck the Interstates layer to make it not visible in the map.

5→ Drag the Counties below the States again to see the state boundaries.

6→ Click one county in Michigan with the Select by Rectangle tool.

6→ Hold down the Shift key and click another county. It is added to the selected set.

6→ Add three more counties using the Shift-click method.

6→ Now hold down the Ctrl key and click one of the counties that are *already selected*. It will be *removed* from the selected set.

Let's efficiently select all the counties in the northern portion of Michigan that lie between Lake Superior and Lake Michigan.

7→ Click and drag a box that goes through as many counties as possible in northern Michigan (Fig. 9.18). It is difficult to get them all.

7→ Use the Shift or Ctrl key to modify the selection until the desired counties are selected.

The default selection type can be modified using the Selection Options.

8→ In the **Map: Selection** group, click the small **Options** button in the lower right corner to open the Selection options.

8→ Examine the Selection options and then focus on the Selection combination mode. *Create a new selection* is the default.

8→ Change the Selection combination mode to *Add to the current selection*. Click OK.

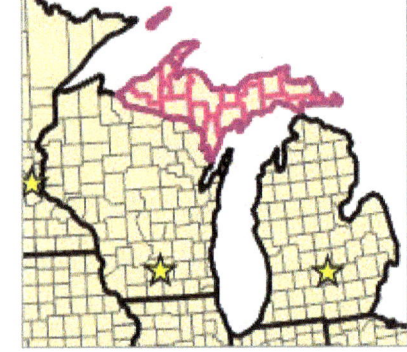

Fig. 9.18. Counties of northern Michigan

Source: Esri

8→ Click a county in southern Michigan. It is added to the selected set. Add more.

8→ Draw a rectangle through several unselected counties to add them as well.

8→ Continue until all of the Michigan counties, northern and southern, are selected.

9→ Open the Selection options again and change the selection combination mode to *Remove from* the *current selection.*

9→ Click on a selected county on the edge of the state (otherwise the difference is not apparent). It is removed from the selected set.

9→ Draw a box around several selected counties to remove them from the set. Continue until the northern counties are selected but the southern ones are not (Fig. 9.18).

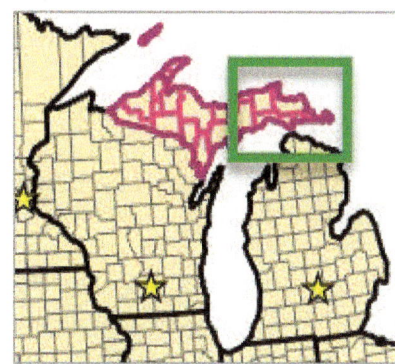

Fig. 9.19. The box
Source: Esri

10→ Set the selection combination mode to *Select from the current selection.*

10→ Draw a box that encloses some of the northern and southern counties, as shown in Figure 9.19.

Notice that the northern counties touching the box were selected, but the southern counties were not because they were not already part of the selected set. Let's use some other selection tools.

10→ Set the selection combination mode back to the default, *Create a new selection.*

10→ Click the **Map: Selection: Clear** button to clear the current selection.

11→ Click the **Map: Selection: Select** button drop-down and examine the other selection tools (Fig. 9.20).

11→ Choose the Circle tool.

11→ Click on the state capital, Lansing, and drag to make a circle (Fig. 9.20). Release the mouse button when satisfied with the circle.

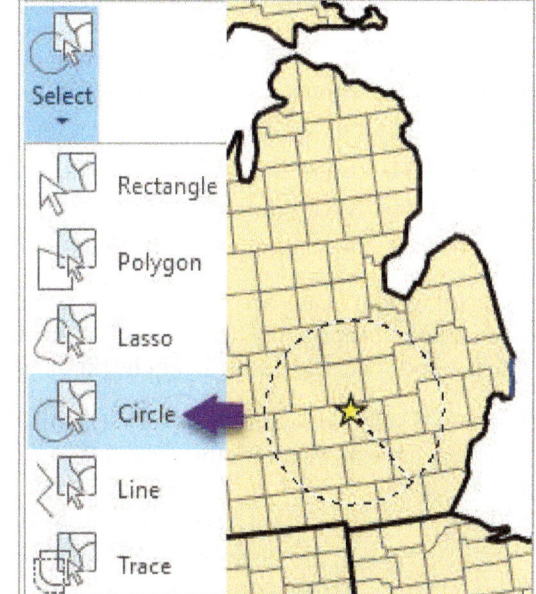

TIP: If the cursor is snapping to Lansing and county corners or edges, use the Snapping button at the bottom of the map to turn snapping off or on as needed.

This tool can be applied with an exact specified radius.

Fig. 9.20. Selecting using a circle
Source: Esri

11→ Click on Lansing again and drag to start the circle.

11→ With the mouse button still down, press the letter *r* (for *radius*) on the keyboard. When the box appears, let go of the mouse button.

11→ Type *100* in the box, select miles (mi) for the units, and press Enter.

12→ Choose the **Map: Selection: Select > Polygon tool**.

12→ Click around the lower peninsula to add the vertices of a polygon that touches all of the southern Michigan counties without including those from other states or the northern peninsula. Double-click to finish the polygon and make the selection.

12→ Experiment with the other selection tools to learn to use them. When finished, clear any selected features and go on with the tutorial.

Selecting by attributes

Recall that the Select Layer By Attribute tool is used to create a selection based on table values. This tool may be opened using the **Map** ribbon or the Geoprocessing pane.

13→ Open the USA bookmark.

13→ In Contents, in the List By Drawing Order panel, turn visibility on for Cities.

13→ Choose **Map: Selection: Select By Attributes**.

13→ In the Geoprocessing pane, set the Layer Name to Cities.

13→ Confirm that the Selection type is set to *New selection*.

13→ Click Add Clause and enter the expression: POP2014 Is Greater Than 1000000 (1 million).

 13→ Use the Verify button to make sure the expression is valid and run the tool.

13→ In Contents, switch back to the List By Selection panel.

TIP: It is helpful to keep the Geoprocessing pane open during the remainder of this tutorial.

1. How many cities in the United States had more than 1 million people in 2014? _____

 14→ In the Geoprocessing pane, click the back arrow to clear the tool settings; then reopen the Select Layer By Attribute tool from the Recent Tools list.

14→ Set the Layer Name to States and keep the *New selection* type.

14→ Enter the expression to POP2014 is Greater Than 5000000 (5 million).

14→ Verify the expression and run the tool.

14→ In Contents, examine the List By Selection panel.

Both cities and states are now selected. States and Cities are separate layers, and their selections are managed separately. Also note that the states were selected even though the checkbox was unchecked. The selectable layer checkboxes only apply to interactive selection.

Let's try a selection with double criteria: all counties that lost population between 2014 and 2010 and that have more males than females.

15→ Use **Map: Selection: Clear** to clear all current selections.

15→ In the Geoprocessing pane, use the back arrow to clear the Select Layer By Attribute tool and reopen it.

15→ Set the layer to Counties and continue to use the *New selection* type.

15→ Enter the partial expression: POP2014 is Less Than.

15→ Click the Fields button and complete the expression by clicking on POP2010. Click Add.

15→ Click Add Clause again and enter the expression: And MALES is Greater Than FEMALES. Click Add.

15→ Verify the expression and run the tool.

2. How many counties with more males than females lost population between 2010 and 2014?

Now from this selection we will select those counties that also have fewer than 10,000 people. Be careful to ensure that the new set of counties is pulled only from the currently selected ones.

16→ In the Select Layer By Attribute tool, change the selection type to *Select subset from the current selection.*

16→ Click the Clear Expression button to erase the old expression.

16→ Enter the expression: POP2014 is Less Than 10000. Verify the expression and run the tool.

3. How many counties remain selected? _____

TIP: It is best to clear and reopen the selection tool if you are changing the layer being selected, so the fields menu can reset. If doing another selection on the same layer, just change the parameters.

Springfield seems to be a popular name for cities. Let's see how many there are.

17→ Use **Map: Selection: Clear** to clear all current selections.

17→ Clear and reopen the Select Layer By Attribute tool.

17→ Set the Layer Name to Cities and use the *New selection* type.

17→ Enter the expression: NAME is Equal To Springfield. (Start typing Springfield in the Values box to help find the name in the list.)

17→ Verify the expression and run the tool. Determine how many cities are selected.

Queries are performed in a language called SQL, used by many databases. It has a specific syntax, such as requiring single quotes around text values. The Add Clause function automatically constructs the input in the correct format. Occasionally it is useful to examine the actual SQL expression or to modify it directly.

18→ In the tool, click the SQL button and examine the expression.

18→ In the SQL expression, replace the text between the single quotes with Las Vegas and run the tool again.

We might be curious how many cities end with the word *City*, such as *Oklahoma City*.

19→ Click the SQL button again to switch back to the Add Clause mode. Clear the expression.

19→ Enter a new expression: NAME contains the text City. Verify and run.

19→ In Contents, right-click the Cities layer and choose Attributes. Scroll through the names of the selected cities in the Attributes pane. Did the query work as expected?

19→ In the Geoprocessing pane, hover the mouse over the gray box containing the expression and click the Edit Clause button.

19→ Replace the City search term with city and click Update. Verify the expression.

No records matched the condition. Databases are often case-sensitive, so it is important to use the appropriate case when constructing queries. It is also important to use case consistently when entering values in a table. Mixing case, like entering Smith and SMITH in the same field, makes selection more difficult.

4. How many city names begin with the word New? Include cities such as New Brunswick, but exclude cities like Newcastle. Check the names in the Attributes pane. _____

5. The answer given for Question 4 includes West New York, which doesn't actually begin with New. Examine the SQL expression and edit it to exclude similar cities from the selection. How many cities meet this more stringent criterion?

The Make Feature Layer tool employs an attribute query to create a new layer, with the option to show only some fields in the table. Imagine developing a report on minority populations in the Mountain subregion of the United States.

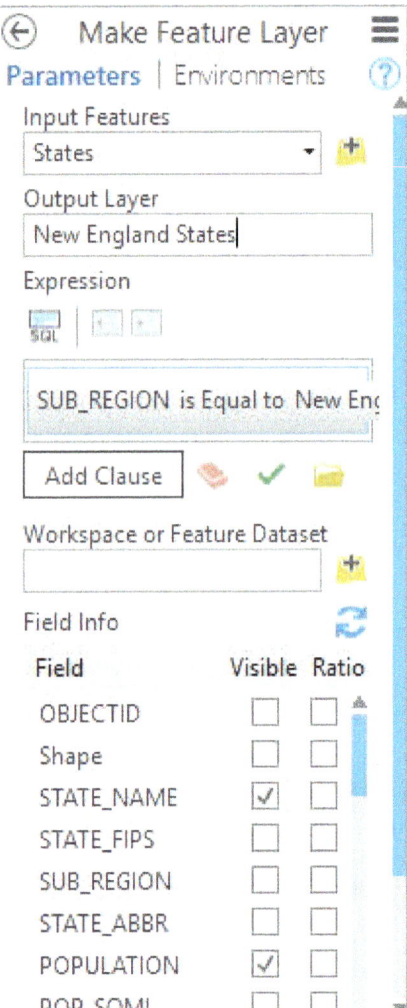

Fig. 9.21. Making a layer from the New England states

Source: Esri

20→ Clear all selections and close the Attributes pane.

20→ In the Geoprocessing pane, open the Make Feature Layer tool (Fig. 9.21).

20→ Set the Input Features to States. Name the Output Layer *New England States*.

20→ Enter the expression: SUB_REGION is Equal to New England.

20→ In the Field Info section, uncheck all fields except STATE_NAME, POPULATION, and the race fields: WHITE, BLACK, etc. (**Hint: Use Shift-Click to select blocks of fields and turn them all on or off with a single check box click.**)

20→ Run the tool.

20→ Open and examine the table for the New England States layer. Close it when finished.

Selecting by location

Next we will try selecting by location. First we will select all the counties that are adjacent to rivers. The features to select (Counties) are the target layer.

21→ In Contents, switch to the List By Drawing Order panel and turn off visibility for the Cities layer. Return to the List By Selection panel when done.

21→ Click **Map: Selection: Select By Location**, or open the Select Layer By Location tool.

21→ Set the Input Feature Layer (target layer) to Counties (Fig. 9.22).

21→ Set the Relationship to Intersect.

21→ Set the Selecting Features (source layer) to Rivers.

21→ Leave the Search Distance blank.

21→ Keep *New selection* as the Selection type. Run the tool.

Next let's select the counties that contain state capitals and save the selection as a new layer.

22→ In the Select Layer By Location tool, change the Relationship to Contains and change the Selecting Features to Capitals.

22→ Run the tool and examine the results on the map.

22→ In Contents, right-click the Counties layer and choose Selection > Make Layer From Selected Features. Rename the new layer *Counties with Capitals*.

Now let's do an example with points and lines by selecting the cities close to interstate highways.

TIP: If a Counties selection layer does not appear, save the project and restart Pro. Repeat the selection and the creation of the layer.

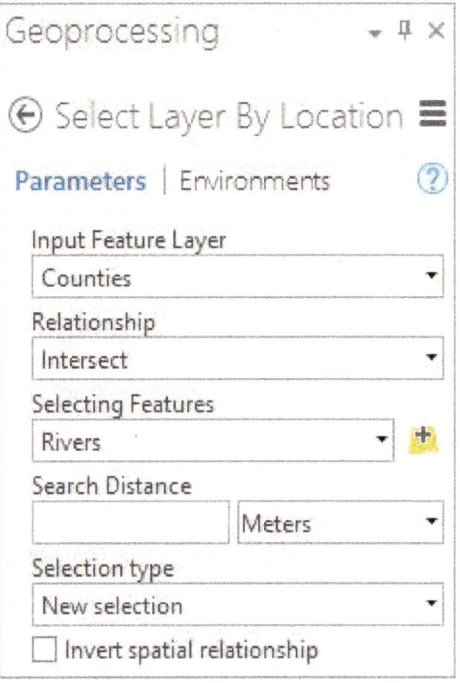

Fig. 9.22. The Select Layer By Location tool
Source: Esri

23→ In the Contents List By Drawing Order panel, turn on visibility for Cities and the Interstates. Return to the List By Selection panel.

23→ In the Select Layer By Location tool, change the Input Feature Layer to Cities and set the Selecting Features to Interstates.

23→ Change the Relationship to *Within a distance* and set the Search Distance to 20 miles. Run the tool and view the results.

Selections on one layer will not be affected by selections on a different layer (the Counties selection is still present). Notice the Selected Features information tab in the lower right corner of the map. The total shows the selected features from all layers. Use Contents to clear the selection of one layer at a time.

23→ In Contents, right-click the Counties layer and choose Selection > Clear Selection.

6. How many US cities are within 20 miles of an interstate highway? _____

23→ Right-click the Cities layer and open the attribute table.

23→ Notice that the bottom of the table says (3641 out of 3905 Selected).

7. Use a handheld calculator, or the one on the computer, to calculate the percentage of cities within 20 miles of an interstate. _____

TIP: Look carefully at less populous states, like Wyoming and Nevada. Sort the cities by ascending population and examine the CLASS field. Does this data set contain *all* cities in the United States? What impact might the completeness of this data set have on the analysis?

In the previous example, we compared *all* the features in one layer to *all* the features in another. Selection honors the selected set in the comparison layer, permitting actions such as selecting rivers that cross a certain state.

24→ Clear all selected features and close the table.

24→ In Contents, in the List By Drawing Order panel, turn off the visibility for Cities, Capitals, and Interstates. Return to the List By Selection panel.

24→ Make States the only selectable layer (it should be the only box checked).

24→ Click the **Map: Selection: Select > Rectangle** tool and select the state of Texas.

25→ In the Select Layer By Location tool, set the Input Feature Layer to Rivers and the Selecting Features to States.

25→ Set the Relationship to Intersect and delete the Search Distance. Run the tool.

8. Which rivers intersect Texas? _____

Spatial queries may also be performed with a single layer, using it both as the target and as the selecting layer. The features comprising the selecting criteria must be selected first. For example, let's select all cities within 400 miles of Kansas City.

26→ Clear all selected features.

26→ In Contents, in the List By Drawing Order panel, turn off visibility for Rivers and turn on Cities.

26→ Open the Select Layer By Attribute tool and set it up to select Kansas City from the Cities layer. Run the tool and check the result.

27→ Open the Select Layer By Location tool.

27→ Set both the Input Feature Layer and the Selecting Features to Cities.

27→ Change the Relationship to *Within a distance* and set the Search Distance to 400 miles. Run the tool.

Attribute and location queries gain power when combined sequentially. Let's select all the rivers that intersect states in the Mountain subregion. We do this by first selecting the mountain states using an attribute query and then selecting the rivers that intersect these states (Fig. 9.23). However, it is often useful to create a temporary layer from a selection so that it can be better visualized or reused in another query. The Make Feature Layer Tool does this easily.

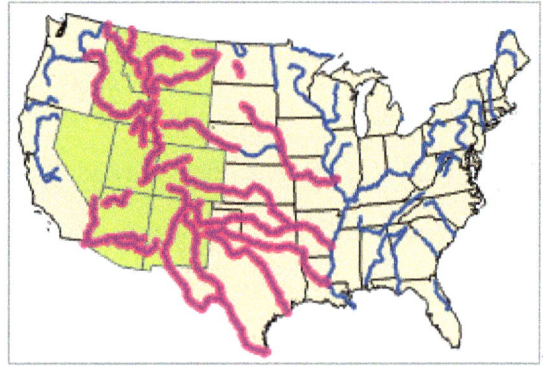

Fig. 9.23. Selecting rivers that cross the Mountain subregion
Source: Esri

28→ Clear all selected features. Turn off visibility for Cities and turn on Rivers.

28→ Open the Make Feature Layer tool.

28→ Set the Input Features to States.

28→ Name the Output Layer Mountain States.

28→ Enter the expression: SUB_REGION is Equal to Mountain.

28→ Leave the fields map as is. Run the tool. Change the Mountain States layer color to see them, if needed.

29→ Open the Select Layer By Location tool.

29→ Change the Input Feature Layer to Rivers and the Selecting Features to Mountain States.

29→ Set the Relationship to Intersect and run the tool.

Notice that many of the rivers extend east to join the Mississippi (Fig. 9.23). Because they pass through the mountain states, they are included. Try selecting only the rivers that are within the mountain states.

29→ In the Select Layer By Location tool, change the Relationship to Within and run the tool again.

Now hardly any rivers are selected. Ideally, one might like to select the *portions* of rivers that fall inside the mountain states. However, Select Layer By Location only works on whole features. To get portions of features, we must use the Clip tool and save the features in a new feature class.

30→ Clear the selected features.

30→ In the Geoprocessing pane, open the Clip (Analysis tools) tool.

30→ Set the Input Features to Rivers.

30→ Set the Clip Features to Mountain States.

30→ Name the output *MountainRivers* and run the tool.

30→ Turn off Rivers to better see the clipped layer.

30→ Save the project.

Combining queries

Combining attribute and spatial queries can provide powerful analysis. Earthquakes and volcanoes create significant hazards for residents. How many people in the United States are at risk, and where do they live?

31→ Choose **Insert: Project: New Map** and rename the new map *Hazards*.

31→ Switch to the USA Map. Select and copy the States, Counties and Cities layers.

31→ Switch to the Hazards map and paste the layers.

31→ Add the quakehis and volcanoes feature classes from Folders\Usa\usdata.

Not all earthquakes and volcanoes are dangerous. First we will create layers for the significantly damaging quakes and the active volcanoes.

32→ In the Geoprocessing pane, open the Make Feature Layer tool.

32→ Set the Input Features to volcanoes and the Output Layer to *Active Volcanoes*.

32→ Enter the expression: KNOWN_ERUP is Greater than 0. Run the tool.

33→ Clear and reopen the Make Feature Layer tool.

33→ Set the Input Features to quakehis and the Output Layer to *Damaging Quakes*.

33→ Enter the expression: MAG is Greater Than 7 Or MMI is Greater Than or Equal to 8. Run the tool.

33→ Remove the original volcanoes and quakehis layers, leaving the selection layers.

First, let's determine how many large cities lie within 50 miles of a dangerous quake or volcano.

34→ Turn on visibility for the Cities layer.

34→ In the Geoprocessing pane, open the Select Layer By Location tool.

34→ Select Cities within 50 miles of Active Volcanoes. (This map represents a large area, so use *Within a distance geodesic* for the Relationship.)

34→ Run the tool and keep it open.

To select the cities close to a volcano OR an earthquake, the Selection type must be set to add the cities close to earthquakes to the current selection.

34→ In the Geoprocessing pane, in the Select Layer By Location tool, change the Selecting Features to Damaging Quakes.

34→ Change the Selection type to *Add to the current selection* and run the tool.

35→ In Contents, right-click Cities and choose Selection > Make Layer From Selected Features.

35→ Rename the new layer *Risky Cities*.

35→ Turn off visibility for the Cities layer.

9. Use the Summary Statistics tool to determine the total number of people in the risky cities.

Not everyone lives in a city, however. Let's try a different approach with counties.

36→ Turn off visibility for the Risky Cities layer.

36→ Repeat the previous steps to select the Counties within 50 miles of an earthquake or volcano (Fig. 9.24).

37→ Make a new layer from the selected features and name it *Risky Counties*.

37→ Give Risky Counties a contrasting color from Counties. Clear the selection.

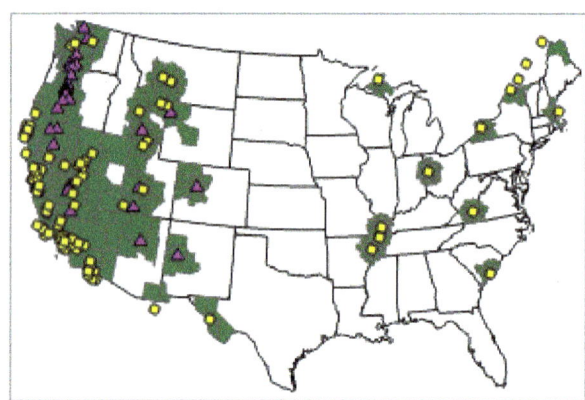

Fig. 9.24. Risky counties (green) near dangerous quakes (yellow) and volcanoes (purple)
Source: Esri

Out west, many counties are large, so the selected risky areas extend further than 50 miles from earthquakes or volcanoes (Fig. 9.24). Buffering and clipping the county areas would better represent the analysis. To buffer both at once, the volcanoes and earthquakes must be merged. We can't combine the table information because the fields don't match, but we can combine the features.

38→ In the Geoprocessing pane, open the Merge tool.

38→ Set the two Input Datasets to Damaging Quakes and Active Volcanoes.

38→ Name the Output Dataset *Dangers*.

38→ Ignore the Field Map. Run the tool.

39→ In the Geoprocessing pane, open the Buffer tool (Fig. 9.25).

39→ Set the Input Features to Dangers.

39→ Name the Output Feature Class *DangersBuf*.

39→ Set the Distance to 50 Miles.

39→ The Planar method is fine since the data are in an equidistant projection.

39→ Set the Dissolve Type to *Dissolve all output features into a single feature*.

39→ Run the tool.

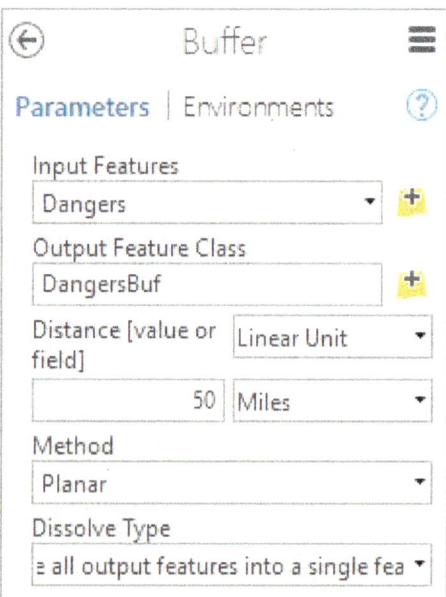

40→ In the Geoprocessing pane, open the Clip (Analysis Tools) tool.

Fig. 9.25. Buffering the dangers
Source: Esri

40→ Set the Input Features to Risky Counties and the Clip Features to DangersBuf.

40→ Name the Output Feature Class RiskyClip. Run the tool.

Finally, a map based on population density portrays the regions that are especially vulnerable.

41→ Turn visibility off for all layers except States and RiskyClip.

41→ Symbolize RiskyClip using a graduated colors symbology based on the POP2014_SQMI field with a Geometric interval classification (Fig. 9.26).

10. Why would summing the population in the RiskyClip layer overestimate the population in the buffers?

Definition queries

A definition query sets a property for a layer that temporarily restricts the layer to features meeting the condition. It is useful for creating layers that are subsets of a feature class.

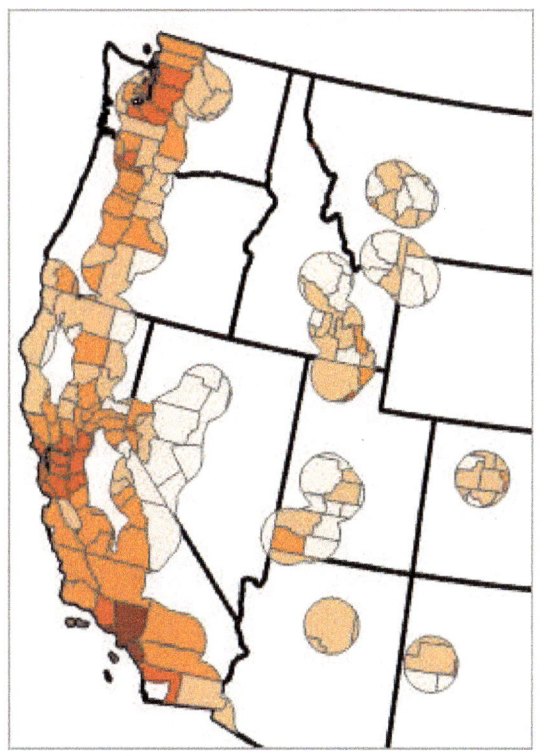

Fig. 9.26. Part of map showing population density of areas close to earthquakes and volcanoes

Source: Esri

42→ Save the project and switch back to the USA Map.

42→ In Contents, turn on the visibility for Capitals.

42→ Open the properties for the Capitals layer and examine the Source settings. The source feature class is named cities.

42→ Switch to the Definition Query settings in the properties. Notice that it contains an expression: CAPITAL is Equal to State.

42→ Click Cancel to close the Layer Properties window without making changes.

The Capitals layer uses the same feature class as the Cities layer. A Definition Query has been created to exclude cities that don't contain "State" in the CAPITAL field. It saves space in the database because it avoids storing a separate feature class for capitals. It also responds to updates of the source. If the percentage of Hispanics in the cities is calculated, the same information will appear in the Capitals layer automatically.

Creating a layer file is a handy way to use the Capitals many times in different maps without having to reset the definition query or symbols each time.

43→ In Contents, right-click the Capitals layer and choose Sharing > Save As Layer File.

43→ Click the Folders entry and navigate into the Usa folder.

43→ Name the file USCapitals and save it.

43→ Switch back to the Hazards map.

43→ Click **Map: Layer: Add Data**. Navigate into Folders\Usa and select the USCapitals. lyrx file. Click OK.

The capitals appear, complete with the yellow star symbol, yet they are still reusing the cities feature class. The Definition Query, as a property of the original layer, is stored in the layer file and continues to suppress cities that are not capitals. Layers, layer files, and definition queries are means to create semipermanent subsets of data without affecting the original feature classes.

TIP: Be sure to Verify when doing definition queries. An error may cause the layer to disappear entirely, which can be confusing.

Exporting data

Sometimes it is more useful to create new feature classes than to manipulate the properties to show a subset. Imagine an analysis that covers only the "Lower 48" states. No field in the table designates this property, so a definition query cannot accomplish it. Furthermore, certain tools (clip, for example) ignore definition queries and will use the entire data set. As a final example, some editing may be needed for the analysis that would be inappropriate for the original data set. In such cases, exporting to create a new feature class is a better option.

44→ Save the project. Click **Insert: Project: New Map**.

44→ Rename the new map Lower 48 and close the USA Map and the Hazards map.

44→ Add the USCapitals layer file to the map.

45→ In Contents, switch to the List By Selection panel and make sure that the Capitals layer is selectable by checking the box.

45→ Click **Map: Selection: Select** and draw a box around the capitals in the lower 48 states.

45→ In Contents, right-click the Capitals layer and choose Data > Export Features.

45→ Name the Output Feature Class L48Capitals and run the tool.

45→ Remove the original Capitals layer from the map.

Let's create a few more lower 48 data sets.

46→ Add the states feature class from the Usa\usdata geodatabase.

46→ Select the lower 48 states and export them as L48States. Remove the original states.

47→ Repeat the process to create an L48Counties feature class.

47→ Clear all selections.

Spatial analysis using queries

Imagine that a national company is looking to develop five new resort properties. The marketing team has determined four criteria to help narrow down the possibilities: (1) low population density, (2) warm climate, (3) close to a highway, and (4) not too close to a city. Since the tools should now be familiar, only general instructions will be given.

48→ Use the Make Feature Layer tool to create a layer of counties with a 2014 population density less than 5 people per square mile. Name the layer Low Pop.

49→ Add the climatestations2005.dbf table from the Usa folder.

49→ In Contents, right-click the climatestations2005 table and choose Display XY Data. Keep the defaults and run the tool.

We will select the stations we need and export them to a new feature class.

49→ Use Select Layer By Attribute on climatestations2005_XYTableT layer to select the stations with an average annual temperature (the Ave_annual field) greater than 60 degrees.

49→ In Contents, right-click the climatestations2005_XYTableT layer and choose Data > Export Features. Name the output warmstations.

49→ Remove the climatestations2005 table and layer.

Not every county has a climate station, so selecting counties that contain warm stations won't work, and using a distance selection would extend too far to the north. Instead, we'll use interactive selection with a polygon to designate the desired counties.

50→ In the Contents List By Selection panel, right-click L48Counties and choose *Make this the only selectable layer* (look at the top of the menu).

50→ Click the **Map: Selection: Select** drop-down and choose Polygon.

50→ Click to enter a polygon that skirts the warmstations points on the north and includes all counties down to the southern border of the United States. Double-click to finish the polygon and make the selection.

50→ The selected set should appear similar to Figure 9.27. It need not be exactly the same.

50→ Make a layer from the selected features and rename it Warm Counties.

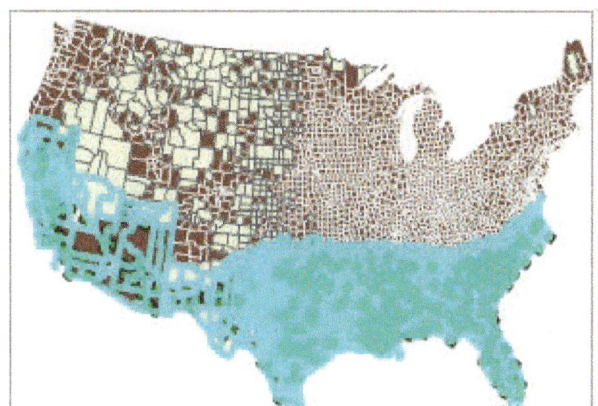

Fig. 9.27. Warm counties selection (cyan)
Source: Esri

51→ Switch to the Contents List By Drawing Order panel and turn off all layers except Low Pop and Warm Counties.

51→ Clear all selections.

51→ Use Select Layer By Location to select the Warm Counties that are identical to the features in Low Pop.

51→ Make a layer from the selected Warm Counties features and rename it Target Counties. Clear the selection.

Next we use a buffer and a clip to find the areas close to major highways. Buffering is an intensive process, so to save time, we will select the roads close to the target counties. Then the Buffer tool will only process the selected roads.

52→ Add the majroads feature class from the usdata\Transportation feature dataset.

52→ Use Select Layer By Location to select the majroads within 35 miles of TargetCounties.

52→ Buffer the majroads layer to 30 miles. Be sure to dissolve all the features. Save the output as roadbuf30.

53→ Clear all selections.

53→ Clip the TargetCounties using the roadbuf30 layer. Name the output *TargetClip*.

Finally, we will exclude the areas that are too close to cities using buffer and erase tools. Again, we will reduce the number of cities processed first.

54→ Add the cities feature class from the usdata geodatabase.

54→ Use Select Layer By Location to select the cities within 25 miles of TargetCounties.

54→ Buffer the cities layer to 20 miles. Be sure to dissolve all the features. Save the output as *citybuf20*.

55→ Use the Erase tool to erase areas of TargetClip using citybuf20. Name the output *FinalTargets*.

55→ Turn off all layers except L48States and FinalTargets.

TIP: The Erase tool requires an ArcGIS Advanced license. Skip that step if you don't have the capability. Or, use the Union tool to combine TargetClip and citybuf20, select the output features for which the field FID_citybuf20 equals −1, and export them as *FinalTargets*. (See Chapter 10 for more about the Union tool).

The FinalTarget areas should look similar to the green and purple areas in Figure 9.28; they may not match exactly.

Of the target areas, the best ones would be close to major airports.

56→ In the Catalog pane, switch to the Portal tab and choose the All Portal icon.

56→ Type the search term **air-ports** and find the feature layer USA Airports. Add it to the map. It contains several different layers organized by airport size. Expand the group layer and find the layer labeled 1,000,000 or more.

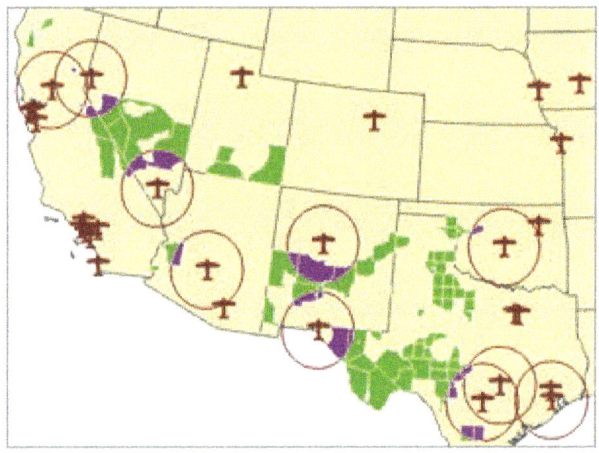

Fig. 9.28. The final target sites (green/purple), the best sites (purple), and the airport buffers

Source: Esri

TIP: If your airport layer does not match the one in the tutorial, try to follow the steps as well as possible. The final result may differ.

57→ Select the airports in the 1,000,000 or more layer that are within 105 miles of the TargetCounties layer.

57→ Buffer the airports to a distance of 100 miles. Name the output *airportbuf100*.

58→ Clip the FinalTargets layer with the airportbuf100 layer. Name the output *BestTargets*.

59→ Clean up the map to show the best targets. They should look similar to the purple areas in Figure 9.28.

This is the end of the tutorial.

→ Save the project and close Pro.

Practice Exercises

Use the data in the mgisdata\Usa folder to answer these questions.

1. How many counties in the United States are named for Abraham Lincoln?

2. How many cities in the United States contain the word *City*? How many end with the word *City*?

3. What percentage of state capitals lies within 25 miles of a river?

4. How many states contain at least one volcano?

5. Congress has awarded FEMA 100 million dollars to help large cities prepare for earthquakes. The cities that qualify for the funding must have more than 500,000 people and be less than 50 miles from one or more earthquakes exceeding 6.0 in magnitude. How many cities are eligible? Which eligible cities are NOT in California?

A conservation group is looking for a place to establish a private bison preserve. The group has developed a set of criteria to narrow down the possibilities. Help them with their analysis.

6. The first step is to target counties that are within the Great Plains states (the stack of states from North Dakota to Texas) or within 100 miles of the Great Plains states. These counties also need to have a population density less than 10 people per square mile. **Capture** a map showing these target counties.

7. The next step is to exclude areas in the target counties that are less than 30 miles from an interstate highway or 30 miles from a city. **Capture** a map showing the remaining areas under consideration.

8. Next, the best area would be within 50 miles of a major river. Determine the areas that also meet this requirement and **capture** a map of them.

9. The preserve must be on private lands. Find the ArcGIS Online feature layer Federal_and_Tribal_Land and use it to exclude such areas. **Capture** a map of the remaining land.

10. With so much area still remaining, the group decides to target areas with an even lower population density, less than two people per square mile. **Capture** a map of the final target areas.

Chapter 10. Joins and Overlay

Objectives

- ▶ Learning the purpose and capabilities of spatial joins
- ▶ Setting up spatial joins based on cardinality and feature type
- ▶ Using map overlay to analyze multiple spatial criteria
- ▶ Understanding differences between spatial joins and overlays

Mastering the Concepts

GIS Concepts

In the most general sense, map overlay occurs whenever two or more input layers are processed to create an output based on some spatial combination of features. GIS practitioners differ widely, however, in which types of operations are included under this heading. The extraction functions discussed in Chapter 9, clip and erase, can be considered queries in one sense, but some classify them as overlay because they can split features that cross each other to create a new output data set. However, other definitions of overlay require the attribute tables of the input feature classes to be combined in the output. Clip and erase do not meet this definition, for they merely copy the attributes of the features being retained to the new feature class unchanged.

Spatial joins meet the general definition of overlay, although they are not usually included in lists of overlay functions, perhaps because they don't split features, even though they combine the tables. The classical functions of intersection and union remain for many the prime examples of overlay, and they represent some of the most commonly cited and used functions in spatial analysis. In this chapter we will explore these different spatial combination techniques without worrying about how to classify them.

What is a spatial join?

Chapter 7 presented attribute joins performed on tables—for example, joining information in a summarized table of earthquake statistics by state to a map of states (Fig. 7.9); the result could be used to create a graduated colors map of damages. This join is based on a common field, the state name, and it has a **cardinality** of one-to-one. The join combines the two tables as if they were one table. The target (or destination) table receives the information from the join (or source) table. A **spatial join** is similar to an attribute join, except that, instead of using a common field to decide which rows in the table match, a spatial relationship between the features is used.

Like attribute joins, spatial joins designate a join feature class and a target feature class. Unlike an attribute join, which appends the join attributes to the existing target table, a spatial join creates a new feature class. It copies the features from the target feature class and appends the attribute information from the join layer (Fig. 10.1). The two original feature classes are unaffected. The target feature class determines the type of features in the output feature class. If an airports feature class is joined to a cities feature class with cities as the target, then the output feature class contains cities.

Spatial relationships for joins

Spatial joins rely on similar spatial relationships (operators) to those used in the Select By Location tool, broadly grouped into categories of intersection, proximity, or containment

(Fig. 10.2). A few of the most commonly used relationships are discussed in detail here. A complete discussion of these relationships and graphic examples can be found in the ArcGIS Pro Help by searching for the topic *Select By Location: graphic examples.*

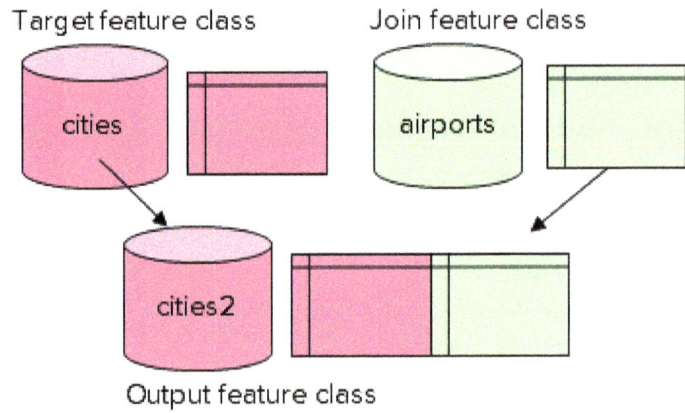

Fig. 10.1. A spatial join copies features from the target feature class and appends attributes from the join feature class.

Spatial joins may be performed on any two feature classes. A user can join points to points, polygons to polygons, lines to points, and any combination of the three geometry types, although the availability of certain spatial operators may be reduced in some cases (e.g., a point feature cannot contain a polygon feature). The output feature class will always have the same feature type as the target feature class.

Intersection	Proximity	Containment
Intersect	Within a distance geodesic	Contains
Intersect 3D	Within a distance	Completely contains
Are identical to	Within a distance 3D	Contains Clementini
Boundary touches	Closest geodesic	Within
Share a line segment with	Closest	Completely within
Crossed by the outline of		Within Clementini
Have their center in		

Fig. 10.2. Spatial relationships used for joins

Because spatial joins are so similar to spatial queries, it is fair to ask how they differ and why someone should use one over the other. The key difference lies in the creation of an output data set with the joined tables. Joins are helpful when it is beneficial to permanently associate the table information—for example, when the distance from one feature to another is stored in the output.

Spatial join examples

Joins with intersect operators

The intersect operator tests whether one feature touches another, such as which zoning category overlaps each parcel in a neighborhood (Fig. 10.3). In this spatial join using the *intersect* relationship, the parcels feature class is the target, and the join feature class is zoning. Each feature class has its own attributes for each feature. When the spatial join is performed, a copy of the target layer is created. Each parcel is tested to determine which zoning polygon intersects it, and the attributes for that zone are attached to the parcel record. The final table includes attribute information for each parcel: the standard parcel attributes plus the zoning information. The output could be used for additional analysis, such as using the area of each parcel and the assigned tax rate to calculate the base land tax for each parcel.

Joins with distance operators

The proximity operators use a distance criterion to link one feature and its attributes to another. Figure 10.4a illustrates a join between airports and cities using the *closest* relationship. The join feature

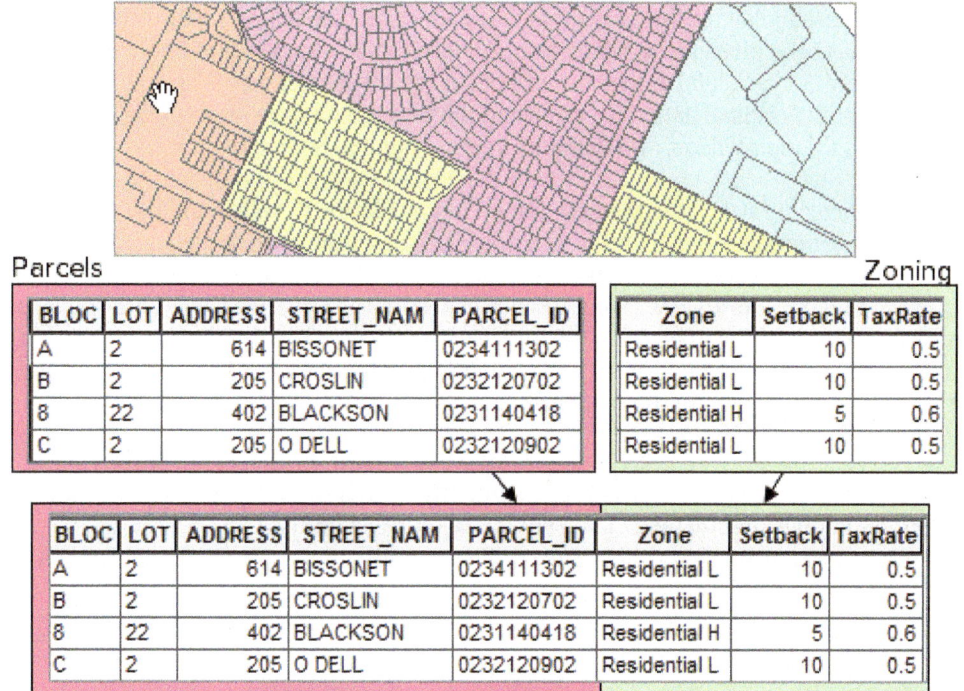

Fig. 10.3. A spatial join using the intersect operator to combine information on parcels and zoning

Source: City of Austin, TX

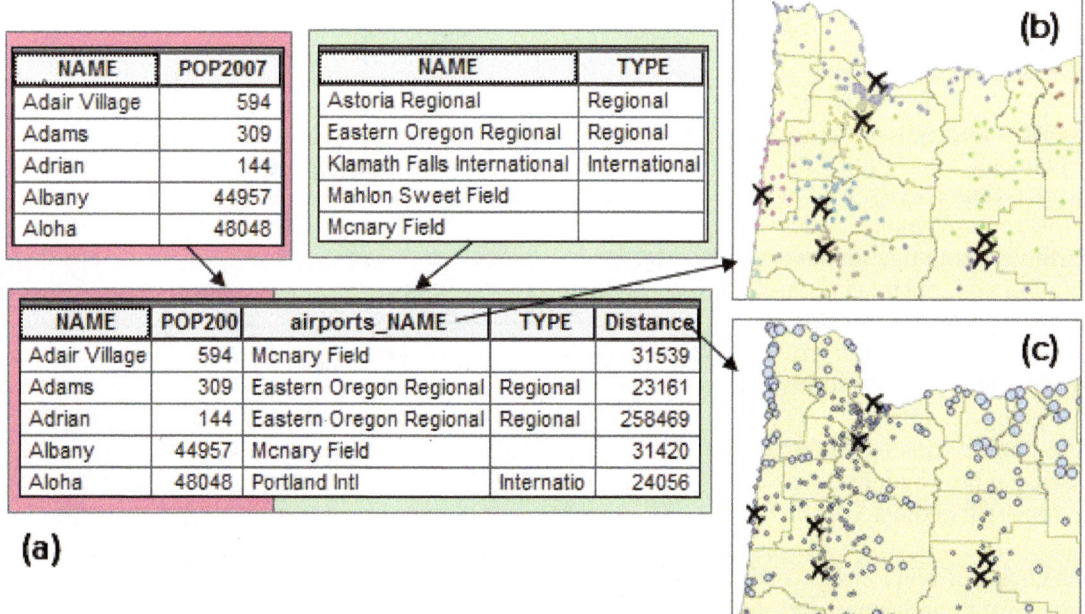

Fig. 10.4. A distance spatial join between cities and airports: (a) joined tables; (b) unique values map based on the airport name; (c) graduated symbol map based on distance

(b, c) Source: Esri

class, airports, has been joined to the target feature class, cities. The output feature class contains cities. Each city has been given the attribute information from the airport that lies closest to it. In a distance join, a new field is added to the output table to record the distance. The output attribute table contains two parts, the original data from cities and the joined data from airports. So McNary Field is the closest airport to Adair Village, and Eastern Oregon Regional is the closest airport to Adams.

Two maps have been created from the new joined layer. Figure 10.4b is a unique values map based on the airport name, so each dot gets a color based on the closest airport. The colors indicate which cities are served by each airport, assuming that people will drive to the closest one. The second map is a graduated symbol map based on the distance field, with the larger circles indicating greater distance from the airport (Fig. 10.4c). The units of the distance field are always given in the stored coordinate system units. These data are in the Oregon Statewide Lambert coordinate system, and the units are meters.

TIP: Always check the original coordinate system after a join to verify the units for the distance values. Also, remember that distance joins should be performed only on projected coordinate systems that preserve distance, or the results will be invalid.

Joins with containment operators

In a join using the containment operator *within*, the records of the feature classes are joined based on whether one feature is inside another. Figure 10.5 shows a point layer containing septic system locations and a polygon layer showing geological units. Imagine that a community has a porous geological aquifer that provides the city water supplies. Extensive development outside city limits has caused the city concern about potential contamination of the aquifer by faulty septic systems. Assessing the threat requires identifying the number of septic systems that occur in the outcrop area of the aquifer.

A spatial join solves this problem perfectly. Septics is the target layer and becomes the output feature class. As shown in Figure 10.5, septic system 800 falls within the Terrace Deposits (Qt) polygon in the geology layer, so the attributes from that geologic polygon are appended to the record. Septic systems 836 and 839 fall within the Navarro Group (Knb) unit and receive that polygon's attributes. The output table will be helpful in assessing how many septics fall on sensitive geological units.

The choice of relationship used from the containment operators will depend in part on the direction of the join. In Figure 10.5, the *within* operator is used because Septics is the target

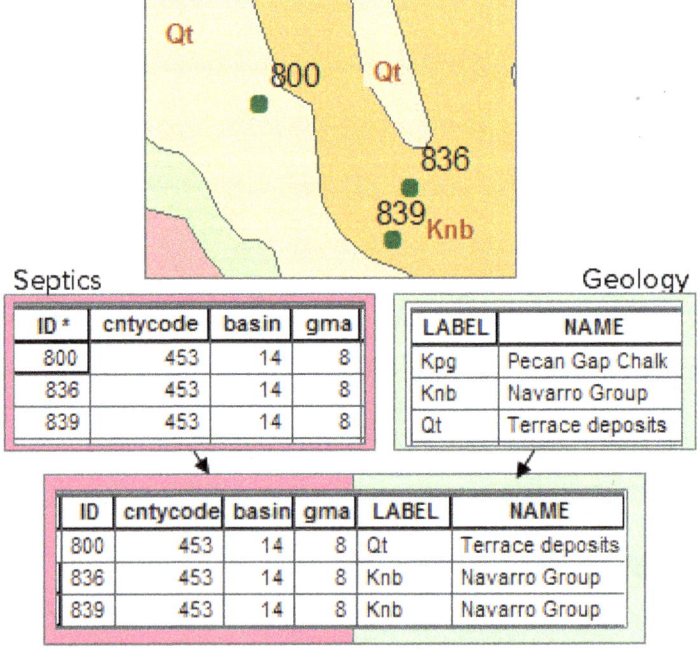

Fig. 10.5. A spatial join gives each septic system the attributes of the geology polygon within which it lies.
Source: USGS, City of Austin, TX

layer, and we wish to identify the polygon that the septic system falls within. If Geology was the target, we would have to ask which septic systems are contained by each polygon.

Depending on the layers used, different operators may generate a similar result. For example, the *intersect* and the *within* operators would usually yield the same result in both the parcel and septic system examples cited. However, differences may occur if the target features lie on or cross boundaries of the join features. In these two examples, this situation would not be expected, but in other situations it would be far more likely to occur (e.g., streams crossing geologic units).

Cardinality in spatial joins

Cardinality is an important issue for spatial joins, just as it is for attribute joins. Each record in the target feature class can only have one row of information from the join feature class appended to it. Before a spatial join is performed, one must assess the potential cardinality between the target features and the join features.

In the first example, assigning zoning to parcels, we specified the *intersect* operator. Since zoning boundaries are usually drawn along road centerlines and rarely intersect parcels, we expect that a parcel will never exist in two different zoning polygons; hence the expected cardinality is one-to-one. However, if we suspected that overlaps might occur due to data errors, we might want to account for that possibility when setting up the join.

In the example with cities and airports, we chose the *closest* operator. Since only one object can be closest to another one, a one-to-one relationship is ensured. We would not need to arrange the join to handle multiple matches.

In the spatial join with septics and geology, we chose the *within* operator. Because a point can never be within two nonoverlapping polygons at once, we again have a guaranteed one-to-one relationship. However, had we chosen to use the *intersect* operator, it would be possible (albeit rare) for a septic point to fall exactly on the boundary between two geology polygons, creating a potential one-to-many relationship. Moreover, if we attempted a join with Geology as the target and specified the *contains* operator, it is quite likely that one geologic unit would contain multiple septic points, again resulting in a one-to-many relationship.

Given that each target record can only be appended to one source record, how do we handle a potential one-to-many cardinality between the target and join features? Two basic approaches are possible. A **one-to-one join operation** specifies that each join feature will be copied once to the target feature class. If the target feature matches more than one record in the join layer, then a **merge rule** will be applied to produce a single record to append, for example, by averaging or summing the numeric values of all the matches to produce a single value. A **one-to-many join operation** creates copies of each target feature until enough are available to append with every matching join feature. This approach must be used cautiously, because it may produce many duplicate copies of each target feature, resulting in very large data sets. It is also often less useful than the one-to-one join operation.

TIP: Be careful not to confuse the cardinality between the target and join features with the type of join operation. *It is common to perform a one-to-one join operation when the cardinality between the target and join features is one-to-many.* Think of the join type terminology as referring to the cardinality between the input target features and the output features, rather than the cardinality between the two input tables.

When a target feature might match multiple join features, it is wise to create merge rules for each field in the join layer. We can call this approach a **merged join**. Consider a spatial join in which

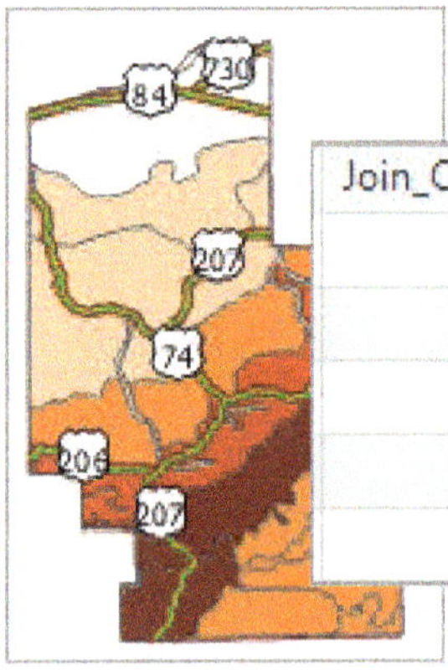

Join_Count	TARGET_FID	HWY	MUID	CLAY	SLOPE	HYDGRP	HYDJOIN
3	638	206	OR148	23.3	23.3	D	D,C,C
3	665	74	OR153	9.6	9.1	B	B,B,B
3	676	207	OR153	9.6	9.1	B	B,B,B
3	681	74	OR160	15.8	5.8	B	B,B,C
2	640	206	OR160	17.9	8.2	B	B,C

Fig. 10.6. Managing merge rules in a one-to-one spatial join with a one-to-many cardinality
Source: USGS, Esri

highways are being joined to soil units in order to assess maintenance costs imposed by clay-rich soils. Each road is likely to cross multiple soil units, a one-to-many cardinality. Figure 10.6 shows a map of highways in Morrow County, Oregon, over a map of the soil clay percentage.

For numeric data, merge rules typically involve statistics, such as calculating the sum or mean from the matching records. The table in Figure 10.6 shows the output table from the merged join. The Join_Count field indicates how many soil units were intersected by each highway feature. Numeric fields, including the clay percentage (CLAY) and the slope (SLOPE), were averaged and appended to the highway record. Hence the first record for Hwy 206 contains clay and slope values that were averaged from the three soil units it crosses.

For text or categorical data, like the map unit identifier (MUID) or the hydrologic group (HYDGRP), the available statistics are limited because such values cannot be averaged or summed. One common approach is to select the first or last value stored in the data table, but this solution produces arbitrary and not very useful results. In Figure 10.6, both the MUID and HYDGRP field contain the first value.

A more complex solution, not available in all GIS programs, would combine the values separated by some sort of delimiter, such as a comma. For example, if a highway crossed three soil units with hydrologic groups B, C, and D, the output field would contain "B,C,D". This option requires creating a new field that is long enough to accept the merged values. In this example, a new field named HYDJOIN, defined with a length of 50, was created to hold the concatenated values. This solution, although convenient, needs to be considered carefully before implementation. If a target feature might match many different join features, then the concatenated values might grow too large for the field.

Classic overlay

One-to-many relationships between target and join features can be difficult to manage, which is one reason why additional tools exist to circumvent these issues. The classic overlay functions were among the first developed for GIS analysis, and they are often presented as examples of what spatial analysis can do. Like a spatial join, these functions combine two layers to create a new output feature class containing information from both of the inputs. Unlike spatial joins, however, features are split if they cross each other, creating new features and enforcing a one-to-one relationship between the attributes from each table, avoiding the problems that occur with one-to-many or many-to-many cardinality situations.

In Figure 10.6, imagine splitting the road whenever it crosses from one soil unit to another. Highway 74 crosses four different polygons, so it would be snipped into four pieces, each of which exists in only one polygon. A one-to-one cardinality between the road sections and the soil polygons is created, allowing the soil units to be joined to the road in every case, with no merging or summarizing needed. Each road in the output table would contain its original attributes plus the attributes of the polygon it crosses.

Assessing geologic hazards, such as landslides or falling rock, provides a good example of overlay analysis. The risk depends primarily on two factors, the slope of the terrain and the strength (competence) of the geological units (sandstone is stronger and less susceptible to sliding than shale). Figure 10.7 shows a slope class map, with the darker colors indicating higher slopes, and a geology map that has been symbolized from low (pale orange) to high (dark orange) competence. The lowest map in the stack shows the overlay result of combining the polygons from the geology and slopeclass feature classes.

During overlay, all boundaries of the original polygons are transferred to the output to create new nonoverlapping polygons, and each new polygon has been given the attributes of the originals, containing information about both geology and slope for the same feature. Each new polygon is considered to have two parents, each of which contributed area and attributes to the child polygon. Consider the selected child polygon (pink) in the output. Its top attributes, outlined in orange, came from the geology parent feature class. The lower attributes, outlined in blue, came from the slopeclass layer.

In Figure 10.7, notice the two fields FID_SDgeology and FID_slopeclass. These fields record the original feature IDs of the parent polygons. In a case where an output polygon has no parent in one

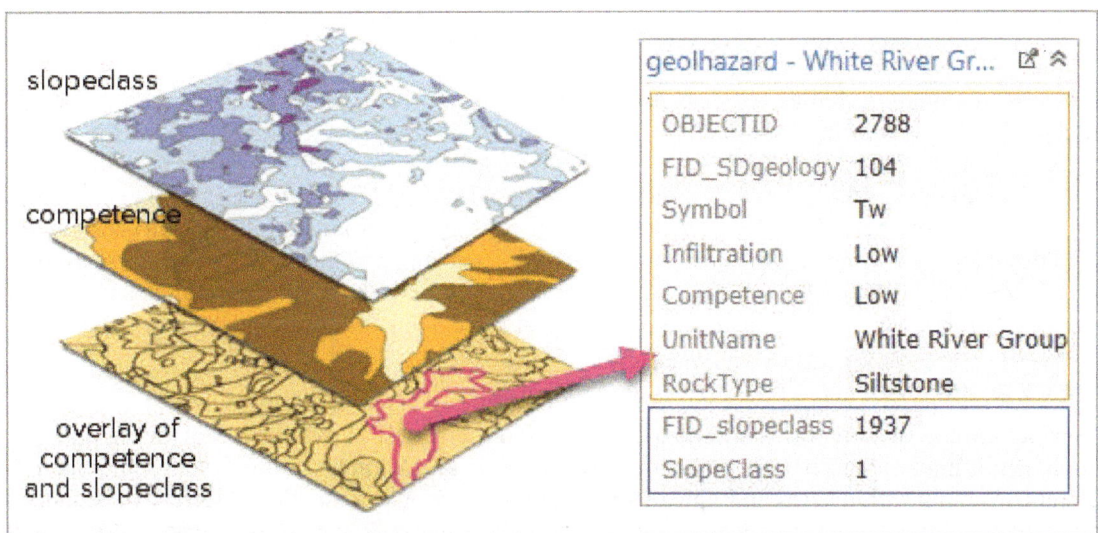

Fig. 10.7. Overlay combines both features and attribute tables of two feature classes
Source: South Dakota Geological Survey

of the input layers, the output field values normally supplied by the parent will be set to <Null>, except the FID_parent value, which is set to −1. This handling allows the user to identify polygons without two parents, which can be useful at times. On rare occasions, it can also be helpful to retain these fields and use them to create a join back to the original feature classes. However, most of the time these fields have little use and could be eliminated to simplify the output table.

To evaluate the geologic hazards, the combinations of attributes that signify risks can be selected and mapped. For example, the combination of high competence and steep slope creates a risk of falling rock, which might then be mapped with roads to find sections that may need signs or mitigation to lower the risk posed to cars (Fig. 10.8).

Intersect and union

Classic overlay includes two functions. **Union** combines two polygon layers, keeping all areas and merging the attributes for both layers. **Intersect** also merges the attributes but retains only the areas (or lines or points) common to both layers, and it may be performed with points, lines, or polygons.

In a union, the primary goal is usually to evaluate combinations of attributes, as demonstrated in the geologic hazard analysis just described. Usually the geographic extent of the two input data sets is the same. If not, areas that existed only in one input would receive attributes only from one of the parent polygons, and it would not be possible to find the combinations of interest.

With intersect, the areas represented by the the input features usually won't match, and the main goal is to find areas of overlap where two or more conditions exist simultaneously. This function provides the foundation of a class of problems known as **suitability analysis**—evaluating a landscape to find which areas best serve a given purpose based on a set of geographic constraints.

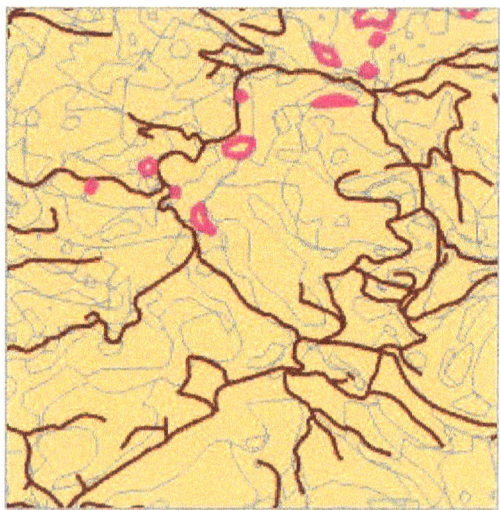

Fig. 10.8. Areas prone to falling rocks (pink) pose a risk when close to roads.
Source: South Dakota Geological Survey, Black Hills National Forest

For example, one could find the potential habitat for a species with specific environmental requirements. Imagine that a rare species of snail inhabits the Black Hills of South Dakota. This snail prefers limey soils in dense coniferous forest, and it is rarely found above an elevation of 1600 meters or below 1200 meters. Each data set (elevation, geology, vegetation) is queried to find the areas meeting the specified condition, and the resulting three layers are intersected to find where all three conditions hold (Fig. 10.9). The habitat polygons in the output contain the full attributes of all three layers, should that information be needed. Such a map would help biologists to create sampling strategies for counting populations, to analyze whether the habitats are interconnected or widely separated, and to make decisions about forest management to protect the snails.

Overlay geometries

The snail habitat demonstrates polygon-on-polygon intersection. However, intersection, unlike union, is not limited to two polygon layers. Intersection can assign attributes from polygons to lines or points within them. In Figure 10.10a, line-in-polygon intersection is used to assign a

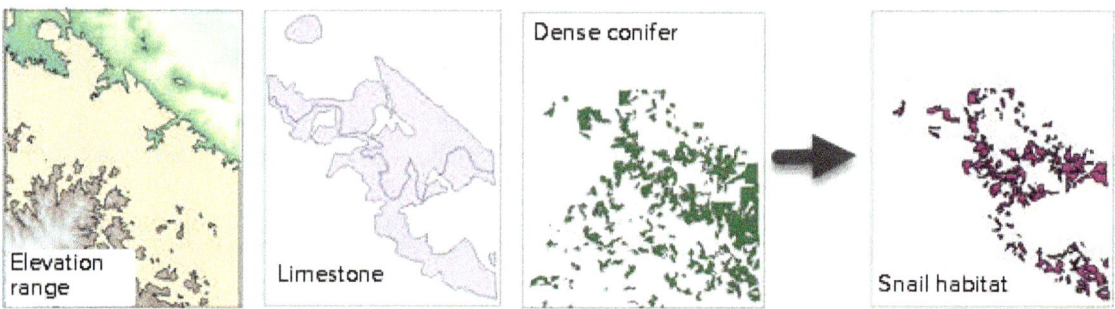

Fig. 10.9. Intersecting elevation, limestone areas, and dense conifer vegetation to identify potential snail habitat
Source: USGS, South Dakota Geological Survey, Black Hills National Forest

geologic unit to each stream reach, useful for estimating stream loss to groundwater. One can also intersect points with polygons, or point-in-polygon intersection. A realtor might have a point feature class of houses for sale and intersect it with polygons representing elementary school districts in order to be able to list the school for each house (Fig. 10.10b). (Note that a point-in-polygon intersection is equivalent to a spatial join using the intersect operator with the point layer as the target).

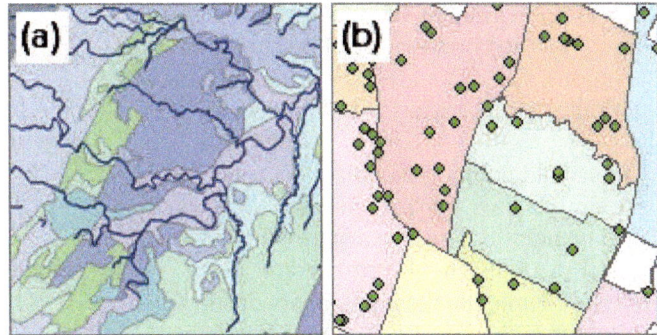

Fig. 10.10. (a) Assign geologic units to streams with line-in-polygon intersection; (b) assign school districts to houses for sale with point-in-polygon intersection. Source: USGS, City of Austin, TX

Whereas the union function requires two polygon input layers, the intersect command is more versatile. Both input and overlay layers may contain points, lines, or polygons. The output geometry may vary, but it cannot exceed the dimensionality of the lowest input.

If the inputs are both polygons, the output may be points, lines, or polygons. In Figure 10.11a, the two circles might represent the drug-free zones around two schools. The polygon output shows the area of overlap; the line output outlines the boundary of the overlap area, and the point output shows where the boundaries of the two circles cross.

If the inputs are both line feature classes, the output may be lines or points (Fig. 10.11b). These lines might represent hiking trails, in which case the line output would find locations where two trails travel together. The point output could find trail intersections that need to have signposts placed on them.

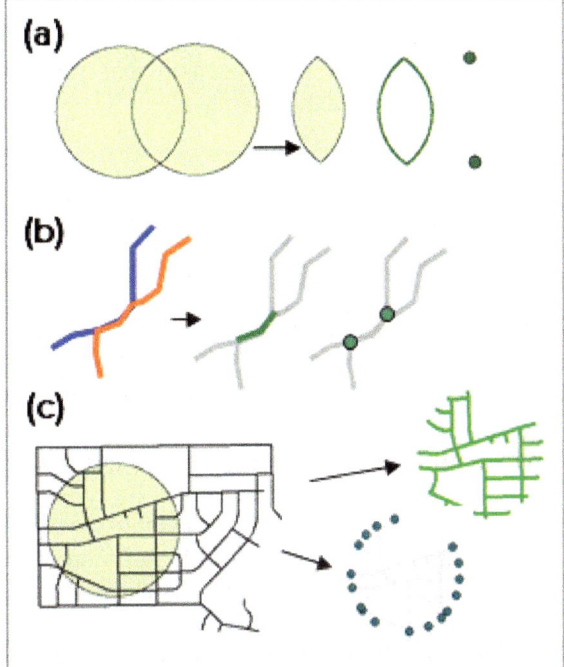

Fig. 10.11. Intersect geometries. (a) Polygons yield polygons, lines, or points. (b) Lines yield lines or points. (c) Polygons and lines yield lines or points.

If the inputs are polygons and lines, the output may be points or lines (Fig. 10.11c). If the circle represents the area around a school where sex offenders are forbidden to live, the line output could provide a list of the streets and address ranges that are off limits. If the city intends to put a sign on every street entering the zone, the points output would estimate the number of signs needed and where they should go.

Intersecting two sets of point data requires the output to be points, but it is not a common analysis, unless someone is looking for duplicate points in two different data sets.

Comparing overlay functions

Figure 10.12 summarizes the different overlay operations using two polygon input layers. Clip and erase do not combine attributes from the input layers; intersect and union do. Clip and erase can accept point and line feature classes for one input, but the boundary input must contain polygons. The output feature classes will have the same geometry as the input classes.

Unions and polygon-on-polygon intersections are similar, so a word about when to use each may be helpful. The primary use of intersect is to find areas where certain conditions overlap. Typically the input layers occupy different regions, as do the elevation range, limestone, and conifer layers shown in Figure 10.9. The goal is to find the areas common to the inputs. Although the attributes from each input are combined in the final table, the attributes themselves are sometimes of little interest. In the snail habitat problem, the pertinent attributes were already preselected using queries. The main goal was developing a map of the overlaps.

In a union, the primary goal is to combine the tables. Generally the input layers all fully occupy the same region, as the geology and slope class layers do in Figure 10.7. It is expected that the entire map area will remain, but that the new features will contain the attributes of both inputs. After a union, the usual next step is to perform attribute queries to find areas with specific combinations of attributes, or to symbolize the combinations to show areas of interest. In the landslide hazard example, we might have extracted the shale and high-slope areas and intersected them to show where the high-hazard areas exist. The hazard is either there or not there. With a union, though, we can analyze different combinations of geology and slope in order to assess potential impacts on slope stability.

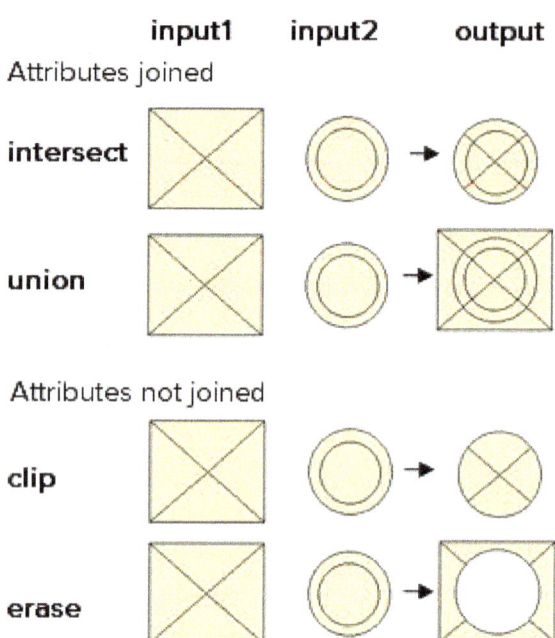

Fig. 10.12. Summary of polygon overlay operations

Union may also be helpful as a substitute for the Erase tool, which requires an ArcGIS Advanced license. Suppose one wishes to erase a square layer using a circles layer (as in Fig. 10.12) but does not have the right license. Instead, a union is performed first, combining all the areas from both layers. However, the output polygons outside the circles have no parent polygon from the circles layer, and as previously described, they would receive a −1 value in the FID_circles field. Selecting the polygons for which FID_circles = −1 and exporting them to a new feature class will yield the same areas as would the result from the Erase tool. The result is not identical, however, because union combines the tables, so the circles and square fields would be included in the output. Erase ignores the attributes of the overlay layer (circles) and yields an output containing only the square layer attributes.

Data quality factors in spatial analysis

Logical consistency

As with spatial queries, overlay and spatial joins can suffer problems from data accuracy and logical consistency (whether data relationships match conditions in the real world). Consider Figure 10.13, which shows census tract polygons and a county boundary (brown line). In the real world, tract and county boundaries should coincide, but these data sets came from two different sources and two difference source scales, and they have not been corrected to match.

In a spatial join, such discrepancies cause matching problems. If these two layers were joined using a *within* or *contains* operator in a spatial join, tracts along the boundary would not actually be within the county (or the county would not contain the tracts), and the output table would receive null values because no match could be identified. If the *intersect* operator were used, a tract might be matched to more than one county, which is patently false and might generate inconsistencies in the output. Solving such issues is usually difficult, in which case the spatial join technique will not be effective.

In extraction or classic overlay, inaccurate boundaries or logical inconsistences may produce small extraneous polygons or lines called **slivers** (Fig. 10.13). These slivers do not represent valid combinations of values and are a nuisance when calculating statistics or performing subsequent operations. It is often desirable to prevent slivers when overlaying. Slivers are especially troublesome when multiple overlay or analysis operations are planned because the errors propagate and multiply with each step.

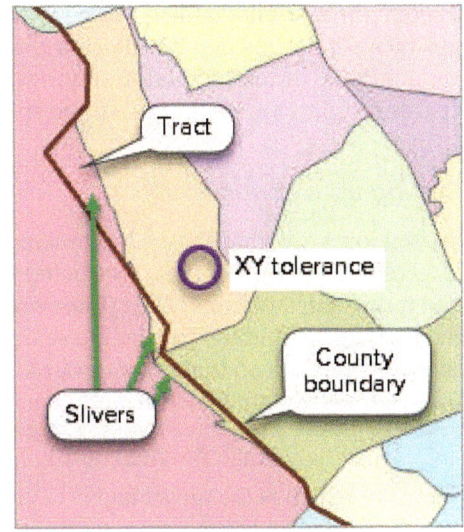

Fig. 10.13. Logical inconsistency between tract and county boundaries
Source: City of Austin, TX

Slivers can be minimized by specifying an **XY tolerance** during an overlay operation. This tolerance specifies the minimum distance between vertices, and it will combine vertices that fall close together. However, the tolerance applies to *all* vertices, and it may degrade the accuracy of the data set. If tolerances are used, caution should be exercised to choose values that are large enough to correct slivers without shifting other vertices by an unacceptable distance.

Consider the purple circle representing an XY tolerance in Figure 10.13. The smaller yellow and green slivers would be eliminated from the output, but the larger pink sliver would not be. Slivers can be removed from an existing data set using the Eliminate tool (an ArcGIS Advanced license is required).

TIP: Always keep in mind that data accuracy and logical consistency issues may negatively affect an analysis, possibly invalidating the results. Be alert for potential problems.

Areas and lengths of features

Some geoprocessing tools (including clip, dissolve, intersect, or union) can affect the areas or lengths of the output features. In geodatabases, the Shape_Length and Shape_Area fields are automatically updated to the new values. However, fields created by the user, such as an ACRES field for parcels, will not automatically be corrected. Users must take steps to update such fields manually, for example by recalculating the acres using a conversion factor based on the units and values in the Shape_Area field, or by using the Calculate Geometry Attributes tool. A table of conversion factors is included inside the back cover of this text. Shapefiles contain no area or length fields that receive automatic updates after geoprocessing, and all such fields must be updated manually using one of these methods.

About ArcGIS

Overlay functions in ArcGIS, including extraction, spatial joins, and classic overlay, are all implemented as tools. Because these tools have similarities, deciding which tool best fits the problem at hand is often the most difficult task. Extraction functions are best when no combination of attribute tables is needed; the output data set will have exactly the same fields as the input but will cover different regions. Spatial joins and classic overlay are both designed to combine attribute tables. When relationships between the features are one-to-one, they produce similar results. When dealing with one-to-many relationships, spatial joins are best for summarizing or merging information while retaining the target features unchanged, whereas classic overlay allows features to be split into pieces.

Setting up a spatial join

Spatial joins are challenging to implement because of the complexity of the options, which are affected by the types of features being joined (points, lines, or polygons), the choice of target layer, the operator being used (intersection, proximity, containment), the cardinality between the join layers, and the cardinality of the join operation itself. It is paramount to clearly visualize the desired result so that the appropriate choices can be made. This series of questions can be followed to help analysts correctly configure the spatial join.

> What should the final output layer or table look like?
>
> Which is the target layer?
>
> Which spatial operator is appropriate?
>
> Which fields do I want in the output table?
>
> Could a target feature match multiple join features? If yes, how should it be handled?

Imagine that a representative from a California congressional district is sponsoring legislation to provide earthquake emergency planning funds, and she is looking for support. She plans to throw a party and invite the representatives from districts with earthquake deaths. She asks her staffer, Jim, to provide names for the invitation list. Jim, grateful for the GIS classes he took, obtains the data sets for the congressional districts and the earthquakes (Fig. 10.14), creates a map to visualize them, and begins to think the problem through.

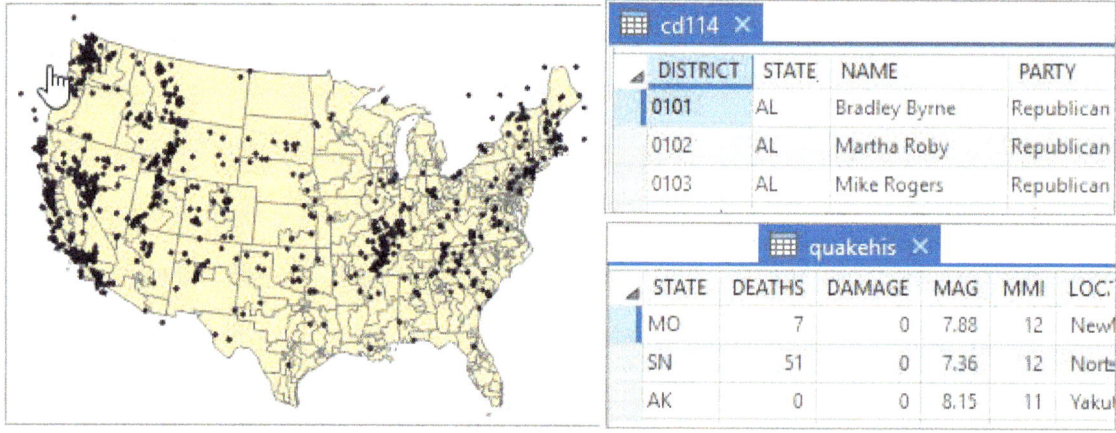

Fig. 10.14. Earthquakes and districts of the 114th Congress
Source: Esri

What should the final output layer/table look like? Jim needs a list of congressional districts, including the names of the representatives. For each district, he also needs to know the number of earthquake deaths.

Which is the target layer? The features in the output layer always match those in the target layer. If districts is chosen as the target, then the output layer will contain districts. If the quakes layer is chosen, then the output layer will contain earthquakes. Jim needs a list of districts in the table, so the districts must be the target.

Which spatial operator is appropriate? The obvious choice is that each earthquake should be assigned to the district in which it occurs. Either the *contains* or the *intersect* operator will work, but in this case, *intersect* is slightly better. It is unlikely, but possible, that an earthquake falls exactly on a district boundary. With *contains*, the earthquake would not be assigned to any district, and its deaths would not be counted. With *intersect*, the earthquake would be assigned to both districts. The deaths would be counted twice, but that would not be an unreasonable solution in this case.

Which fields do I want in the output table? At a minimum, Jim needs the district identifying information, the representative name, and the sum of the earthquake deaths.

Could a target feature match multiple join features? Yes, a district can contain many earthquakes. Jim doesn't want multiple copies of districts in the output; he wants a single district with the summed deaths. Thus, the one-to-one join operation with merge rules is the best solution.

Now that the questions are answered, Jim is ready to perform the join. He opens the Spatial Join tool and sets it up as shown in Figure 10.15. The Target Features, Join Features, Output Feature Class, and Join Operation are quickly filled in, using the answers to the questions. He hesitates for a moment to decide whether to uncheck the Keep All Target Features box. Checking it means that districts without earthquakes will be included in the output feature class. Jim decides that he will keep all districts for now. If his boss decides to invite a few extra people to the party, he will still have the names.

The Field Map takes more thought. Jim keeps the most relevant fields from the districts table (DISTRICTID, STATE, NAME, and PARTY) for use when he prepares the invitation list. For the earthquake fields, he chooses the DEATHS field and applies the Sum Merge Rule. Just because he is curious, he also specifies the DAMAGE field with the Sum rule. Finally Jim selects the Intersect operator and runs the tool.

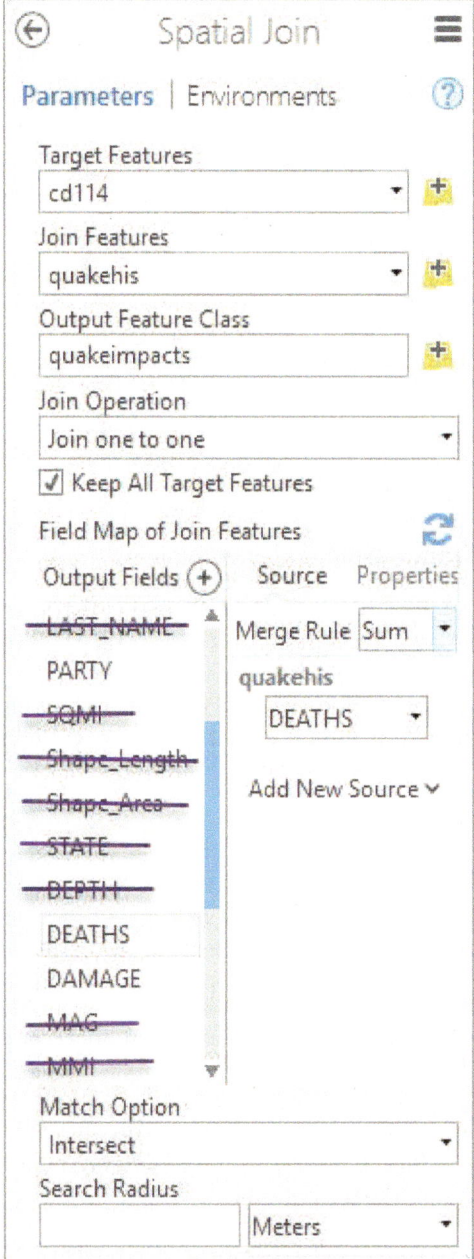

Fig. 10.15. Finding the earthquake deaths in congressional districts

Source: Esri

Figure 10.16 shows the output table. The Join_Count field indicates the number of earthquakes found in the district. Notice that the first three records have a count value of zero (they contain no earthquakes), so the summary fields contain <Null> values. If the staffer had unchecked the Keep All Target Features box, these three districts would have been dropped from the table. The next four districts had earthquakes but no deaths or damage. However, Don Young's Alaska district had 125 deaths.

In creating the new table for the output, each feature receives a new FeatureID (FID) to uniquely identify it in the table. The TARGET_FID field records the FID of the feature in the

Join_Count	TARGET_FID	DISTRICT	STATE	NAME	PARTY	DEATHS	DAMAGE
0	1	0101	AL	Bradley Byrne	Republican	<Null>	<Null>
0	2	0102	AL	Martha Roby	Republican	<Null>	<Null>
0	3	0103	AL	Mike Rogers	Republican	<Null>	<Null>
1	4	0104	AL	Robert B. Aderholt	Republican	0	0
1	5	0105	AL	Mo Brooks	Republican	0	0
4	6	0106	AL	Gary J. Palmer	Republican	0	0
1	7	0107	AL	Terri A. Sewell	Democrat	0	0
29	8	0200	AK	Don Young	Republican	125	31100000

Fig. 10.16. Output table for the spatial join
Source: Esri

original target layer. Occasionally this information is useful in tracking down a particular feature from the input layer.

As the final step, Jim selects the districts with at least one earthquake death and produces a table of people to invite, getting a total of 28 representatives. His boss wants more people at the party, so she directs Jim to invite all the reps from California, whether their districts had earthquake deaths or not. Grateful that he kept all districts in the output, Jim adds the California reps to the selection and produces a final list with 65 names.

Summary

▶ A spatial join combines the records of two feature tables based on the location of the features evaluated by a spatial operator. A new feature class is created by a spatial join.

▶ A cardinality of one-to-many between the target features and the joined features requires a one-to-many join operation or a one-to-one join operation with merged fields.

▶ Joins based on a proximity operator may include adding a field for the distance to the output table, measured in map units of the target layer.

▶ Map overlay resembles a spatial join, but it splits features when they partly overlap. This function enforces a one-to-one relationship between features when their attributes are joined in the output table.

▶ Map overlay comes in two basic types. A union keeps all the features from both layers. An intersect keeps all the features that are common to both input layers.

▶ Spatial analysis should be performed with layers having projected coordinate systems that do not distort distances. Using layers with a geographic coordinate system may yield incorrect results.

▶ Logical inconsistencies, such as tracts and counties having slightly different boundaries, may cause spatial joins to generate incorrect results or create troublesome slivers during an intersect or union operation.

▶ The Shape_Area, Shape_Length, and Shape_Perimeter fields in geodatabase feature classes are stored and updated automatically. Other area-based or length-based fields, such as Area, Acres, or Road_km, are not automatically updated.

▶ Lengths and areas of features stored in shapefiles must be updated manually.

Important Terms

cardinality	merged join	sliver	union
intersect	one-to-many join operation	spatial join	XY tolerance
merge rule	one-to-one join operation	suitability analysis	

Chapter Review Questions

1. What primary characteristics distinguish a spatial join from selecting by location? From the intersect function?

2. What options may be used to handle a one-to-many cardinality in a spatial join?

3. If a point feature type is joined to a polygon layer, with the points as the target layer, what will the feature type of the output layer be?

4. Joining points to polygons may be performed using either the intersect or containment operator, with very similar results. In what situation will the results be different, and how will they differ?

5. Explain the difference between cardinality and join type. Describe why a one-to-many cardinality is often handled using a one-to-one join.

6. For the following spatial join problems, answer the series of questions in the text and then state the type of join that should be used.

    (a) Determine the number of windmills in each county in Wyoming.
    (b) Find the closest city to each rest area on an interstate.
    (c) Determine the FEMA floodplain designation(s) for each agricultural parcel in a county.
    (d) Use grocery stores and block population points to estimate the number of people likely to shop at each store.

7. Describe the difference between intersect and union, including the purposes for which they are used.

8. Which different types of outputs are possible (points, lines, polygons) when performing intersect and union?

9. Why is logical consistency important when overlaying data sets with spatial joins? When using intersect or union? Can these problems be reduced? How?

10. Explain why feature lengths or areas are a concern following overlay, and describe what should be done to mitigate those concerns.

Expand your knowledge

Expand your knowledge by reading these sections of the ArcGIS Pro Help. These sections include a new location, the Tool Reference tab next to the Help tab in the Help.

> Help > Data > Data Types > Tables > Spatial joins by feature type
> (and the subsequent sections involving joins and spatial joins)
> Tool Reference > Tools > Analysis Toolbox > Overlay toolset
> Tool Reference > Environments

Mastering the Skills

Teaching Tutorial

The following examples provide step-by-step instructions for doing basic tasks and solving basic problems in ArcGIS. The steps you need to do are highlighted with an arrow →; follow them carefully.

> → Start ArcGIS Pro and open the ClassProjects\OverlayAnalysis project. Remember to save the project often.

Understanding spatial joins

TIP: At the word **STOP**, pause and set up the problem using the questions presented in the Concepts section. Then read on to see if you analyzed the problem correctly.

The Austin map view shows the geology and water wells for Austin, TX. For each well, we would like to know the geological unit associated with it. **STOP** and think it through.

The desired output is a table of wells containing a field with the geological unit. Therefore, Wells is the target layer. We want to join attributes from the geologic unit that the well is *within*. Since wells can fall within only one polygon, the cardinality is many-to-one, and no merge rules are required. In the rare case that a well falls exactly on a geologic boundary, it will be ignored.

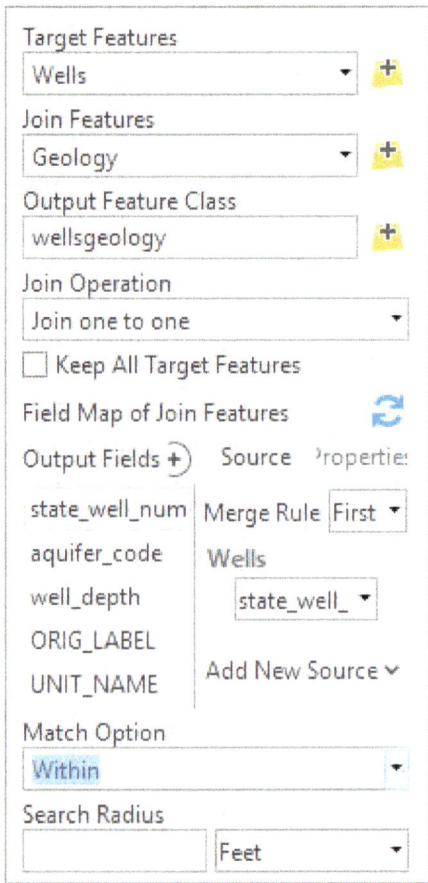

TIP: Set up the windows to keep the Geoprocessing pane open throughout the tutorial.

1→ In the Geoprocessing pane, open the Spatial Join tool (or select it from the **Analysis: Tools** scroll box).

1→ Set the Target Features to Wells and the Join Features to Geology (Fig. 10.17).

1→ Name the output *wellsgeology* and save it in the home geodatabase.

1→ Select *Join one to one* for the Join Operation.

1→ Uncheck the box to Keep All Target Features to eliminate wells outside any geologic unit. Don't run the tool yet.

The wells feature class has many attribute fields we don't need. To simplify the output table, we will remove them from the output feature class.

2→ Scroll through the Output Fields in the Field Map and find the switch point between the well fields and the geology fields. (**Hint**: the case changes.)

2→ Click the county_code field name in the wells section of the table to select it; then click the red X to delete it.

Fig. 10.17. Parameters for joining the wells and geology
Source: Esri

2→ Delete all the wells fields except: state_well_number, aquifer_code, and well_depth. (Use the Ctrl-key to select multiple fields and then delete them together. To reset the field map in case of an error, click the blue reset button.)

2→ Delete all the Geology fields except ORIG_LABEL and UNIT_NAME.

2→ No merge rules are needed, so do not set any Merge Rules.

2→ Set the Match Option to Within. The tool should appear as in Figure 10.17. Run it.

TIP: It helps to be familiar with the contents of the tables before making decisions about what to keep and what to delete. If necessary, examine the tables before starting a join.

3→ In Contents, right-click the wellgeology layer and choose Attribute Table. Examine the fields.

Let's find out how many of the wells are on limestone units.

3→ Use the Select Layer By Attribute tool to select wells from the wellsgeology layer with the expression: UNIT_NAME contains the text Limestone.

1. How many wells are situated on limestone? _____

3→ Clear the table selection and close the table. Turn visibility off for all layers except Creeks and Soils.

Managing cardinality options

The substrate of a creek affects its connection with groundwater. A creek on high-infiltration soils will provide more recharge than one on low-infiltration soils. We assigned geology information to wells using a spatial join; we can also do it for creeks and soils. **STOP** and think it through.

At first glance, this problem appears similar to the previous one. We want a table containing creeks with a field indicating the soil unit and the infiltration value. Thus, Creeks is the target layer and the *within* operator can be used.

4→ In the Geoprocessing pane, open the Spatial Join tool. Set the Target to Creeks and the Join Features to Soils. Name the output *creeksoilswithin*.

4→ Use the *Join one-to-one* operation. Uncheck the box to Keep All Target Features.

4→ In the Field Map Output Fields, from Creeks, delete WATERSHED, Shape_ Length, AREA, and PERIMETER.

4→ In the Field Map Output Fields, from Soils, delete Rowid_, MUID_1, and STATE. Scroll past the soil attributes and delete the three Shape_ fields at the bottom.

4→ Set the Match Option to Within and run the tool. Keep the tool open.

5→ Open the creeksoilswithin table and examine the fields.

5→ Give the creeksoilswithin symbols a different color than Creeks.

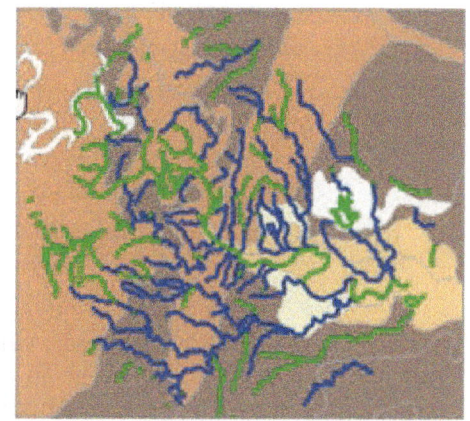

Fig. 10.18. Creeks before (blue) and after (green) the *within* spatial join

Source: City of Austin, TX

The output shows that we overlooked an important issue (Fig. 10.18). Any creek that crossed a soil boundary was not within a single soil unit, so it was dropped from the output. The joined feature class contains fewer creeks than the original, 306 instead of 381. The *within* operator did not produce a good result, leaving many creeks unjoined.

Perhaps we should use the *intersect* operator instead. However, the cardinality between creeks and soils is potentially one-to-many, so we should specify merge rules. The numeric fields are easily managed by taking the mean values. The soil identification MUID is a text field; for it we will use the Join rule and add a new text field to contain the concatenated values.

6→ The Spatial Join tool should still be open with the same settings. Change the output name to *creeksoilint* and the Match Option to Intersect.

6→ In the Field Map Output Fields, select the first numeric soil attribute field, AWC. Under the Source heading to the right, set the Merge Rule to Mean.

6→ Use Shift-Click to select CLAY and all the fields underneath it, and set the Merge Rule to Mean for all of them at once.

TIP: The main Geoprocessing pane keeps track of previously run tools. Use the drop-down arrow by a tool name to open a recently run tool, complete with its parameter settings.

Next we need to add a new field to contain the concatenated MUID values.

7→ In the Field Map section (Fig. 10.19), click the + sign next to Output Fields (1) to add a new field. Replace the NewField name with *MuidJoin* (2).

7→ To the right, click on the Properties heading (3). Enter *MuidJoin* as the Alias and change the Length to *100*.

7→ Click the Source heading (4). Set the Merge Rule to Join and enter a comma for the Join Delimiter.

7→ Click Add New Source (5). In the drop-down, click the Soils table to view its fields and select MUID. Click Add Selected at the bottom of the field list.

7→ The Field Map should appear as in the bottom half of Figure 10.19. Run the tool.

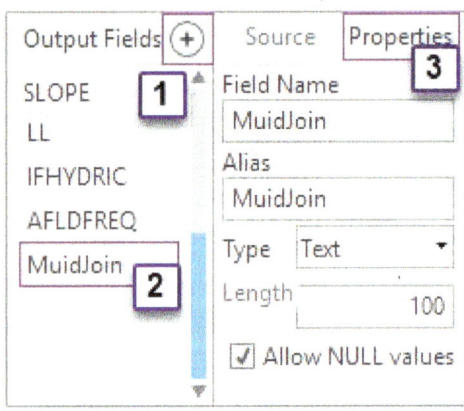

8→ Open the creeksoilint attribute table. Right-click the Join_Count field heading and sort it in Descending order to put the creeks with multiple matches at the top.

8→ Examine the fields, particularly the new MuidJoin field containing the concatenated labels.

This time every creek was assigned a value averaged from its soil matches. The MUID field contains the first match found; the MuidJoin field records all the matches. Incidentally, creating the MuidJoin field was not required unless we needed to keep all the MUID values; we could have omitted that step if we didn't care about them.

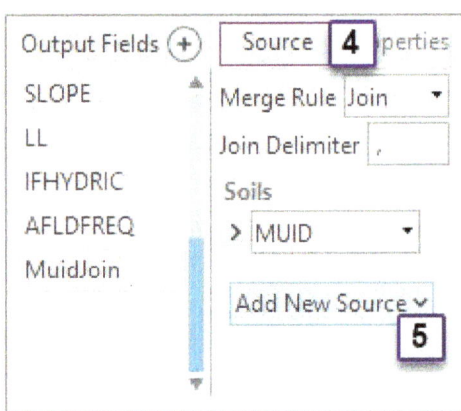

Fig. 10.19. Creating a join merge rule with a new field

Source: Esri

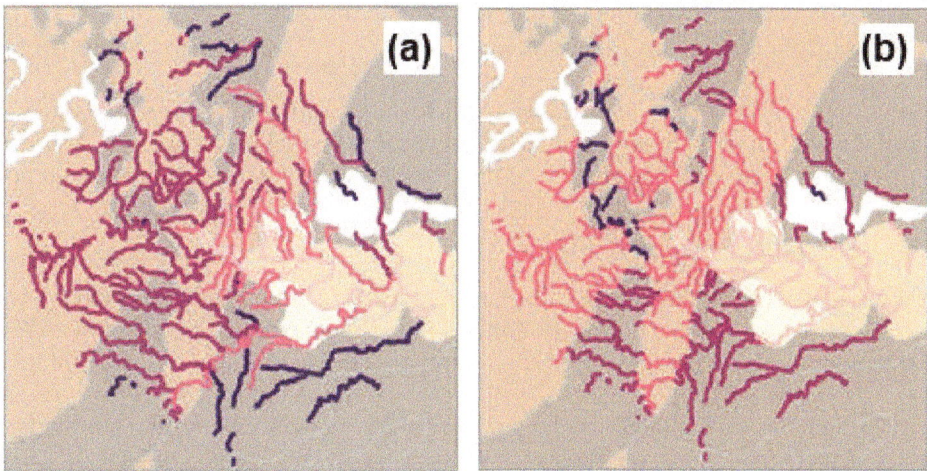

Fig. 10.20. Creeks symbolized by soil infiltration rate (a) after a merged spatial join and (b) after intersecting
Source: City of Austin, TX

8→ Symbolize the creeksoilint layer using a graduated colors map based on the HYGRP field with 2-pt. lines, as in Figure 10.20a. The darker colors show slower infiltration.

Another way of dealing with a one-to-many cardinality between the target and join features is to specify that multiple copies of the target features be created.

9→ In the Spatial Join tool, change the output name to *creeksoilint* many. Change the Join Operation to *Join one to many.*

9→ The merge rules will not be needed or used, but it is easier to leave them than to change them. Run the tool.

9→ Open the creeksoilsintmany table.

9→ The TARGET_FID field contains the original polygon IDs from Creeks. Sort it in Ascending order and examine the records for FIDs 3 and 7.

We now have 488 streams instead of 381. New copies of streams 3, 7, and others were created to match all of the soil polygons they crossed. Although a good example of how a one-to-many join type works, the result is not very helpful in this case.

As the creek on soils example makes clear, spatial joins are complicated when one-to-many relationships exist between the target and joined features. The merged solution worked best, but long creeks were given one average value even though they might cross soils with widely different infiltration values. The Intersect tool provides a better solution: it eliminates the cardinality issue by splitting the creeks to create a one-to-one correspondence between each stream segment and the soil unit that contains it.

10→ In the Geoprocessing pane, open the Intersect tool.

10→ Click the drop-down box for Input Features and select Creeks.

10→ Click the new drop-down box underneath and select Soils.

10→ Name the output feature class *creekintersect* and run the tool.

11→ Symbolize the creekintersect layer (Fig. 10.20b) in the same way as creeksoilint.

11→ Compare the two layers, and also examine the tables.

11→ Save the project before going on.

Cardinality is an important consideration when performing spatial joins, and a join may not be the best tool for a given task. However, let's return to more spatial join examples. Merged spatial joins may be problematic with lines and polygons, but they work well with points and polygons.

Merged point to polygon joins

Consider a watershed, the collection area for surface water. Runoff that comes from a watershed with many people is likely to have more pollutants than runoff from a watershed with fewer people. Let's analyze which watersheds in the Austin area are at greatest risk for polluted runoff.

12→ Remove the creeksoilswithin, creeksoilint, creeksoilintmany, and creekintersect layers from the map. They are still saved in the home geodatabase.

12→ Close any open tables, and turn off all layers except Watersheds and Block Population.

12→ Open the table for the Block Population layer.

Block Population contains points representing the centroids of block groups, the smallest unit for which the Census Bureau summarizes population data. Each point represents numbers of people and households. We will use these points to determine the number of people in each watershed. **STOP** and think it through.

We need a list of watersheds with the total number of people in it. Watersheds is the target layer. One watershed can contain many block points, so the cardinality between the target and join features is one-to-many. A merged one-to-one join operation is the best choice. Since we want to know the total number of people, we need to request the Sum merge rule for population.

13→ Close the table.

13→ In the Geoprocessing pane, open the Spatial Join tool.

13→ Set the Target Features to Watersheds and the Join Features to Block Population. Name the output watershedpop.

13→ Use *Join one to one* for the Join Operation and uncheck the box to Keep All Target Features. Some watersheds have no points, so we don't need to keep them.

13→ In the Field Map, keep only DCM_CODE and DCM_NAME from Watersheds.

13→ Keep only POP2010, HOUSEHOLDS, and HSE_UNITS from Block Population. Select them and set the Merge Rule to Sum.

13→ Set the Match Option to Contains and run the tool.

14→ In Contents, in the List By Drawing Order panel, drag the watershedpop layer just below the Block Population layer.

14→ Open the watershedpop table and examine the fields.

The Join_Count field records how many block points were found in the watershed. The population and household fields now contain sums for all block points contained by the watershed.

15→ Close the table and turn off the Watersheds layer.

15→ Symbolize the watershedpop layer with a graduated colors map based on the POP2010 field.

15→ The population values are influenced by polygon area, so normalize the classification using the Shape_Area field. The map should look similar to Figure 10.21.

Distance joins

Consider an analysis of post office usage. One might assume that, generally, people will go to the closest post office. For each street, it would be nice to know which post office to go to. **STOP** and think it through.

We want a list of streets with a field indicating the closest post office, so Streets is the target layer. The *closest* operator is appropriate. Since only one post office can be "closest," this is a one-to-one relationship, so no merge rules are needed.

16→ Turn off all layers except the Streets and Post Office layers.

16→ In the Geoprocessing pane, use the back button to clear and reopen the Spatial Join tool.

16→ Set the Target Features to Streets and the Join Features to Post Office. Name the output streetpost.

16→ Keep *Join one to one* for the Join Operation, and keep all target features.

16→ Keep all Output Fields. Set the Match Option to Closest and name the Distance Field PODistanceFt. Run the tool.

17→ Open the streetpost attribute table and examine it.

2. Which field contains the post office names? _____

17→ Symbolize the streetpost layer with a unique values map based on the FACILITY_N field, so that each street is symbolized by the post office it is closest to (Fig. 10.22).

A distance join also creates a distance field recording how far each street lies from the nearest post office (as the bird flies, not driving distance). We know that the units of the distance field we specified (PODistanceFT) are in feet because the input feature classes have coordinate systems with x–y units in feet.

3. How many streets are more than two miles (10,560 feet) from a post office? _____

18→ Clear any selected features and close the table.

18→ Symbolize the streetpost layer with a graduated colors map based on the PODistanceFt field (Fig. 10.23).

A different proximity question might be: How many people are within walking distance (0.5 mi) of a recreation center? **STOP** and think it through.

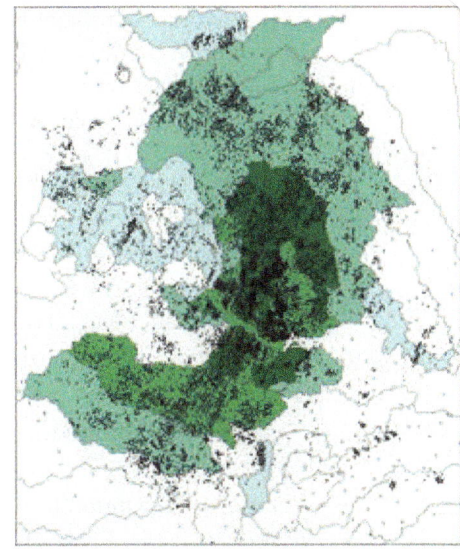

Fig. 10.21. Watershed population density
Source: Esri, City of Austin, TX

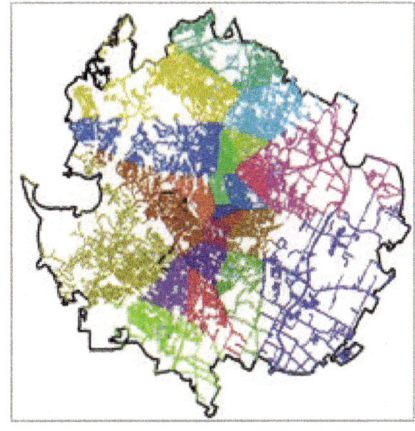

Fig. 10.22. Post office service areas
Source: City of Austin, TX

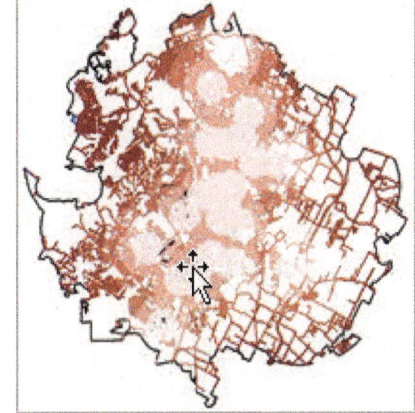

Fig. 10.23. Distance of streets from the nearest post office
Source: City of Austin, TX

We want a list of recreation centers with sums of the nearby population. The target is the centers, and the operator is *Within a distance*. The cardinality will be one-to-many, and we want to set a Sum merge rule for the population.

19→ Turn off all layers except Block Population and Rec Centers.

19→ Click the back arrow to clear and reopen the Spatial Join tool.

19→ Set the Target Features to Rec Centers and the Join Features to Block Population. Name the output *centerwalkpop*.

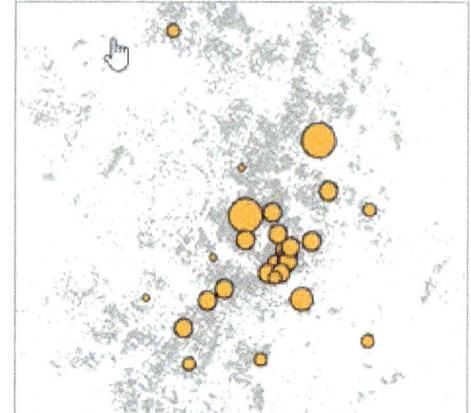

19→ Use *Join one to one* for the Join Operation and keep all target features.

19→ Delete all output fields except FACILI-TY_N, POP2010, and HSE_UNITS.

19→ Specify a Sum Merge Rule for POP2010 and HSE_UNITS.

19→ Set the Match Option to *Within a distance* and set the Search Radius to 0.5 miles. Run the tool.

20→ Examine the output table.

20→ Symbolize the centerwalkpop layer with a graduated symbols map based on the POP2010 field (Fig. 10.24).

Fig. 10.24. Recreation centers by population within walking distance
Source: City of Austin, TX

4. How does Spatial Join differ from Select Layer By Location for this problem?

Spatial joins are a versatile and powerful analysis tool. However, as we saw in the creek and geology example, sometimes the ability to split features that cross each other provides a more elegant solution, which brings us to the classic overlay tools of intersect and union.

Using intersect for habitat analysis

Intersect excels at sieve analysis, or finding where combinations of different conditions exist, such as the problem described earlier concerning habitat for the Black Hills snail. The habitat will be defined by three criteria: on a limestone geologic unit, in dense coniferous forest, and between the elevations of 1200 and 1600 meters.

21→ Close any tables and save the project. Close the Austin map and open the Sturgis Area map.

The first step is to create layers that contain polygons where each condition holds. We will use an attribute query on Geology to place the limestone units in a separate layer.

22→ In the Geoprocessing pane, open the Make Feature Layer tool.

22→ Set the Input Features to SDgeology and name the Output Layer *Limestone*.

22→ Enter the expression: RockType is Equal to Limestone. Run the tool.

TIP: Using the Make Feature Layer tool, although not necessary for processing, is a good practice. It keeps the selection available if a procedural question arises or if a tool must be rerun.

Next, we will make a layer for the dense conifers. This query has three conditions because we must select both ponderosa pine (TPP) and white spruce (TWS), plus we want only the dense stands. We must be careful to execute them in the right order.

23→ Turn off all layers except Vegetation.

23→ Clear and reopen the Make Feature Layer tool. Set the Input Features to Vegetation and the Output Layer to Conifers.

23→ Enter the expression: COV_TYPE is Equal to TPP.

23→ Click Add Clause again and enter: Or COV_TYPE is Equal to TWS.

23→ Click Add Clause one more time and enter: And DENSITY96 is Equal to C.

23→ Click the SQL button and insert parentheses around the first two conditions to ensure that they get executed first. The SQL expression should appear as: (COV_TYPE = 'TPP' Or COV_TYPE = 'TWS') And DENSITY96 = 'C'.

23→ Verify the expression and run the tool. If executed correctly, the selected features should match those shown in Figure 10.25.

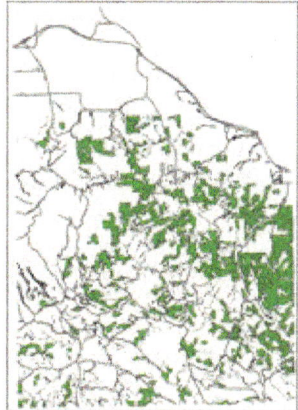

Fig. 10.25. Selecting dense conifers

Source: Black Hills National Forest

Notice that we are not choosing to alter the field maps in the Make Feature Layer tool. In sieve analysis, the spatial regions are usually the focus, rather than the attributes.

24→ Turn off all layers except the Elevation Range layer, the Limestone layer, and the Conifers layer.

The elevation range has already been prepared. However, the intersection will be faster and the output simpler if we use the Dissolve tool to remove the internal boundaries in the conifer areas.

24→ In the Geoprocessing pane, open the Dissolve tool (Fig. 10.26).

24→ Set the Input Features to Conifers and name the Output Feature Class *DensConifDisslv*.

24→ Set the Dissolve Field to Density96. Leave the Statistics fields blank.

24→ Uncheck the box to *Create multipart features*. Run the tool.

Now we are ready to find the areas common to all three layers using Intersect.

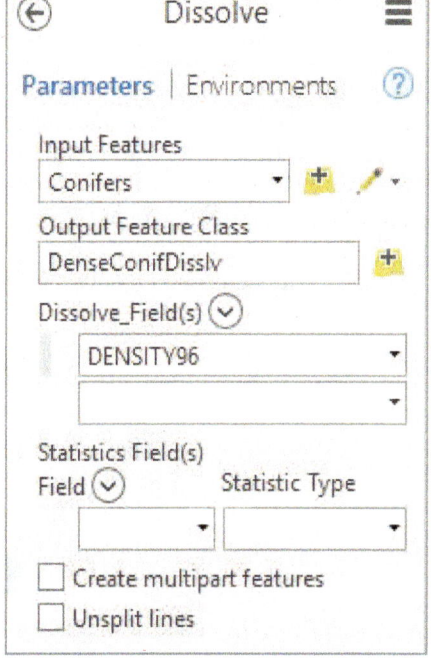

Fig. 10.26. Dissolving the conifers

Source: Esri

TIP: Most students have an Advanced license, which allows multiple layers to be intersected at once. With a Basic or Standard license it must be done in two steps. First intersect the Limestone and Conifers layers; then intersect the output with the Elevation Range layer.

25→ In the Geoprocessing pane, open the Intersect tool (Fig. 10.27). Examine the Help pop-ups.

25→ Under Input Features, click the first drop-down box and set it to Limestone.

25→ Click the next Input Features drop-down box and set it to DenseConifDisslv. Set the third box to Elevation Range.

25→ Name the Output Feature Class SnailHabitat.

25→ Set the Attributes To Join option to *All attributes except feature IDs.*

25→ The remaining options can be left to their defaults. Run the tool.

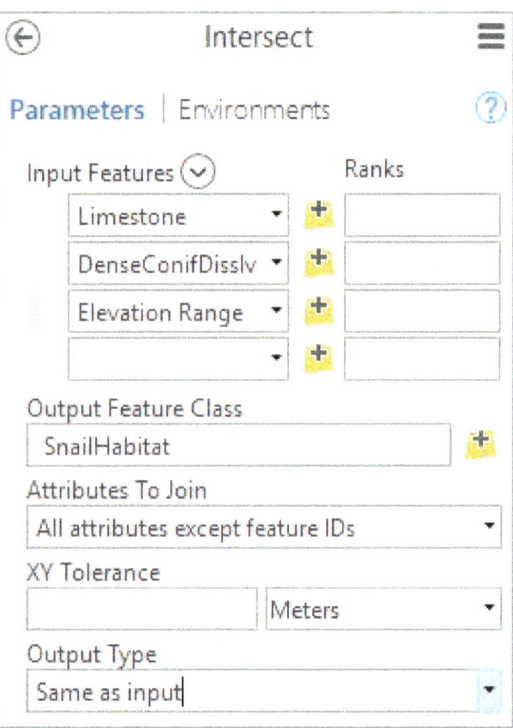

Fig. 10.27. Intersecting to find snail habitat
Source: Esri

TIP: The Feature ID, or FID, is a unique integer assigned to every feature for tracking. During an intersection, new FID values are assigned to each feature. The option we selected eliminated the original FID values from the output table, since usually they serve little purpose.

Overlay of lines in polygons

The snails have a three-week breeding season in early June. During this period they seek the open areas offered by roads, and many get crushed. The Forest Service wants to consider closing primitive roads that traverse through snail habitat during the breeding season to lessen the number of crushed snails. They need to assess which roads would be affected by making a map with the proposed road closures. This process involves a line-in-polygon intersection.

26→ In Contents, turn visibility off for all layers except SnailHabitat and Roads.

26→ Use the Select Layer By Attribute tool with the Roads layer to select the Primitive roads using the expression: TYPE is Equal to PR.

27→ Open the Intersect tool. Set the Input Features to the Roads layer (with the primitive road selection) and the Snail-Habitat layer.

27→ Name the output *ProposedClosures* and run the tool. Clear the selected features.

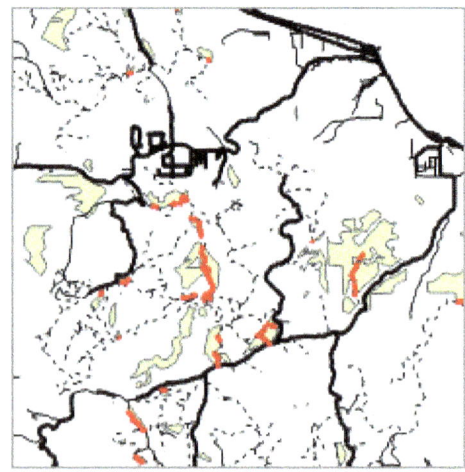

28→ Create a map similar to Figure 10.28, with the closure roads highlighted in red and the other roads in black.

28→ Save the project.

Fig. 10.28. Some of the proposed road closures
Source: Black Hills National Forest

Combining overlay with buffers

To prevent snail death on the major roads, which cannot be closed, a Forest Service biologist has suggested thinning the tree stands within 200 meters of the roads to make these areas less attractive to snails and thus to keep them away from roads. We can prepare a map showing the stands to be cleared and determine the total area of snail habitat that would be eliminated.

29→ Use Select Layer By Attribute to select the primary and secondary roads from the Roads layer, using the expression: TYPE is Equal to P Or TYPE is Equal to S.

30→ Open the Buffer tool. Set the Input Features to Roads and name the output *roadbuf200*.

30→ Set the Distance to *200* and select Meters for the units.

30→ Change the Dissolve Type to *Dissolve all output features into a single feature* and run the tool.

TIP: By default the Buffer tool does not dissolve boundaries between buffers, creating messy overlapping output. Usually it is best to dissolve during a buffer.

Now intersect the road buffers with the snail habitat to determine the potential areas to be cleared.

31→ Open the Intersect tool. Choose SnailHabitat and roadbuf200 as the Input Features.

31→ Name the output *ProposedThin* and run the tool.

32→ Clean up the map display to show the results (Fig. 10.29).

The final step is to determine the area of snail habitat that will be eliminated by the thinning. Geodatabases automatically keep track of the areas, lengths, and perimeters of features.

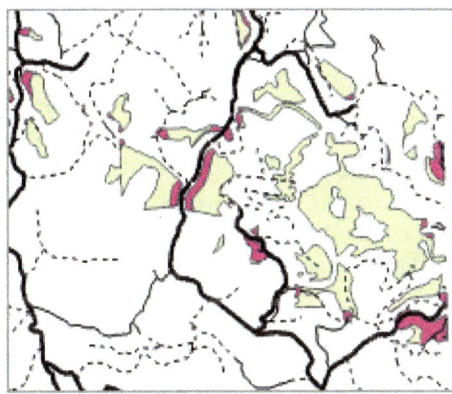

Fig. 10.29. Portion of map showing the proposed thinning areas
Source: Black Hills National Forest

33→ Open the Summary Statistics tool. Set the Input Table to ProposedThin and name the Output Table *ThinArea*.

33→ Set the Statistics Field to Shape_Area and the Statistic Type to Sum. Run the tool.

33→ Open the ThinArea table and examine the results.

The units of the Shape_Area field will be in the same coordinates as the map units of the feature class coordinate system, in this case, square meters.

TIP: To convert square meters to square kilometers, divide by 1 million (1,000,000).

5. What is the total area of the proposed thinning in square kilometers? _____

TIP: Shapefiles do not automatically maintain area or length fields. Values in these fields may not be correct if geoprocessing actions have been performed since the fields were created. The fields may be updated in using the Calculate Geometry Attributes tool.

Investigating relationships with union

Intersect is valuable for finding areas where known criteria exist simultaneously. Union can be used to investigate relationships between two types of information. The Forest Service wishes to analyze combinations of slope and geology to predict geologic hazards such as landslides.

34→ Close all tables and turn off all layers except SDgeology.

34→ Add the slopepct feature class from the Features feature dataset of the mgisdata \BlackHills\Sturgis geodatabase.

34→ Open the Union tool. Choose SDgeology and slopepct for the Input Features. Name the Output Feature Class *GeolSlope.*

34→ Make sure Attributes To Join is set to *All attributes.* We will need the original Feature-ID's this time. Run the tool.

In the output, notice that all areas of both input layers are included, even though the slopepct layer does not cover as much area as the SDgeology layer does.

35→ Click the **Map: Navigate: Explore** tool.

35→ Click on several of the small polygons in the lower half and examine the attributes.

35→ Click several of the large polygons near the top and examine those attributes.

The large upper polygons received no attribute information from slopepct because the layer does not extend there. When an output polygon is missing data from one layer, the original FID value from the layer is set to −1. (That is why we kept the FID values in this case.) Let's eliminate these areas from further consideration and also hide the fields we don't need.

36→ Close the Pop-up window and open the Make Feature Layer tool.

36→ Set the Input Features to GeolSlope and name the Output Layer *Valid Combos.*

36→ Add the expression: FID_SDgeology is does Not Equal −1 And FID_slopepct does Not Equal −1.

36→ In Field Info, uncheck visibility for all fields except: Symbol, Competence, UnitName, RockType, and SLOPE_PCT. Run the tool.

37→ Remove the GeolSlope layer and examine the Valid Combos table. Close the table when finished.

The top areas of concern for sinks (collapsed areas caused by rock dissolution) include limestone areas with low slopes. We will select the polygons that meet these criteria.

38→ Clear and reopen the Make Feature Layer tool. Set the Input Features to Valid Combos and name the Output Layer *Sink Hazard.*

38→ Enter the expression: RockType is Equal to Limestone And SLOPE_PCT is Less Than 20. Run the tool.

38→ Give the Sink Hazard layer a yellow symbol (Fig. 10.30).

Areas at risk for rockfall incidents include stiff rocks in steeply sloping areas.

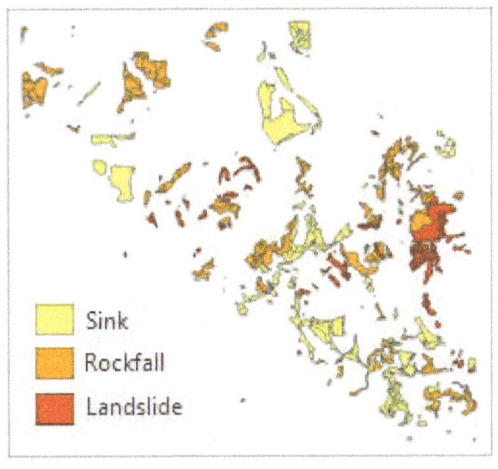

39→ Clear and reopen the Make Feature Layer tool. Set the Input Features to Valid Combos and name the Output Layer *Rockfall Hazard.*

39→ Enter the expression: Competence is Equal to High And SLOPE_PCT is Greater Than 40. Run the tool.

39→ Give the Rockfall Hazard layer a medium orange symbol.

Finally, the landslide hazard areas contain weaker rocks with high slopes.

Fig. 10.30. Geologic hazard areas

40→ Clear and reopen the Make Feature Layer tool. Set the Input Features to Valid Combos and name the Output Layer *Landslide Hazard.*

40→ Enter the expression: Competence is Equal to Low And SLOPE_PCT is Greater Than 40. Run the tool.

40→ Give the Landslide Hazard layer a dark orange symbol.

41→ Create a map showing only the three hazard areas, similar to Figure 10.30.

Rockfall is especially hazardous when close to roads.

42→ Intersect the Roads layer with the Rockfall Hazard layer. Name the output *Rock-HazardRoads.*

42→ Symbolize RockHazardRoads to clearly see the affected roads.

42→ Save the project before going on.

Working with slivers and tolerances

We will now demonstrate the problem of slivers and how to prevent them using tolerances. Performing a union on data sets with different source scales illustrates this problem nicely: we will use the congressional districts and counties in California as an example.

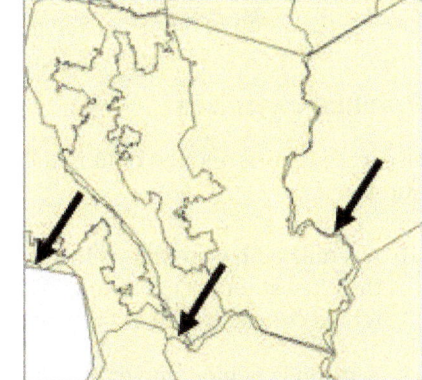

43→ Close the Sturgis Area map and open the Tolerances map, showing the San Francisco Bay area of California. Compare the boundaries for cd114 and counties.

44→ Open the Union tool. For the Input Features, select cd114 first and then counties (this order gives better default symbols for the output).

44→ Name the output *union0* and run the tool.

Notice that many slivers occur in the output (Fig. 10.31). The district and county boundaries coincide in real life, but

Fig. 10.31. Slivers from the union
Source: Esri

these two data sets come from different sources, and the boundaries aren't identical. To specify a processing tolerance, it would be helpful to know the typical width of the slivers.

45→ Zoom in to a region with typical slivers.

45→ Click the **Map: Inquiry: Measure** drop-down and choose Measure Distance.

45→ In the Distance pop-up, click Options > Distance Units > Meters.

45→ Measure across several slivers to estimate their widths.

45→ Click **Map: Navigation: Explore** again to close the Measure pop-up.

A typical sliver width is about 1000 meters. A tolerance is a radius, so we should use about half the measured value. Let's repeat the union with a tolerance of 500 meters.

46→ The Union tool should still be open. Change the output name to *union500* and enter *500* in the XY Tolerance box. Run the tool.

46→ Turn the union500 layer off and on a few times to compare the boundaries with the union0 layer.

46→ Zoom to the Bay Area bookmark.

Notice that some slivers are still present, and the district boundaries have lost detail. The XY Tolerance was not large enough to fix all the slivers, but it is already affecting the precision of the output. Nevertheless, we will try a larger tolerance.

47→ In the Union tool, change the output name to *union1000* and change the XY Tolerance to *1000* meters. Run the tool and compare the output to the other layers.

Now the generalization of the boundaries is obvious, and a few stubborn slivers remain. At this scale, it may appear that the boundaries have been severely compromised, but keep in mind that we are zoomed in far beyond the source scale of the districts, about 1:5,000,000.

47→ Zoom to 1:5,000,000. The union looks better.

This example not only demonstrates the use of tolerances to reduce the impact of slivers, but it also illustrates why trying to combine data from very different source scales tends to give disappointing results. The boundaries of these two original data sets were so different that it was difficult to eliminate all slivers and maintain reasonable coordinate accuracy.

This is the end of the tutorial.

→ Save the project and close Pro.

Practice Exercises

Use spatial joins and the data in the mgisdata\Austin\Austin geodatabase to answer the following questions.

1. A tennis club is implementing an after-school tennis program. Use a spatial join to determine the closest tennis court to each school. What is the average and median distance of schools to tennis courts, in miles? Create a histogram of distances (in miles) and **capture** it.

2. A study is being done to evaluate staffing levels in the police districts. Determine the number of people (using the blockpop points) living in each district. Map and chart the results and discuss them. Can you explain why some of the districts have such low numbers of people? (**Hint**: compare the blockpop points to the base map.)

3. The University of Texas at Austin received a grant for water quality education. Each school will study aspects of the watershed in which it is located. Create a list of schools, each with its designated watershed. Sort the list alphabetically by school name and **capture** the results for the first 10 schools, with the grade range and watershed name. Which watershed has the most schools, and how many does it have?

4. A runner's club is analyzing access to parks in Austin, assuming that runners will go to the closest park and not run to parks more than one mile away. Determine the potential usage at each park based on the block population closer to that park than any other, but not more than 1 mile away.

5. Determine the number of wells situated in each census tract in Austin. **Capture** maps showing the wells per 1000 people and the mean well depth in each tract.

Use the Austin data and classic overlay (intersect/union) to answer the following questions.

6. Septic systems are commonly used in low-density housing areas, but the impacts on aquifers may be of concern. Map the areas in Austin where the census tract population densities are less than 1000 people per square mile and where the geologic unit names contain Limestone or Terrace or alluvium.

7. The clay fraction of soils as well as road use may affect the frequency of street repairs. How many street miles in Austin have more than 50% clay and a road class of 5 or less? **Capture** a map showing these roads.

8. Water quality and risks of gasoline spills may be affected by the prevalence of road arteries in a watershed. Using the arteries feature class, determine the number of road miles in each watershed. Which three watersheds have the highest mileage? **Capture** a map showing the mileage for each watershed.

9. Perhaps the total of street miles is a better measure than the arteries. Repeat the watershed analysis using the streets feature class and compare the results. List the three watersheds with the highest mileage, and **capture** a map showing the mileage of all watersheds. What significant issue impacts the quality of this result?

10. The impact of flooding on Austin would depend on the flood frequency and the density of the population. Use the tracts and soils feature classes to calculate a flood impact index based on the population density times the flood frequency, scaled from 1 to 10. What aspect of the input data most limits the applicability of this map?

Chapter 11. Raster Analysis

Objectives

▶ Learning different raster analysis functions and techniques

▶ Using Boolean overlay and weighted overlay to analyze suitability

▶ Applying distance, density, surface, and neighborhood functions

▶ Converting between raster and vector data models

▶ Controlling the analysis environment when using Spatial Analyst

NOTE: This chapter requires **Spatial Analyst**, an optional extension to ArcGIS. It must be installed and licensed for you to be able to perform the functions described.

Mastering the Concepts

GIS Concepts

The availability of two different data models, raster and vector, provides flexibility for data storage and analysis. Neither data model is intrinsically superior; they both have areas in which they excel and areas in which they are at a disadvantage. The GIS professional who has a grasp of both tools holds the keys to developing the most efficient and accurate analysis.

Raster data

The analysis functions available for rasters number in the hundreds and include sophisticated applications. This chapter introduces basic concepts for analyzing rasters, but the topic spans a greater breadth of theory and technique than we have time to explore. The serious student of raster analysis will find additional material in the software documentation for Spatial Analyst.

Recall that spatial data come in two forms (Fig. 11.1). Discrete data, such as soil types or land use, represent spatial objects and form distinct objects or regions on the map. Continuous data, such as elevation, vary more smoothly over a range of values and form a surface or field. The vector data model best stores discrete data. Rasters easily store and analyze continuous data.

Like vector data, rasters have a defined coordinate system with a projection and a datum. The raster is georeferenced by specifying the x-y coordinates of the center of the upper-left pixel and the cell size (the height/width of the pixels). Usually pixels are square. A 30-meter raster has pixels composed of 30 m by 30 m squares. The cell size is specified in the coordinate system's map units, and meters or feet are typically used.

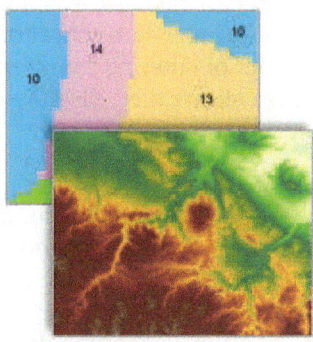

Although one can store a raster in a GCS, this practice is not common because the distortion gives an unsatisfactory view of the data. Furthermore, many raster analysis functions incorporate distance and angle formulas that will give inaccurate results when applied to a raster in a GCS or in a projection with significant distortion. Therefore, raster data are normally stored in whatever projected coordinate system gives the best and most undistorted

Fig. 11.1. Discrete land use data and a continuous digital elevation model
Source: USGS

view of the data. The main exception occurs for data providers, such as the US Geological Survey, that provide topographic or other data in GCS format. Such data should be projected before they are used.

Raster analysis

Raster processing is fundamentally different from vector processing. Vector analysis manipulates *x-y* coordinates. Raster processing operates on stacks of aligned cells and can be very efficient. This chapter describes the most commonly used functions, but many more are available.

Map algebra

A raster is an array of numeric values arranged in rows and columns. Several stacked rasters representing different attributes can be analyzed using cell-by-cell formulas based on mathematical, logical, or Boolean operators. This technique is called **map algebra**. Consider the stack of two input rasters in Figure 11.2. These rasters can be added together cell by cell to produce a new raster with cells containing the sum of the corresponding cells in the inputs. If the cells of the input rasters are not already aligned, then they must be resampled. Resampling, discussed in Chapter 6, may modify the values and accuracy of a raster, so ideally the user creates and manages the raster analysis such that the need for resampling is minimized.

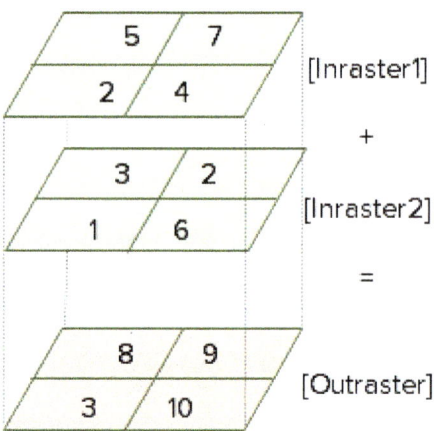

Fig. 11.2. Map algebra addition

The map algebra expression for the operation in Figure 11.2 would appear as [Inraster1] + [Inraster2] with the raster names enclosed in brackets. Imagine a raster of precipitation in inches. The expression [Precip] * 2.54 produces an output raster with precipitation in centimeters. Expressions can include multiple rasters, functions, and values. A more complex expression to predict erosion potential as a function of rainfall, slope, soil erodability, and vegetative cover might read:

[Precip] * 2 + [Slope] * 4 / ([Erosion] – [Vegcover]).

Map algebra also includes logical functions that can return **Boolean rasters**, which have values of true (1) or false (0) indicating where some condition holds. Figure 11.3 shows the result of entering an expression [Elevation] > 1400; the purple area shows the cells that meet this criterion.

Boolean overlay

Overlay is commonly applied to **suitability analysis**, in which a landscape is evaluated to find which areas best serve a given purpose based on a set of factors. The snail habitat problem in Chapter 10, which finds ideal habitat based on the factors of elevation, vegetation density, and geology, is a classic example of overlay. Raster overlay is especially well adapted to suitability problems. First, it can incorporate conditions from continuous data sets, such as elevation or slope, which cannot be easily stored as vectors. Second, the process is faster because corresponding cells are simply added or multiplied

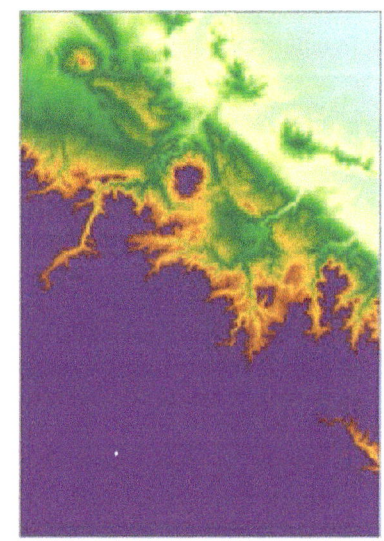

Fig. 11.3. The purple area shows the pixels where Elevation > 1400 meters.
Source: USGS

together, instead of splicing the existing polygons and creating new ones. Third, raster overlay can process multiple layers more easily than vector overlay.

Boolean overlay is similar to vector overlay, but it uses map algebra with Boolean rasters and operators. Chapter 9 introduced Boolean operators, which evaluate two input conditions, A and B, to determine a result that is true or false. In raster analysis, this function is performed on a cell-by-cell basis.

Figure 11.4 demonstrates the Boolean operators applied to raster analysis. The circles represent input rasters A and B, in which the cells contain true values inside the circles and false values outside. The squares represent two 4 × 4 input rasters, and the third column details the

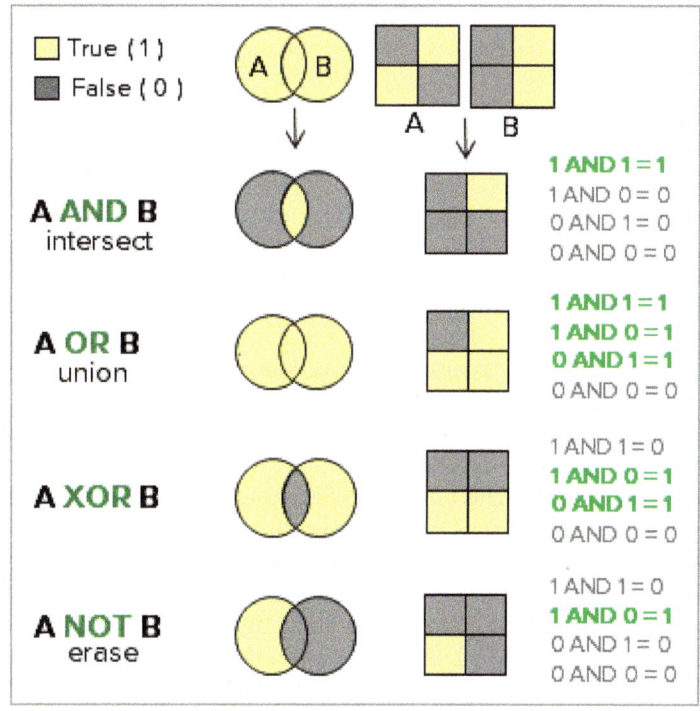

Fig. 11.4. The Boolean operators applied to rasters

operator logic. The AND operator finds the areas common to both inputs and is the equivalent of the vector intersect function. The OR operator result combines the areas of both and matches the union function. NOT is equivalent to the erase function and is the only Boolean operator that is not commutative. The XOR operator has no equivalent tool for vectors (although the same result could be achieved with intersect followed by an erase).

Imagine using Boolean overlay to create a habitat map for lodgepole pine trees based on two criteria: areas above 1500 feet in elevation and with more than 60 centimeters of precipitation annually. In Figure 11.5a, the Boolean raster shows where the precipitation conditions for lodgepole are favorable. Areas where the precipitation is greater than 60 cm have the value true

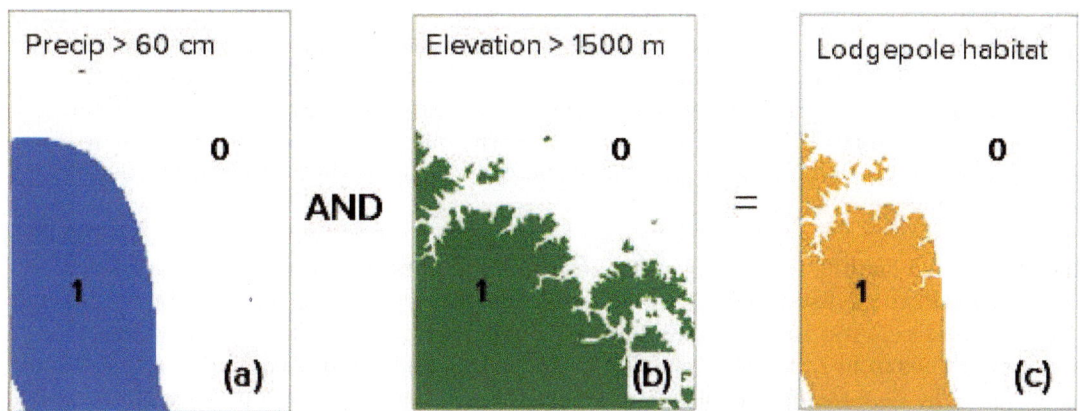

Fig. 11.5. Boolean overlay: (a) a Boolean raster in which the value 1 (true) indicates where precipitation is > 60 cm; (b) a raster indicating elevations > 1500 meters; (c) the result of multiplying the first two rasters, showing where both conditions are true

(1), and areas labeled false (0) have precipitation <= 60 cm. Figure 11.5b shows the favorable conditions based on elevation, where the cells containing 1 have elevations > 1500 meters.

To find the lodgepole habitat, we apply the Boolean operator AND to the input rasters. If both inputs are 1, the output cell is 1. If either or both of the input cells has a 0, then the output will be 0. Figure 11.5c shows the output raster with the potential habitat areas identified as cells with the value 1 (true).

The Boolean AND operates the same way as multiplication of the two input rasters. Examine the expressions 1 AND 1, 1 AND 0, etc. in Figure 11.4, and notice that substituting multiplication (*) for AND gives the same result. Thus, either AND or multiplication could be used to find the lodgepole habitat. Since both of these operators are symmetrical, meaning that the order of inputs does not matter, they can be applied to three or more rasters as easily as two.

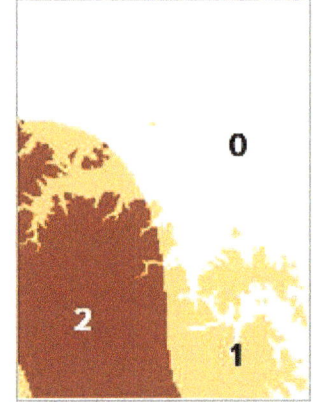

Map *addition* provides another way to model the habitat—to *rank* the probability of finding lodgepole. In this model, if both elevation and precipitation are ideal, then lodgepole is most likely to occur. If only one is ideal, then it might occur but is not as likely. If neither condition is ideal, then lodgepole is unlikely. To implement this model, we add the two input maps instead of multiplying them. The final raster in Figure 11.6 has three possible values: 0 if neither condition is true, 1 if either condition is true, or 2 if both conditions are true. Thus, 2 indicates the highest probability of finding lodgepole, 1 a moderate probability, and 0 a low probability. Of course, we would wish to test our model of lodgepole habitat suitability by testing to see if we predicted where lodgepole is actually found.

Fig. 11.6. A Boolean addition overlay

Weighted overlay

Boolean overlay has several drawbacks for suitability analysis. First and greatest, it demands that all variables be portrayed as true/false, when in reality it makes little sense that a slope of 10.0 is all right for a building site, but 10.1 is not. It is often more realistic to consider a gradation of values going from less ideal to ideal. Second, it assumes that all input layers are of equal importance. Yet in deciding on a building site, perhaps being close to an existing road is more important than slope because the cost of developing a long driveway is greater than doing some extra bulldozing to prepare the site.

Weighted overlay analysis addresses these two drawbacks. Like Boolean overlay, it develops a set of K rasters (R_1, R_2, ..., R_k) specifying the suitability of a set of conditions. Instead of a simple true/false, however, each raster is ranked into ordinal values 1 to N from worst to best, where N is any positive integer. Each raster is also assigned a weight W_k between 0 and 1, with all the weights summing to 1 ($W_1 + W_2 + ... + W_k = 1$). The output is then calculated as $W_1R_1 + W_2R_2 + ... + W_kR_k$, yielding a raster with values between 1 and N. High values in the output indicate a higher suitability.

Consider siting a landfill, for example (Fig. 11.7). Four factors (K = 4) might be considered: slope, soil infiltration, distance from streams, and distance from roads. A suitability index value of N = 3 is desired. First, one must **reclassify** each raster, converting its original values to a set of new values from 1 to 3. The result is four index rasters, one for each factor, as shown in Figure 11.7. The final raster is achieved by multiplying each index raster by its weight and adding them together. The final raster contains floating-point values between 1 and 3, with the highest values (and darkest colors) indicating the most suitable areas. Weighted overlay provides a more flexible and sophisticated approach to suitability modeling than Boolean overlay. Its main disadvantage is that assigning the suitability ranks and weights is usually a subjective process,

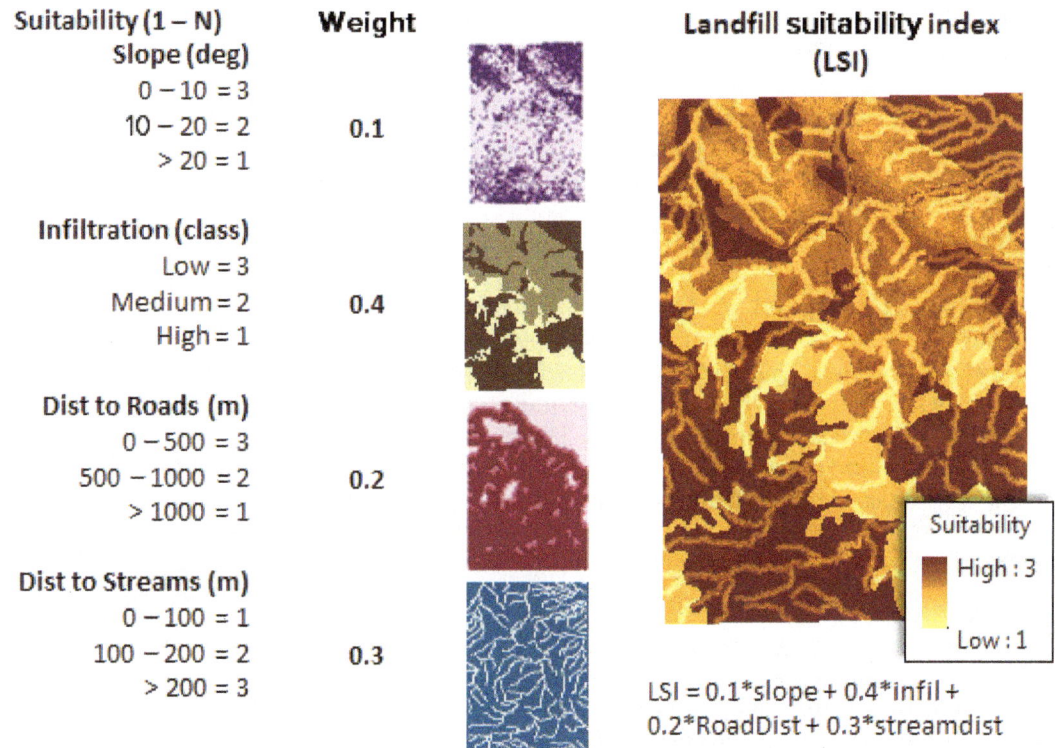

Suitability (1 – N) Slope (deg) 0 – 10 = 3 10 – 20 = 2 > 20 = 1	Weight 0.1

Infiltration (class)
Low = 3
Medium = 2   0.4
High = 1

Dist to Roads (m)
0 – 500 = 3
500 – 1000 = 2   0.2
> 1000 = 1

Dist to Streams (m)
0 – 100 = 1
100 – 200 = 2   0.3
> 200 = 3

Landfill suitability index (LSI)

Suitability
High : 3

Low : 1

LSI = 0.1*slope + 0.4*infil + 0.2*RoadDist + 0.3*streamdist

Fig. 11.7. Weighted overlay to determine suitability for a landfill site. The dark colors in each raster indicate the most suitable areas.

and the final model is heavily dependent on the analyst's choices.

Distance functions

Distance functions calculate a continuous range of distances from a specified set of objects. The **Euclidean distance** function produces a raster in which each cell represents the shortest distance from the objects, for example, streams (Fig. 11.8a). A related output is the Euclidean *direction* raster (Fig. 11.8b), which records the direction of the closest object in angular degrees with north equal to 0. At any given point, these two grids would show the distance from the closest stream and the direction to travel to get to it.

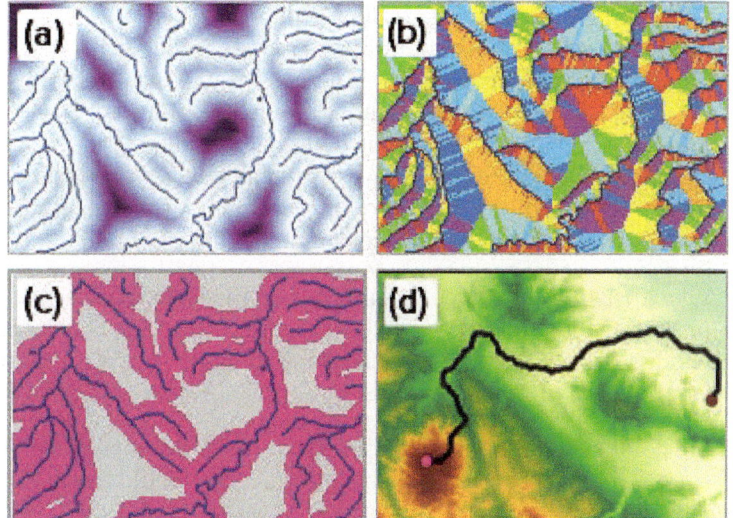

Fig. 11.8. (a) Straight-line distance; (b) distance-direction raster; (c) raster buffers; and (d) least-cost path between two points
Source: USGS, Black Hills National Forest

The Euclidean distance function is also the means for creating buffers as raster data sets. Once a distance surface is generated, it is a simple matter to identify the distances that are greater than

343

or less than a specified buffer value, to produce a Boolean raster with a 1 value inside the buffer and a 0 value outside (Fig. 11.8c).

Cost-distance functions consider more complex and flexible approaches to travel. Consider a mountain with two points on opposite sides. The shortest path to the opposite point goes over the mountain. However, it is easier to go around the base, even though the path is longer, due to the effort (cost) of climbing over the mountain. Slope is one kind of cost. Heavy vegetation, marshes, and off-limits areas are other factors that increase the cost of travel.

In a cost-distance analysis (Fig. 11.8d), a raster is first prepared that estimates the cost of travel over a surface. A single cost factor, such as slope, could be used, or the raster could combine different types of costs. This cost raster is used to develop a cost direction raster, which tells a traveler which directions incur the least cost, and a cost accumulation raster to keep track of the total effort expended as the traveler moves. These rasters can be combined to calculate the least-cost path between two points, indicating the easiest journey rather than the shortest. In Figure 11.8d, a slope raster was used to designate the cost, and a least-cost hiking trail was calculated.

Density

A density function creates rasters showing the frequency of objects within a search radius specified by the user. The function visits each cell of the output raster, counts the points falling inside the search radius centered on that cell, and assigns the count to the cell. An attribute field may be specified to weight the counts. Figure 11.9 shows a point density map of city populations in the United States. Without the population attribute, the function would count the number of cities; with the attribute, it counts the number of people. The counts are given per unit area, as in people per square mile or cities per square kilometer. A large search radius produces a smoother map but will show less detail than a smaller radius. Choosing the search radius can take some experimentation, but

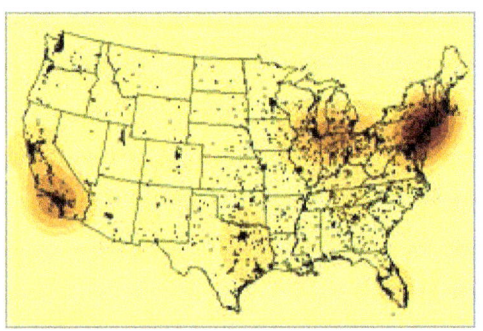

Fig. 11.9. Density of population from cities using a 500-mile search radius
Source: Esri

it should be large enough so that every cell will have at least one feature within the search radius; to meet this condition in Figure 11.9, a radius of several hundred miles would be required because of the low number of cities in the Mountain subregion.

A line density function counts line occurrences within the circle instead of points, and it weights them by length and optionally by attribute. This function could be used to calculate road density within a national forest. An attribute field might give stronger weights to primary roads and a lesser weight to primitive roads.

The point density function assumes that the entire weight value is present at each point, but this is not always true. A city cannot have 100,000 people living at a single x-y location. A kernel density function spreads the population value smoothly over a circular area, with the maximum value in the center and decreasing values around it. The kernel density function provides a more realistic model of the population distribution, and the resulting raster will appear smoother than the result obtained with the point density function. The search radius value has less impact on the kernel density function, although larger radii will produce more generalized rasters.

Interpolation

Interpolation takes measured values at points and distributes them across a raster, estimating the values in between the measurements. A common example involves taking precipitation measurements at weather stations and interpolating the precipitation values in between stations to create a continuous precipitation surface. Figure 11.10 shows the interpolation of the total precipitation measured at the stations over the Black Hills region, with the watersheds superimposed on the image.

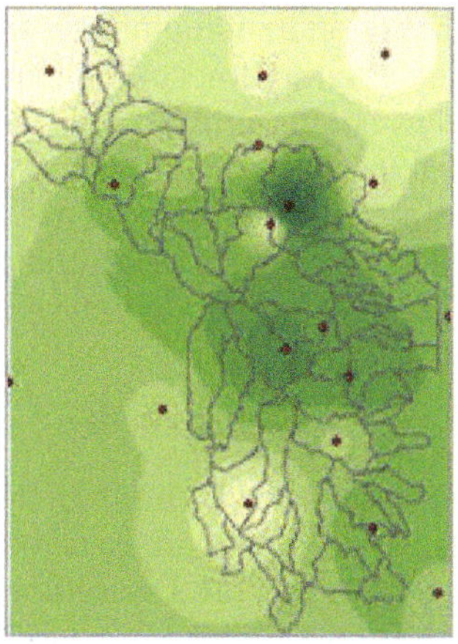

Interpolation algorithms are based on an important geographic principle known as **Tobler's law**, which states that values measured closer together are more likely to be similar than values that are measured at greater distances. It is also called the first law of geography, for many geographic data sets exhibit this behavior. Continuous raster data sets, such as elevation or temperature, can use many more colors than a typical polygon classification because of the underlying structure of the data according to Tobler's law.

Four different interpolation methods are common: inverse distance weighted (IDW), splines, trends, and kriging. Each method uses different assumptions and different means to obtain the result. More information about these four interpolators can be found in the Help documentation for Spatial Analyst. Interpolation is a complex topic and the subject of entire textbooks. GIS professionals should become familiar with the different methods in order to understand the benefits, drawbacks, and assumptions of each one.

Although interpolated and density maps appear similar, the functions work differently. Density functions count occurrences within a radius and divide by the area. Interpolation estimates a value for each raster cell based on nearby measurements.

Fig. 11.10. Rainfall raster interpolated from weather station measurements
Source: USGS, National Climatic Data Center

Surface analysis

Surface analysis includes a set of functions for calculating properties of a surface, such as slope and aspect. Although most commonly applied to elevation data (Fig. 11.11a), these functions can be used on any type of continuous raster. Using these functions on a DEM could help identify an ideal location for a cabin, for example. A **slope** map (Fig. 11.11b), showing the steepness of the terrain, could help identify sites flat enough to build on. An **aspect** map (Fig. 11.11c), showing the compass direction of the slope, could locate south-facing hillsides to give the cabin a warm southern exposure. A **hillshade** map (Fig. 11.11d), which represents the surface

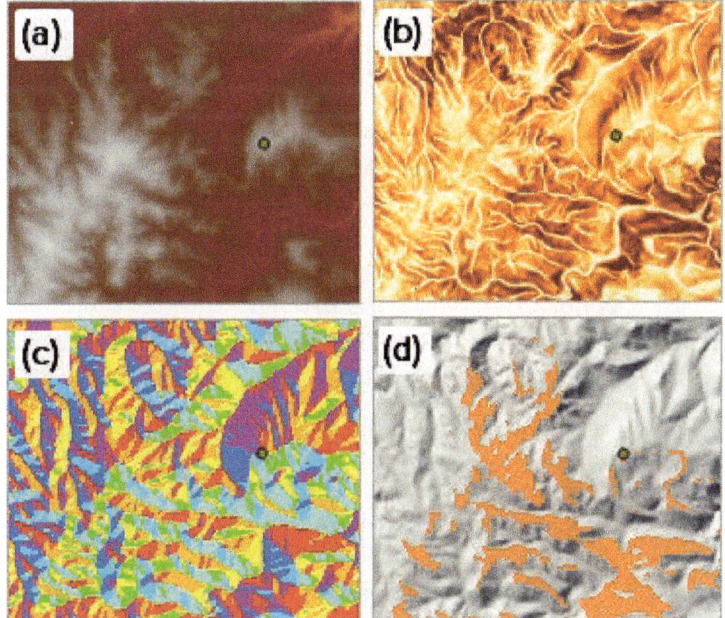

Fig. 11.11. Surface functions: (a) elevation raster; (b) slope; (c) aspect; (d) the view (orange) from an observer at the green point, shown over a hillshade raster
Source: USGS

as though an illumination source were shining on it, can help the reader visualize the landforms. The illumination source can be varied by **azimuth**, or the direction in degrees (0° to 360° with 0° as north), and the **zenith angle**, or the angle from 0° to 90° above the horizon. A **viewshed** analysis (orange areas in Fig. 11.11d), which determines what parts of a surface can be seen from specified viewpoint(s), could help in choosing the site with the best view. Finally, a **cut/ fill** function calculates the differences between two input surfaces and is useful for developing estimates of the amount of material to be added or removed from a building site.

Neighborhood statistics

Neighborhood functions operate on an area of specific size and shape around a cell, known as a neighborhood. A moving window passes over the raster and calculates a statistic from all the cells in the neighborhood, records the result, and moves on. Neighborhoods may have different shapes, including squares, circles, rectangles, and annuli. The statistical functions include minimum, maximum, range, sum, mean, standard deviation, variety, majority, minority, and median. Two types of neighborhood functions are used: block functions and focal functions.

A **block function** moves the window such that it never overlaps the previous window, but it produces a set of side-by-side adjacent windows that are independent of each other. Once the statistic is calculated for the window, every cell in the neighborhood receives the same value before the window moves on. Every cell is used only once. A block function can be used to resample a raster from a higher to a lower resolution, as might be needed to adapt a high-resolution land cover data set for a coarse climate modeling grid.

A **focal function** centers the window on a cell, called the target. It calculates the statistic for the neighborhood and assigns the value to the target cell; then it repeats the process at the next adjacent cell. Consider a 3 × 3-cell averaging window (Fig. 11.12). For each target cell, the algorithm finds the average of the nine cells in the window and places it in the target cell of the output raster. The window then moves to the next cell. An edge cell is still evaluated but will have fewer cells in the neighborhood; the cell in the upper-left corner of Figure 11.12 has only four values to average: the target and its three neighbors. A **filter** is another name for a focal function.

Figure 11.13 shows the difference in output between a block and focal majority function applied to a land cover raster (Fig. 11.13a). In both cases, a 4 × 4-cell neighborhood containing 16 cells

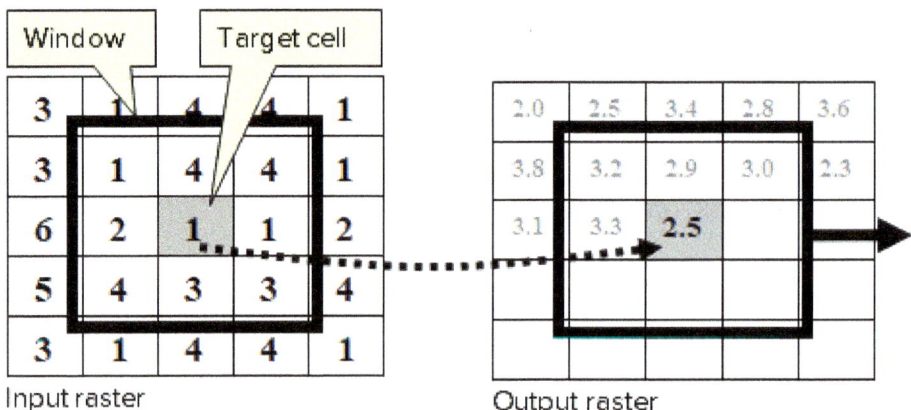

Fig. 11.12. A 3 × 3 averaging window averages the nine cells in the window and places the result in the target cell of the output raster.

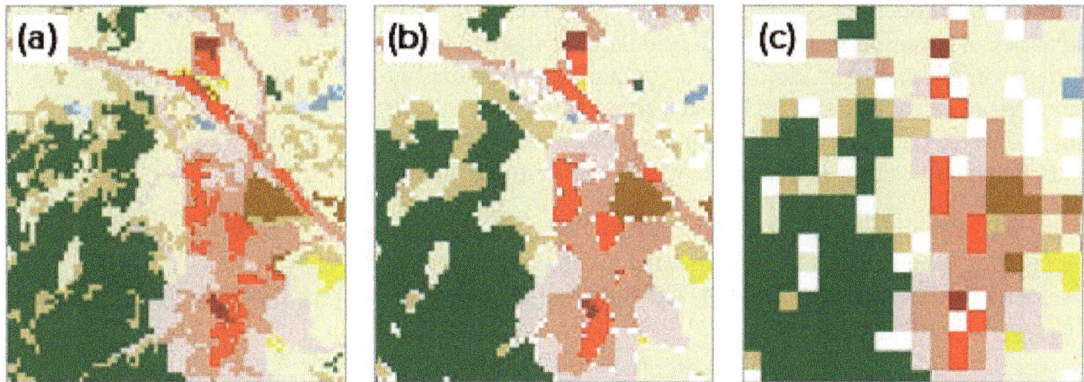

Fig. 11.13. A focal function versus a block function: (a) original land cover raster; (b) after a 4 × 4 focal majority function; (c) after a 4 × 4 block majority function. The white cells show where no majority was found.
Source: USGS

was used, and the statistic is the most frequent land use value found in the neighborhood. The white output cells occur where no majority was found. In Figure 11.13b, the focal function evaluated each cell with its 15 immediate neighbors and then moved to the next adjacent cell. In Figure 11.13c, the block majority function evaluates the 16 cells and then moves to the adjacent group of 16. Both generalize the original data set, but to a different degree.

Zonal statistics

The **zonal statistics** function is sometimes considered another type of neighborhood function, but the neighborhoods are not a fixed size or shape. Instead, they are defined by another layer, called the zone layer. A **zone** is a region of a raster or a polygon feature class that shares the same attribute value. A zone might be a watershed, a parcel, or a soil type. The attribute might be a name, a category, or a unique identifier, such as a parcel-ID or even the table ObjectID.

The function gathers all the cells from an input raster that fall inside each zone and calculates cell statistics for the zone. For example, if a watershed layer defined the zones and the input raster contained slope, then zonal statistics would calculate the average slope for each watershed. Figure 11.14a shows a set of watersheds displayed on a hillshade map; it is apparent which watersheds have more variable topography and a higher average slope. When the zone function is executed, the slope values are averaged for each watershed.

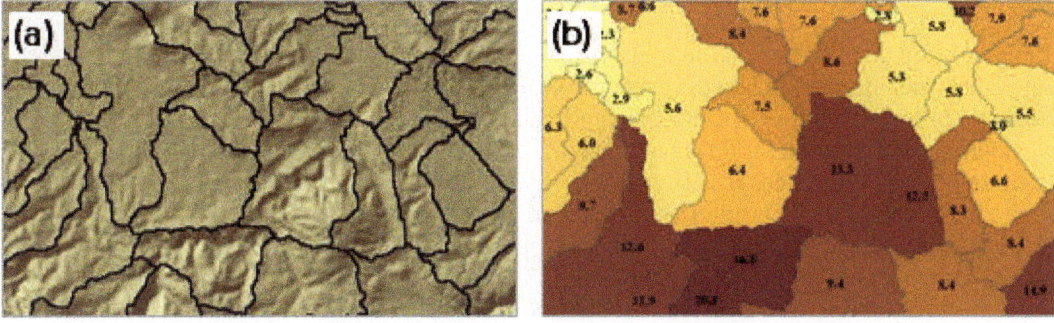

Fig. 11.14. Using zonal statistics to find the average slope of watersheds: (a) the zone raster shown over a hillshade; (b) the watersheds colored by and labeled with the average slope
Source: USGS, Black Hills National Forest

Zonal statistics functions can produce two types of output, rasters and tables. A raster output places the calculated statistical value into every cell in the zone. The table output contains the zone attribute field plus one or more statistics for each zone. This table could be joined to the original watershed polygons using the zone field and used to map the average slope of the watersheds (Fig. 11.14b).

Zones need not be contiguous like the watershed zones, however. The geology map in Figure 11.15 contains many unconnected areas with the same geological unit. The map has only six zones, one for each color.

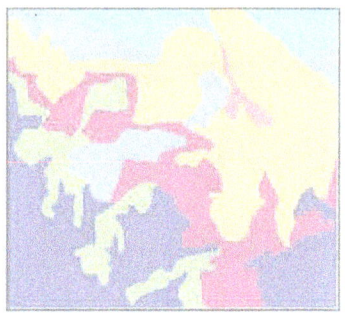

Fig. 11.15. This geology map has six zones.
Source: South Dakota Geological Survey

The zonal statistics function can also assign values from a surface to line or point data, such as finding the average slope along a stream section or attaching an elevation value from a DEM to a layer of wells. Each line or point must have a unique name or ID value that identifies it as a distinct zone.

Reclassify

The **reclassify** function changes the values of a raster according to a scheme designed by the user, such as classifying a slope map into three regions of low, medium, or high slope, or ranking suitability factors in a weighted overlay (see Fig. 11.7). For a map with only a few values, such as land use codes, the user might specify a new value for each code individually. Perhaps a town has annexed a nearby planning unit that used different land use codes and must convert one set of codes to another; the code 143 becomes 347, and so on. With continuous data covering a range of values, such as slope, the values are typically classified into a discrete number of ranges. Classification uses a tool similar to the one described in Chapter 2 for classifying attribute data.

About ArcGIS

Using Spatial Analyst

Spatial Analyst is an extension to ArcGIS and must be purchased separately. It contains a set of tools for ArcToolbox under the heading Spatial Analyst Tools (Fig. 11.16). The tools are organized by function type. Spatial Analyst tools are fully integrated into the geoprocessing environment. They can be run from the toolbox or the command line, used to build models in Model Builder, and incorporated into scripts.

The tools are well documented in the Help files, in many cases including complete discussions of the concepts behind each tool. Users interested in interpolation, for example, will find extensive information on, and references to, the various interpolation methods (IDW, kriging, trending, and splining). Reading about unfamiliar tools before implementing them is highly recommended.

Spatial Analyst Tools

- Conditional
- Density
- Distance
- Extraction
- Generalization
- Groundwater
- Hydrology
- Interpolation
- Local
- Map Algebra
- Math
- Multivariate
- Neighborhood
- Overlay
- Raster Creation
- Reclass
- Segmentation
- Solar Radiation
- Surface
- Zonal

Fig. 11.16. Spatial Analyst tools
Source: Esri

Environment settings

The Spatial Analyst tools obey the Environment settings like any other tool. Several settings are particularly useful (Fig. 11.17).

The **Workspace** settings specify the default input and output locations for data sets. Spatial Analyst produces many output rasters that may be intermediate steps toward a solution. By default, these are stored in the project geodatabase. However, specifying a separate geodatabase to hold the outputs helps keep them separated from permanent data collections and simplifies cleanup later.

The **Output Coordinates** setting specifies the output coordinate system. It is usually best left to the default, Same as Input. Projecting rasters consumes time and memory, and it may cause system overload when large rasters are being processed. Moreover, resampling during projection may affect the resolution and accuracy of the original data. Projecting rasters should be done only when necessary, preferably as a separate and conscious step in which the parameters are carefully chosen, not as a forgotten background process.

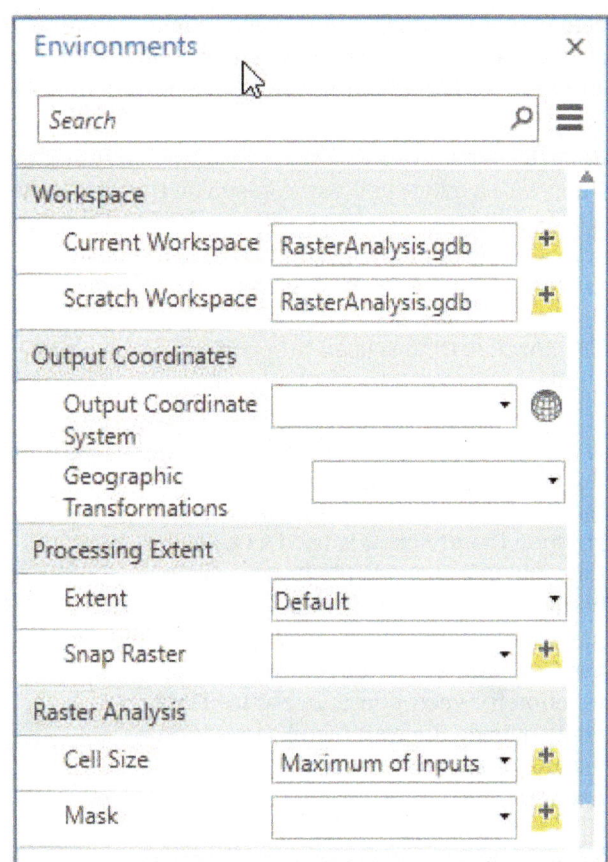

Fig. 11.17. Key Environment settings for raster analysis
Source: Esri

The **Processing Extent** controls the rectangular area used to preselect features or raster cells to be used in a tool. By default, all areas of the input data sets are used. With rasters, this setting is particularly important because a raster must have a rectangular extent. If two data sets represented by the black squares in Figure 11.18 are input to a tool, the output raster might contain the entire blue area even though the cells outside the black lines will be NoData. The default is the Intersection of Inputs option (the yellow area containing all cells with valid values from both inputs). The extent may be set to match the display area, or to be the same as a specific data set to which everything else should conform. This last option is usually the best because it minimizes resampling as analysis steps unfold.

The Snap raster setting instructs the program to exactly match the corner of the first cell to an existing raster, and it can be used to ensure that cells in the output always align, reducing the need for resampling in later steps.

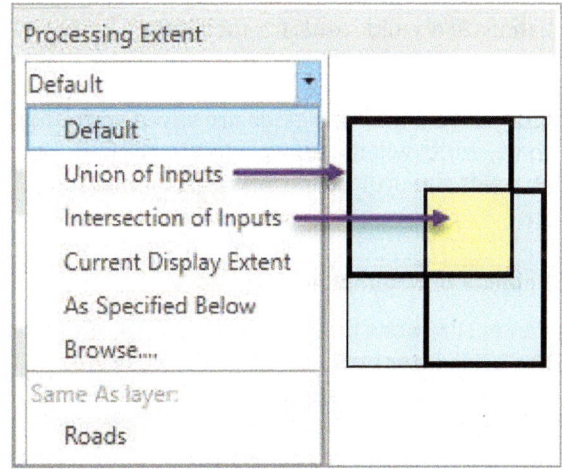

Fig. 11.18. Processing extent options
Source: Esri

The **Cell Size** setting controls the output resolution of rasters. It can be set to a specific value or to match an existing data set. The default is the maximum of the inputs. For example, if two inputs have cell sizes of 50 meters and 30 meters, the output will have 50-meter resolution, and the

30-meter raster will be resampled to match (with potential impacts on the data values). The cell size is always measured in coordinate system units and is typically in meters or feet. If a raster is stored in a GCS, however, the units will be degrees.

Some tools ask for a cell size even if the environment is set. In this case, the cell size box will be filled with a suggested value based on the map units and the analysis extent. The suggested size will usually create a raster with a sufficient, but not excessive, number of cells. If the suggested size is far different from what is expected or wanted, then double-check the extent settings and the map units before proceeding.

The **Mask** setting is used to specify a polygon feature class or raster that controls which cells will be processed. A **mask** raster contains NoData cells outside the region of interest, and all output cells there will be given NoData values. If a polygon feature class is used as a mask, then any cells outside the feature class boundary will receive NoData values.

Using a mask is similar to using the clip function for vector data. In Figure 11.19, areas outside the watershed have NoData instead of elevation after the mask is applied.

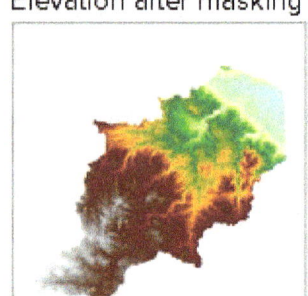

Fig. 11.19. NoData cells in a mask raster become NoData cells in the output raster.
Source: USGS, Black Hills National Forest

Some tools allow the user to specify different methods for handling NoData values. If the tool is set to *Ignore NoData values*, then the tool tries to complete the computation using the values it has available; this is usually the preferred and default option. The other option is to assign NoData to the output cell if the input cells contain one or more NoData values. In other words, if any input values are NoData, the tool assumes that it does not have enough information to calculate a valid result for that cell.

TIP: Environment settings are saved with projects and persist until changed. They are reset to the defaults when a new project is created. Use the Environments panel on a tool to temporarily override the project settings for that one tool, that one time.

Impacts of resampling

Raster functions treat layers as arrays of numbers, which can be processed as a stack. To perform raster analysis, all of the input rasters must have the same cell size and extent so that the cells in each stack align. If a tool receives inputs with different cell sizes or extents, the rasters will automatically be resampled to match each other. By default, all rasters are resampled to the most coarse cell size present in the input rasters.

If rasters with different coordinate systems are input to a tool for analysis, the output raster will be given a coordinate system based on this order of precedence: (1) the coordinate system specified by the Environment settings and (2) the coordinate system of the first input to the tool. Spatial Analyst will automatically project the rasters to match, and the projection will often include resampling.

The automatic resampling and projection process is invisible to the user and seems convenient, but the pitfalls of relying on it can be significant. First, the automatic process may be degrading the accuracy of a data set without the user being aware of it, potentially leading to inconsistencies or even errors in the results. Second, the default settings used in the automatic process may not be optimized for the data. When resampling a DEM, the bilinear interpolation option produces the best result, but the automatic process uses the nearest neighbor option and may introduce troublesome discontinuities in the surface. Third, the order of processing may lead to problems. Consider an input raster that needs both clipping and projecting. Because projecting is a computer-intensive process, it is best to clip first and then project, but the automatic process may cause the tool to fail by projecting first and overwhelming the computer.

Before beginning a raster analysis, it is best to determine a standard coordinate system, extent, and cell size that will be consistently used throughout. If an input data set does not match the selected parameters, it is wise to project, clip, or resample the data manually beforehand so that the appropriate parameters can be specified instead of letting them default to the automatic processes. Once analysis begins, using appropriate environment settings will ensure that the extents, cell sizes, and projections remain consistent throughout the analysis, which will eliminate the need for resampling and the problems that may occur as a result. If the raster tools are failing unexpectedly, the solution often lies in reviewing the inputs and environments and ensuring that the inputs and environment settings are carefully managed.

Summary

▶ Rasters consist of spatial data stored as individual cells in an array and can represent discrete features or continuous fields of information.

▶ Rasters have coordinate systems just like other spatial data. Many raster calculations are based on distance and area functions, so it is important to store rasters in a projected coordinate system with minimal distortion.

▶ Map algebra treats rasters as arrays of numbers that can be added, multiplied, and otherwise manipulated on a cell-by-cell basis. Arithmetic, logical, Boolean, and bitwise operators and functions offer a wide array of options for analyzing rasters and scalars.

▶ Raster overlay is generally faster and better incorporates continuous data than vector overlay for suitability analysis. Boolean overlay uses true/false rasters and Boolean operators. Weighted overlay uses reclassification and weights for greater flexibility in modeling suitability.

▶ Distance functions measure distances from a set of objects, and they can be used to calculate costs and paths associated with traveling over a surface.

▶ Density functions measure the number of objects occurring within a specified search radius and report it in features per unit area.

▶ Raster interpolation estimates a surface based on measurements at isolated locations.

▶ Statistical measures can be calculated for rasters based on single cells, within a specified neighborhood, or within large, irregular zones.

▶ Rasters can be reclassified and given new values as needed.

▶ Resampling occurs automatically when rasters with different cell rasters are analyzed together, but it can degrade the quality of a data set. Users should try to minimize resampling when possible by aligning raster data before analysis.

▶ The analysis environment controls certain settings for all new rasters created during a session. The settings include the working directory, the extent and cell size of output rasters, and the analysis mask.

Important Terms

aspect	filter	reclassify	weighted overlay
azimuth	focal function	slope	zenith angle
block function	hillshade	Spatial Analyst	zonal statistics
Boolean overlay	interpolation	suitability analysis	zone
Boolean raster	map algebra	surface analysis	
cut/fill	mask	Tobler's law	
Euclidean distance	neighborhood function	viewshed	

Chapter Review Questions

1. Before any map algebra expression can be executed, what must be true of the resolution and structure of the input rasters?

2. Imagine you have Boolean rasters for 10 different criteria regarding the siting of a landfill. If the rasters are added together, what is the potential range of values in the output raster? What is the potential range of values if the rasters are multiplied?

3. Explain the advantages of using weighted overlay instead of Boolean overlay.

4. Why is it best to store rasters in the same coordinate system in which they will be analyzed? Which types of projections work best?

5. Compare the range of values encountered in a raster showing distance to streams and a Boolean raster showing buffers around the streams.

6. Describe Tobler's law. Give an example of a data set that likely follows this law and another data example that likely does not.

7. When should a kernel density function be used instead of a simple density function?

8. Focal and zonal statistics functions are similar except for how the neighborhoods are defined. Explain this difference.

9. Imagine that a parcels feature class is used as the zone layer for a zonal statistics function. Explain how the results would differ if the zone attribute used were (1) the parcel-ID number or (2) the land use designation.

10. List and describe the five Environment settings that can be helpful in managing raster analysis. When should these be checked and set?

Expand your knowledge

Expand your knowledge by reading these sections of ArcGIS Help.

> Help > Analysis and geoprocessing > Spatial Analyst > Spatial Analyst basics
> Help > Analysis and geoprocessing > Spatial Analyst > Map algebra
> Tool Reference > Tools > Spatial Analyst toolbox (skim the variety of tools)

Mastering the Skills

Teaching Tutorial

The following examples provide step-by-step instructions for doing basic tasks and solving basic problems in ArcGIS. The steps you need to do are highlighted with an arrow →; follow them carefully.

> → Start ArcGIS Pro and open the ClassProjects\RasterAnalysis project. Remember to save often.

To become acquainted with raster analysis, we will use rasters to solve the snail habitat suitability problem from Chapter 10 (Fig. 11.20). Certain Environment settings are especially important for raster analysis, including the default cell size and extent, so that all inputs and outputs have the same cell grids and no resampling will be needed.

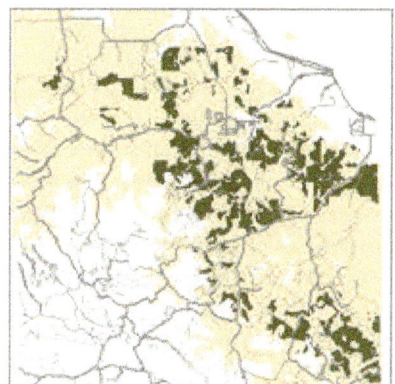

Fig. 11.20. Snail habitat determined by vector analysis
Source: Black Hills National Forest

> 1→ Click **ANALYSIS: Geoprocessing: Environments**.
>
> 1→ Note that the Current and Scratch Workspace are set to the home geodatabase.
>
> 1→ Set the Processing Extent: Extent setting to *Same as layer: Elevation*. The coordinate boxes update to show the specific *x-y* values being used.
>
> 1→ Set the Processing Extent: Snap Raster to Elevation.
>
> 1→ Change the Raster Analysis: Cell Size setting to *Same as layer Elevation*.
>
> 1→ Keep the Raster Analysis: Mask setting blank. Click OK.

TIP: It is NOT a good practice to change the output coordinate system during raster analysis because it will cause resampling of the input data. Usually one should leave the output coordinate system blank for raster processing.

In Chapter 10, we predicted snail habitat from vector layers of geology, vegetation, and elevation. We will now perform the same analysis by creating three Boolean rasters, each showing one of the ideal snail conditions, and then multiplying them to find the habitat.

Performing a Boolean overlay

The snails prefer elevations from 1200 to 1600 feet. We need to create a Boolean raster in which the cells meeting the elevation criteria have a value of 1 and the rest have a value of 0. The Reclassify tool is used to convert one set of raster values to another.

TIP: The Spatial Analyst and 3D Analyst extensions share some tools, so two options may appear for a tool search. In this tutorial, always choose the Spatial Analyst tool.

> 2→ Arrange the Geoprocessing pane so that it will remain open during the tutorial.
>
> 2→ In the Geoprocessing pane, open the Reclassify tool (Fig. 11.21).

2→ Set the Input raster to Elevation.

2→ The Reclass field will default to Value, the field containing the values stored in the raster. This is usually the field to use when processing rasters.

With numeric data, we need to specify ranges, in this case: 0–1200, 1200–1600, and 1600–2000. It is helpful to create the desired number of classes first and then edit the ranges.

3→ Click the Classify link and enter 3 for the number of classes.

3→ Click in the first Start box and change the value to 0. Click Tab to go to the End box and enter 1200. Tab to the New box and enter 0.

3→ Continue using Tab to enter the ranges shown in Figure 11.21, including 0 for the New NoData value.

3→ Name the output *ElevRange* and run the tool.

TIP: When reclassifying, the minimum and maximum values in the raster will be automatically entered in the first Start box and the last End box. It works best to ensure that the start and stop values are below and above these automatic values. Also, to create a true Boolean raster with only 0 and 1 values, it is best to change the NoData values to 0.

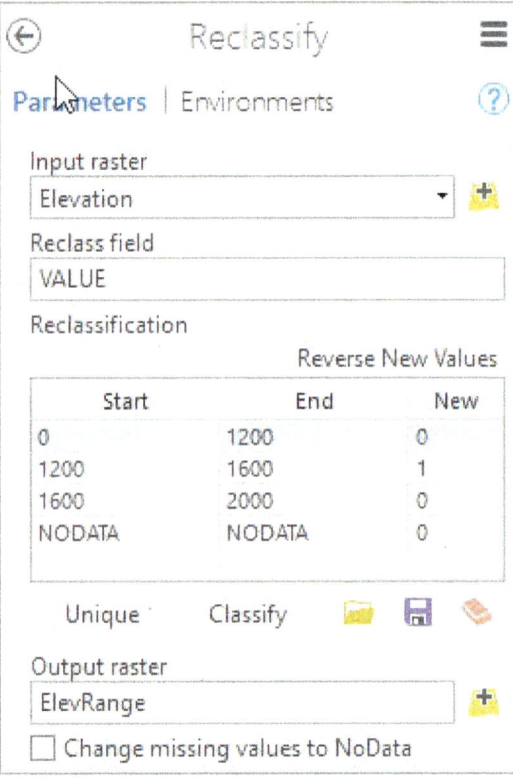

Fig. 11.21. Reclassifying elevation
Source: Esri

3→ Right-click the 0 value symbol for ElevRange and set its color to Gray 10%, which helps us to visualize which areas meet the criteria.

The new raster should appear as in Figure 11.22a (your color may be different). This is a Boolean raster with two values: 1 where elevation is in the desired range (green) and 0 where it is not (gray).

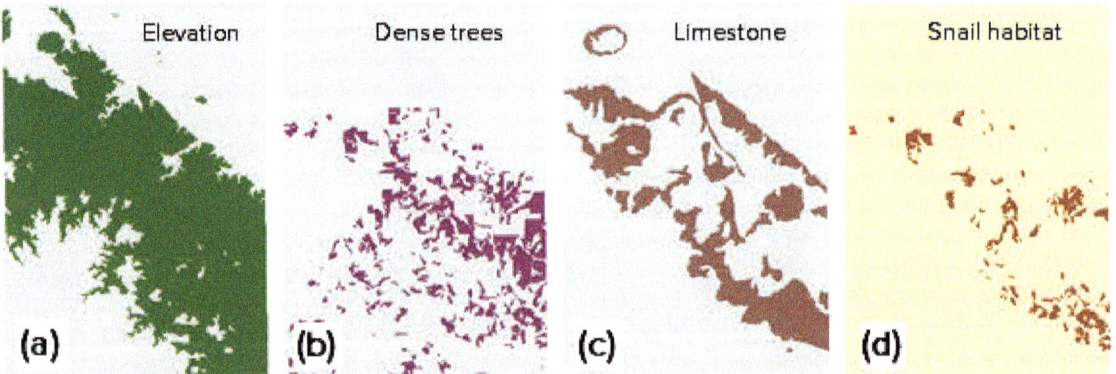

Fig. 11.22. Boolean overlay to determine snail habitat

Next, we will use the Vegetation polygon layer to create a Boolean raster showing areas of dense trees. Rasters only store one attribute value, so we must specify which field in the Vegetation table will supply the data. We will use the DENSITY96 text field.

> 4→ Turn off all layers except Vegetation.
>
> 4→ Open the Feature to Raster tool. Set the Input Features to Vegetation and the Field to DENSITY96. Name the output *vegdensity* and run the tool.
>
> 4→ Turn off the Vegetation polygon layer to see the raster (Fig. 11.22b).
>
> 4→ Right-click the vegdensity layer and choose Attribute Table.

The vegdensity raster table is shown in Figure 11.23. A raster stores only numbers (shown in the Value field), determined by arbitrarily assigning an integer to each unique text value in DENSITY96. The table documents the integer assigned to each density class, although the layer uses the text field in the legend. The Count field indicates how many cells have each value. For example, this raster contains 75,914 cells of density class C.

OBJECTID	Value	Count	DENSITY96
1	1	147840	0
2	2	80898	A
3	3	75914	C
4	4	97386	B

Fig. 11.23. The vegdensity table
Source: Esri

1. Use the cell size (30 m) to calculate the area of class C in km^2. _____

> 5→ Close the table and open the Reclassify tool.
>
> 5→ Set the Input raster to vegdensity and the Reclass field to DENSITY96.
>
> 5→ Change the New value to 1 for C, and the rest to 0, including NODATA.
>
> 5→ Name the output *DenseTrees*. Run the tool.

Notice that the output symbols are labeled 0/A/B and C, instead of 0 and 1. When the attribute table contains a text field, the raster renderer uses it by default. To remember that we are using Boolean layers, let's reset the symbols to show the actual raster values, not their text labels.

> 6→ Open the Symbology pane for the DenseTrees layer. It uses a Unique Values map, which is fine, but change the Value field to Value instead of DENSITY96.
>
> 6→ Set the 0 color to Gray 10% to enable us to better visualize the desired areas (Fig. 11.22b).

Finally, we will convert the Geology layer to a raster. Because there are so many different geologic units, we will select the desired one before converting. Areas outside the selected polygons will be given a NoData value.

> 7→ Turn off all layers except Geology.
>
> 7→ Open the Select Layer By Attribute tool and select the limestone from the Geology layer using the expression: RockType is Equal to Limestone.
>
> 8→ Open the Feature to Raster tool. Select Geology as the Input features and set the Field to RockType. Name the output *Limestone* and run the tool.
>
> 8→ Open the Limestone attribute table.

NoData values are treated differently than 0 values during Boolean overlay; they do not even show up in this table. To ensure consistent processing during a Boolean overlay, and to avoid having both 0 and NoData values in the final result, we should reclassify NoData values to 0.

9→ Close the table and open the Reclassify tool.

9→ Set the Input raster to Limestone and set the Reclass field to Value.

9→ If necessary, type 1's across the first row and hit tab.

9→ In the NODATA row, change the New value to 0.

9→ Name the output *LimestoneBoo* and run the tool.

10→ Turn off Geology to see the new raster.

10→ Change the 0 value symbol in LimestoneBoo to gray (Fig. 11.22c).

We are ready to overlay. We have three Boolean rasters, each containing 1 where the habitat condition holds and 0 where the condition is absent (Fig. 11.22a–c). After multiplying them, areas that meet all three conditions will be assigned the value 1, and all other cells will be 0.

11→ Open the Raster Calculator tool (Fig. 11.24).

11→ In the Rasters list, double-click the raster names and double-click the * operator in the Tools list to enter the expression: "ElevRange" * " LimestoneBoo" * "DenseTrees"

11→ Name the output *SnailHab* and run the tool.

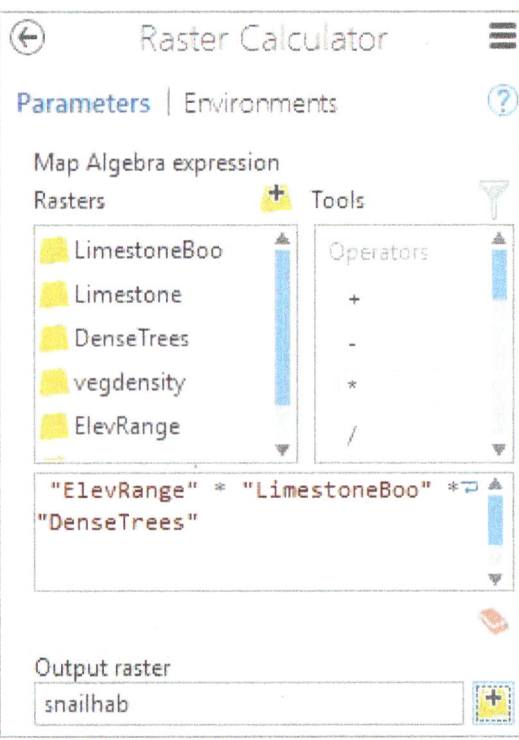

Fig. 11.24. Boolean overlay with the Raster Calculator

Source: Esri

12→ Turn off all layers except for SnailHab.

12→ Zoom to the extent of the SnailHab layer and compare it carefully with Figure 11.22d (the colors may be different).

In Figure 11.22d, the snail habitat has the value 1 and appears brown. It is the raster equivalent of the snail habitat calculated in Chapter 10 by intersecting vector features. With rasters, though, we can choose to add them to get a ranking of suitability, rather than just a simple yes or no.

13→ In the still open Raster Calculator tool, change the * operators to + operators, changing the expression to: "ElevRange" + "LimestoneBoo" + "DenseTrees"

13→ Name the output *SnailHabrank* and rerun the tool.

13→ Open the Symbology for SnailHabrank and change the color scheme to a light to dark green color ramp.

This time the best habitat receives the value of 3, and other areas are ranked by how many of the preferred conditions occur there (Fig. 11.25).

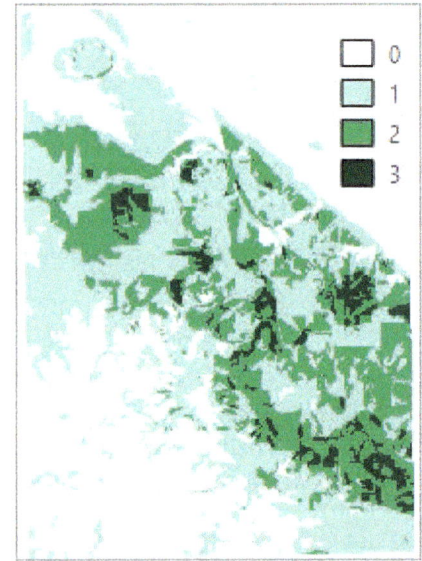

Fig. 11.25. Ranked snail habitat

Distance functions and buffers

Vector analysis uses buffers to evaluate proximity to features, such as areas within 200 meters of a road. Distance analysis using rasters is more powerful and flexible, partly because rasters can store continuous distance information, whereas vectors are limited to portraying discrete objects. Let's explore some of the raster distance functions.

14→ Use Ctrl-click to collapse the legends and turn off all layers except for Roads.

14→ Open the Euclidean Distance tool (Fig. 11.26).

14→ Set the Input data to Roads and name the output distance raster *RoadDist*.

14→ Important: Leave the maximum distance blank.

14→ Accept the default cell size of 30 meters.

14→ Specify an Output direction raster, *RoadDir*. Run the tool.

15→ Turn off RoadDir for the moment to examine RoadDist.

15→ Open the Symbology pane for the Road-Dist layer and change the Method to the Natural Breaks (Jenks) classification.

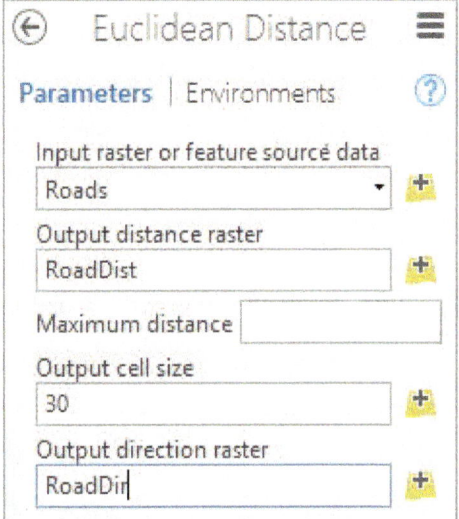

Fig. 11.26. Euclidean Distance tool
Source: Esri

Normally, equal interval is the appropriate classification for distance rasters, but in this case the roadless area in the upper right is skewing the distance measurements and the symbols.

16→ Click the **Map: Navigate: Explore** tool and click on cells of the raster to examine the values they contain.

Each cell of the RoadDist raster represents the distance of that cell to the nearest road (Fig. 11.27a) in the coordinate system units, meters.

16→ Turn on the RoadDir raster and examine the values in its legend. Click on several cells to examine the values. Close the Pop-up window when finished.

The RoadDir raster cells record the direction to the closest road (Fig. 11.27b) in degrees from 0 to 360, with 0 (and 360) indicating north. For any location in the map area, these two rasters indicate the distance to the closest road and the travel direction to get there.

Chapter 9 introduced buffers. The Raster Calculator can quickly create raster buffers in the form of a Boolean raster with 1 inside the buffer and 0 outside it (Fig. 11.28). Imagine that we want buffers showing areas within 300 meters of the roads.

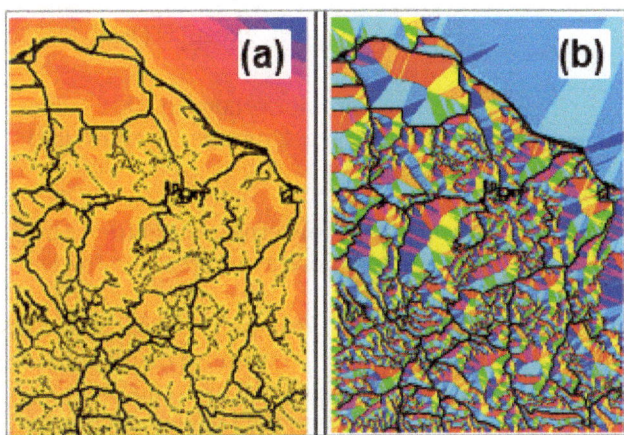

Fig. 11.27. The (a) RoadDist and (b) RoadDir rasters
Source: Black Hills National Forest

17→ Open the Raster Calculator tool.

17→ Enter the expression: "RoadDist" <= 300.
(Note that "<=" is a single operator entered from its own line, not by entering first "<" and then "=".)

17→ Name the output *Roadbuf300* and run the tool.

We can also create multiple ring buffers with rasters. As an example, consider the snail habitat again. Because snails tend to get crushed on roads, the best habitat is farther away from roads. Biologists have determined that the snails rarely travel more than 400 meters during their lifetimes. We can rank the habitat based on distance using 200-meter ring buffers.

18→ Open the Reclassify tool. Set the Input raster to RoadDist and the Reclass field to Value.

18→ Click the Classify link. Set the number of classes to four.

18→ In the Reclassification section, enter 0 in the first box, then use the Tab key to enter the remaining ranges as shown in Figure 11.29.

18→ Name the output *Ringbuf200* and run the tool.

Now we have a raster with 1 signifying the least desirable distance and 4 the best. Let us multiply the ring buffers with the snail habitat raster. Recall that habitat has a value of 1 and non-habitat a value of 0, so multiplying by Ringbuf200 will yield a new raster with values from 1 to 4, but only in the snail habitat.

19→ Open the Raster Calculator tool and enter the expression:

"SnailHab" * "Ringbuf200".

19→ Name the output *DistHabitat* and run the tool.

19→ Turn off all layers except DistHabitat and roads.

19→ Change the color scheme of DistHabitat so that 1 is red, 2 orange, 3 yellow, and 4 green (Fig. 11.29).

This map might raise new concerns—very little of the habitat lies in the safer regions away from the roads.

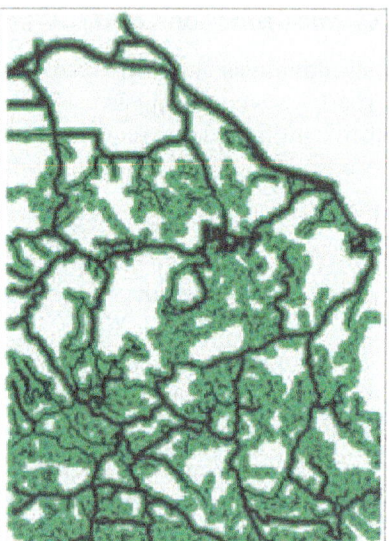

Fig. 11.28. Boolean road buffers
Source: Black Hills National Forest

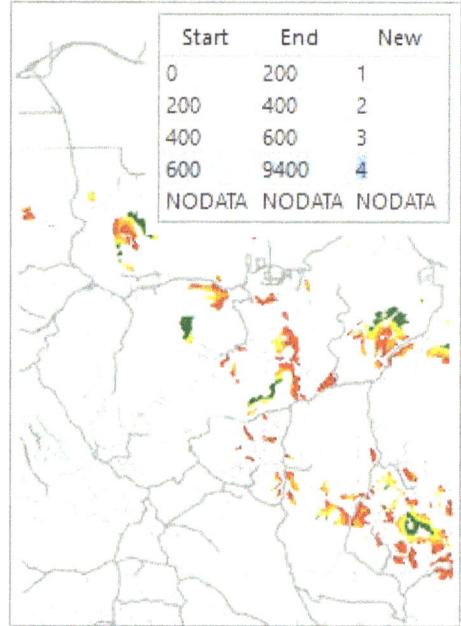

Start	End	New
0	200	1
200	400	2
400	600	3
600	9400	4
NODATA	NODATA	NODATA

Fig. 11.29. Ranked habitat ranked by distance to roads
Source: Esri, Black Hills National Forest

It is important to note that the classified display of a distance raster (Fig. 11.27a) looks similar to buffers, but do not be fooled. A distance raster contains continuous distances, whereas a buffer raster contains discrete values of 0 or 1. One cannot create buffers by changing the symbology of a distance raster. Even though they may <u>appear</u> as buffers, the values are still distances.

20→ Compare the legends and rasters of the Ringbuf200, Roadbuf300, and RoadDist layers.

Nor can one create buffers using the maximum distance feature in the Euclidean Distance tool. This is a common misconception.

21→ Open the Euclidean Distance tool. Set the input features to Roads, the output to *MinDist*, and the maximum distance to *300*. Don't specify a direction raster; it isn't needed. Run the tool.

21→ Turn off all layers except MinDist and Roadbuf300. Zoom in and compare them carefully. Compare their legends in the Contents pane as well.

Superficially, the MinDist raster looks like buffers because only the cells close to roads have values. However, the cell values contain distances. The true Boolean roadbuf300 has 1 in the buffer cells. To turn MinDist into buffers, it must still be reclassified. So setting the maximum distance does not save any effort, and the output will not work correctly for Boolean overlay.

Topographic functions

Topographic functions can powerfully transform elevation data into useful products. In this section, we will use them to find low-slope, sunny areas as prospective sites to build a winter ski cabin. The Contents pane of the current map is getting crowded with layers, so let's start fresh with a new map.

22→ Save the project and close the Sturgis map window.

22→ Choose **Insert: New Map** and change its name to *Topo Functions*. Turn off the Topographic basemap.

22→ Add the raster Folders\BlackHills\Sturgis\topo30m to the map.

The slope function calculates, in layperson's terms, how steep the surface is. In mathematical terms it calculates the rate of change of the surface.

23→ Open the Slope tool (Fig. 11.30). Set the Input raster to topo30m. Name the output *Slope30m*.

23→ Keep the defaults for the Output measurement (Degree) and the Z factor (1). Run the tool.

23→ Turn off topo30m and examine Slope30m.

Let's find the areas that are flat enough to build a cabin, with slope less than 10 degrees.

24→ Open the Raster Calculator tool and enter the expression: "Slope30m" < 10. Name the output *LowSlope*. Run the tool.

24→ Set the 0 values of LowSlope to light gray and the 1 value to a nice color.

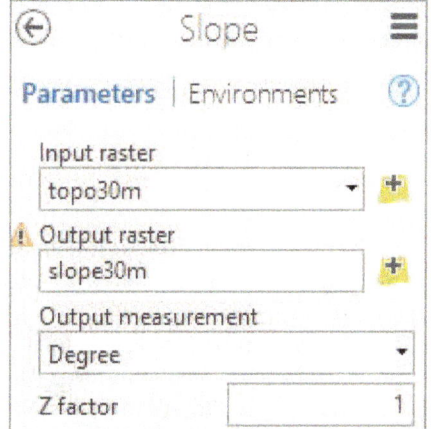

Fig. 11.30. The Slope tool
Source: Esri

To find sunny areas, we need an aspect raster. Aspect is the direction a slope faces—measured in degrees from north, with true north being 0 (and 360) degrees, east 90 degrees, and so on.

25→ Open the Aspect tool. Set the Input raster to topo30m, and name the output *Aspect30m*. Run the tool.

2. The aspect map is automatically colored and labeled according to compass direction. Which color indicates a northern aspect? _____ A southern aspect? _____

Let's find the areas with a warm southerly exposure, where the aspect is between 130 and 230 degrees (50 degrees either side of south).

26→ Open the Raster Calculator tool.

26→ Carefully enter this expression, including the parentheses:

("Aspect30m" > 130) & ("Aspect30m" < 230)

26→ Name the output *Southerly* and run the tool.

TIP: The parentheses are required to force the calculator to evaluate the two conditions (aspect > 130) and (aspect < 230) before it executes the Boolean AND (&).

Now we do a Boolean overlay to find potential cabin sites.

27→ Clear and reopen the Raster Calculator tool and enter the expression: "LowSlope" * "Southerly". Name the output *CabinSites* and run the tool.

27→ Compare the results with Figure 11.31.

27→ Collapse all legends, and turn off all layers.

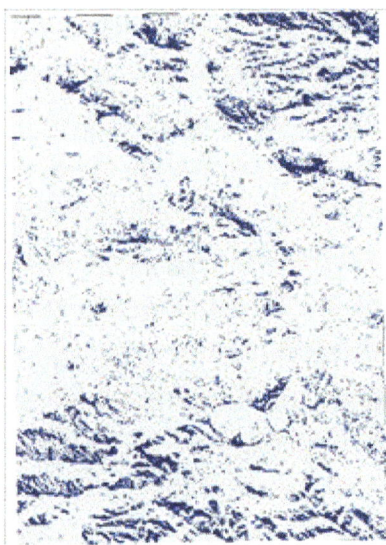

Two other surface functions are worth investigating. The hillshade function creates a raster that mimics the illumination of the surface from a light source at a specified direction (**azimuth**) and altitude (**zenith angle**), both measured in degrees. The result looks similar to what a person might see when flying over the surface.

28→ Open the Hillshade tool. Set the Input raster to topo30m, and name the output *Hillshd30m*. Leave the other settings to their defaults. Run the tool.

Fig. 11.31. Cabin sites

Compare the elevation and hillshade rasters, and notice how the hillshade brings out fine topographic details. It makes an excellent basemap.

The Viewshed function determines what areas are visible from a set of points. It is used for applications such as placing fire watchtowers or determining whether timber clear-cut areas are visible from a lookout. We will use it to determine the visibility offered by three fire towers.

29→ Add the towers feature class from the Features feature dataset in the Folders\Black-Hills\Sturgis geodatabase.

29→ Open the Viewshed tool. Set the Input raster to topo30m and the observer features to towers. Name the output *TowerView* and run the tool.

3. Examine the attribute table of towerview. What values does the raster actually store? Use the tool Help to determine what the values mean.

TIP: As rasters accumulate, one must wait while each draws. For best performance, get in the habit of turning off or removing any rasters not currently in use.

Neighborhood functions

A neighborhood function examines a target window in a raster and calculates statistics from cells in the window. Consider the idea of habitat variety.

30→ Add the Landuse.lyrx layer file from the Folders\BlackHills folder. It refers to a landcover raster in the Sturgis geodatabase.

Some areas are homogeneous, with only one or two land cover units, but other areas are more variable. We can analyze the variety of land cover types in different areas using neighborhood functions, such as finding the variety of land cover types within a 10 × 10-cell moving rectangle.

31→ Open the Focal Statistics tool (Fig. 11.32).

31→ Set the Input raster to landcover and name the output Lcvarfoc.

31→ Choose a Rectangle neighborhood. Enter 10 for both Width and Height and set the Units type to Cell.

31→ Select Variety for the Statistics type and run the tool.

31→ These are ordinal data, so open the Symbology pane for the Lcvarfoc layer and change the color scheme to a monochromatic color ramp.

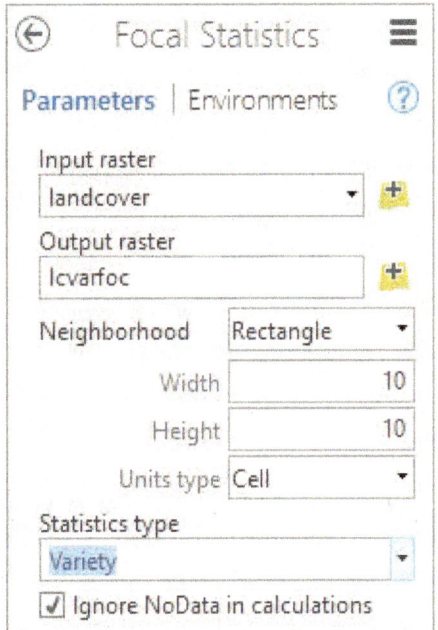

Fig. 11.32. Focal Statistics tool
Source: Esri

A value of 1 in the lcvarfoc raster indicates that only one land cover type is found within a 10 × 10-cell (300 m × 300 m) area; it is homogeneous. A value of 9 indicates that nine different types were found. This raster gives an objective estimate of landscape heterogeneity.

A block statistics function works much in the same way except that the target window, instead of moving cell by cell, moves to a completely new area each time.

32→ Open the Block Statistics tool.

32→ Use the same inputs as for the Focal Statistics tool, but name the output Lcvarblock. Run the tool.

32→ Change the color scheme to match Lcvarfoc and compare the two rasters. Then turn them both off and collapse the legends, leaving only landcover turned on.

The neighborhood functions can use different statistics. The focal majority function, for example, does a good job of simplifying a raster for conversion to features. This land cover data set (Fig. 11.33a) was generated from raster satellite data, but perhaps a vector version is needed. In its original form it would make a complex set of polygons with many tiny spots and stringers, and it needs simplifying before conversion. The Majority Statistic tool examines the target window and selects the land cover value that occurs most often. We will use the focal version to get a smoother result. The larger the window, the greater the generalization.

33→ Open the Focal Statistics tool. Set the Input raster to landcover and name the output Major5.

33→ Specify a 5 × 5-cell Rectangle window and the Majority statistic. Run the tool.

When simplifying, a smaller rectangle done two or three times usually works better than a larger rectangle, so run the majority filter once more.

33→ In the Focal Statistics tool, set the input raster to Major5 and name the output Major5x2. Leave the other parameters as is and run the tool again.

We can import symbols from a prepared layer file to Major5x2 to better compare the original and final rasters.

34→ Open the Symbology pane for Major5x2. Click Options and select Import…

34→ Navigate to the BlackHills folder and select the LanduseNumbers.lyrx file.

34→ Compare Major5x2 (Fig. 11.33b) with landcover (Fig. 11.33a).

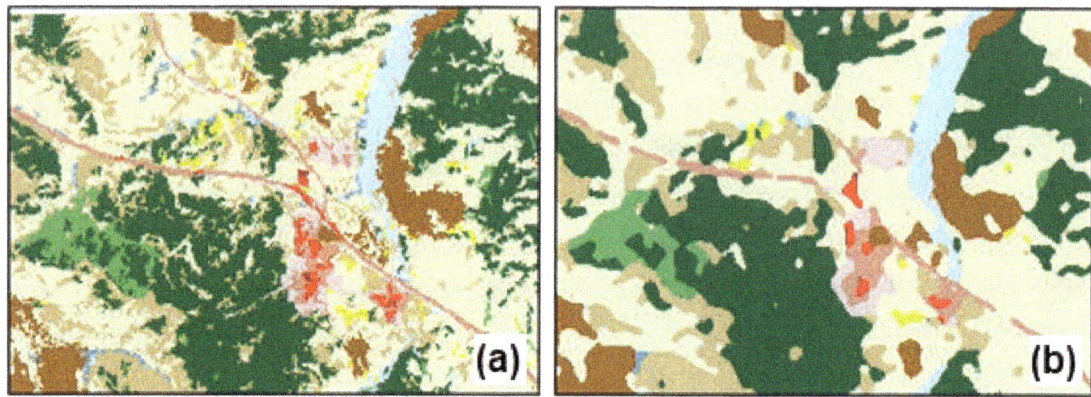

Fig. 11.33. (a) Original land cover and (b) land cover after two passes of a 5 × 5 majority filter
Source: USGS

Notice the simpler appearance of the new map. It will convert to polygons more easily. However, the original resolution and accuracy of the data have been generalized. Keep this fact in mind when interpreting results based on the new data.

Converting rasters to features

Since we've done the simplification, let's go ahead and convert the land cover to polygons.

35→ Open the Raster to Polygon tool.

35→ Set the input raster to Major5x2 and the Field to Value.

35→ Name the output LandcoverPoly and check the box to *Simplify polygons*. (Otherwise, the polygon edges will follow the cell boundaries.) Run the tool.

The field specified, Value, is transferred to the polygon table as an attribute named *gridcode*. Unfortunately, the text field containing the names of the land cover classes, which was present in the original landcover raster, was omitted when the neighborhood majority tool was run. However, we can copy the field from the original landcover table using a join.

36→ Open the LandcoverPoly table and examine the gridcode field.

36→ Open the Add Join tool. Set the Layer name to LandcoverPoly and the Input Join Field to gridcode.

36→ Set the Join Table to landcover and the Output Join Field to Value. Run the tool.

37→ In Contents, right-click the LandcoverPoly layer and choose Data > Export Features.

37→ Set the Input Features to LandcoverPoly and the Output Feature Class to LandcoverFinal. Run the tool.

37→ Open the LandcoverFinal attribute table and confirm that it contains the joined fields.

38→ Remove the LandcoverPoly layer and close the remaining table.

38→ Open the Symbology pane for LandcoverFinal and display it using a unique values map based on the Covertype field. There is no way to copy a raster colormap to a polygon data set, unfortunately.

The SnailHab raster from the first analysis will serve to demonstrate another potential issue when converting rasters to polygons.

39→ Collapse and turn off all layers.

39→ Add the SnailHab raster from the home geodatabase, RasterAnalysis.

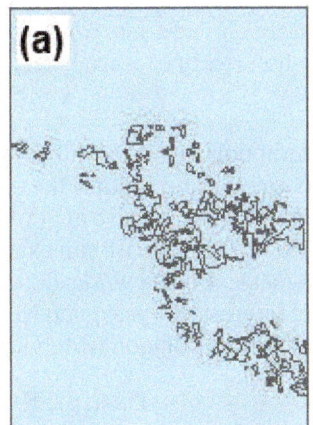

Notice that the raster has 0 and 1 values. If we convert this raster, both 0 and 1 areas will be converted to polygons (Fig. 11.34a), even though we only want the polygons with 1 value. We should first use Reclassify to convert the 0 values to NoData.

40→ Open the Reclassify tool. Set the Input raster to SnailHab and the Reclass field to Value.

40→ In the New column, change the output value for 0 to NODATA and the output value for 1 to 1 (Fig. 11.34b).

40→ Name the output *SnailHabOnly* and run the tool. Turn off SnailHab.

	Value	New
(b)	0	NODATA
	1	1
	NODATA	NODATA

41→ Open the Raster to Polygon tool. Set the Input raster to SnailHabOnly, and set the Field to Value. Name the output *SnailHabPoly* and run the tool.

41→ Examine the polygons. Then save the project.

Fig. 11.34. Snail habitat polygon conversion

Interpolation and zonal statistics

Interpolation calculates a raster surface from point measurements. For example, climate stations in the Black Hills record precipitation. In between the stations, we don't know what the precipitation is, but we can estimate it at each cell to create a precipitation raster.

42→ Close the Topo Functions map and open the Climate Stations map.

42→ Open the Climate Stations table and examine the fields. This data set contains monthly precipitation values in centimeters for 1997, with annual totals.

42→ Close the table.

This data set covers a greater extent than the others, and keeping the default cell size at 30 will generate rasters with large storage requirements. Nor do we need that much resolution at this scale. We will reset the processing extent and increase the default resolution.

43→ Click **Analysis: Geoprocessing: Environments**.

43→ Set Processing Extent: Extent to *Same As layer Climate Stations* and delete the Snap Raster, Elevation.

43→ Enter *200* for the Raster Analysis: Cell Size setting. Click OK.

We will create a raster with an estimate of annual precipitation from the Sum field.

44→ Open the IDW tool (Fig. 11.35). Set the Input points to Climate Stations and the Z value field to SUM.

44→ Name the output *AnnPrecipcm*. Keep the other default parameters and run the tool.

4. Is the precipitation raster an integer or a floating-point raster? _____ Examine its properties to check the answer.

TIP: Using interpolation functions properly requires understanding the settings and using them appropriately. Spend some time with the Help and/or textbooks when ready to learn more.

Although we needed all the stations to get the best possible interpolation, the estimate worsens farther from the stations. We also are only interested in areas inside the watersheds. We can extract part of a raster using a mask. A mask works like a clip, keeping the areas of interest and assigning NoData to the others. Either rasters or polygon features may be used as masks.

45→ Open the Extract by Mask tool.

45→ Set the Input raster to AnnPrecipcm.

45→ Set the mask data to the feature class Watersheds.

45→ Name the output *PrecipMask* and run the tool.

46→ Turn off AnnPrecipcm.

46→ Use the Symbology pane for the PrecipMask raster to display it using the Stretch method with a brown to green color ramp (Fig. 11.36a).

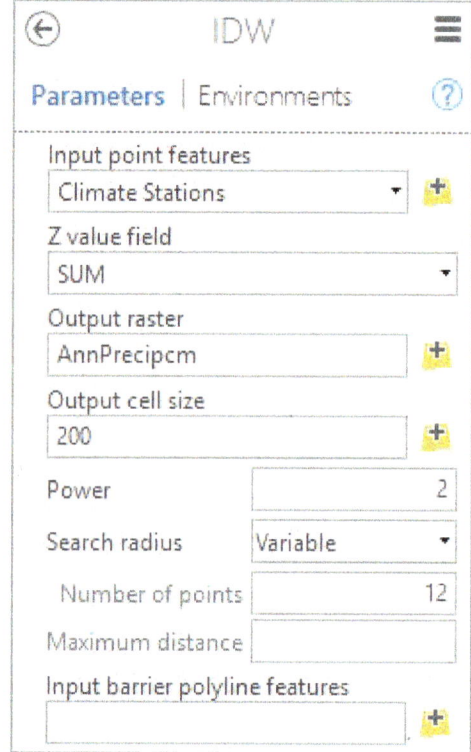

Fig. 11.35. The IDW tool
Source: Esri

The Raster Analysis Environment settings include a mask setting to ensure that every new raster has NoData in the space outside the mask area.

46→ Choose **Analysis: Geoprocessing: Environments**. Set the Raster Analysis: Mask setting to Watersheds. Click OK.

We are doing a hydrology study and wish to calculate the total volume of precipitation received by each watershed. We start by calculating the volume, V, of precipitation per cell in cubic kilometers. The depth of received water in the cell is PrecipMask* 0.00001 cm/km. The area of each cell is 200 × 200 m². Therefore, $V(km^3)$ = PrecipMask (cm) * 0.00001 km/cm × 0.2 km × 0.2 km (Fig. 11.36b).

47→ Open the Raster Calculator tool and enter the expression:

"PrecipMask" * 0.00001 * 0.2 * 0.2

47→ Name the output *PrecipVolkm* and run the tool.

The output raster values range from about 1.6×10^{-5} to 3.5×10^{-5} km³. Next, if we add all the cells in the watershed, we get the total volume of water in the watershed. Zonal statistics will sum the cell volumes in one raster (PrecipVolkm) based on zones defined by another raster or feature class (Watersheds). In the zone feature class, we must choose an attribute field that holds the unique values that define the zones. We will use the field SHEDNAME.

48→ Open the Zonal Statistics tool (Fig. 11.37a).

48→ Set the input zone data to Watersheds and the Zone field to SHEDNAME.

48→ Set the Input value raster to PrecipVolkm.

48→ Name the output *ShedPrecipVol* and set the statistic to Sum. Run the tool.

49→ Open the symbol properties of ShedPrecipVol and give it a light to dark blue color ramp (Fig. 11.37b).

In the output raster, each cell in the watershed contains the summed volume over the watershed, from about 0.005 km³ to 0.16 km³. The Zonal Statistics as Table tool provides an alternative way to calculate and view the totals.

50→ Open the Zonal Statistics as Table tool.

50→ Fill out the first three boxes as before (Fig. 11.37a).

50→ Name the output table *ShedVolTable*.

50→ Instead of the single SUM statistic, choose All. Run the tool.

51→ Find the standalone ShedVolTable table, open it, and examine the statistics for the watersheds.

The output table lists more information, including the cell count, the area, and statistics about the water volumes (min, max, etc.). This table could be joined back to the Watersheds layer if needed.

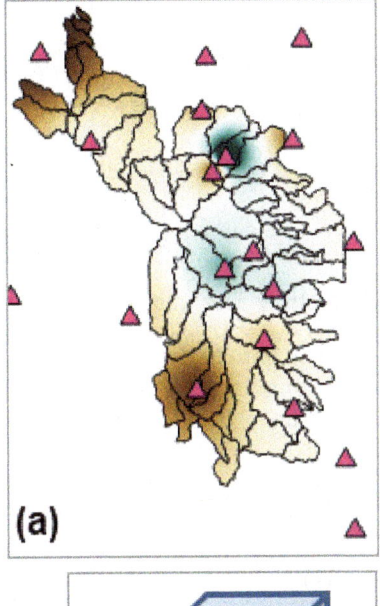

(a)

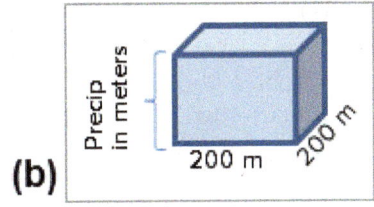

(b)

Fig. 11.36. Precipitation raster after running Extract by Mask
Source: Black Hills National Forest

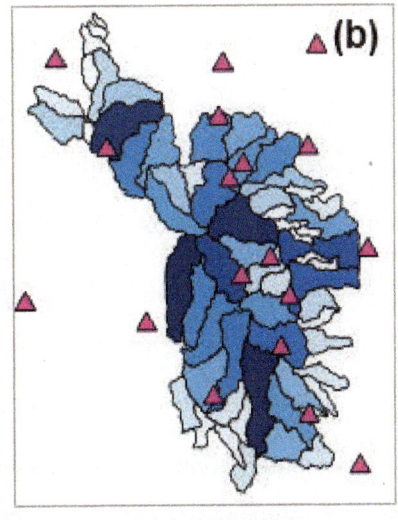

Fig. 11.37. Zonal statistics tool and output
Source: Esri

5. What is the name of the watershed that has the largest volume of water? What is the volume?

TIP: Zones can also be defined by points or lines, enabling one to determine the elevations of a set of summits or the average slopes of individual streams.

This is the end of the tutorial.

$\rightarrow$ Save the project and close ArcGIS Pro.

Practice Exercises

Use the data in the BlackHills\Sturgis geodatabase to answer the following questions. Reset the Environment settings for Cell Size to 30 meters and the Extent to Default; also verify that the Snap Raster and the Mask are not set.

Imagine that a biologist is looking for a suitable area to reintroduce bighorn sheep to the Black Hills. The site needs to be forested, must contain steep slopes, and must be on Forest Service land.

1. Create a Boolean raster showing areas where vegetation density is moderate to high (DENSITY96 is B or C) and the land is owned by the Forest Service (OWNER is NFS). **Capture** the map.

2. Create a Boolean raster showing the areas that lie beyond 300 meters from a primary or secondary road (TYPE = P or S). **Capture** the map.

3. Use Block Statistics on the slope30m raster to calculate average slopes over 10 × 10-cell (300 m × 300 m) areas. Then create a Boolean raster showing where the average slopes are greater than 15 degrees. **Capture** the map.

4. Use the rasters produced in Exercises 1 to 3 to create a Boolean raster showing the potential reintroduction areas. What is the total area of these potential sites? **Capture** the map.

5. The site must be at least 3 km² in area to be considered for the reintroduction site. How many potential sites have the minimum area? Create a helpful map of the sites meeting this criterion for the final selection. (**Hint:** Convert the sites to polygons.) Looking at the map, what other factors might be used to select the final site?

A study needs to compare the erosion potential of different watersheds. Erosion risk is increased by high slope and decreased by thick vegetation canopy and high soil infiltration.

6. The canopy raster contains the forest canopy percentage. Create a raster named CanopyEffect, using the formula (100 – canopy)/100. **Capture** the map.

7. Convert the SDgeology feature class to a raster based on infiltration. Reclassify the low, medium, and high values to 5, 3, and 1 respectively. Divide the resulting raster by 10 to create an InfilEffect raster ranging from 0.1 to 0.5. **Capture** the map.

8. Calculate the erosion index as slope30m * CanopyEffect * InfilEffect. **Capture** the map.

9. Determine the mean erosion index for each watershed in wshds2c. **Capture** the map.

An example of weighted Boolean overlay to find a landfill is portrayed in Figure 11.7.

10. Perform the landfill analysis using the same approach, but use the maximum index value N = 5 instead of N = 3 and assign your own weights. Use the figure as a template to document your ranges, ranks, and weights, including a map of each input raster, the index equation, and the final map. **Capture** the final diagram.

Chapter 12. Sharing GIS

Objectives

▶ Understanding GIS services and data sharing in ArcGIS Online

▶ Sharing workflows with ModelBuilder or Tasks

▶ Developing comprehensive metadata to document shared items

▶ Using ArcGIS Online to drive mobile mapping and survey applications

Mastering the Concepts

GIS Concepts

The power of GIS increases when it is shared, whether it is with a project team, colleagues, a far-flung team in multiple locations, or the general public. Being able to distribute not only data, but also processes and workflows is an increasingly important part of geospatial work, and the software and network systems that support sharing are developing at an increasing pace. This chapter presents techniques that expand geospatial work beyond the desktop and local server, including publishing data to ArcGIS Online, staging a multiuser field data collection effort using smartphones or tablets, leading colleagues through a workflow using Tasks, and sharing geoprocessing procedures using ModelBuilder and Python scripting.

Web GIS

From the time that desktop computers became inexpensive and powerful, most GIS work was done with the software installed on desktop computers. The data usually resided on the same computer or on a file server in a local area network. The data were often stored in proprietary formats recognized only by the software that created them. However, the urge to share information, the expansion of the Internet, the popularity of smartphones and other mobile devices, and the desire to spread GIS beyond the scope of a few trained professionals have all driven the development of an alternate model, known as Web GIS.

Web GIS has the same familiar components as desktop GIS (hardware, software, stored data, etc.), but those components may be spread across thousands of miles instead of residing in a single office or campus. A typical Web GIS includes servers, resources, and clients, all connected through the Internet (Fig. 12.1). **Resources** are information designed to be shared with others, such as web maps or data services. A **client** is a computer or other device that requires information or services and requests them over the Internet. Clients may include desktop computers, laptops, smartphones, tablets, or other servers. Clients have software that enables them to make information requests to the server. The client software might be a simple web browser, or it might be a custom application designed to work with geospatial data such as a mobile phone app or a specialized program. The **web server** computer runs software that allows it to receive requests from clients and to deliver the requested resources.

Resources may reside on the same computer as the server, but often they are stored on additional servers, or **GIS servers**, that deliver the information while the web server focuses on communicating with the clients. In this scenario, the web server receives a request from the client and forwards it to the computer with the geospatial resources, which accesses its data, performs any actions required, and delivers the result back to the web server for transmission to the client.

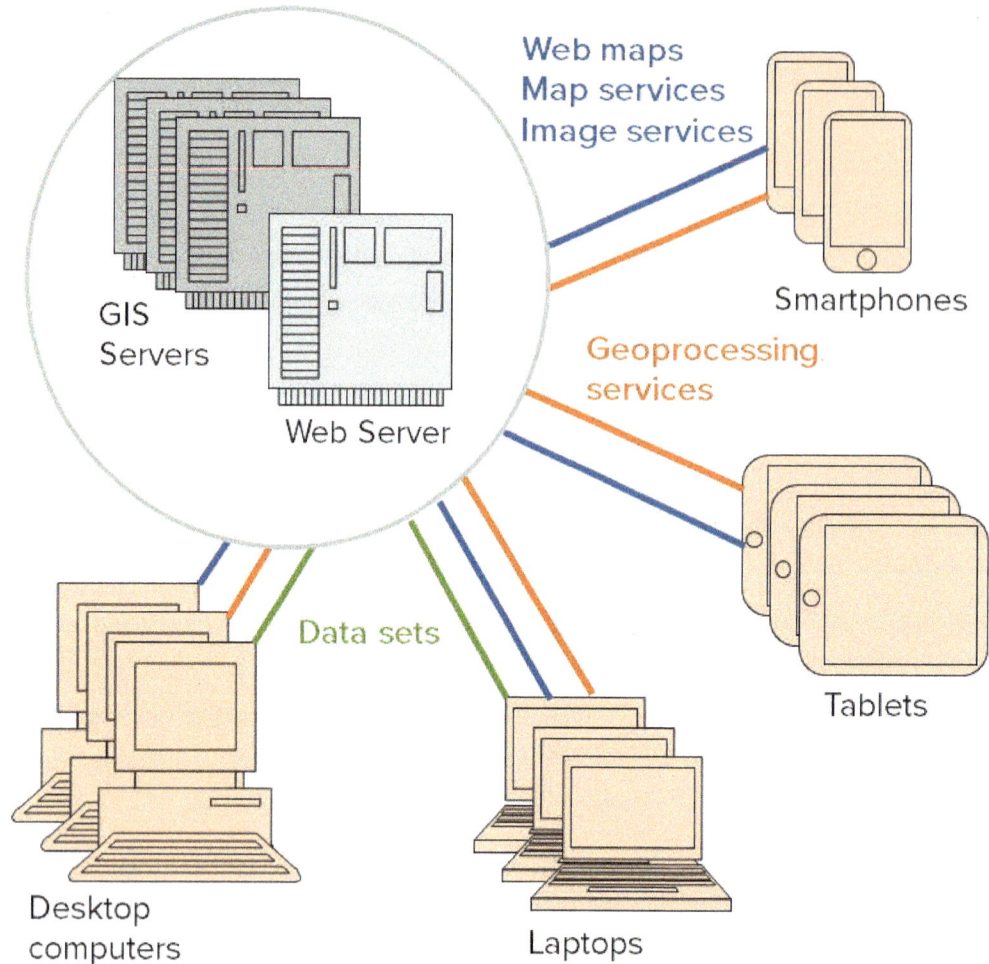

Fig. 12.1. A typical Web GIS configuration

Imagine that Sally is using a smartphone to get directions to a party. She opens an app on her phone (the client) and types in the address. The client transmits the address to the web server, which forwards it to the GIS server, which parses the address, compares it to its stored database of streets, and determines the location on a map. Geospatial analysis services then calculate how to get from Sally's current location to the address, returning a set of directions, a route, and an image of the street map. This information is relayed through the web server back to the client app so it can show the route map and present the directions. As Sally drives with the app open, the client and web server periodically communicate to update Sally's location and recalculate the route and directions if Sally takes a wrong turn.

This description is simplified, but it illustrates the success of this architecture, and its beauty, in the fact that Sally does not need the entire world street map on her phone (where it would leave little room for music or videos). The data reside on the GIS server, and it only needs to provide a small amount of information about Sally's immediate surroundings. Also, if a street error is found and fixed, the revised data are simply downloaded the next time they are accessed; the updates need not be transmitted to millions of phones.

Clients may be characterized as thin clients or thick clients. A thin client has few geospatial capabilities and relies on the GIS server to do most of the work and return a simple result, such as an image of a map or a list of nearby restaurants. A web browser is an example of a thin client.

A thick client can perform geospatial data handling on its own. A desktop GIS software program is the ultimate thick client, capable of doing more than the GIS server. Most mobile apps fall in between. In the previous example, the GIS server provides the data and determines the route, whereas the client tracks the current location against the route.

GIS services are typically consumed through a **web map**, which is a map that uses only GIS services rather than locally stored data. Users can search through thousands of data services and combine them in new maps (mashups). Web maps can be created by some types of software, such as the ArcGIS Online web application or ArcGIS Pro. Even more applications can consume these maps, including smartphones and tablets.

Resources are identified and located by a unique address known as the Universal Resource Locator, or URL; for example, http://www.esri.com is the URL for a web page. The URL for a map service hosted by ArcGIS Online might appear as http://maps1.arcgisonline.com/ArcGIS /rest/services/USGS_Earthquake_Faults/MapServer. A map service URL must be opened by a GIS client; a standard web browser cannot interpret it.

Sharing workflows

Web GIS is capable of sharing more than data or maps. A workflow is a series of data management or analysis steps used to create a final product from a set of inputs. The mapping of snail habitat, introduced in Chapter 10 and reiterated in Chapter 11, provides an example. Three inputs (elevation range, limestone areas, and dense conifer areas) were processed to yield a map of potential snail habitat. In both cases, each step was manually performed. A colleague, to perform the same analysis, would need to repeat the same steps.

Some geoprocessing sequences need to be repeated many times by the same person or by multiple people. The Black Hills, the home of the snails, is continually being logged. The Forest Service needs to repeat the snail habitat analysis at regular intervals to track long-term changes. As timber companies submit maps detailing logging plans, each plan can be evaluated for its impact on the habitat. Ensuring that the process is repeated in the same way each time is critical. Of course, the author of the process could write a series of instructions, but the document might be lost or misunderstood.

Several benefits accrue to encoding the steps as a stored workflow. First, the process is documented with the right information needed for someone else to repeat it. Second, the analyst can run the process automatically in a fraction of the time needed to execute the steps manually. Third, workflows can be iterated or nested together to quickly accomplish more.

GIS systems typically enable workflow sharing through macros, models, or scripting. A **macro** records a series of steps as the user performs them, allowing them to be repeated. Some macros may be edited afterwards to fix a misstep or add capabilities. A **model** generally provides a visual canvas to string together a series of steps graphically (see Fig. 12.6), setting the parameters for each step and defining inputs and outputs. A **script** is a software program written in a programming language but able to call functions from the GIS software program, such as running the Buffer tool. Employing a programming or scripting language, such as Perl or Python, expands the sophistication and complexity of the workflows by allowing them to test conditions, make decisions, and communicate with the person running the script. Once a macro, model, or program is created, it can be distributed to others who need to accomplish the same task in a quick or standardized way.

Sharing and metadata

As we have learned, metadata is "data about data," allowing users to discover, access, and understand geospatial data sets and services. Providing metadata is a part of creating GIS data sets. It is a professional obligation when data will be distributed to clients or the public. Best management practices dictate that at least basic metadata should also be prepared for in-house data so that critical information about the source data and processing will not be lost.

Shared data and workflows should also be well documented so that others can use them effectively and give appropriate credit to the authors. When searching for data in a GIS clearinghouse or within a repository like ArcGIS Online, it is much easier to review a summary about the data set than to download it first and then explore it to find out whether it meets the intended need. Moreover, the information published in the item's description is often the first clue as to the authority and professional quality of the data set.

One basic purpose of metadata is providing enough information for users to decide whether a data set has sufficient quality for a particular application. Publishers should address six key issues of data quality in metadata: (1) lineage, (2) positional accuracy, (3) attribute accuracy, (4) logical consistency, (5) completeness, and (6) temporal accuracy.

Lineage documents the original source of the data set and records the processing steps that occurred between the source and the final product. Questions to be answered include who collected the data, how and why they were collected, what was the original scale or accuracy, and how they were converted into digital form.

Positional accuracy is an assessment of how closely features in the data set correspond to their locations in the real world. Positional accuracy may be affected at several stages in the development of a data set: (1) errors present in the original data collection, (2) errors associated with transformations or georeferencing, and (3) errors associated with subsequent processing, such as performing intersection with an XY tolerance. Positional accuracy is typically reported in terms of the average displacement distance between the data set features or pixels and the true positions.

Attribute accuracy assesses how well the values in the attribute fields represent the true values. Does the label on Rapid Creek represent its real name? Does the 40% crown cover value of a tree stand reflect the actual density of the crown foliage? Because there are so many different kinds of attributes, assessing accuracy is complex, variable, and unguided by any sort of standard methodology. However, any information that can be provided about known errors or inconsistencies in the attributes can be helpful to the user.

Logical consistency is a measure of how well the stored features represent the objects and spatial relationships in the real world. This aspect of data quality is another complex and difficult area that is rarely tested in full. However, it is a good idea to characterize the internal logical consistency of a data set and how well it fulfills the topological rules that have been defined for it. Polygons should not have gaps or overlaps, lines should intersect at nodes and not cross each other, and so on. Steps taken during editing to create the correct relationships or steps applied afterward by testing relationships with formal topology tools can be reported.

The **completeness** of a data set refers to how well it has captured every possible object in the real world. What is the smallest size of a wetland that was mapped? Did the department of transportation digitize only public roads in its GPS survey, or are private roads also included? Did it include gravel and dirt tracks or only paved roads? Does the table of oil and gas wells in the state include every known well, or have some historical records been lost?

Temporal accuracy concerns the time period for which a data set is considered valid. Some geospatial data, such as a map of greenness condition or a NEXRAD radar rain map, represent ephemeral conditions valid for a few minutes to days. Population changes rapidly but is usually referenced to a census year. Land use data change slowly as cities grow, and they may be reasonably accurate for several years. Geology, major road networks, rivers, and international boundaries change less often. For data that will change significantly over the lifetime of the data set, it is important to designate how often the data provider updates the information.

Metadata standards

The short and simple **Item Description** introduced in previous chapters (Fig. 12.2a) can be used to quickly document an ArcGIS data set. However, GIS professionals who produce data for a client, or for the public, usually create standards-based metadata that can be used across different GIS platforms. Metadata standards are developed by agencies such as the **Federal Geographic Data Committee (FGDC)** and the **International Organization for Standardization (ISO)**. A **metadata standard** defines the specific information fields that should be included in metadata and defines which ones are optional and which are mandatory. It also specifies the format and organization of the information provided, so the metadata are transportable from system to system.

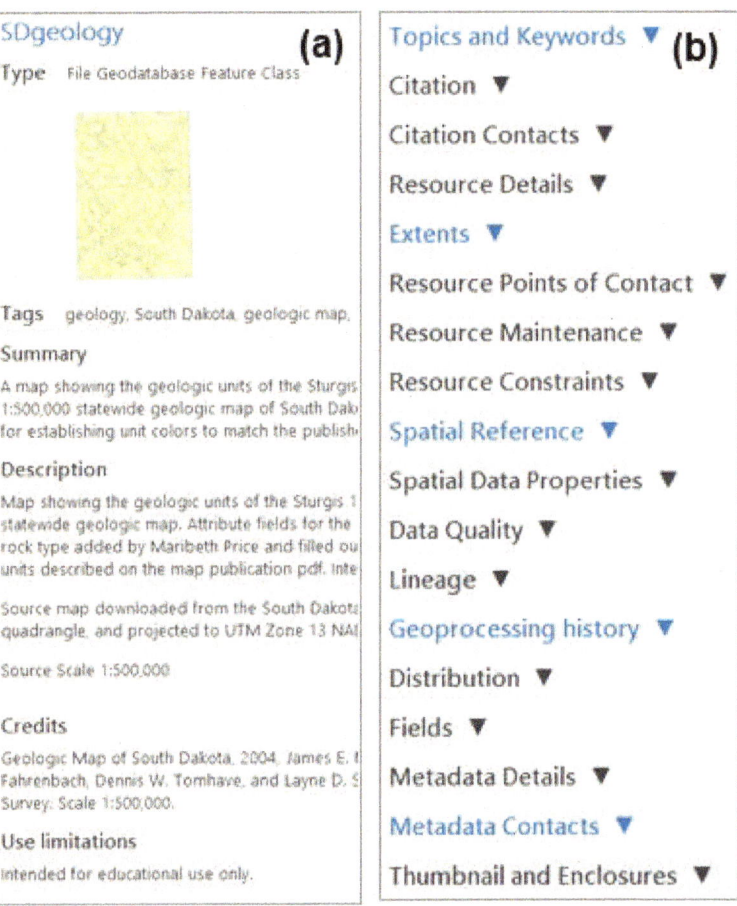

Fig. 12.2. (a) Streamlined and (b) complete standards-based metadata
Source: Esri

Standards-based metadata contain a wealth of information. Figure 12.2b shows the section headings (the complete metadata would require many pages to show). Resource refers to the data set being documented, the Citation documents the original source (such as the formal citation for a published paper map), the Lineage describes the data sources and processing steps, and the Distribution tells a user how to obtain a copy. It even contains metadata about the metadata, such as who developed them and which standard was used.

Currently, metadata standards in the United States are in a state of flux. For many years, users in the United States have employed the **Content Standard for Digital Geographic Metadata (CSDGM)** developed by the FGDC. Companies that produce data for federal agencies are required to provide metadata that complies with the CSDGM standard. International organizations are likely to use a different standard called ISO 19115. The United States and Canada geographic data committees jointly developed a customized standard known as the North American Profile of ISO 19115, or NAP. It has been adopted by Canada. The FGDC has formally endorsed the ISO 19115 standard, and federal agencies are being encouraged to move to it, although this process may take years. Thus, users are likely to encounter both CSDGM and NAP metadata. Much of the information contained in the two standards is similar, but the organization is different, and the cross-walk is not perfect or simple. More information and helpful publications on both standards can be found at the FGDC web site http://www.fgdc.gov/metadata and publications list (http://www.fgdc.gov/metadata/metadata-publications-list).

In ArcGIS products, metadata is stored and edited in an internal format associated with each data set. Users who are not required to produce metadata in a specific standard can simply create and consult the metadata within the ArcGIS internal framework. That framework permits users to specify a **metadata style**, which controls how the internal information appears and is edited. Users may set the metadata style to match the type of metadata they use most often (Fig. 12.3). When formal metadata meeting a particular standard are required (as when delivering data to a client), the internal metadata can be translated and saved in a standard-compliant XML file.

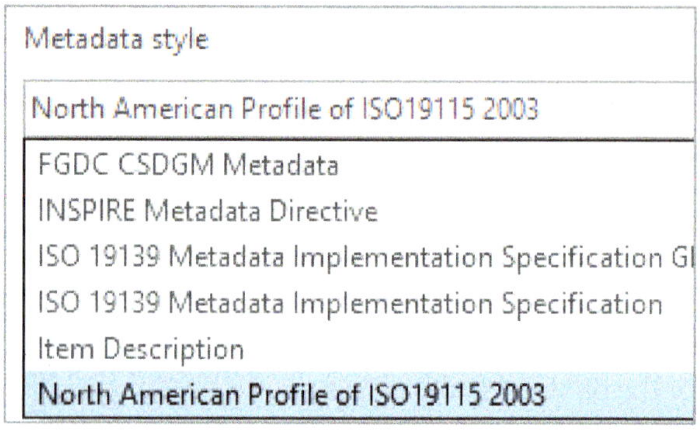

Fig. 12.3. Metadata styles in ArcGIS
Source: Esri

The streamlined Item Description is the default style in ArcGIS Pro. The Item Description is easy to create and to understand, and although it falls far short of a complete metadata record, it is much better than nothing. However, data users may not always find sufficient information about a data set to understand its characteristics or evaluate it for a given purpose. In this case, one of the more extensive styles may be chosen to display the additional information (if it was filled out by the data creator).

As advocated in Chapters 5 and 6, creating an Item Description for each data set downloaded or final data set created should not be neglected. An Item Description can be created for any item type in ArcGIS, not just geographic data sets but also models, tables, scripts, and more. This minimum Item Description is also required for every data set shared on ArcGIS Online.

About ArcGIS

About ArcGIS Online

ArcGIS Online (AGOL) is a cloud-based platform that provides GIS data and geoprocessing services as well as an environment to share maps and data with other users. AGOL provides services to many types of clients, including web browsers, desktop software packages like ArcGIS Pro, and smartphone and tablet apps.

AGOL is based on the capabilities of a geospatially smart data server software package known as ArcGIS for Portal, designed to store and distribute GIS data and services over the Internet. Previous names for this product were ArcGIS for Server and ArcIMS (Internet Map Server). Setting up and maintaining an ArcGIS Portal is a challenging task beyond the capabilities of most GIS users. AGOL is essentially a platform that interfaces with an army of cloud-based ArcGIS Portal installations, allowing its users to host GIS data on the Internet without needing their own GIS servers. AGOL can also organize and access data from servers outside the cloud, hosted by organizations or agencies that have created their own Portal installations.

Most AGOL activities require a sign-in regardless of the type of client being used. AGOL offers free individual public accounts to introduce users to the system, although these accounts are limited in what they can do and access. However, the bulk of AGOL services are structured around an **organizational account**, which is required to fully access the functionality and data sets available. An ArcGIS Online organization is set up by an administrator for multiple users. A company might set up an organization to allow its employees to create and share data related to their work, and it might also present a public face to allow citizens access to some of its data sets. Many local governments have AGOL organizations that serve this role.

ArcGIS Pro is designed to integrate directly with the capabilities of ArcGIS Online, and it can access and manage the resources saved in an AGOL account. On program start-up, the user will be asked to supply an AGOL organizational login name and password, and the account must be configured by the AGOL administrator to include an ArcGIS Pro license. Universities with an Esri™ GIS site license may have one or more AGOL organizations run by the university or its departments. This type of account will most likely be provided to a student to run ArcGIS Pro.

AGOL organizations are not free; they are assessed an annual fee based on the number of users. Activities by members of the organization also consume service credits. Each annual subscription includes a certain number of credits, but if the organization runs out, more must be purchased. Most users' daily actions consume few credits, but certain types of analysis or publishing can burn through an organization's entire credit stock in a single day. (Most functions that use a significant number of credits will provide a warning and an estimate of the credits that will be consumed.) Users are often assigned credit limits and may be blocked from certain capabilities until they have the training they need to avoid wasting credits out of ignorance.

Content within AGOL is categorized as open, subscriber, or premium content. Open content may be used by anyone with a public account. **Subscriber content** is only open to users with an organizational account. **Premium content** requires an organizational account and also consumes credits.

TIP: Esri™ offers a 60-day free trial that includes an organizational account for ArcGIS Online.

Optimizing services

A major challenge of web-based data includes delivering large volumes of data through a network connection. With vector data, two main strategies are used. When actual features are being transferred, the trick is to restrict the geographic area so that fewer features are needed or that features are only shown when a specified scale range is given. For complex maps based on many different vector data sets, the features are displayed and symbolized with the desired properties, and then a snapshot is taken to convert the map to a simple raster **tile**, as shown in Figure 6.2f in Chapter 6. Visually, the map appears identical, but the raster tile is much quicker to transmit. These snapshots might be rendered dynamically at the time the user wants to view the data, or they may be generated ahead of time and stored until they are needed, a process known as **caching**.

The familiar USGS topographic paper maps are helpful in visualizing the idea of tiles and caching. To create detailed printed maps for the United States, the USGS developed several meshes defined by equal values of degrees (1 × 1 degree, 0.5 × 0.5 degree, 15 minutes, 7.5 minutes), defining the extent of each map (Fig. 12.4). For each mesh, it designated a scale that allowed the map to fit on a standard paper size, each more detailed than the last (1:250,000, 1:100,000, 1:48,000, 1:24,000). The resulting quadrangles are analogous to tiles, and the effort to create and print each quadrangle represents caching. The fact that the cache is composed of paper maps (rather than PNG images) and was stored in file cabinets (instead of a hard drive) does not negate the similarity. This analogy helps one appreciate that representing a large area like the United States at a detailed scale of 1:24,000 requires an enormous number of tiles and a staggering amount of data. Even with automated computer processing, a modern digital cache of world data for multiple detailed scales can take weeks to create.

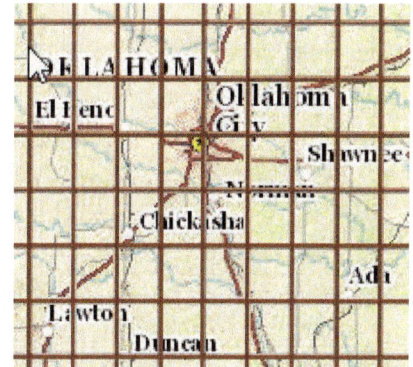

Fig. 12.4. Some 15-minute quadrangles (tiles) near Oklahoma City.
Source: USGS

The sizes of the tiles and the collection of scales generated together constitute the **tiling scheme**. The USGS quadrangle tiling scheme is one example, but ArcGIS Online and other map services like Bing and Google Earth use a common tiling scheme that covers the globe and designates standard tile sizes and scales for caching. There are display benefits to using matching tiles, so users are encouraged to use this standard scheme for their own caches, although users with special requirements (and expert knowledge) can establish other schemes if needed.

With raster data, the service provider typically uses both tiling and pregenerated pyramids to minimize the number of pixels sent through the network. The zoom scale determines the resolution of the raster tiles that will be transmitted, and the extent determines which tiles will be needed.

Types of ArcGIS Online services

Many different types of data services, both vector-based and raster-based, are provided by ArcGIS Online. Although knowledge of the data types is not necessary to search for and use the data, as demonstrated by our use of AGOL data in the previous tutorials, this knowledge helps in finding the most appropriate data sets for a given purpose, as well as understanding the capabilities of the data and how it may be used. It is also critical to learning how to publish data to ArcGIS Online.

A summary of the different data types is presented in Table 12.1, followed by a discussion of each type. The items in the table are the most commonly used services available for Pro, although ArcGIS Online supports other formats for its web interface as well. In the discussion of service types that follows, pay attention to which services provide actual geospatial data as features or

Table 12.1. Types of GIS services available in ArcGIS Online

File type	Description
Feature layers	Feature layers serve point, line, or polygon features and their attributes
Imagery layers	Imagery layers serve raster data products like satellite images, aerial photography, and elevation in an analyzable form
Map image layers	A collection of map cartography based on vector data, dynamically rendered and served as image tiles.
Tile layers	A collection of web-accessible tiles that reside on a server; pre-rendered for rapid display; includes many basemaps.
Elevation layers	A collection of cached elevation image tiles in a compressed format, suitable to show terrain in 3D scenes
Web tools	A geoprocessing function than runs on the GIS server rather than using local GIS software capabilities
Scenes	Symbolized 3D spatial content for visualizing a geographic area in three dimensions; may include 2D and 3D layer services
Web maps	A map based solely on GIS services; may include other types of services listed in this table
Packages	Data uploaded to ArcGIS Online for sharing with others; includes layer, map, project, rule, or geoprocessing packages
Scene layers	A collection of 3D objects and z-values, including points, 3D objects, integrated meshes, or point clouds
Tables	A set of rows and columns where each row represents a feature; may contain locations but are not typically drawn on the map

raster values, which can be analyzed and manipulated, and which present only pictures of spatial data that can only be viewed. New types of services may be added over time. For a current list and detailed descriptions of service types, consult the Help in ArcGIS Online.

A **feature layer** is a vector-based data set, shared as individual point, line, or polygon features. Feature layers may be slow to render, especially if the service contains many features. The advantage is that users can specify the drawing symbols to use, review the information in the attribute tables, and use the features in an analysis, just as if the features were stored locally. If configured to allow it, feature layers can also be edited by any type of client, including the AGOL web interface, ArcMap, and ArcGIS Pro. Feature layers also form the basis for shared editing of data sets from different locations, or collecting field data using smartphones or tablets running the Explorer, Collector, or Survey123 app.

An **imagery layer** is designed to serve satellite imagery or aerial photography as actual cell values, which can be manipulated and analyzed just as local rasters may be. Imagery layers can be symbolized as the user wishes and incorporated into raster processing tools.

A **map image layer** dynamically (in real time) draws a symbolized vector-based map using the server and then renders a snapshot of the map as tiles to deliver them quickly to the client, usually in the PNG image format. These layers often use visible scale ranges to present different data sets at various scales as appropriate (states vs. counties, for example). Some map image layers include the ability to click features to identify their attributes.

A **tiled layer** may be based on either vector or raster information and is similar to a map image layer except that the tiles are created ahead of time and stored on the server until they are needed, that is, they are cached. This type of service is efficient and fast, but the user cannot modify how the map is symbolized, and features cannot be identified. The ArcGIS Online basemaps are examples of this service type. Vector tile layers are a recently added service type. They include style settings that may, with some effort on the part of the user, be modified to present the layer with different fonts, colors, or symbols.

A **scene layer** is a cached web layer that is optimized for rapid display of 2D or 3D data in a three-dimensional display. An **elevation layer** is a specific and often-used type of scene layer that provides compressed elevation data for rapid 3D display.

Web maps (2D) and **web scenes** (3D) are collections of service layers assembled and pre-symbolized to communicate geographic information. When added to Pro or the AGOL web interface, one or more feature, map, tile, or scene layers may be included.

A layer *package* assembles a layer with its symbology and other properties, as well as its data, into a single file that can be shared. Packages can also be created for maps (including all its layers), scenes, geoprocessing services, and even for projects.

GIS servers can also provide analysis tools in addition to data. A **geoprocessing service** makes available certain computations and functions so that users can perform the designated analysis through a web site, even if they do not have GIS software installed. These services may operate on either vector or raster data, depending on the task.

Two other ArcGIS Online products expand its capabilities beyond the custom client. A **web application** is a web page designed to work in a standard browser, but configured to consume a web map and to allow the viewer to interact with it. Esri™ has also popularized a particular type of web application known as the **story map**, which presents one or more interactive maps along with descriptions, photos, videos, and other multimedia items. Esri™ has developed a selection of templates that can be used to quickly create and customize a web application or story map with little to no web programming experience. Story maps have become a popular way to share geospatial content, and they can be beautiful and inspiring (https://storymaps.arcgis.com/en/).

Not all GIS services are equally valuable. Many professional and government agencies publish services, and these tend to be authoritative and well-documented sources of data. Ordinary users can also share and publish maps and data in ArcGIS Online, and the quality can be variable. It is up to the user to evaluate a data set and make sure that it meets the quality standards required. It is important to look at the source of the data: is it published by a well-known agency or a random user? The thoroughness of the metadata can be an important clue to the professionalism and skill level of the data provider. Poorly documented data sets tend to be published by beginners or sometimes by qualified analysts who are "trying things out" and don't intend the data set to be a lasting or widely used commodity. Although it is fine to use such data for fun or for a class project, one should think twice about using questionable data in a project that affects real-world issues in business, safety, property, or legal affairs. Regardless of the source, the user should also understand the limitations and potential errors or inconsistencies in any data set used. Metadata is designed to provide this information and guide users to make sound choices.

Sharing data in ArcGIS Online

Sharing GIS data in AGOL can be done two ways. The first is to create feature layers, image layers, tile layers, and other files and share them directly as hosted services. Hosting services in AGOL requires an organizational account and specific publishing permissions set by the organization administrator. Publishing hosted data sets may be restricted by the administrator because the process may consume a great many credits. Feature layers are fairly innocuous as long as they don't have too many features. Hosted tile layers, however, require knowledgeable configuration to prevent the caching process from burning through excessive credits.

Even without publishing permissions, any user with a public or organizational AGOL account can still share data in the form of a **package**, a single file that includes symbols, data, and additional items needed to completely re-create a layer, map, scene, or project on someone else's computer. Packages may be saved to a local computer or network drive, or they can be uploaded and shared in ArcGIS Online to reach a wider audience. AGOL packages cannot be displayed using the ArcGIS Online web interface, or by tablets and smartphones, but they can be used by ArcMap and ArcGIS Pro. Some AGOL layers appear as both map services and layer packages; choosing the package will download the data, taking some additional time at first, but it may improve performance in the long run. Because packages include the actual data, they can often be exported and saved in alternate locations, and so they are a potential source of data when building geodatabases.

Preparing data for sharing

Publishing a feature service or layer package is more complex that taking the current map and popping it into ArcGIS Online. Proper preparation is needed to ensure that the final result looks and works well. The publishing process includes an Analyze step that reviews the publishing configuration and identifies potential issues. Some issues are severe and must be fixed before publication can be completed; others may ignore best practices or result in sub-par performance, but the publication can proceed if desired. Some best practices are listed here.

- Basemaps cannot be included in a service or package. The basemaps already exist in AGOL; there is no reason to include them again.

- The map should have a default map extent so that the viewer is immediately taken to the area encompassed by the map.

- Keep the number of features in a feature layer to a minimum. Layers with thousands of features will be slow, especially if they contain complicated polygons. To share large collections of features, consider using a package or a map image layer instead. It may be advantageous to generalize features before hosting them or to provide group layers with scale ranges.

▶ Set visible scale ranges to restrict the display of layers to the appropriate scales. If planning a service that covers a wide range of scales, set the scale ranges to be consistent with the tiling scheme.

▶ Decide whether the data need to be editable, who may view them, and who may edit them. Set these parameters during or after publication.

▶ Provide quality metadata and good search terms.

▶ If caching tile layers, be careful when selecting the scales to include.

When publishing tile layers in ArcGIS Online, care must be taken in selecting the cache scales because the data extent and the chosen scales will determine the total size of a cache and the number of credits consumed in generating it. The user will be advised of the estimated size before publishing. If the size is less than 100 MB, the credits consumed will not be excessive. To generate larger caches, talk to the AGOL administrator first to ensure that the caching parameters are correctly configured and that enough credits are available.

Tiles may also be cached locally, using a desktop computer's processing power instead of the cloud computer in AGOL, and uploading the tiles afterwards. This method significantly reduces the number of credits consumed, although the user should still keep an eye on the size. Creating caches with superfluous scales is a waste of time and space.

Sharing features for mobile applications

Esri™ provides several different options for collecting geospatial data using a mobile device such as a smartphone or tablet (Fig. 12.5), including the mobile apps Explorer for ArcGIS, Collector for ArcGIS, and Survey123. Each app is free to download, but using Collector and Survey123 requires an organizational AGOL subscription. Mobile devices are typically not as accurate as a dedicated GPS, having typical accuracies of 5 to 10 meters (examine the sidewalk and landscaping boundaries in Fig. 12.5). However, they are simple to learn and adequate for many purposes. Add-on devices can be purchased, at prices from several hundred to several thousand dollars, to improve the accuracy of the mobile device.

ArcGIS mobile data collection applications rely on hosted feature layers. Before data are collected, a feature layer service must be created and published. A web map is also required to package the hosted layers for use by the mobile application.

Because mobile data collection is based on shared cloud services, multiple field workers can collect and edit information in the same shared data set. Moreover, this information is updated live, so as one field worker collects new features in one area, the others will see those features appearing on their mobile maps in real time. They can tell when one area has already been covered and move to a different location, preventing duplication of effort. The field work leader can track which sections of the map are

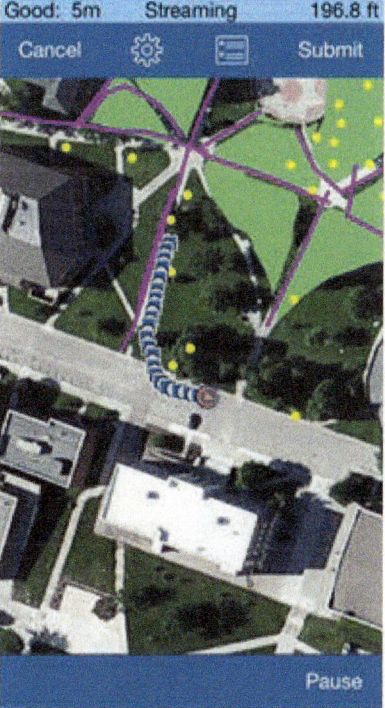

Fig. 12.5. Mapping sidewalks with Collector
Source: Esri

377

complete and send idle crew members to new areas. Workers don't need to upload the data back in the office once the field day is complete.

When preparing the hosted feature service, you should create coded domains for the attributed fields, providing pick-lists for the field workers. This practice saves time on data entry, avoids typing errors, and enforces the use of a consistent terminology. When collecting information about trees, for example, typing "coniferous" each time would quickly become tedious. Moreover, without a list, users may make typing errors, use abbreviations, or use case inconsistently, which complicates queries later. Domains and the rest of the feature layer schema should be carefully planned and implemented before publishing the service; schema changes are much more difficult to implement afterwards.

Certain mobile applications, such as the Collector app, can perform disconnected data collection when no cell or wireless service is available, as may be the case when collecting data in remote areas. This feature is also handy to reduce the demand for cellular data when working in an area with service. Features and attributes generally do not consume large quantities of one's monthly data limit, but if photos are being collected as part of the field work, then the data requirements increase dramatically and may start to have a real impact on a user's data allotment. With disconnected editing, the data and photos are stored locally until the user is back in range of a wireless server and can manually sync the data and photos to the hosting server without using cellular data transmission. However, disconnected editing will not allow a field team to view live edits as they are collected because the new data will only appear on the server after the work has been synced.

Sharing workflows

ArcGIS Pro provides three main ways to share workflows: models, scripts, and tasks.

ModelBuilder provides a graphic canvas to string tools together and execute them in sequence. Figure 12.6a shows a model to map the snail habitat analysis from Chapter 10. The blue ovals

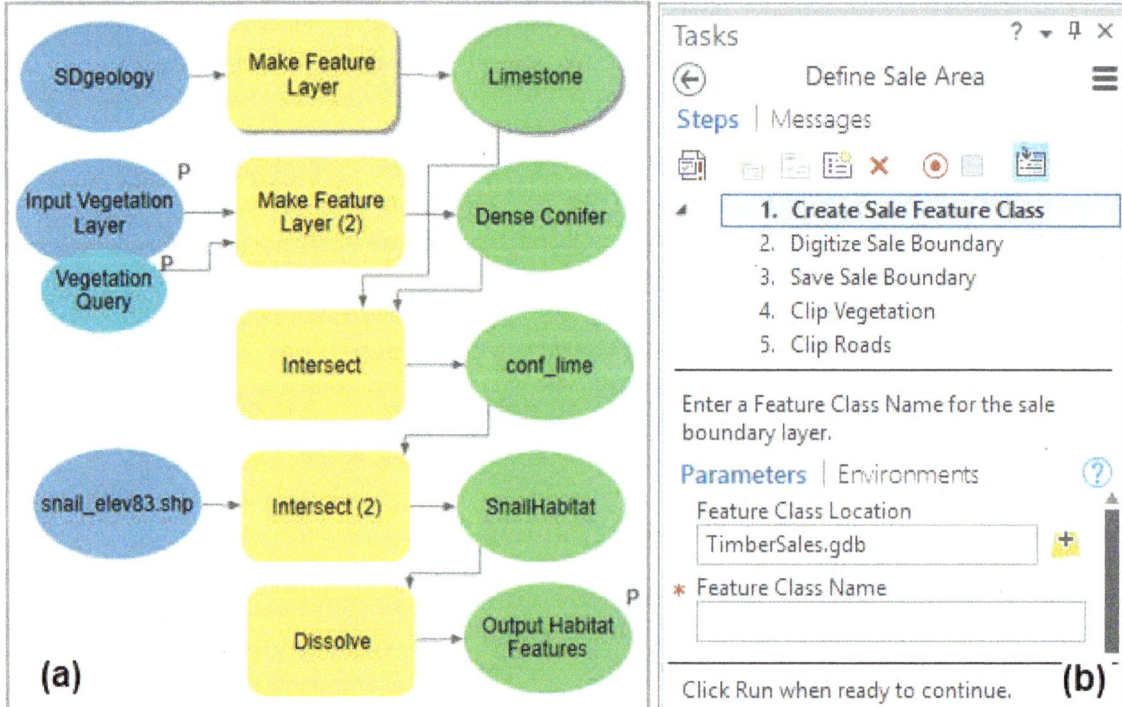

Fig. 12.6. Sharing workflows: (a) a snail habitat model; and (b) a task to define a timber sale area
Source: Esri

represent data inputs, the yellow rectangles are tools, and the green ovals are output data. Data and tools are connected to create a flow, and each tool's parameters are set and saved so that the model can be run automatically. Once a model is created, it can be run as many times as desired. The tool settings can be altered to produce different results as needed. Models are stored in a project's toolboxes and can be run as if they are tools.

More advanced models can incorporate decision structures, conditional statements, and iteration functions if needed. For even more flexibility, models can be saved as computer programs. A **script** is a computer program, written in a programming language such as Jscript, VBScript, or Perl, which has been enabled to include geoprocessing tools or commands. Esri™ has adopted the open-source language Python as its recommended scripting platform. A copy of Python is installed with ArcGIS, and all Esri™ scripting examples and supplemental tools are written in Python.

Models and scripts are ideal for automatically executing a series of geoprocessing steps, but sometimes a procedure requires interaction with a user to provide spatial input or decisions, such as defining the region of interest before the steps are executed, or performing a set of prescribed edits. A **task** is a series of steps that reside within a project, designed to document a process and step the user through it with detailed instructions. Figure 12.6b shows a task that prompts a user to digitize a polygon representing the area of a timber sale, and then it steps through clipping a series of layers to the boundary while allowing the user to specify the right input layers and name each output. Tasks assist workflows in many ways. They help nonexperts to complete an analysis the same way that an expert would. They speed repetitive tasks by eliminating keystrokes, opening the next tool automatically, and providing appropriate default values. They record processes that are done infrequently, enabling a user to do it quickly and correctly even if months have elapsed since the last execution. They can be shared with many different users to establish a consistent method of operation.

A **task item** is a container within a project that can hold one or more tasks. Each task is composed of a number of steps (as in Steps 1 through 5 of the Define Sale Area task in Fig. 12.6b). When creating tasks, each step may be manually configured by selecting a tool or command and assigning parameters, or the creator may use a Record function to record a series of steps as a process is performed. Multiple task items may be stored within a project.

Summary

▶ A web GIS may spread the standard GIS components across long distances instead of concentrating them in a single office or network.

▶ GIS services provide data or analysis over the Internet; these services may be consumed by a variety of devices with different software and diverse capabilities.

▶ Workflows and geoprocessing sequences may be shared with others using models, macros, or scripts.

▶ Metadata documents six issues related to data quality so that users can locate and evaluate data sets to determine if they are fit for a given purpose.

▶ Metadata may range from a streamlined Item Description to full-fledged professional metadata prepared according to national or international standards. The thoroughness and quality of the metadata are often important clues to the authority and value of a data set.

▶ ArcGIS Online is a cloud-based environment for creating and sharing GIS data, applications, workflows, and knowledge. It revolves around organizational subscriptions maintained by various institutions that provide accounts and service credits to their users.

▶ ArcGIS Online offers a variety of service types that run the gamut from actual features or pixels that can be analyzed to tiled representations of complex maps. Three-dimensional scenes can also be created and shared.

▶ Anyone with an AGOL account can share a package, but hosting GIS services requires publishing privileges in an ArcGIS Online account or in an institutional server installation set up using Portal for ArcGIS.

▶ Publishing quality services that perform well requires forethought, careful documentation, and attention to best practices. For cached tile layers, the generation scales must be carefully selected to avoid consuming excessive credits.

▶ Data collection using mobile devices relies on hosted feature layers in AGOL or an institutional Portal. Several different mobile applications are available.

Important Terms

ArcGIS Online	imagery layer	package,	temporal accuracy
attribute accuracy	ISO (International	layer	tile
caching	Organization for	map	tiled layer
client	Standardization)	project	tiling scheme
completeness	Item Description	positional accuracy	web application
CSDGM (Content	lineage	premium content	web map
Standard for Digital	logical consistency	resources	web scene
Geographic Metadata)	macro	scene layer	web server
elevation layer	map image layer	script	
feature layer	metadata standard	story map	
FGDC (Federal Geographic	metadata style	subscriber content	
Data Committee)	model	task	
geoprocessing service	ModelBuilder	task item	
GIS server	organizational account		

Chapter Review Questions

1. In what ways is Web GIS similar to traditional GIS? How do they differ?

2. Explain what a client is. What types of clients can access GIS resources? How do they differ from each other?

3. How does a web map differ from the many maps used throughout the tutorials in this book?

4. Describe different methods for sharing workflows with others.

5. Briefly summarize the six different data quality issues addressed by good metadata.

6. What types of accounts are available in ArcGIS Online? Compare their capabilities.

7. How does a Web GIS limit the amount of data that needs to be transmitted to the user, in order to ensure optimum performance? Discuss for both vector and raster data.

8. A logical grouping of ArcGIS Online services might include feature layers, imagery layers, and packages in one group, and map image layers and tile layers in another. Explain the basis of this grouping.

9. Describe the similarities and differences between tasks and models.

10. Why are domains especially useful when preparing for data collection using mobile devices?

Expanding your knowledge

Visit https://storymaps.arcgis.com/en/ to learn about story maps and to view examples.

Read through these sections of the ArcGIS Pro Help to learn more about sharing GIS data.

> Help > Share your work (all sections)
> Help > Workflows > Tasks
> Help > Analysis and geoprocessing > ModelBuilder
> Help > Share your work

Log in to ArcGIS.com using a web browser and explore the Help there.

Mastering the Skills

Teaching Tutorial

The following examples provide step-by-step instructions for doing basic tasks and solving basic problems in ArcGIS. The steps you need to do are highlighted with an arrow →; follow them carefully.

> → Start ArcGIS Pro and open the ClassProjects\TimberSales project. Remember to save often.

Geoprocessing with ModelBuilder

This section shows how to use ModelBuilder to record sequences of geoprocessing steps. When a timber company submits a logging plan in the form of a map of the final vegetation after logging, the national forest personnel evaluate the plan, including how it may affect the snail habitat. A model to run the habitat analysis would save time and ensure that the analysis is repeated the same way each time.

> 1→ Create a new map and name it *Snail Model*.
> 1→ Choose **Project** > **Options** and click the Geoprocessing section. Make sure the box is checked to *Allow geoprocessing tools to overwrite existing data sets*.
> 1→ Click OK and click the back arrow to return to the project.

Creating and running a model

The project has a toolbox for storing models and tools related to the project.

> 2→ In the Catalog pane, expand the Toolboxes entry to see TimberSales.tbx.
> 2→ Right-click the TimberSales.tbx toolbox and choose New > Model.

A view named Model opens and the **ModelBuilder** ribbon is shown. Take a moment to examine the groups and buttons on the ribbon.

> 2→ Click **Modelbuilder: Model: Properties**.
> 2→ Enter *SnailHabitat* for the Name of the model.
> 2→ For the Label, enter *Snail Habitat Analysis*. Make sure the box is checked to *Store tool with relative path*.
> 2→ Click OK and then click **Modelbuilder: Model: Save**.

TIP: Storing relative pathnames for a model makes it work even if the project folder is copied to a new location or shared online as a project package. Uncheck this box if the data used in the model were fixed to a specific network location like an organization's file server.

In ModelBuilder, chains of data and tools are created, arguments are set for the tools, and the steps are then run. We begin by selecting the geology and vegetation characteristics conducive to snail habitat and making layers from them.

3→ In the Catalog pane, from the mgisdata \BlackHills\Sturgis \Features feature dataset, drag the SDgeology and vegetation feature classes to the model canvas.

3→ Also drag the snail_elev83.shp shapefile from the mgisdata\Black-Hills folder onto the model canvas.

3→ Click an oval to select and move it. Arrange the three data sets in a vertical stack (Fig. 12.7).

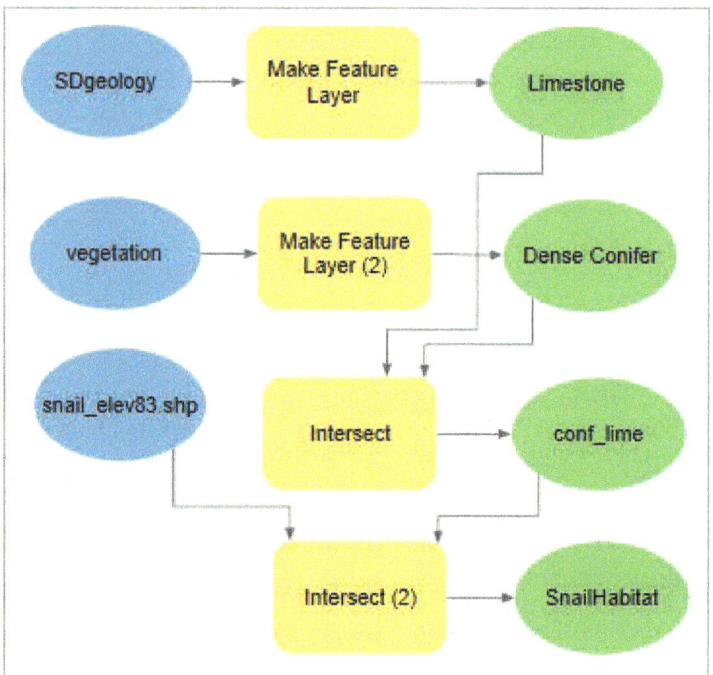

Fig. 12.7. The Snail Habitat Analysis model

4→ In the Geoprocessing pane, search for the Make Feature Layer tool and drag it onto the model canvas, placing it next to the SDgeology oval.

4→ With the tool selected, click on the blue SDgeology oval and drag toward the Make Feature Layer box to connect the two. Choose Input Features from the pop-up box.

4→ Right-click the Make Feature Layer tool in the model canvas and choose Open.

4→ Name the Output Layer *Limestone*, as shown in Figure 12.7.

4→ Enter the selection clause: RockType is Equal to Limestone. Verify it and click OK.

TIP: Gray boxes indicate that a tool is missing valid parameters and is not ready to run. Colored boxes indicate that the step is ready to execute. A drop shadow indicates that the step has been executed.

5→ From the Geoprocessing pane, drag another copy of the Make Feature Layer tool onto the model canvas and place it next to the vegetation oval.

5→ Connect the vegetation oval to the Make Feature Layer (2) tool and select Input Features as the parameter.

5→ Right-click the tool to open it. Name the Output Layer *Dense Conifer*.

5→ Enter an expression with these three clauses: (COV_TYPE is Equal to TPP Or COV_TYPE is Equal to TWS) And DENSITY96 is Equal to C.

5→ Click the SQL button and add parentheses to match the preceding expression.

5→ Verify the expression and click OK.

TIP: Click on a model shape to select it. Use click-and-drag boxes or the Shift key to select multiple shapes. Click and drag selected shapes to move them around on the model canvas or to resize them. Use these techniques to arrange the model in neat patterns.

Recall that intersecting more than two feature classes requires an ArcGIS Advanced license. To ensure that this tool will work for anyone, we will perform the intersection in two steps, two feature classes at a time.

6→ In the Geoprocessing pane, search for the Intersect tool and drag it to the model canvas.

6→ Connect the green Limestone oval to the Intersect tool and choose Input Features from the pop-up.

6→ Also connect the Dense Conifer oval as Input Features.

6→ Right-click the Intersect tool on the model canvas and open it.

6→ Name the Output Feature Class *conf_lime*. Click OK.

7→ Drag another copy of the Intersect tool from the Geoprocessing pane.

7→ Connect both snail_elev83.shp and conf_lime to the Intersect tool as Input Features.

7→ Open the Intersect (2) tool and name the output feature class *SnailHabitat*. Click OK.

The model should look similar to Figure 12.7. It's okay if yours is arranged differently, as long as the connections are correct. We are ready to run it, but we should save it first.

8→ Choose **Modelbuilder: Model: Save**.

8→ Choose **Modelbuilder: Run: Validate** first to test the model.

8→ If no errors are shown, click **Modelbuilder: Run: Run**.

8→ Close the message box when the model finishes running.

Drop shadows appear behind each oval and box to indicate that the step has been executed, but the output will not appear on the map automatically.

8→ Right-click the green SnailHabitat oval and choose Add to Display. The layer appears in the Contents pane.

8→ Switch to the Snail Model map to view the output. Zoom to the layer if needed.

We should dissolve the interior lines in the snail habitat by adding a new step to the model.

9→ Switch back to the Model view.

9→ In the Geoprocessing pane, search for the Dissolve tool and add it to the model.

9→ Connect the SnailHabitat oval to the Dissolve tool as Input Features.

9→ Open the Dissolve tool. Set the Output Feature Class to *SnailHabitatDiss* and the Dissolve_Field to OWNER.

9→ Uncheck the box to *Create multipart features* and click OK.

9→ Save the model and then click Run. Close the message box.

Notice that only the last Dissolve step was executed. Completed steps are not executed again (unless they were opened and changed).

10→ In Contents, right-click the SnailHabitat layer and choose Remove.

10→ Right-click the SnailHabitat oval and click Add to Display to uncheck it.

10→ In the Model view, right-click the SnailHabitatDiss oval and choose Add to Display.

10→ Switch to the Snail Model map to view the new result. Save the project.

Changing and rerunning a model

The saved model provides a record of the inputs and parameters used to derive the output. It can also be edited and run again to try different solutions. A new study has indicated that the snails also do well on another rock type known as felsic intrusives.

11→ Switch back to the Model view.

11→ Right-click the Make Feature Layer input box for the geology query and open it.

11→ Add the clause: Or RockType is Equal to Felsic intrusives.

11→ Verify the query and then click OK to close the tool.

Notice that the model boxes lose their shadows (except the unchanged vegetation query), indicating that they need to be run again with the new conditions.

11→ Save the model and then click the Run button to run the model again.

11→ Switch to the map and examine the SnailHabitatDiss output.

11→ Save the project.

Creating a tool from a model

Models can be set up to run as tools by turning the inputs into parameters. This approach is helpful when the same series of steps is to be run with different inputs each time.

12→ In the Model view, right-click the blue vegetation oval and choose Parameter.
    A small P appears to the right of the blue oval, indicating that it is now a parameter.

12→ Right-click on the SnailHabitatDiss green oval and make it a parameter.

12→ Save the model and close the Model view.

12→ In the Catalog pane, expand the TimberSales.tbx toolbox and double-click the Snail Habitat Analysis model.

The model is opened as a tool in the Geoprocessing pane, with input boxes for the two parameters. Examine the titles above the boxes, vegetation and SnailHabitatDiss. The titles are taken from the names assigned to the ovals in the model. Let's rename them so they make more sense to a user.

13→ In the Catalog pane, right-click the model and choose Edit.

13→ Right-click the vegetation oval and choose Rename. Edit the name to Input Vegetation Layer and click Enter.

13→ Rename the SnailHabitatDiss oval Output Habitat Features.

13→ Save the model and close it.

13→ In the Catalog pane, double-click the model to open it as a tool.

TIP: Double-clicking or choosing Open will open the model as a tool to be run. To open the model canvas, right-click the model and choose Edit.

There is a more recent vegetation feature class in the Sturgis geodatabase called veg2006. We'll run the model again using the new data to see if the habitat has changed.

14→ In the Geoprocessing pane, in the Snail Habitat Analysis tool, click the Browse button for the Input Vegetation Layer and add the veg2006 feature class from the Sturgis geodatabase.

14→ Set the Output Habitat Features to SnailHab06. Run the tool.

14→ Compare the SnailHabDiss and SnailHab06 layers, noting the loss of habitat.

We can also set individual arguments as parameters. Imagine that we wanted to be able to evaluate less than ideal but still viable habitat by choosing different vegetation types.

15→ In the Catalog pane, right-click the Snail Habitat Analysis tool and choose Edit.

15→ Right-click on the Make Feature Layer (2) box associated with the vegetation query and choose Create Variable > From Parameter > Expression. A new blue input oval appears for the expression.

15→ Rename the new oval Vegetation Query. Move or resize if needed.

15→ Right-click the Vegetation Query oval and choose Parameter. A P appears next to it (Fig. 12.8).

15→ Save the model and close it.

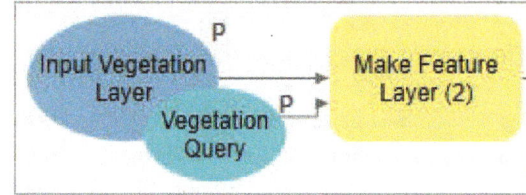

Fig. 12.8. Making the vegetation query a parameter

Now let's run the model again. Suppose we want to evaluate the potential habitat if all moderately to densely forested areas are considered, not just conifers.

16→ In the Catalog pane, double-click the Snail Habitat Analysis model to open it.

16→ Set the input and output feature classes to vegetation and SnailHabAllC.

16→ Hover on one of the COVER_TYPE clauses and click the red X to delete the clause. Delete the other one, leaving only the DENSITY96 clause.

16→ Add a new clause: Or DENSITY96 is Equal to B.

16→ Run the tool again. Compare the outputs. Save the project.

This brief introduction shows only a small part of what ModelBuilder can do. To learn more, read about ModelBuilder in the Pro Help files.

TIP: Models can be saved as scripts in the Python programming language as a starting point for creating full-fledged programs with more advanced capabilities.

Sharing workflows with Tasks

Tasks are used to record the steps of a repeatable workflow to ensure that it is done the same way each time. We will start by running an existing task.

17→ Close the Snail Model map and open the Timber Sale 15A map.

17→ In the Catalog pane, expand Tasks and double-click Timber Sale Maps.

The Timber Sale Maps task item opens in a Tasks pane. (A *task item* is a container for storing tasks.) This one contains a single task named Define Sale Area, which will walk through defining the area of a proposed timber sale and clipping several layers to it.

TIP: If a project has no existing task items in which to create a task, or a new task item is needed, right-click the Tasks entry in the Catalog pane and choose New Task Item.

Running a task

17→ In the Tasks pane, mouse over the Define Sale Area task and click the blue arrow to start it.

The task opens to the first step, Create Sale Feature class. Beneath the step, some instructions tell what to do (*Enter a Feature Class Name for the sale boundary layer*). Beneath it, the Create Feature Class tool has been embedded in the Tasks pane. At the bottom is another set of instructions telling how to proceed to the next step and a bar to show progress through the steps.

17→ Enter Sale15A for the Feature Class Name. Click the Tab key.

17→ Examine but do not change the other tool parameters.

17→ At the bottom of the Tasks pane, click Run to proceed to the next step.

The second step, Digitize Sale Boundary, starts automatically, saving mouse clicks.

18→ Follow the instructions to digitize a timber sale polygon, staying within the green National Forest area. An example is shown in Figure 12.9.

18→ After finishing the polygon, click Next in the Tasks pane to continue.

18→ Follow the instructions for Step 3 to save the sale boundary.

The next steps prompt the user to clip three different input layers to the timber sale area. Name the clipped outputs with descriptive names.

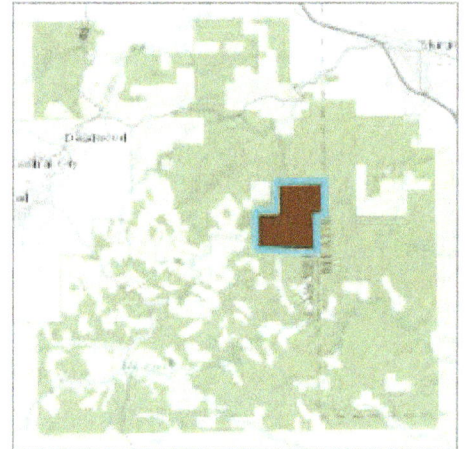

19→ Follow the directions in Step 4, naming the clipped output Sale15A_Veg.

19→ Continue following the task until it is complete. Name the clip outputs Sale15A_Roads and Sale15A_Streams.

19→ The task is done. Save the project.

Fig. 12.9. Suggested timber sale area
Source: Esri

Creating a task

After this demonstration of how tasks work, it is time to create one. This new task will identify the stands of timber with commercial value for logging, based on tree size and density.

 20→ In the Tasks pane, click the Options button and choose Edit in Designer.

A Task Designer pane opens, which is used to modify the properties of whatever is currently selected in the Tasks pane. We will refer to it as the Designer pane, for short.

20→ Mouse over the icons in the Tasks pane to see what they are called.

 20→ In the Tasks pane, click the New Task button.

20→ In the Designer pane, name the task Find Logging Areas and enter the description *Finds the areas of timber to be logged based on tree size and density.*

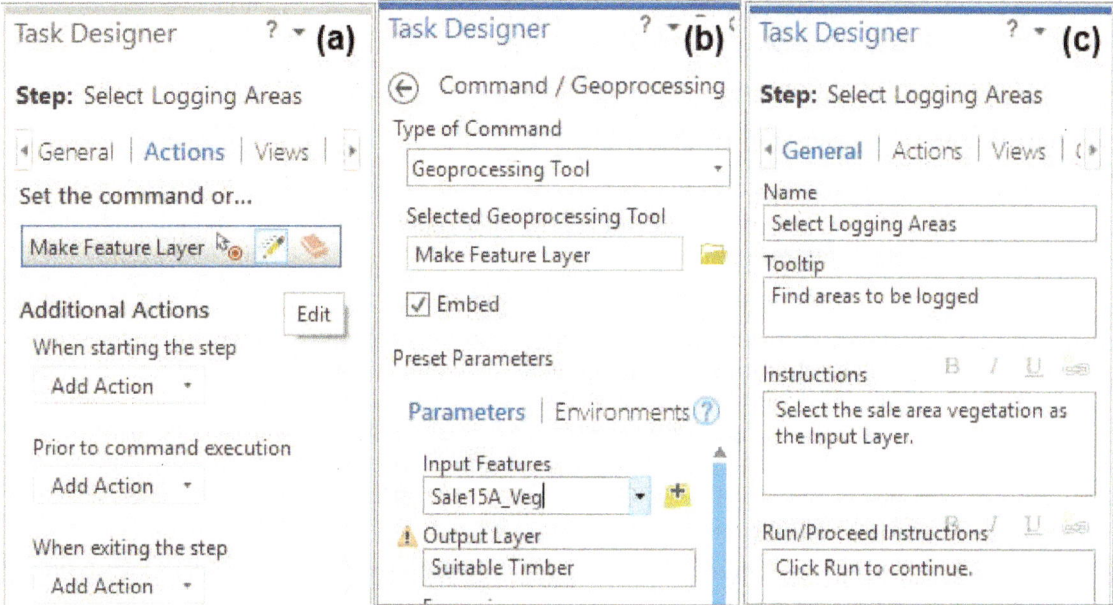

Fig. 12.10. Configuring the Select Logging Areas step
Source: Esri

 21→ In the Tasks pane, click the New Step button.

21→ In the Designer pane, switch to the Actions tab (Fig. 12.10a).

 21→ Under *Set the command or geoprocessing tool,* mouse over the <none> bar and
choose the Edit icon (pencil).

21→ For Type of Command, choose Geoprocessing Tool. A search box opens.

TIP: If the search box does not open automatically, or if it closes unexpectedly, click the Folder
icon under Selected Geoprocessing Tool to open it again.

22→ Search for and select the Make Feature Layer tool. Click OK.

22→ Keep the Embed box checked (Fig. 12.10b). This setting will show the tool inside
the Tasks pane when it runs.

22→ Select Sale15A_Veg as the Input Features.

22→ Name the Output Layer *Suitable Timber.*

22→ Click Add Clause and enter the expression: SSTAGE96 is Equal to 4B Or SSTAGE96
is Equal to 4C Or SSTAGE96 is Equal to 3C.

22→ Verify the expression and click Done.

23→ In the Designer pane, switch to the General tab (Fig. 12.10c). Enter *Select Logging
Areas* for the step Name. Enter the remaining information as shown in Figure 12.10c.

23→ Use the circled Information buttons to review the Step Behavior options.

23→ Select Auto Proceed for the behavior of this step.

The next step will prompt the user to symbolize the map using a command. We want the step to
launch the Symbology pane automatically.

24→ In the Tasks pane, click the New Step button.

24→ Switch to the Actions tab in the Designer pane.

24→ Mouse over the <none> bar and choose the Edit icon.

24→ For Type of Command, select Command. Search for *Symbology*.

Examine the commands. So many are listed, it is difficult to choose the right one. Let's try a different approach.

25→ Click Cancel and click the back arrow. Make sure the Sale15A_Veg layer is selected in the Contents pane.

25→ Mouse over the <none> bar again and choose Record.

25→ Click the Feature Layer: Appearance: Symbology button. The step action is entered and the recording stops automatically.

26→ Switch to the General tab in the Designer pane.

26→ Change the step name to *Symbolize Logging Areas*.

26→ Enter the tooltip *Symbolize areas to be logged*.

26→ For Instructions, enter *Create a unique values map based on the SSTAGE96 field*.

26→ For the Run/Proceed Instructions, enter *Click Finish*.

26→ Select Auto Run for the Step Behavior.

26→ Close the Designer pane and save the Project to save the task edits.

Now we will run the task just created.

27→ In the Tasks pane, mouse over the Find Logging Areas task and click the blue arrow.

27→ Follow the instructions to run the task.

We will create one more task to find the protected areas that cannot be logged. In this task, we will demonstrate how to record steps instead of configuring them manually.

28→ In the Tasks pane, choose Options > Edit In Designer.

28→ Click New Task.

28→ In the Designer pane, enter the Name *Find Protected Areas* and enter the description *Finds the areas of a timber sale where logging is prohibited*.

 29→ In the Tasks pane, click the Record Commands button.

29→ In the Geoprocessing pane, search for and open the Buffer tool. A new step named Buffer is automatically added to the task.

29→ Select Sale15A_Streams for the Input Features.

29→ Enter *StreamBuffers* for the Output Feature Class.

29→ Set the Distance to *100* and select Meters for the units.

29→ Set the Dissolve Type to *Dissolve all output features into a single feature*.

29→ Run the tool.

TIP: The Task Designer pane must be open while recording. Place it side by side with the Geoprocessing pane before proceeding.

30→ Click the back arrow to return to the main Geoprocessing pane.

30→ Open the Buffer tool again. Another buffer step is created.

30→ Set the Input Features to Sale15A_Roads.

30→ Name the Output Feature Class *RoadBuffers*.

30→ Set the Distance to *50* meters.

30→ Set the Dissolve Type to *Dissolve output features into a single feature.*

30→ Run the tool.

 30→ In the Tasks pane, click the Stop Recording button.

Recording steps has one drawback: the parameters entered into the tool are not saved. To make these tasks as easy as possible, we should set the appropriate tool parameters and provide default names for output files. Inputs, since these will differ each time a user runs the task, will usually be left blank. Let's preset the parameters for the two buffer steps.

31→ In the Tasks pane, select the first Buffer step.

31→ In the Designer pane, switch to the Actions tab.

31→ Mouse over the Buffer button and select the Edit icon.

31→ Enter StreamBuffers for the Output Feature Class. Set the Distance to *100* meters and the Dissolve type to *Dissolve all...* Click Done.

32→ Repeat the previous tutorial step to edit the parameters of the second Buffer step, setting the output to *RoadBuffers*, the Distance to *50* meters, and the Dissolve type to *Dissolve all...*

Finally, we describe the steps for the user and set the appropriate step behaviors.

33→ In the Tasks pane, click the first Buffer step.

33→ Switch to the General tab of the Designer pane and name it *Buffer Streams.*

33→ Enter a tooltip, *Buffer streams for protected areas.*

33→ Enter the instructions, *Select the timber sale streams for the Input Features.*

33→ For Run/Proceed Instructions, enter *Click Run to continue.*

33→ Set the Step Behavior to Auto Proceed.

34→ In the Tasks pane, select the second Buffer step.

34→ In the Designer pane, enter a Name, Tooltip, and other items as previously, replacing references to streams with roads.

34→ Set the Step Behavior to Auto Proceed.

Having to reset the parameters makes recording less efficient, so we will enter the last two steps manually. The next step requires no user input, so it will be set to run automatically.

 35→ In the Tasks pane, click the New Step button.

35→ In the Designer pane, open the Actions tab. Mouse over the <none> bar and choose the Edit icon.

35→ Choose Geoprocessing Tool. Select the Union tool and click OK.

35→ Enter StreamBuffers and RoadBuffers for the Input Features.

35→ Name the Output Feature Class *NoLogBuffers*.

35→ Click Done.

36→ Switch to the General tab in the Designer pane. Name the step *Combine Buffers* and enter a tooltip.

36→ For Instructions, enter *Wait while the buffers are combined.*

36→ For Run/Proceed, enter *Next step starts automatically.*

36→ Select Automatic for the Step Behavior.

37→ In the Tasks pane, click the New Step button.

37→ Switch to the Actions tab. Mouse over the <none> bar and choose the Edit icon.

37→ Choose Geoprocessing Tool. Select the Dissolve tool and click OK.

37→ For the Input Features, type the name *NoLogBuffers.* (A red x appears because the previous step has not been run yet; ignore it.)

37→ Name the Output Feature Class *Sale_Protected.* Click Done.

38→ Switch to the General tab. Name the step *Dissolve Buffers* and enter a tooltip.

38→ For Instructions, enter *Edit the Output Feature Class name to uniquely identify this timber sale's protected areas.*

38→ For Run/Proceed, enter *Click Run to continue, and then Finish to complete the task.*

38→ Set the Step Behavior to Auto Proceed.

38→ Close the Task Designer pane and save the project. Always save before testing a task, in case of an error.

TIP: Editing an action may reset the step behavior to Manual. Make it a habit to check the step behavior after editing an action.

39→ In the Contents pane, turn off all layers except Suitable Timber.

39→ Switch back to the Tasks pane, showing the Timber Sale Maps task item with its three tasks (Fig. 12.11).

39→ Mouse over the Find Protected Areas task and click the blue arrow to start the task.

39→ Run through the task as a test to make sure it works.

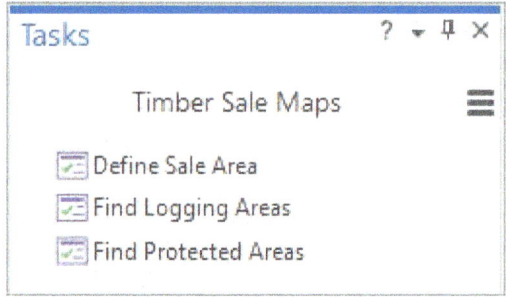

Fig. 12.11. The Timber Sale Maps task item
Source: Esri

Just like data sets, models and tasks should have metadata filled out by the creator. Even if they won't be shared, the information may be a useful reminder the next time the creator uses them. We won't fill out the metadata now, but take a quick look at the formats.

40→ Switch to the Catalog view.

40→ In Contents, under Project, select the Toolboxes entry and double-click the TimberSales.tbx toolbox that appears in the view to open it.

40→ Click the model and examine the metadata. Model metadata is somewhat different than metadata for a data set. See the Help Tool Reference for examples.

40→ Examine the metadata for the task item.

40→ Save the project.

TIP: To save and share a task item, expand the Tasks entry in the Catalog pane and choose Export > Save As.

Examining metadata

Sharing data incurs a responsibility to create quality metadata. In ArcGIS Online or in other Internet sources, the completeness of the metadata is often the primary clue to the quality and value of a data set. Users may find it necessary at times to explore the standards-based metadata before accepting a data set as a valid contribution to a project.

41→ Make sure the Catalog view is open.

41→ Select Folders in the Contents pane. In the Catalog view, use double-clicks to navigate inside the mgisdata\Usa\usdata geodatabase.

41→ Click the quakehis layer to select it, and scroll through the metadata to the end.

By default, the Catalog view shows only a few fields of metadata, the **Item Description**. To view more extensive metadata, if it is available, the setting must be changed.

42→ Choose **Project: Options** and view the Metadata section.

42→ Change the metadata style to *North American Profile of ISO 19115 2003*. Click OK and click the back arrow to return to the project.

42→ To update the metadata view, click the data set above quakehis and then click on quakehis again.

42→ Scroll down again and notice the additional information.

Styles may be based on a standard such as ISO 19115. ArcGIS stores the metadata information internally in its own format. The metadata style controls which information items are displayed and edited. The metadata can be exported to the standard format if needed.

TIP: The large-font section headings may be collapsed (side arrow) or expanded (down arrow). Use these headings to hide or view the information as directed in the next steps.

43→ Start by collapsing all the headings beneath Scale Range so that the organization of the 18 sections is apparent (Fig. 12.12).

43→ Expand the Topics and Keywords section and skim through the information.

43→ Close it when finished by clicking the heading again.

TIP: In the following steps, when you are finished examining each section, close it again.

43→ Examine the Citation section.

1. What was the publication date for this data set?

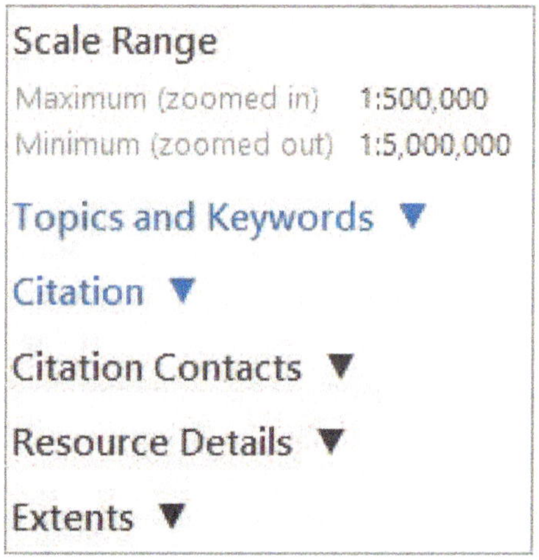

   _____

43→ Examine the Citation Contacts section.

2. Who originated this data set?_____ Who is the publisher? _____

44→ Examine the Resource Details section.

44→ Examine the Resource Points of Contact to find out who to contact with questions about this data set.

Scale Range

Maximum (zoomed in)   1:500,000
Minimum (zoomed out)   1:5,000,000

Topics and Keywords ▼

Citation ▼

Citation Contacts ▼

Resource Details ▼

Extents ▼

Fig. 12.12. Some of the metadata sections
Source: Esri

44→ Examine the Resource Maintenance section to find out how often this data set is updated.

Let's look more closely at where this data set came from.

44→ Expand the Lineage section and look at the process steps.

3. How many process steps are listed? _____ When did they occur? _____

4. What source was used to compile this data set? _____

5. What is the original scale (scale denominator) of this data set? _____

TIP: In Chapter 7, we were puzzled why some earthquakes prior to 1960 had magnitudes and some didn't. These metadata suggest the reason: several sources were combined to make this feature class, and not all of them included magnitude estimates for older earthquakes.

44→ Examine the Data Quality section, noticing in particular the five data quality reports and how they compare to the data quality issues discussed in this chapter.

6. Examine the Distribution section and determine how one obtains a copy of these data.

7. Examine the Metadata Details and Metadata Contacts sections to learn who produced the metadata and what standard was used to create it.

8. Examine the Fields section to find is the earthquake depth unit of measurement.

Clearly, it takes a lot of effort to document data sets to this level of detail. Organizations may have one or more people dedicated to metadata production, and such people are often in high demand. Learning to interpret and create standards-based metadata are valuable skills for geospatial professionals.

Many data sets in the United States have the older FGDC CSDGM style metadata, which cannot be viewed in Pro until it is converted or imported. The NLCD land cover data downloaded from The National Map in Chapter 6 came with metadata stored in a separate XML file.

45→ Use **Insert: Project: Add Folder** to add a connection to the gisclass\ClassProjects folder containing the CraterLake project.

45→ Navigate into the CraterLake.gdb geodatabase and examine the metadata for Landcover. It is empty.

45→ Navigate into the CraterLake\MyDownloads folder and find the NLCD2011_LC_N42W120.xml file and click to select it.

45→ Read the message about the FGDC metadata and click View Content. Your web browser should open to show the metadata in its native XML format.

The metadata from the .xml file can be imported to the Landcover raster.

46→ In Catalog view, return to the CraterLake geodatabase and select your Landcover raster.

46→ Click **Catalog Metadata: Import.**

46→ Click the Browse button for the Import metadata from box, navigate to the NLCD2011_LC_N42W120.xml file, and select it. Click OK.

46→ Set the type of metadata to import as FGDC CSDGM. Click OK. Examine the imported metadata.

Sharing a layer package

ArcGIS Online can be used to share data with others in an organization or anywhere else. One of the simplest methods is the layer package, which compiles the layer settings and the data into a simple file that can be uploaded to ArcGIS Online.

47→ Save the current project and open the ClassProjects\CraterLake project.

47→ Close all maps in the project except the map named Crater Lake.

47→ Turn off all layers except the group layer Crater Lake Geology and zoom to its extent.

A friend is planning to visit Crater Lake and wants to explore the geology, so we will share the Crater Lake Geology group layer as a package. When sharing data, it is important to provide metadata and give credit for the source. If the layer metadata have been completed, the Summary and Tags will be copied when the metadata are shared. We can consult the original information that came with the geology units feature class.

48→ Open the Catalog view and navigate to the Folders\CraterLake\CraterLake home geodatabase. Examine the metadata for the geologyunits feature class.

48→ Copy the Item Description information from the Tags section through the Use Limitations section.

48→ Paste the metadata into a word processor blank document. This step makes it easier to transfer the information.

49→ Switch back to the Crater Lake map view.

49→ Open the properties for the Crater Lake Geology <u>group layer</u> and click the Metadata section.

49→ Copy and paste the Tags, Summary, Description, Credits, and Use Limitations from the word processor document to the metadata sections.

A new process step was performed for this data set, so it is appropriate to update the metadata with that contribution.

50→ In the Summary section, add this text to the bottom of the summary: *Symbolized and uploaded to ArcGIS Online by <your name> on <today's date>.*

50→ Scroll to the bottom and click the Generate Thumbnail button. Click OK to save the metadata.

50→ Minimize the word processor document to save it for later.

It is a good practice to organize the AGOL account content using folders; otherwise it quickly becomes difficult to find things. Take a moment now to create a folder to contain these practice data. Folders can only be created and managed from the web version of ArcGIS Online.

TIP: Although ArcGIS Online allows one to create a folder name with spaces, Pro will not write to the folder. It is never wise to put spaces in a folder name.

51→ In a web browser, go to ArcGIS.com and sign in using the same user name and password that you use for Pro.

51→ Near the top of the window, click the Content tab (Fig. 12.13).

51→ In the Folders box on the left, click the New folder button.

51→ Enter a name for the folder, without spaces, such as *ProPractice*, and click OK.

51→ Minimize the browser for now, rather than closing it.

52→ In Pro, in the Catalog pane, switch to the Portal tab and select the first icon, My Content, to see the content from your AGOL account.

52→ If the new ProPractice folder is not visible, click the Options button (widen the pane if it isn't shown) and select Refresh.

Now we are ready to share the group layer.

53→ Switch to the Crater Lake map.

53→ With the Crater Lake Geology group layer selected in Contents, click **Share: Package: Layer** (or right-click the group layer in Contents and choose Share As Layer Package).

53→ Choose to *Upload package to Online account* (Fig. 12.14).

53→ Accept Crater Lake Geology for the package Name

53→ Click the Location Browse button and select the Portal\My Content\ProPractice folder just created. Click Open.

54→ Check the Item Description Summary and Tags to make sure they contain the metadata you provided.

54→ For now, do not check any additional *Share with* boxes, so that the package is only visible to you. This setting can be changed later if needed.

54→ Click Analyze. Pro will review the settings and identify any issues that would prevent the layer from publishing properly. This data set should not have errors.

54→ Click Package. It will take a couple of minutes to finish. After it finishes, close the Package Layers pane.

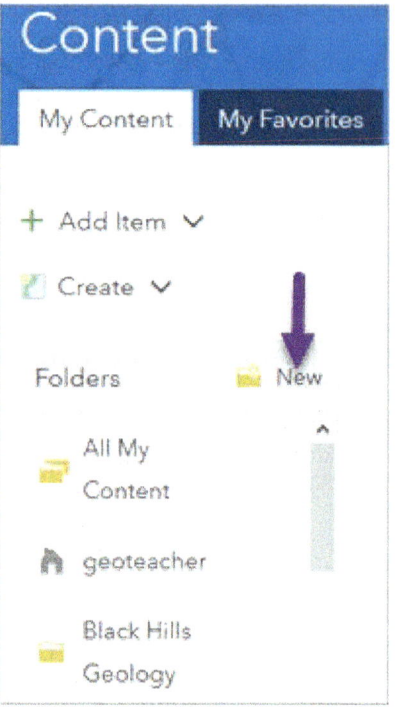

Fig. 12.13. Adding a new folder to AGOL content
Source: Esri

The layer package will be placed in your AGOL content. (By the way, don't actually share any of these practice items with the public. AGOL does not need hundreds of identical Crater Lake packages and services posted by students doing this tutorial.)

TIP: When an AGOL account contains folders, only the folders are visible by default. To see other data sets, click the green *Find more items…* text at the bottom of the pane.

55→ In the Catalog pane, click the Options button and choose Refresh.

55→ Double-click the ProPractice folder to open it and see the layer package.

55→ Hover over the package to see the brief pop-up description.

56→ Switch to the Catalog view.

56→ In Contents, select the Portal\My Content folder.

56→ Navigate to the ProPractice folder and select the Crater Lake Geology layer package to view the complete metadata.

56→ Save the project.

Publishing services

A trip to Crater Lake could benefit from a smartphone-accessible map that draws the geology and the park map. A geology feature layer would be slow, so a tile layer would give the best performance and can include both the geology features and the park map. When publishing map services, it is helpful to have a separate map containing just the desired layers.

57→ Use **Insert: Project: New Map** to create a new map and name it *Publish*.

57→ Switch to the Crater Lake map view. In Contents, right-click the Crater Lake Geology group layer and choose Copy.

57→ Switch back to the Publish map and paste the group layer in it. The metadata already created for the layer will be copied as well.

57→ In the Catalog pane's Project tab, look in the CraterLake home folder for the georeferenced crlamap1GR.jpg image, and add it to the map.

This time, we need to provide metadata for the map itself and not just its layers. We can use the information copied to the word processor document earlier, but we also need the credits for the park visitor map.

58→ Switch to the Catalog view and open the metadata for the crlamap1GR.jpg image.

58→ Read the description and then copy the Credits section.

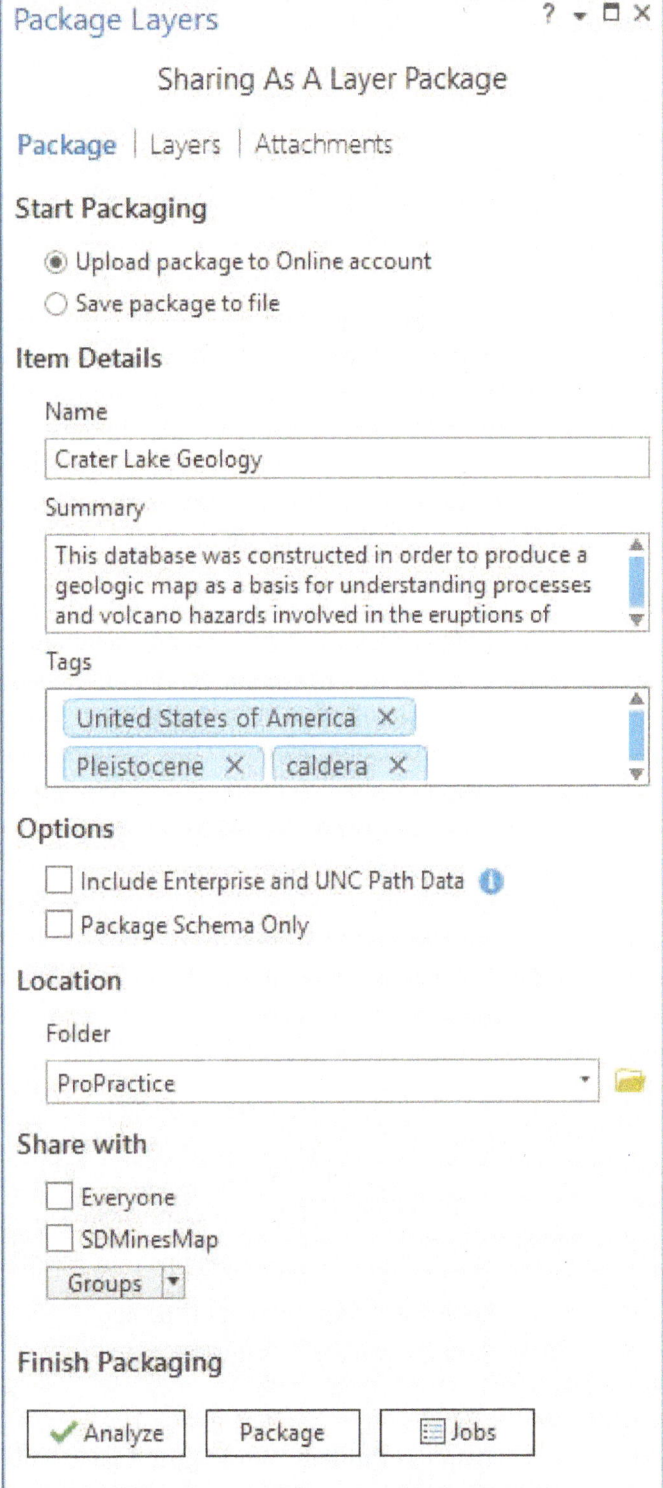

Fig. 12.14. Creating and sharing a layer package

Source: Esri

58→ Switch back to the Publish map.

58→ In Contents, double-click the Publish heading to open the map properties. Click the Metadata section.

58→ Title the map *Crater Lake Mobile Map.*

58→ Scroll to the Credits section and past the visitor map credit.

58→ Open the word processing document from earlier and copy the Tags and Credits to the map metadata properties.

Now we will add our own information describing the published map.

59→ In the Summary section, describe the purpose of the map (e.g., to provide geologic and park information for visitors with smartphones.)

59→ In the Description section, briefly describe the two data sets (the geology units and the Park Service map) included in the map. Use information from the word processor document if it helps.

59→ Scroll to the bottom and generate a thumbnail.

59→ Click OK to save the metadata.

Next, it is important to establish some basic properties to ensure good performance of the map.

60→ Double-click the Publish map name to open the map properties again.

60→ Open the Extent section. By default the map will show the extent of data in all layers, in this case the crlamap1.jpg, which is fine.

60→ Open the Coordinate Systems section. Web maps perform best in the WGS 1984 Web Mercator Auxiliary Sphere coordinate system, so set it to this CS.

60→ Click OK to save the map settings.

61→ In Contents, select the Crater Lake Geology group layer and use the **Layer Appearance: Effects** transparency slider to set it to about 50%.

61→ In Contents, right-click and remove the Topographic basemap.

For hosted tile layers, large scales generate the most tiles and will have the primary effect on the size of the cache. We must choose the scales carefully.

61→ Examine the current scale. Zoom in until the geology subunits appear with their labels and the park map starts to look pixilated. Note the scale.

61→ In Contents, click the Classes and Labels layer and examine the **Feature Layer Appearance** and the **Labeling** ribbons.

The detailed geology symbols are set to appear at 1:24,000, and the labels appear at 1:10,000.

61→ Zoom back to the extent of the map.

61→ Save the project. It is always smart to save before publishing.

62→ In Contents, select the Publish map entry and click **Share: Share As: Web Layer > Publish Web Layer**. A pane will open.

62→ Enter the name, *CraterLakeMobile_xxx*, where xxx is your initials (service names within an organization must be unique). No spaces are permitted in this name.

62→ Set the Layer Type to Tile.

62→ Click the Location Folder dropdown and select the ProPractice folder.

62→ For now, keep the default sharing, which is private. It can be changed later to share the map with others.

63→ Click the Configuration tab (Fig. 12.15). Keep the default tiling scheme.

63→ Examine the Levels of Detail and the Estimated cache size.

The suggested scale range encompasses Level 10 (1:577,791) to Level 16 (1:9,028) and would generate a cache approximately 53 MB in size.

63→ Place the cursor in the middle of the range slider and click and drag it to a new position. The entire range moves.

63→ Place the cursor on one end of the range and click/drag to change it independently from the other end.

63→ Experiment with sliding the scale range to higher and lower values. Watch the Level and Scale values change, and observe the effect on the estimated cache size.

A decision is needed here. The 1:9028 scale will show the labels, but the cache will be larger and will consume more credits. Since this is only a practice exercise, we will choose a smaller cache.

63→ Set the range from Level 11 to Level 15 (Fig. 12.15) with a cache size of 12 MB.

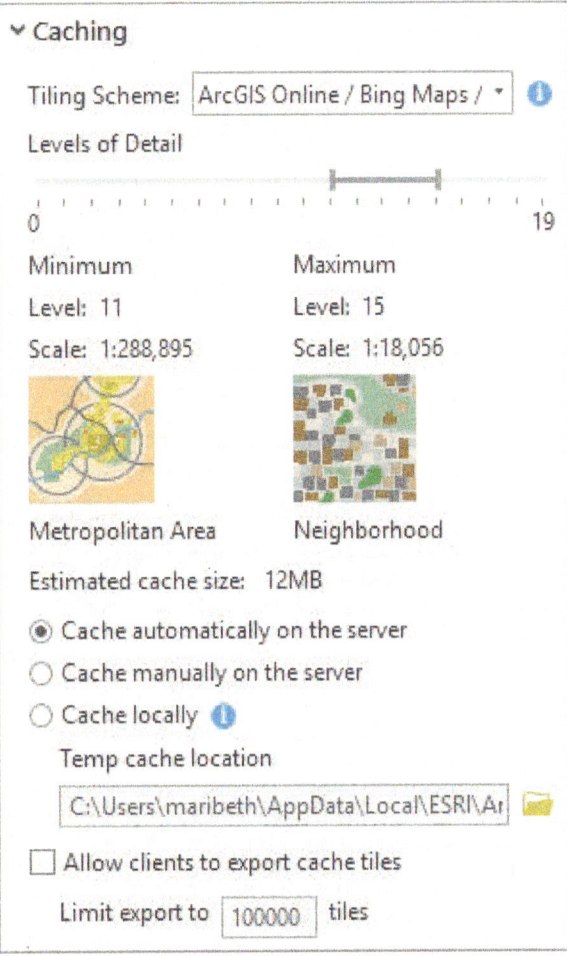

Fig. 12.15. Selecting the levels of detail for the cache
Source: Esri

Finally, analyze the publication scheme for potential issues.

64→ Keep the default to *Cache automatically on the server.*

64→ Click the Analyze button at the bottom.

64→ No errors or warnings should appear; if they did, they would need to be cleared before publishing. See the Pro Help.

64→ Click Publish. It will take a few minutes. A successful report should appear at the bottom of the Share As Web Layer pane when it is finished.

64→ Close the Share As Web Layer pane after seeing the report.

TIP: Click the Jobs button in the Share ribbon to keep track of a large caching process. One can continue to work in Pro while a cache is being generated.

65→ After the cache is generated, in the Catalog pane, switch to the Portal panel and navigate to the ProPractice folder. Use Options to Refresh the view.

65→ If the CraterLakeMobile service is not shown, click Find more items… at the bottom of the pane, open and close the folder, and refresh again (or again).

Using the service on a smartphone with the ArcGIS app requires a web map based on the hosted service, so we should create one and share it now. Web maps must only use map services and contain no local data.

65→ Use **Insert: Project: New Map** to create a new map and name it *Visit Crater Lake Geology*.

65→ In the Catalog pane, find the CraterLakeMobile tile layer and drag it into the map.

65→ Zoom in until the park map and geologic units appear. This will be the map extent.

A new shared map requires new metadata and new properties.

66→ In Contents, open the map properties for Visit Crater Lake Geology.

66→ Open the Extent section and set the extent to Custom extent. Choose Calculate from: Current visible extent.

66→ Click the Metadata section. Title the metadata *Visit Crater Lake Geology*.

66→ Enter several tags, a summary, and a description for the map. Give yourself credit, but don't forget to acknowledge the USGS and the Park Service.

66→ Click Generate Thumbnail.

66→ Click the Coordinate Systems section and make sure the map coordinate system is set to WGS 1984 Web Mercator Auxiliary Sphere.

66→ Click OK.

Now it is time to share the map.

67→ Save the project.

67→ In Contents, select the Visit Crater Lake Geology map entry and choose **Share: Share As: Web Map**.

67→ Name the map *Visit Crater Lake Geology* and save it in the ProPractice folder

67→ Keep the default settings for now.

67→ Click Analyze. No warnings should appear.

67→ Click Share.

67→ When it finishes, in the Catalog pane, open the ProPractice folder. If necessary, click the Options > Refresh button to see the new map.

Now that it is a web map, it can be viewed by mobile devices using the following procedure. The instructions are general, since every mobile operating system is different.

→ Download the ArcGIS Explorer or Collector app on the mobile device.

→ Open the app and sign in with your AGOL/Pro login name and password.

→ Find the Visit Crater Lake Geology map and explore it.

TIP: For Help on specific apps and operating systems, log in to the web browser version of ArcGIS Online and search for "mobile".

This is the end of the tutorial. The next optional section shows how to prepare a feature layer to collect information in the field using a smartphone or tablet.

→ Save the project.

Collecting data on a mobile device (optional)

Viewing a map on a mobile device is powerful, but imagine adding the ability to collect data using the built-in GPS. Mobile editing is based on hosted feature layers nested within a web map. In this next section, we will build a set of hosted layers to collect objects from a college campus or park. For this exercise, start a new project.

68→ Click the New Project button on the Quick Access toolbar at the top of the window.

68→ Name the project *MobileCollection* and save it in the gisclass\ClassProjects folder.

68→ Use **Insert: Project: New Map** to create a new map. Rename it *Hometown*.

68→ Use the **Map: Inquiry: Locate** tool to zoom to a local public area, such as a campus or a park, to be your field area.

69→ Adjust the zoom so that the entire field area fills the screen.

69→ Open the map properties Extent settings. Choose Custom Extent and select to calculate it from the current visible extent.

69→ Click the Coordinate Systems settings and verify that the map is set to WGS 1984 Web Mercator Auxiliary Sphere.

We will be publishing this map, so take a moment to fill in the metadata.

69→ Click the Metadata settings. Title the map *Hometown Mobile Collection*, assign some tags, and write a brief summary and description. Give yourself the credit for the map. Generate a thumbnail.

69→ Click OK to save all of the settings.

To ensure that all data created for this project are in the same coordinate system, we will set it in the project environment settings.

70→ Choose **Analysis: Environments**. Click the drop-down button for Output Coordinate System and select Current Map [Hometown]. Click OK.

TIP: If you are collecting data for an established database that uses a different coordinate system, such as UTM or State Plane, set the map and environment settings to that coordinate system.

Next we create feature classes to record objects and areas likely occur on a campus or park: point objects such as trees and light poles, sidewalk centerlines, and area features such as athletic fields and parking. We will create them in a feature dataset. The tutorial will use Sioux Park in Rapid City; name your feature dataset appropriately and substitute as needed in the instructions that follow.

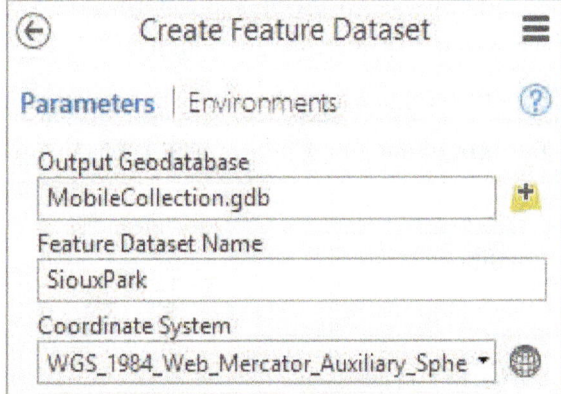

TIP: Although a feature dataset is not required, it can be used later to create a formal topology to assist with editing dangles, gaps, and overlaps that are common when collecting GPS data.

70→ In the Geoprocessing pane, open the Create Feature Dataset tool (Fig. 12.16).

Fig. 12.16. Creating a feature dataset
Source: Esri

399

70→ Leave the default output geodatabase and coordinate system.

70→ Type a name without spaces for the data set, such as *SiouxPark*, and run the tool.

71→ In the Geoprocessing pane, open the Create Feature Class tool.

71→ Click the Feature Class Location browse button and navigate inside the MobileCollection geodatabase.

71→ Select the SiouxPark feature dataset (or what you named it) and click OK.

71→ Type *Objects* for the Feature Class Name.

71→ Set the Geometry Type to Point.

71→ Leave the other settings to their defaults and run the tool.

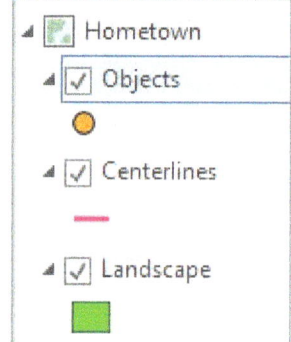

72→ In the open tool, change the Feature Class Name to *Centerlines* and the Geometry Type to Polyline. Run the tool again.

72→ Repeat the previous step to create a polygon feature class named *Landscape*.

To improve the visibility of the features on a smartphone screen on a sunny day, the symbols should be bold.

Fig. 12.17. Layers for collection

Source: Esri

73→ Open the Symbology pane for each new layer and select a large and bright symbol with good contrast (Fig. 12.17).

Data collection is faster and more consistent when domains are used to provide pick-lists. Table 12.2 shows some suitable domains for collecting information from a park or campus.

We will create the domains for the geodatabase first. (These domains will work for most campus or park areas. Add additional code values to the lists if desired or needed.)

74→ In the Catalog view, find and right-click the MobileCollection geodatabase, choosing Domains.

74→ Click in the first empty Domain Name box and enter *Things*.

74→ Tab to the Description box and enter a description, such as *Things found in a park*.

74→ Accept Text as the Field Type and Coded Value Domain as the Domain Type.

74→ Leave the Split and Merge policies set to their defaults.

Table 12.2. Domains for the MobileCollection geodatabase

Domain name	Field type	Domain type	Coded values
Things	Text	Coded	Tree, Bench, Lightpole, Sign, Sculpture, Other
ConditionClass	Text	Coded	Good, Fair, Poor
LineCategories	Text	Coded	Sidewalk, Street, Trail, Powerline, Other
SurfaceMaterials	Text	Coded	Concrete, Asphalt, Gravel, Dirt, Grass, Matting, Bark, Sand, Other
LanduseClass	Text	Coded	Parking, Athletic field, Pool, Landscaping, Playground, Game court, General use, Other

74→ Click in the Codes section. For each Code, use the first five letters (in lowercase) of the value listed in Table 12.2, and use the complete value listed for the Description (Fig. 12.18).

75→ Click in the next empty Domain Name box and enter the information and codes for the *ConditionClass* domain. Use the same method for entering the codes and descriptions as in the previous step. Be sure each code does not exceed five letters.

75→ Continue until all of the domains in Table 12.2 have been entered (Fig. 12.19).

75→ Click **Domains: Changes: Save** to save the domains.

75→ Close the Domains: MobileCollection view. Open the Hometown map.

Code	Description
tree	Tree
bench	Bench
light	Light pole
sign	Sign
sculp	Sculpture
other	Other

Fig. 12.18. Codes for the Things domain
Source: Esri

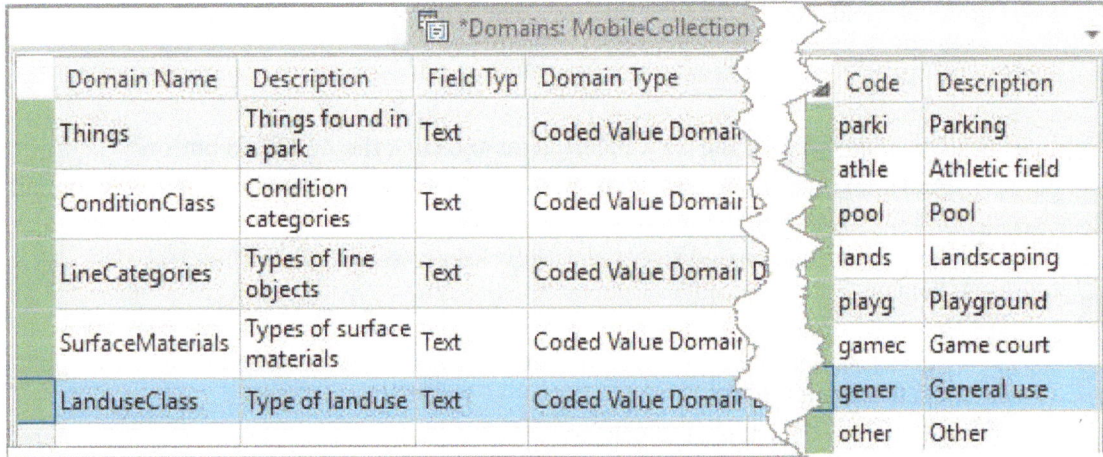

Fig. 12.19. Domains for the MobileCollection geodatabase, ready to be saved
Source: Esri

Next we create the fields for each feature class, assigning the domain as needed. We will also assign default values for certain fields to speed data entry. One expects, for example, that concrete sidewalks in good condition will be the most common Centerline feature.

76→ Open the table for the Objects layer and click the Add Field button.

76→ In the first empty Field Name box, enter *ObjectType*. Leave the alias blank.

76→ Set the Data Type to Text and the Length to 5.

76→ Click in the Domain box and select Things from the list of domains.

76→ Click in the Default box and select Tree as the default.

77→ Click to add another new field. Name it *Condition*, give it a Text field type, set the length to 5 characters, and assign the ConditionClass domain. Set the default to Good.

77→ Add a field named *CheckDate* with a Date field type and no domain.

78→ Add a field named Notes with a Text field type and a length of 255. The fields should appear as in Figure 12.20.

78→ Choose **Fields: Changes: Save** and then close the Fields: Objects view.

Field Name	Alias	Data Type	Domain	Default	Length
OBJECTID	OBJECTID	Object ID			
Shape	Shape	Geometry			
ObjectType	ObjectType	Text	Things	Tree	5
Condition	Condition	Text	ConditionClass	Good	5
CheckDate	CheckDate	Date			
Notes	Notes	Text			255

Fig. 12.20. Fields for the Objects feature class table
Source: Esri

79→ Open the table for the Centerlines layer and click the Add Field button.

79→ Add a text field named *Linetype* with 5 characters and the LineCategories domain. Set the default to Sidewalk.

79→ Add a text field named *Material* with 5 characters and the SurfaceMaterials domain. Set the default to Concrete.

80→ Add a text field named *Condition* with 5 characters and the ConditionClass domain. Set the default to Good.

80→ Add a date field named CheckDate.

80→ Add a text field named *Notes* with 255 characters.

80→ Choose **Fields: Changes: Save** and close the Fields: Centerlines view.

81→ Open the table for the Landscape layer and click the Add Field button.

81→ Add a text field named *Areatype* with 5 characters and the LanduseClass domain. Set the default to General use.

81→ Add a text field named *Material* with 5 characters and the SurfaceMaterials domain. Set the default to Grass.

82→ Add a text field named *Condition* with 5 characters and the ConditionClass domain. Set the default to Good.

82→ Add a date field named *CheckDate*.

82→ Add a text field named *Notes* with 255 characters.

82→ Choose **Fields: Changes: Save** and close the Fields: Landscape view.

82→ Close all of the tables and save the project.

TIP: It is wise to carefully design and create the fields, domains, and default values ahead of time. Schema changes are harder to implement after the service has been published.

Next we must prepare the feature classes and map for publishing adhering to the common best practices. You already set the default map extent and entered metadata for the map.

83→ Open the layer properties for Objects.

83→ Click the General settings and set the Out beyond (minimum scale) to 1:500,000.

83→ Click the Display settings and set the Display field to ObjectType. Click OK.

84→ Open the layer properties for Centerlines. Set the same minimum scale and choose Linetype for the default label field.

84→ Open the layer properties for Landscape. Set the same minimum scale and choose AreaType for the Display field.

85→ Remove the Topographic basemap from the map. (Don't be surprised that the display is blank; no features have been created yet in the feature classes.)

85→ In Contents, select all three layers (hold the Shift key down). Right-click one of them and choose Sharing > Share As Web Layer to open that tool.

TIP: To have friends or classmates help collect the data, use a unique map name and distinctive searchable tags instead of the generic ones given here, and set the sharing options to an organization, a group, or Everyone as needed to give them access.

85→ Enter a distinctive name for the service (without spaces), such as SiouxParkMapping.

85→ Enter a brief summary and distinctive search tags.

85→ Select Feature for the Layer Type.

85→ Choose ProPractice for the Location Folder.

85→ If desired, turn on additional Sharing Options.

85→ Click the Configuration tab in the tool.

85→ Check the box to Enable editing and allow editors to *Add, update, and delete features* (Fig. 12.21).

85→ Do not Enable Sync for this exercise (unless the chosen park has no cell service).

85→ Check Export Data to allow others to download copies of the feature layers, if desired. It is not required for this exercise and can be changed later if needed.

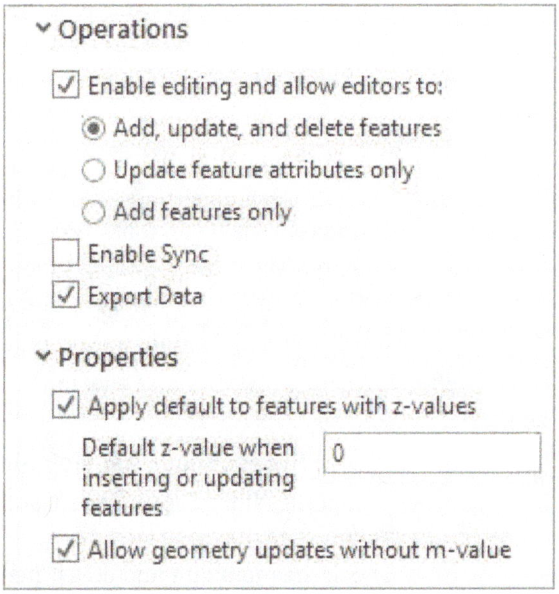

Fig. 12.21. Configuration settings for the hosted features

Source: Esri

TIP: When checked, the Enable Sync box allows certain applications that support it, such as Collector, to collect data offline when no cell service is available. However, it negates the live editing aspect in which field workers can view each other's collection efforts in real time.

86→ Click Analyze to begin testing the layer before publishing.

86→ A medium severity error appears because the layers do not have editing templates.

Although one could proceed with publishing, allowing the templates to be created automatically, we will fix the problem manually to show how to make changes in the middle of publishing. Leave the Share tool open while addressing the error to avoid losing the information already entered in the tool.

86→ Click **Edit: Features: Create** to open the Create Features pane. This step creates a template for each layer in the map.

86→ Close the Create Features pane.

86→ In the Share As Web Layer tool, click Analyze again. This time no errors are found.

86→ Click Publish and wait.

So far only the feature layers have been published. Recall that a web map is required to access the information in a mobile app, and it must use the published feature layers, not the local layers shown in the current map.

87→ Click **Insert: Project: New Map** and rename it *Sioux Park Mapping*.

87→ In the Catalog pane, switch to the Portal tab showing the My Content panel. Navigate inside the ProPractice folder.

87→ If needed to see the new service, use Options/Refresh and/or Find more items....

87→ Find the SiouxParkMapping feature layer and add it to the map.

87→ Use the Locate tool or zoom to the park area chosen. (With no features yet, the feature layers do not have valid extents, so Zoom to Layer won't work this time.)

87→ Choose **Map: Layer: Basemap > Imagery** for the map.

88→ Open the map properties.

88→ Click the Extent settings and set it to Custom extent, choosing to calculate from the current visible extent.

88→ Click the Metadata settings. Enter search tags and metadata. Make the tags distinctive if you are sharing this map with others.

88→ Scroll down and create a thumbnail. Click OK to close the map properties.

88→ Save the project before sharing.

89→ Choose **Share: Share As: Web Map**.

89→ Accept the Name *Sioux Park Mapping* and select ProPractice for the Location folder.

89→ Check the metadata and edit if needed.

89→ Check more Sharing Options if you plan to share the map with others.

89→ Click Analyze. No errors should appear. Click Share.

89→ Close the Share As Web Map pane when done. Save the project.

Use the Explorer or the Collector app to collect data on a smartphone or tablet. One must go to the park to actually collect data, but for now, make sure you have the app and can find the map.

> → Download the Explorer or Collector app on the mobile device.
>
> → Open the app and sign in with your AGOL /Pro login name and password.
>
> → Find the Sioux Park Mapping map within the app.

TIP: Each mobile operating system is different. To get help with using the app, log in to the web browser version of ArcGIS Online Help and search for "mobile".

This is the end of the tutorial.

> → Save the project and close Pro.

Practice Exercises

Imagine that an Oregon realtor's association is suddenly receiving many requests for parcels suitable for growing a particular specialty crop, which tends to do best under these conditions:

- ▸ rocktype1 is basalt or andesite
- ▸ drainage is at least 3 and less than or equal to 4
- ▸ slope is at least 10 and less than or equal to 30%

The association found soils and geology feature classes (in the oregondata geodatabase) to use to identify these areas. Implement a model and a task to help them find suitable properties.

1. Start a new project. Include your name or initials in the project name and the names of the maps in the project. Add a connection to the mgisdata\Oregon folder. Copy the geology feature class from the OregonStateGeology geodatabase and the soils feature class from the oregondata geodatabase to the project home geodatabase. Then delete the folder connection. (This step will reduce the size and complexity of the project package when it is shared.)

2. Use ModelBuilder to construct a model to conduct this analysis for Oregon. Be sure to create quality metadata documenting the criteria and identifying the model author.

3. Construct a task that allows a user to create a polygon showing a potential parcel area, enter different criteria for drainage and slope, and receive a clipped map of the areas within the polygon that meet the criteria. Be sure to include good instructions. Document the task with good metadata. (**Hint**: Use the Task Designer to examine the Define Sale Area task from the tutorial to see how it was done.)

4. Share the project as a project package (which will include the model and task). Check with your instructor about which sharing options to use.

5. Use the hosted feature layers and map from the tutorial to collect some data from the local park or campus that you published. Include at least 15 features of each feature type (point, line, and polygon). **Capture** a map showing the collected data.

Glossary

absolute pathname a file pathname that starts at the drive letter

adjacency a spatial condition that quantifies whether one feature touches another

affine transformation a process to translate, rotate, or skew an image to fit a new coordinate system

aggregate to combine measurements from one set of spatial units to a larger unit using a regular or hierarchical scheme, such as adding county populations to determine the state populations

alias an alternate name displayed for a field in a table that does not have to follow length and character restrictions

alpha a color parameter that defines the opacity or transparency of a color

annotation labels created from map features and stored separately for detailed editing

append a function that combines features from two different feature classes

ArcGIS Online a cloud-based platform for creating and sharing GIS maps, data, and applications

archive a single file that stores multiple files within it, often in compressed form, for easier transfer and downloading

area-weighted average resampling raster resampling technique that takes an area-weighted mean of the original cells to provide a value for the new resampled cell

ASCII an international code used to store simple text letters, numbers, and symbols; synonymous with "text file"

aspatial data data entries that are not tied to a location on the earth's surface

aspect the direction of steepest slope at a location on a surface

attribute accuracy a measure of the type or frequency of errors occurring in the attributes of a data set

attribute field a column in a table containing information about spatial features

attribute query an operation to extract specific records of information from a file based on values in the records

attribute table a table containing rows of information for features

attributes information about map features stored in columns of a table

azimuth a measurement of compass direction in 360 degrees with 0 = north

azimuthal projection a map projection onto a flat plane that is tangent or secant to the earth's surface

background areas and objects that are low in the visual hierarchy

band a single array of values stored in a raster, which may store one array or many

band stack a set of rasters with the same extent and cell size, stored together as a single unit; see also *multiband raster*

base heights elevations used to drape 2D map features on a 3D surface for visualization in a scene

bilinear resampling a resampling method that interpolates a new cell value based on the four closest values

binary numeric data stored in base 2 as a series of ones and zeros; a computer's native data storage format

block function an analysis function that moves over a raster in adjacent, nonoverlapping neighborhoods

block resampling a raster resampling scheme that aggregates only whole pixels into the new grid, such as converting a four-by-four block into one new pixel

Boolean an adjective describing data, operators, values, or files that contain or manipulate only true or false values

Boolean operators operators (AND, OR, NOT, XOR) that evaluate a pair of true or false propositions and return a true or false result

Boolean overlay evaluating rasters with Boolean operators to identify areas where combinations of specific conditions exist

Boolean raster a raster with cells that contain a 1 to indicate true or 0 to indicate false regarding a specified condition

branding repeating colors, symbols, fonts, images, or other design elements across multiple publications to foster recognition

buffer the delineated area within a specified distance of a feature

byte a unit of data storage in base 2 containing eight zeros or ones and holding a number from 0 to 255

caching manufacturing tiled snapshots of a map ahead of time and storing them on a GIS server for rapid delivery over the Internet

capture using the Alt-PrntScrn keys to place a copy of the active window on the clipboard for pasting into another program

cardinality the numeric relationship between matching records in two tables: one-to-one, one-to-many, many-to-one, or many-to-many

cartographic generalization simplifying map features in order to construct clearer and less cluttered maps

case field an attribute field used to group records in a table before calculating separate statistics for each group

categorical data data that place objects into unranked groups; examples are land use and geology data

cell a square data element in a raster corresponding to one value representing conditions on the ground

central meridian the central longitude of a map projection for which the x coordinate equals zero

chart map a map showing several different attributes in chart form, with one chart for each feature

choropleth map a map in which each feature, such as a state, is colored according to the values in a data field

classification assigning features to two or more groups based on numeric values in an attribute field

classified a raster display method that divides values into two or more groups based on their numeric values

client a device with software that allows it to communicate with an Internet server providing data or processing services

clip to remove features and portions of features that lie outside of the features of another layer

cloud a computing system composed of warehouses of computers and hard drives managed by a company that rents processing power and disk space to clients

cluster tolerance a defined distance used in topological editing, causing vertices to be made coincident if they are closer together than the tolerance distance

CMYK a method of specifying color using mixtures of cyan, magenta, yellow, and black; primarily used in printing and publishing

coded domain a rule that permits only certain specific values to be assigned to an attribute, such as land use codes

coincident boundary an identical boundary shared by two adjacent polygons

colormap a set of defined colors matched to specific image pixel values, which determines how the image will appear

completeness a measure of how well a data set has captured the entire area or all representatives of a feature class

compression mathematical or statistical schemes used to reduce the storage size of a file or files

conic projection a map projection derived by projecting latitude-longitude values on a paper cone covering a sphere

connectivity a property of linear features when they are connected to each other via junctions

connotation a typical emotional reaction elicited by certain colors or symbols

containment the property of one feature including another in whole or in part

Contents pane part of the Pro GUI that lists the objects being worked with, such as layers in a map or design elements in a layout

contextual ribbon in the Pro GUI, a strip of commands that only appears when the appropriate object is selected

continuous data that take on a variety of values and that change rapidly across a data set, such as elevation

contrast the difference in hue, saturation, or value between adjacent colors

convention a symbol for mapping that is commonly chosen and understood to have particular meaning

coordinate pair a single pair of x and y values indicating position in a planar coordinate system

coordinate space the range of x and y values onto which maps are plotted

coordinate system (1) a specified range of x-y values onto which a map is plotted; (2) the definition of a coordinate space used by a map layer, including the ellipsoid, datum, and projection

core ribbon in the Pro GUI, a strip of commands that is nearly always visible and accessible

coverage the spatial data format created for, and used by, Arc/Info

CSDGM (Content Standard for Digital Geospatial Metadata) a metadata standard developed by the FGDC and commonly used in the United States

cubic convolution resampling a resampling method that interpolates a new cell value from the 16 closest input cells

cut/fill a function that determines the difference between an initial and final topographic surface

cylindrical projection a map projection derived by projecting spherical latitude-longitude values onto a cylinder wrapped about a sphere

dangle a line feature that fails to connect to another line feature

database a data construct designed to store information as tables

data raster a raster that contains measured values, such as elevations, that can be used for analysis; contrast with *picture raster*

datum a combination of an earth ellipsoid and a reference point to reduce mapping discrepancies

dBase a database program whose file format has been adopted for the shapefile data model and tables in ArcGIS

Define Projection tool a tool to guide the user through the task of assigning a coordinate system to a spatial data layer

defined interval a classification method in which the user specifies a size range for all the classes

definition query an operation to set a map layer to display only the features whose attributes meet specific criteria

degrees the measurement units used in the spherical coordinate system; a circle has 360 degrees

DEM see *digital elevation model*

destination table the table that receives data from another table during a join operation; previous terminology for a *target table*

digital elevation model (DEM) a raster array of values representing elevations at the earth's surface

digital raster graphic (DRG) a scanned image of a USGS topographic map

digitize to convert shapes on a paper map to a digital map layer by entering vertices

discrete refers to data that represent distinct spatial objects such as wells, roads, or counties

display units the units in which the current x-y location of the cursor on the map is reported

dissolve to combine features when they share the same value for an attribute

divergent color set a set of colors with a neutral middle value that grades to increasingly saturated colors of different hues above and below

division units the units in which a map scale bar is measured and drawn, such as miles or kilometers

division value the length of one section of a scale bar as given in division units, such as 100 km

divisions the number of sections given to a scale bar

domain a rule that determines the values that may be entered into an attribute

dot density map a map representing attribute values by a proportional number of randomly placed dots

double-precision a floating-point numeric value stored using 16 significant figures

DRG see *digital raster graphic*

draping laying a 2D map layer over an elevation surface in order to generate a 3D visualization

dynamic constraints an editing capability that controls aspects of the next vertex placed when entering a sketch, such as the distance, direction, or location

dynamic labels labels determined from an attribute and automatically placed on a map each time features are drawn and redrawn

edge a shared boundary between two features being edited using topology

edge snapping ensuring that new features are automatically connected to the edges of existing lines or polygons

elevation layer a web service that provides compressed elevation data for rapid display of 3D scenes

ellipsoid a spheroidal volume with unequal axes, used to approximate the shape of the earth in map projections

end snapping ensuring that new features are automatically connected to the ends of existing line features

enterprise GIS a long-term GIS project developed by a large organization and involving many people over a long period of time

Environment settings program-level or tool-level settings that affect how tools run or set characteristics of the output, such as cell size or coordinate system

equal interval a classification method in which the user specifies a number of classes that have equal size ranges

erase an overlay function that removes features lying inside the external boundary of another polygon feature class

Euclidean distance the straight-line distance between two points

event layer a map layer of points created from a series of coordinate pairs in a table

export to create a new data file from all or a subset of features in an existing one, often while converting it to a different format

expression a statement containing field names, values, and/or functions used to extract records in a query or calculate values in a table

extension an additional set of tools, commands, and functions that can be purchased separately to extend the capabilities of ArcGIS products

extent the range of x-y coordinates displayed in a map or stored in a data layer

extent rectangle the range of x-y coordinates occupied by the features in a data layer

extraction functions or commands that pull out certain records or pixels based on field values or a designated spatial area

extrusion extending a flat feature into the vertical dimension by an amount specified in an attribute field (such as building heights) or defined for the layer as a whole

false easting an arbitrary x-coordinate translation applied to a map projection, usually to ensure that all values are positive

false northing an arbitrary y-coordinate translation applied to a map projection, usually to ensure that all values are positive

feature a spatial object composed of one or more *x-y* coordinate pairs and having one or more attributes in a single record of an associated table

feature ID (FID) a unique number assigned to every feature in a spatial data file and used for identification and tracking

feature class a set of similar objects with the same attributes stored together in a spatial data file

feature dataset a set of feature classes in a geodatabase that share a common coordinate system and can participate in networks and topology

feature layer a GIS data service that delivers point, line, or polygon features over an Internet connection

feature template a set of properties and attributes that stores all the information needed to create new features for a layer when editing

feature weight the priority assigned to a layer when determining which features may be drawn on top of others

FGDC (Federal Geographic Data Committee) an organization that promotes and establishes standards for the exchange of data

FID see *feature ID*

field a single column of information in a data table

field definition parameters specified when creating an attribute field, including type of data, field length, precision, and scale

field length the maximum number of characters that can be stored in a text attribute field

Fields view in the Pro GUI, an interface for modifying the properties of a table's fields, such as the name, alias, or visibility

filter a moving window applied to a raster data layer that calculates new values for the center pixel based on some function of the values in the window

flat file database a database that stores data in simple files

focal function raster analysis functions that determine new values for a target cell based on values in a moving neighborhood around the target

folder connection (1) a link in ArcGIS Pro that points to a folder with GIS data and serves as a shortcut to frequently used folders; (2) a link to a DBMS allowing data to be transferred

foreground objects and area that appear higher in the visual hierarchy

generalize see *generalization*

generalization simplifying a feature in order to store it or display it at smaller scales than its source scale

GCS see *geographic coordinate system*

geocoding the matching of a location stored in a table to a spatial point feature based on a reference spatial data layer, most often applied to converting addresses to locations

geodatabase a data model developed for ArcGIS 8 that employs recent database technology for storage and implements rules and topology

geographic coordinate system (GCS) a spherical coordinate system of degrees of latitude and longitude that is used to locate features on the earth's surface

geoid the shape of the earth as defined by mean sea level and affected by topographic and gravitational factors

geometric accuracy the accuracy with which the shape and position of features are represented

geometric interval a classification method that bases the class intervals on a geometric series in which each class is multiplied by a constant coefficient to produce the next higher class

geoprocessing analysis of spatial data layers, such as dissolving, intersecting, and merging

geoprocessing history a log of the geoprocessing operations and parameters kept by ArcGIS Pro during a session

geoprocessing service an analysis tool provided by a GIS server in which the processing is done on the server rather than on the client's device

georeferenced a spatial data layer that is tied to a specific location on the earth's surface for display with other data

geospatial adjective describing software, data, models, or other constructs that constitute, use, or manipulate geographic information

GIS server a computer configured to provide GIS data or analysis over the Internet

graduated color map a map that divides numeric data from a polygon feature class into classes based on value and displays the classes with different colors

graduated symbol map a map that divides numeric data from a line or point feature class into classes based on value and displays the classes with different size or thickness of symbols

graphic text text placed on a map layout that is not associated with a feature attribute

graticule grid marks of latitude and longitude placed on a map boundary

grid (1) specific raster format native to GRID and Spatial Analyst; (2) generic term for a raster data set

ground control points a set of points that match easily identifiable locations on two different data layers to enable georeferencing of one layer to another

hierarchical database a database that stores information in tables with permanent links between them

hillshade a raster that displays the brightness variation of a surface as if it were illuminated by a light source at a specified azimuth and zenith angle

histogram a graph showing the number of values from a data set that occur in evenly spaced ranges known as bins

home geodatabase the geodatabase stored within a project, which serves as the default location for geoprocessing output

HSV a method of specifying color based on hue, saturation, or value

hue (in HSV color model) the shade of a color, such as red, green, or blue, measured on a 0 to 360 scale

image a raster data layer, usually referring to a raster that displays brightness values, as in a photograph

imagery layer a GIS data service that delivers raster cells or pixels over an Internet connection

import to bring a copy of a data set into a geodatabase, sometimes including a conversion from a different data format

interactive query or **interactive selection** extracting one or more features by manually picking them from the screen with selection tools

Interchange file a file used to store a coverage for file transfer; has an extension of .e00

interpolation to calculate values at locations between known measurements; to populate a raster with values extrapolated from a known set of point values

intersect an operation to overlay two spatial data layers and find the areas common to both while discarding areas unique to either

intersection the property of two features touching each other in whole or in part

interval data values that follow a regular scale but have no natural zero point, such as degrees Celsius or pH

ISO (International Organization for Standardization) an international body that approves standard industry practices, including metadata standards

item description brief style of metadata offering basic information about a data set

Jenks method a way to classify numeric data into ranges defined by naturally occurring gaps in the data histogram

join the temporary combination of data from two tables based on a common attribute field or location

join table the table that provides the information to be appended to another table in a join; replaces the previous terminology *source table*

key an attribute field that is used to extract or match records in a table

latitude a spherical unit measuring angular distance north or south from the equator

latitude of origin the reference latitude of a map projection where the y value is zero

layer a reference to a feature class and its associated properties

layer file a file that stores a pointer to spatial data along with information on how to display it

layer package a file with a symbolized map layout and necessary data included for use by someone else; used to share map layers

layer properties settings that control how a data set is to be displayed or viewed

layout the specification for a map page, including the map frames, legend title, scale bar, and so on; stored in a project

layout file a file that contains all of the properties, settings and data links needed to re-create a layout

line a spatial feature composed of a string of x-y coordinate vertices and used to represent linear features such as streets

lineage the original source and processing steps behind the production of a digital geospatial data set

locator a reference data set and rules used to find locations based on addresses, city and state combinations, or other data not already in x-y coordinates

lock file a small file created in a data set's folder to indicate that the data set is in use

logical consistency how well the spatial relationships between features in a data set represent the real-world relationships; an aspect of topology

logical expression a statement composed of field names, operators, and values that specifies criteria used to select records or values from a layer or table

logical operators functions such as >, <, or = that compare values and generate a true or false result

longitude a spherical unit measuring angular distance east or west from the Prime Meridian

lossy compression a compression scheme that drops information from the original data and, once completed, cannot be reversed to re-create the original file

macro a series of program steps, often recorded by a user, that allows someone to quickly execute the steps automatically

mantissa the part of an exponential number that stores the significant digits

map (in ArcGIS Pro) a collection of data layers, labels, and other elements assembled to portray a geographic area

map algebra a system that permits calculations and operations on entire raster arrays, such as adding two rasters together

map elements objects placed on a map layout, such as titles, legends, scale bars, north arrows, images, and charts

map extent the range of x-y values of the area being displayed in a map

map frame a representation of a project map placed within a layout, linked to the properties of the layers in the original map but with an independent scale and extent

map image layer a GIS data service that dynamically renders a map based on vector or raster data and delivers a raster snapshot of the map to a client

map package a file with a symbolized map and necessary data included for use by someone else; used to share maps

map scale the ratio of feature size on a map to its size on the ground

map topology temporary spatial relationships developed between features during editing to facilitate the editing of features with common nodes or boundaries

map units the units of the coordinate system in which a map is stored or displayed

mask a raster layer applied during analysis to nullify unwanted cells, such as those outside a study area boundary

MAUP see *modifiable areal unit problem*

measurement grid a data frame grid that provides measured x-y coordinate system values around the perimeter of the frame

merge (1) to combine two or more map features into one feature; (2) to combine two or more data layers into a single layer

merge rule used in spatial joins to specify how to combine attributes when multiple join features match a single target output feature

merged join a spatial join that combines attributes from multiple join features that match a single output feature, using a merge rule

metadata information stored about data to document their source, history, management, uses, and more

metadata standard a set of requirements laying out the types of information and organization to be used for creating metadata

metadata style a setting that controls how the internal ArcGIS metadata are displayed and edited

model a sequence of steps or calculations used to convert raw data into useful information; a scheme used to understand and predict processes in the real world based on the manipulation of data

ModelBuilder a graphical interface used to combine existing tools to create new tools and scripts

modifiable areal unit problem (MAUP) statistical and visual issues caused by aggregating measured data using arbitrary areal units such as political boundaries

multiband raster a set of rasters with the same extent and cell size, stored together as a single unit; see also *band stack*

multipart feature a single feature composed of nontouching units, such as a single state feature composed of the seven Hawaiian islands

multipatch feature a three-dimensional feature constructed of simple rectangular or circular facets

NAD see *North American Datum*

nearest neighbor resampling a resampling method for discrete data that grabs the closest value to a cell center

neatline a line used to enclose one or more map elements in a rectangle

negative space blank areas in a graphic or map design composition

neighborhood function calculates a value for a target cell or feature based on surrounding values from a defined region

netCDF the Common Data Form, a standard data format for storing three-dimensional information, especially space-time information

NoData a special value used to designate that a data value is absent or unknown

node the beginning or end x-y pair of a line feature

nominal data values that name or identify an object, such as a street name

normalize to divide the values of an attribute field by the total of the field or by the values in another field

North American Datum (NAD) a combination of a spheroid and reference point that is used to minimize map distortion in North America

object ID (OID) unique number identifying a row in a table or a feature in a geodatabase feature class

oblique projection a map projection in which locations on a sphere are projected to a cylinder or cone of paper at an arbitrary angle

OID see *object ID*

one-to-many join operation a spatial join that creates enough output target features to combine attributes with every matching join feature

one-to-one join operation a spatial join that creates only one target output feature for each target input feature; merge rules are used to specify how to summarize or combine the attributes of multiple matching join features

ordinal data data values that indicate a rank or ordering system

organizational account an ArcGIS Online account provided by an organization that maintains its own instance of ArcGIS Online and provides accounts to users

origin the (0,0) point of a coordinate system

orthographic projection a map projection in which locations on a sphere are projected onto a planar surface

overlap a spatial condition that quantifies whether one feature covers all or part of another feature

package a single file containing all the elements of a layer, map, or project, created in order to share them with others, typically on ArcGIS Online

pane a part of the Pro GUI used to present commands and settings for working with objects in the program

panel a collection of commands or settings · within a pane, accessed by selecting a specific icon within the pane

parameter (1) specific value associated with map projections that define how the map appears; (2) a variable that serves as an input to a model

path or **pathname** a list of the folders that must be traversed to locate a particular file, such as c:\mgisdata\usa\states.shp

picture raster a raster that portrays a visual snapshot of geographic information, rather than analyzable data like elevation; contrast with *data raster*

pixel a square data element in a raster corresponding to one value representing conditions on the ground

pixel depth the number of bits or bytes allotted to store a single cell of a raster

planar topology an association of feature classes in a feature dataset, established by rules regarding the spatial relationships between features, such as not overlapping each other

point a one-dimensional feature defined by a single x-y coordinate pair

point cloud a data set composed of a multitude of x-y-z locations, often used to store and visualize three-dimensional objects or surfaces

point snapping ensuring that new vertices are automatically connected to existing point features

polygon a closed, two-dimensional area feature defined by three or more x-y coordinate pairs

portal a connection that accesses GIS services over the Internet

positional accuracy a measure of the likelihood that features on a map are actually in the locations specified

precision the number of digits or significant figures allotted to store a numeric value

premium content in ArcGIS Online, content that is only available to users with an organizational account; it also consumes service credits

Prime Meridian the line of zero longitude on the earth, passing through Greenwich, England

project (noun) a collection of data, tools, models, tables, and other items for a geographic area; the basic working file structure used by ArcGIS Pro

project (verb) to convert a three-dimensional geographic coordinate system to a two-dimensional planar one

project file the data storage file that contains and tracks all the elements of information stored within a project; has an .aprx file extension

project folder a filesystem folder that contains the project file, the home geodatabase, and the other items associated with a project

Project tool a tool that converts a feature class from one coordinate system to another

project package a file containing a complete project with data, maps, layouts, and other items; used to share a project with others

projection a mathematical transformation that converts spherical units of latitude and longitude to a planar x-y coordinate system

proportional symbol map a map that displays attribute values with marker or line symbols that are proportional in size relative to the value of the feature

proximity the property of one feature being close to another feature

public account in ArcGIS Online, a free account that gives access to a subset of the data and services available

pyramids a set of rasters with different resolutions that is calculated from a raster and used to speed displays at smaller scales

quantile a classification method that divides the data into the specified number of quantiles so that each class has the same number of features

query an operation to extract records from a database according to a specified set of criteria

range domain a rule that stipulates the largest and smallest values that can be stored in an attribute field

raster a data set composed of an array of numeric values, each of which represents a condition in a square element of ground

ratio data data having a regular scale of measurement and a natural zero point, such as precipitation or population

read access granting permission to a user to view a file but not to change or save it

reclassify to replace sets or ranges of values in a raster with different sets or ranges of values

record a row in a table containing one object or feature

rectify to rotate, resize, or warp an image to match a map base using a selected set of ground control points

reference grid a data frame grid that provides lettered and numbered squares on a map for use with an index of the features

reference latitude the latitude of origin; see also *latitude of origin*

reference scale the scale at which text or symbols appear at their assigned size

relate a temporary association between two tables based on a common field whereby fields may be selected based on whether they match selected records in the other table

relational database a database that stores information in tables and constructs temporary relationships between them

relative pathname path to a file that starts in the current folder

resampling to change the resolution of a raster using a defined strategy to convert values from one cell grid to another

resolution (1) the ground area represented by one cell value in a raster; (2) the default storage precision of a vector data set

resources generic term used to refer to data, analysis, maps, scenes, or other content available from a server over the Internet

RGB a method of specifying color using mixtures of red, green, and blue on scales of 0 to 255 each

RGB composite an image displayed by assigning one band of brightness information to each red, green, and blue color gun in a display monitor

root mean square error (RMSE) the root mean squared differences between a set of original and translated points

RMSE see *root mean square error*

rubbersheeting a process to deform a raster to fit it to a new coordinate system

Rule of Joining each record in the destination table must have one and only one matching record in the source table

rule of thirds guideline to place important parts of a graphic composition at the intersections of lines dividing the page in thirds

saturation (in HSV color model) the intensity of a color measured on a scale of 0 to 100

scale (1) the ratio of the size of features in a map to their size on the ground; (2) the number of decimal places allotted to an attribute field for storing numbers; (3) a factor used to convert raster decimal values to integers for more efficient storage

scale range the range of scales for which a data layer will be displayed, set by the user to avoid clutter, or the display of layers at inappropriate scales

scene a three-dimensional geographic view achieved by draping map data onto an elevation surface

scene layer in ArcGIS Online, a scene available as a GIS service over the Internet

schema the layout plan of a geodatabase, including the tables, the fields, the domains, and the relationships

schema changes changes of properties that affect the underlying storage structure of the feature class or geodatabase, rather than the cosmetic properties of layers

script a program that includes calls to ArcGIS commands and functions

secant projection a map projection in which spherical coordinates are projected onto a surface that intersects the sphere along two great circles

segment the connector between two adjacent vertices; usually a straight line, but it may also be specified as one of several types of curves

segment constraints editing guides that work to control the next vertex of a sketch, such as its distance, direction, angle, or location

selection type a parameter that controls what happens to previously and newly selected objects when a new selection is made

shapefile the spatial data model developed for, and used by, ArcView 3 and later versions

single-precision a floating-point numeric value stored using eight significant figures

sketch a provisional figure created during editing; when finished, it becomes a feature in a feature class

Sketch menu a context menu with segment constraints, accessed by right-clicking *off* a sketch during editing

slice to divide the values in a raster into a specified number of even classes

sliver small polygon created during map overlay due to slight boundary differences in the inputs; usually considered an error

slope the drop in elevation for a specified horizontal distance, expressed as an angle or a percentage

snap tolerance a distance set during editing such that new vertices close to an existing feature are snapped to it

snapping ensuring that features within a specified distance are automatically adjusted to meet at exactly the same location to avoid gaps between features

source data (1) a spatial data file that provides the features for a map layer; (2) the original information used to develop a spatial data set

source scale the original analog scale or resolution at which a digital data set is converted to digital form

source table the table that provides the information to be appended to another table in a join; previous terminology for a *join table*

space-time cube a three-dimensional data set that combines *x-y* data with time as the third axis

spaghetti model a model that stores spatial features as a series of *x-y* coordinates and does not store topological relationships between features

Spatial Analyst a program extension that is used to analyze raster data

spatial data information that is tied to a specific location on the earth's surface

spatial join a function that combines the attributes of features in two layers based on a spatial relationship

spatial operators a set of functions that evaluate spatial relationships, including intersection, containment, and proximity, between sets of features

spatial query an operation to extract records from a data layer based on location relative to another data layer

spatial reference the complete description of how the spatial data are stored for a feature class that includes the coordinate system, the X/Y Domain, and the precision

SPCS see *State Plane Coordinate System*

spheroid ellipsoid; see also *ellipsoid*

SQL see *Structured Query Language*

standalone table a table of information not linked to spatial data features

standard deviation a classification scheme in which the class breaks are based on the standard deviation values of the data being mapped

standard parallel parameter of a map projection indicating the latitude(s) at which the projection surface lies tangent or secant to the sphere

State Plane coordinate system (SPCS) a group of projections defined for different regions of the United States and designed to minimize map distortions

stereographic projection a map projection in which locations on a sphere are projected onto a planar surface

story map a combination of interactive maps with text, photographs, or videos and served as a web application

stretch to spread the values of an image to cover the entire range of symbols available, often supplemented by ignoring the tails of the distribution; see also *stretched*

stretch geometry an editing option to make gradual adjustments to connected features when moving nodes or edges in a topology

stretched a display method that spreads the data values over the entire range of symbols available; see also *stretch*

Structured Query Language (SQL) an established syntax for creating logical expressions to extract records from a database according to specified criteria

style a collection of map symbols and colors that are stored together and used together

subdivisions the number of units into which a single division of a scale bar is split, usually appearing on the left end

subscriber content in ArcGIS Online, content that is only available to users with an organizational account

suitability analysis evaluating a landscape to determine which areas are best fit for a given purpose, based on a set of factors

surface analysis functions designed for application to rasters representing a three-dimensional surface, such as elevation

table data stored as an array of rows and columns, with each row representing an object or a feature and each column representing an attribute or a property of that object

table view in the Pro GUI, an interface for viewing and manipulating the information in a table

tangent projection a map projection in which spherical coordinates are projected upon a surface that lies tangent to the sphere along one line of latitude

target table the table that receives data from another table during a join operation; replaces the previous term *destination table*

task a stored sequence of geoprocessing steps with detailed instructions to allow a user to interactively follow a designated procedure

task item a container in a project used to store one or more tasks

temporal accuracy a determination of the time period for which a data set is considered valid

thematic accuracy the degree to which attribute values represent the true properties in the real world

thematic raster a raster that contains categorical or nominal data values, such as land use codes or soil types

tile a rectangular subset of a larger raster, usually stored as a regularly spaced non-overlapping set, to speed the transmittal of GIS services

tiled layer in ArcGIS Online, a map that has been split into regularly spaced tiles, rendered ahead of time, cached on a server, and delivered as a GIS service; includes basemaps

tiling scheme parameters specifying the sizes of map tiles and the multiple scales for which they will be generated when creating a tiled layer

TIN see *triangular irregular network*

Tobler's law measurements taken at locations close together are more likely to be similar than measures taken at locations further apart

tool a computer program designed to perform one specific function

toolbox a collection of tools in ArcGIS Pro, organized by function

topological model a data model that stores spatial relationships between features in addition to their x-y coordinates

topology the spatial relationships between features, such as which are connected or adjacent to each other

transformation conversion of one geographic coordinate system (GCS) and datum to another

transverse projection a map projection in which spherical coordinates are converted to locations on a cylinder or cone tangent to the sphere along a line of longitude

triangular irregular network (TIN) a data model for storing surfaces as triangular facets with varying orientations

union an operation to create a new feature or set of features by combining all the areas from two input layers

unique values map a map in which each attribute value is assigned its own symbol

Universal Transverse Mercator (UTM) a family of map projections defined for 60 zones around the world and based on a transverse cylindrical projection

UTM see *Universal Transverse Mercator*

value (in HSV color model) the darkness or lightness of a color on a scale of 0 to 100

vector a spatial data storage method in which features are represented by one or more pairs of x-y coordinate values forming points, lines, or polygons

Venn diagram a diagram used in set theory to portray sets and subsets

vertex (pl. **vertices**) a point at which the segments of a line or polygon feature change direction

Vertex menu a short context menu for manipulating individual vertices, accessed by clicking *on* the sketch during editing

vertex snapping ensuring that new features are automatically connected to the vertices of existing line features

vertices see *vertex*

view in the Pro GUI, a window in the central display area that contains a project item being worked with, such as a map or a table

viewshed the area on a three-dimensional surface that is visible from a specified point or set of points representing observer(s)

visual center where the center of a graphic composition appears to be—about 5% above the geometric center

visual hierarchy the order in which objects in a design are perceived

voxel a rectangular solid representing a volume in space, a three-dimensional pixel

web application a web page that incorporates interactive display and analysis rather than simply presenting information; used in ArcGIS Online to deliver interactive maps on a web page

Web GIS a system with similar capabilities to those of desktop GIS, but the components are connected by the Internet and may be thousands of miles apart

web map an interactive map based solely on Internet data services and accessed from many clients, including desktop software, web pages, and mobile devices

web scene an interactive 3D scene based solely on Internet data services and accessible from many clients, including desktop software, web pages, and mobile devices

web server a computer configured to communicate with clients and respond to requests over the Internet

weighted overlay an analysis that uses a set of reclassified and ranked condition rasters and multiplies them with weights to calculate a suitability index

window in the Pro GUI, a container that organizes panes and views within a constrained box on the screen; windows may contain multiple open panes or views as quickly accessible tabs

world file a text file containing the georeferencing information for a raster

write access granting permission to a user to view and edit and save a file

XY tolerance a setting used for geoprocessing that defines the minimum allowable distance between vertices

X/Y Domain the maximum allowable range of values stored by a feature class

zenith angle an angular measure of the distance of an object above the horizon from 0 to 90 degrees

zipped refers to a set of files that have been compiled into a single file and compressed in order to make file transfers and downloads easier

zonal statistics raster analysis functions that calculate statistics from one raster for zones defined by a different raster

zone the combined areas of a raster that share the same attribute value, such as all areas with the same land use code

Selected Answers

Some problems can be done several ways. Your answers may sometimes be slightly different.

Because of the small size, legends and other items are omitted from most map answers. Your submissions should be better; check with your instructor to determine what you need to submit.

Ch. 1 Tutorial

1. 5412 hectares

2. Points: vents; line: faults and bathymetry; polygon: floorgeology, geologyunits, lake, ParkBoundary, topocliparea; rasters: dem30m and hillshade10m

3. No, it did not update.

4. USA Mean Temperature (2016) [imagery layer]. US Geological Survey on ArcGIS Online. URL: https://landscape3.arcgis.com/arcgis/rest/services/USA_Mean_Temperature /ImageServer [August, 2017]. NOAA provides the data; USGS hosts it.

5. 36 counties, avg. pop. 108,964

6. Wheeler, 1460 and Multnomah, 759,000

Ch. 1 Exercises

Answers will vary.

Ch. 2 Tutorial

1. ELEVATION is interval; TYPE is categorical.

2. Click FEATURE LAYER: APPEARANCE: Drawing: Symbology, or click the layer symbol in the Contents.

3. I for Interstates, U for US highways, S for state highways

4. Ratio data; a graduated color map

5. Interval data; stretched or classified

6. Ordinal data; unique values with monochromatic ramp

Ch. 2 Exercises

Answers will vary. Pay attention to colors, symbols, legend text, formatting of legend values, and choice of classification methods for each map.

Volcanic Hazards frame

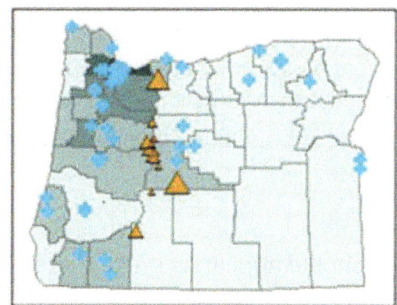

Source: Esri

Agriculture frame

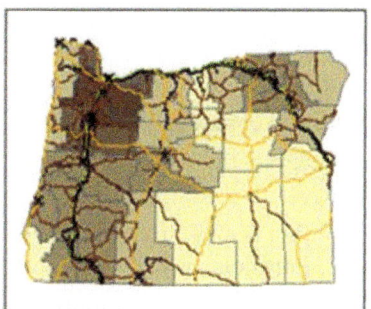

Source: Esri

Housing frame

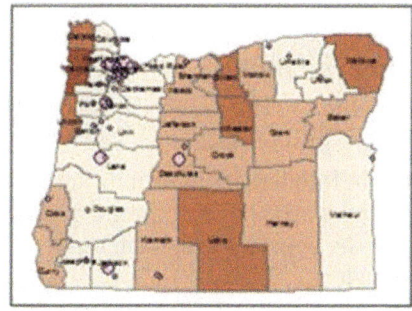

Source: Esri

Physiography frame

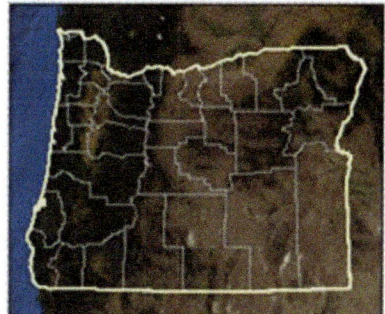

Source: Esri

421

Ch. 3 Tutorial

1. A map is a container for layers; a map frame places a map in a layout and has the same layer properties but independent extent.
2. Left to right: subdivision, division, division units

Ch. 3 Exercises

1. The map should appear as the one shown in the question.
2. Answers will vary.

Ch. 4 Tutorial

1. GCS_WGS_1984, degrees
2. About –80° and 25°
3. Both units are decimal degrees.
4. Mercator units are meters.
5. South America mostly negative; Asia mostly positive
6. Central meridian is –96° and the latitude of origin is 40°. Parallels are 20° and 60°.
7. They are not overwritten because they are being stored in different geodatabases.
8. NAD 1983 South Dakota State Plane

Ch. 4 Exercises (Hints)

1. Examine the Coordinate Systems setting in the layer properties.
2. Examine the Coordinate Systems setting in the layer properties.
3. Locate Mt. Rainier, zoom way in, and use the map properties to modify the display units for the coordinate system readout in the map view.
4. Open the map properties to view the coordinate system details.
5. Open the map properties to view the coordinate system details.
6. Open the map properties to view the coordinate system details.
7. Open the map properties to view the coordinate system details.
8. Display and label the UTM and State Planes from mgisdata.
9. In the map properties, start with a State Plane zone and modify it.
10. Examine layer properties, guess the correct coordinate system, and modify the label so it will be properly displayed.

Ch. 5 Tutorial

1. Layer names are cosmetic, and file naming rules don't apply.
2. Not found: coverages, database connections, AGOL feature layers, packages, and CAD drawings.
3. Contents pane shows layers with cosmetic settings; Catalog pane shows stored data sets.
4. One is for vector data; one is for rasters.
5. Different sources with different source scales.
6. Clipping would replace portions of the original boundaries and degrade the geometric accuracy.
7. From reading the metadata on the web site.

Ch. 5 Exercises

Answers will vary.

Ch. 6 Tutorial

1. NAD 1983 UTM Zone 10N; meters
2. The value of 180; comprises the lake
3. Number 6, ignimbrite

Ch. 6 Exercises

Answers will vary.

Ch. 7 Tutorial

1. About 39 million Blacks and 50 million Hispanics
2. South Atlantics has the most Blacks; Pacific has the most Hispanics
3. 190 Democratic districts
4. Standalone table
5. DISTRICTID field
6. one-to-many; states is the target table; districts is the join table
7. 21 reps from New England
8. Many western stations are in the mountains and are thus cooler.
9. They only contain partial information instead of decimal degrees.
10. STATION_NAME and STATION

Ch. 7 Exercises (Hints)

1. Add table to Pro and export to database.
2. Examine the number of records in the table and sort by opening date.

3. View metadata in Catalog view.

4. Use the Frequency tool.

5. Create a per capita field and join the states to stores in order to calculate the values.

6. Use joins to combine the standalone frequency tables to spatial data so they can be mapped.

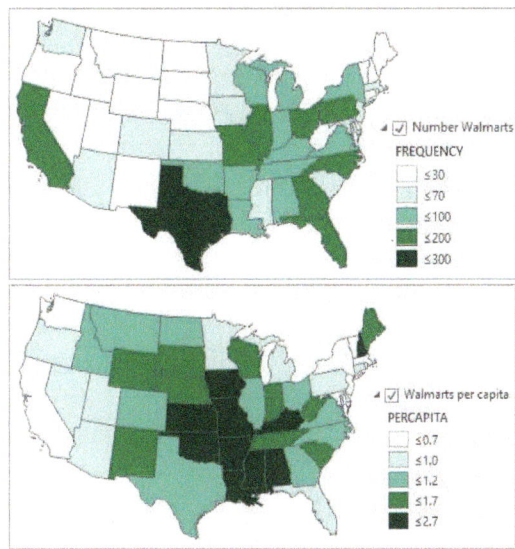

Source: Esri

7. Create a scatter chart.

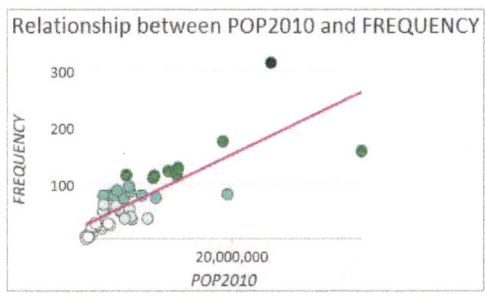

Source: Esri

8. Use summary statistics.

9. Use the layer properties to set the time settings and explore with the time slider.

10. Answers will vary.

Ch. 8 Tutorial

1. NAD 1983 State Plane Texas Central

2. Schema changes are being made.

3. Source scale is 1:200,000; about 1200 ft.

Ch. 8 Exercises

Answers will vary. Pay attention to snapping and completing adjacent polygons properly.

Ch. 9 Tutorial

1. 9 cities

2. 303 counties

3. 133 counties left

4. 36 cities contain "New"

5. Remove the first % wildcard in SQL; then 34 cities remain

6. 3641 cities within 20 miles of interstate

7. 93% of cities

8. Brazos, Canadian, Pecos, Red, and Rio Grande

9. 43 million people in the risky cities

10. It counts people in portions of counties removed by the clip.

Ch. 9 Exercises (Hints)

1. Use attribute selection

2. Use attribute selection

3. Use location selection

4. Use location selection

5. Select by attributes and then by location with appropriate selection type settings

6–9. See maps.

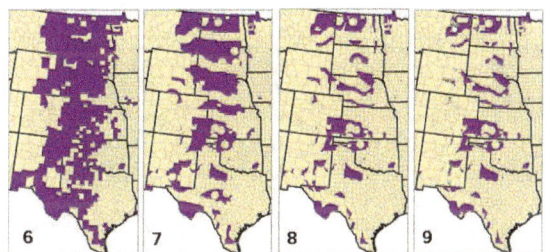

Source: Esri

10. See map.

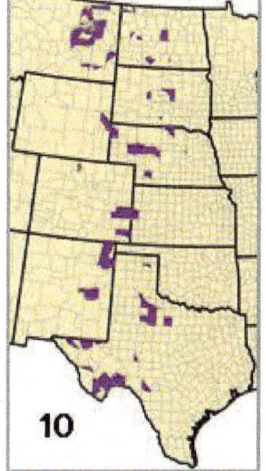

Source: Esri

Ch. 10 Tutorial

1. 591 out of 1360 wells
2. FACILITY_N
3. 16,477 streets
4. Selecting by location only selects points; it cannot sum the populations for the centers.
5. 2.4 km²

Ch. 10 Exercises (Hints)

1. Join closest with a one-to-one join type. Create and calculate a distance field and chart the results.
2. See map.

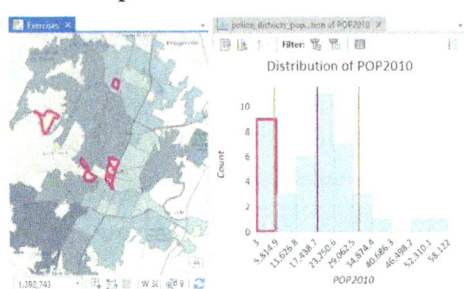

Source: Esri, City of Austin

3. Join watersheds to schools.
4. See map.

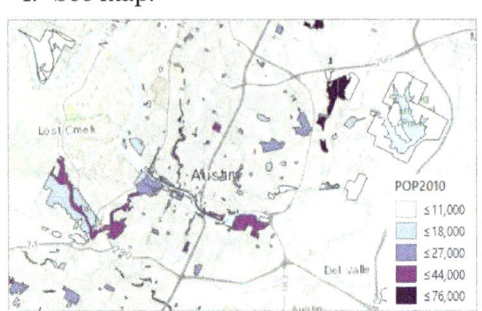

Source: Esri, City of Austin

5. See maps.

Wells per 1000 people

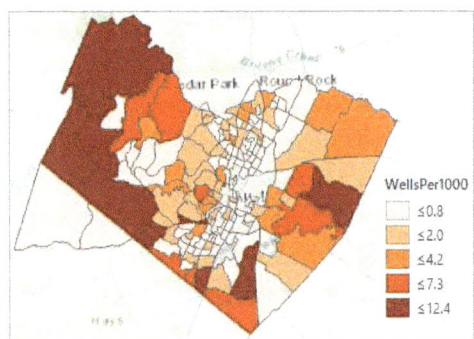

Source: Esri, City of Austin

Well depths

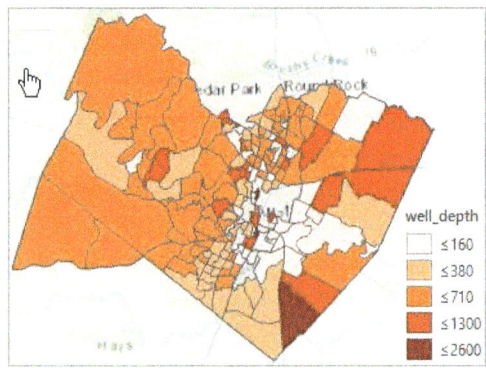

Source: Esri, City of Austin

6. See map.

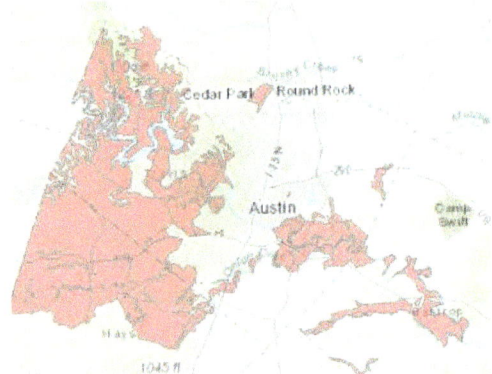

Source: Esri, USGS, City of Austin

7. See map; selected roads in purple

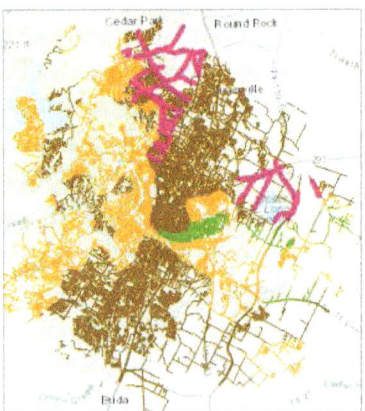

Source: Esri, City of Austin

Ch. 10 Exercises, (Hints) con't.

8. See map.

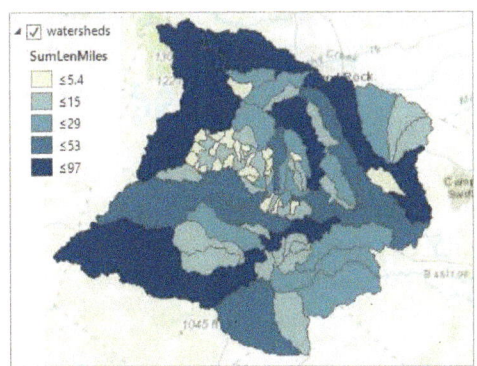

Source: Esri, City of Austin

9. See map.

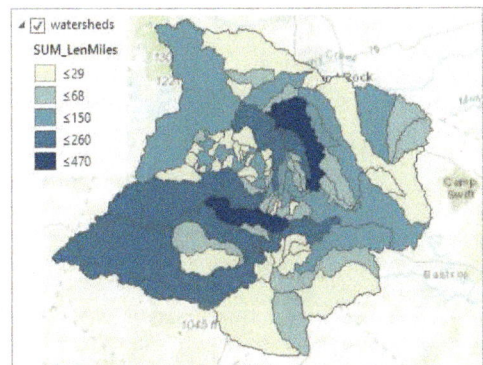

Source: Esri, City of Austin

10. See map.

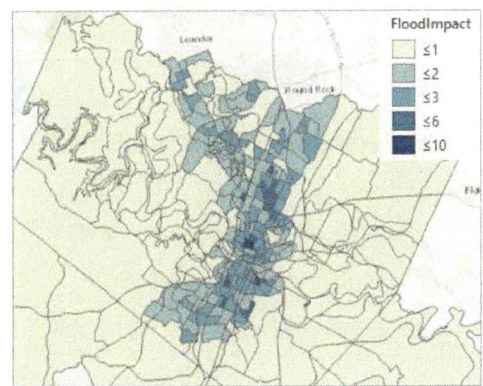

Source: Esri, City of Austin

Ch. 11 Tutorial

1. About 319 km² of class C

2. North is red; south is light blue.

3. Stores 0–3, indicating the number of points from which the cell can be seen

4. Floating-point raster because the precipitation values have decimals

5. Pleasant Valley, about 0.166 km³

Ch. 11 Exercises

1. See map.

2. See map.

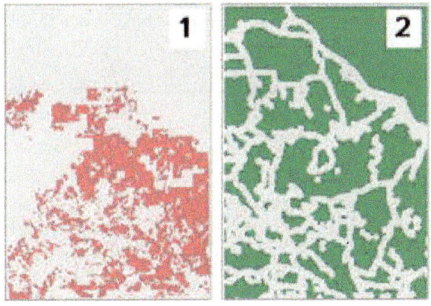

Source: Black Hills National Forest

3. See map.

4. Area is 40.4 km². See map.

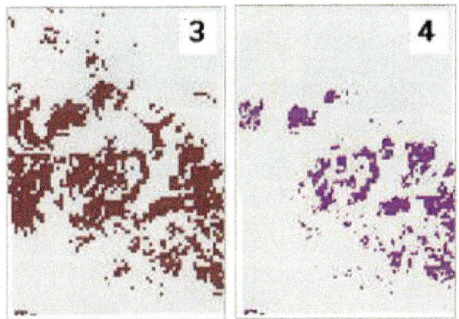

Source: Esri, Black Hills National Forest, South Dakota Geological Survey

5. See map; best sites in green.

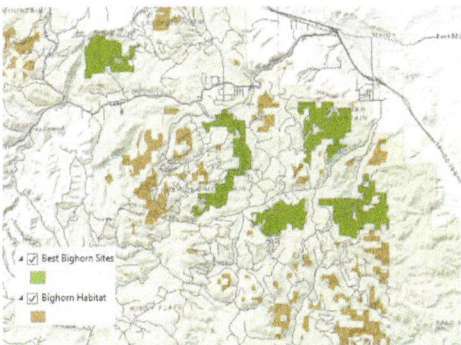

Source: Esri

Ch. 11 Exercises, con't.

6–9. Answers will vary; see maps for examples.

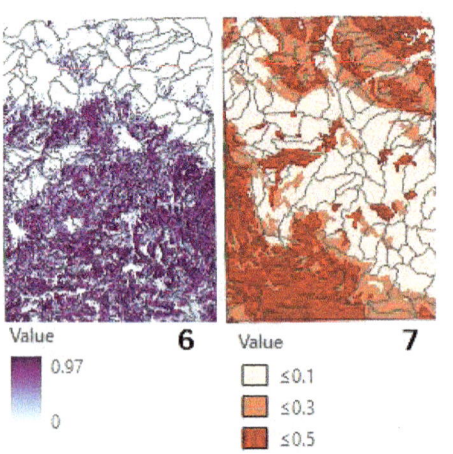

Source: Black Hills National Forest

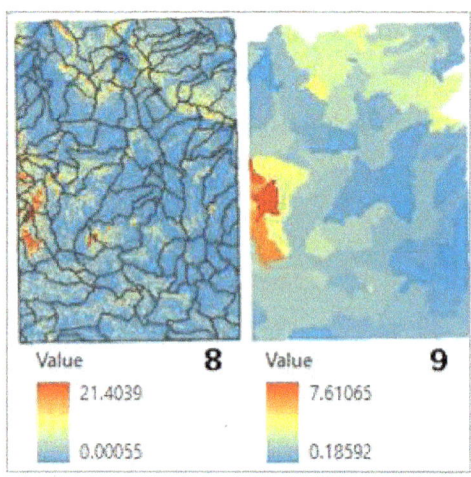

Source: Black Hills National Forest

10. Answers will vary, but they should appear somewhat similar to Figure 11.7.

Ch. 12 Tutorial

1. Publication date is 6/30/2010.
2. Originated by National World Atlas and USGS; published by esri
3. Two steps in 2004 and 2006
4. The National Atlas
5. Source scale 1:2 million
6. Contact esri to purchase the software.
7. Produced by ArcGIS Content Team; follows NAP of ISO 19115 2003
8. Depth unit is kilometers.

Ch. 12 Exercises

Answers will vary.

Areas in Oregon that meet the criteria.

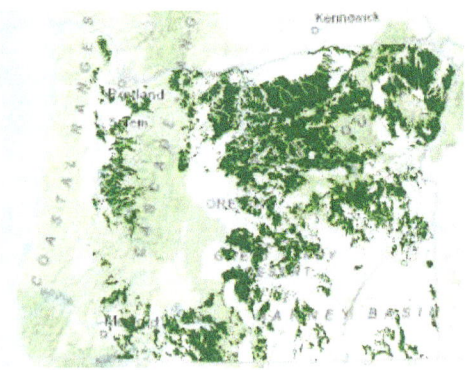

Source: Esri

Index